QUEUEING THEORY

A Linear Algebraic Approach

Lester R. Lipsky

Macmillan Publishing Company
NEW YORK

Maxwell Macmillan Canada
TORONTO

Maxwell Macmillan International Publishing Group
NEW YORK OXFORD SINGAPORE SYDNEY

Editor: John Griffin

Production Supervisor: Leo Malek

Production Manager: Aliza Greenblatt

Text Designer: Erikson/Dillon Art Services

Cover Designer: Natasha Sylvester

Illustrations: Erikson/Dillon Art Services

This book was set in Computer Times Regular 10/11, by the American Mathematical Society, was printed and bound by Book Press. The cover was printed by Lehigh Press.

Macmillan Publishing Company
866 Third Avenue, New York, New York 10022

Macmillan Publishing Company is part
of the Maxwell Communication Group of Companies.

Maxwell Macmillan Canada, Inc.
1200 Eglinton Avenue East
Suite 200
Don Mills, Ontario M3C 3N1

Library of Congress Cataloging-in-Publication Data

Lester R. Lipsky
Queueing theory : a linear algebraic approach / Lester R. Lipsky.
p. cm.
Includes index.
ISBN 0-02-370962-9
1. Queuing theory. 2. Algebra, Linear. I. Title.
T57.9. L56 1992
519.8'2—dc20 91-11790
CIP

Printing: 1 2 3 4 5 6 7 8 9 Year: 2 3 4 5 6 7 8 9 0 1

DEDICATION

This book is dedicated to my wife, Jacqueline Ristin Lipsky, my lifelong friend, who, after all these years of taking second place to school, thesis, kids, tenure, research, tenure (a second time), and for the last three years, *the book*, still remains married to me. It is also dedicated to our parents, Morris (Pod)Lipsky, Ray Polonski Lipsky, Saul (Zvo)Ristin, and Helen Birnberg Ristin.

A Path to Discovery

"Theories of the known which are described by different ideas, may be equivalent in all their predictions and are hence scientifically indistinguishable. However, they are not psychologically identical when trying to move from that base into the unknown. For different views suggest different kinds of modifications which might be made. Therefore, a good scientist today might find it useful to have a wide range of viewpoints and mathematical expressions of the same theory available to him. This may be asking too much of one man. Then new students should as a class have this. If every individual student follows the same current fashion in expressing and thinking about the generally understood areas, then the variety of hypotheses being generated to understand the still open problems is limited. Perhaps rightly so, · · · BUT if it is in another direction, who will find it?"

- Richard P. Feynman

"So spoke an honest man, the outstanding intuitionist of our age and a prime example of what may lie in store for anyone who dares to follow the beat of a different drum."

- Julian Schwinger

From a special issue on Richard Feynman (who died on 15 February 1988) in: PHYSICS TODAY, February 1989. Feynman's quote (slightly paraphrased here) was taken from his Nobel Lecture in June 1965.

[Note: Feynman and Schwinger shared the Nobel prize with S. Tomonaga in 1965 for their work on Quantum Electrodynamics in the late forties. Working independently, and using radically different methods, they ended up with mathematically equivalent theories. Schwinger and Tomonaga were the "mainstreamers", but everyone calculates using Feynman's method to this day.]

PREFACE

> *"'Necessity is the mother of invention' is a silly proverb. 'Necessity is the mother of dodges and temporary fixups' is much nearer to the truth. The basis of the growth of modern invention is science, and science is almost wholly the outgrowth of intellectual curiosity."* – Alfred North Whitehead

At least 50 worthwhile books on queueing theory have been written in the last 35 years. Two or three times as many books have been published in which queueing theory plays an important part. Most of these books, even the older ones, are still useful for understanding at least some part of the subject. Why, then, should yet another book be published? The answer, simply, is that there is no book (or even collection of papers) that covers *intermediate queueing theory* using what I call LAQT (linear algebraic queueing theory). There are in fact only two books which use a linear algebraic approach, both by Marcel Neuts [NEUT81] and [NEUT89], and both of them are written for experts in the field. I waited five years for someone to write a book that could be used for a first or second course in the subject (never do anything if someone else is going to do it), but to no avail. So in 1988 I started to write it myself.

The reason that LAQT should become familiar to novices as well as to those who are already knowledgeable in intermediate and advanced queueing theory is that any problems which can be cast into a matrix-vector format can easily be adapted to make use of the high-speed parallel and vector processors available today. Also, many problems in queueing theory that traditionally are solved by unrelated mathematical techniques can now be solved in a consistent integrated fashion. This allows for better physical insight. But, most important, many system performance measures which are normally ignored because of their computational and formulational difficulties can be dealt with easily in LAQT. Some examples are: properties of the *busy period*, departure processes, *first-passage times*, *residual* times, distinctions between what an observer sees and what a customer sees, and compound processes in general. Each of these topics is treated here without requiring prior knowledge of the reader. This book makes the following claim. "Any problem that can be solved for exponential servers can somehow be extended to treat nonexponential servers." Of course, it remains to be seen whether the future will vindicate this optimism.

Many decisions had to be made before this book could be written. First, who is the intended audience? There are a half a dozen disciplines which claim queueing theory as one of their "bread-and-butter" techniques. Applied probability, computer science, electrical engineering, management science, operations research, systems engineering, and even physics lay claim to various parts of this subject as their own, each with its own terminology. Since I dabble in all these fields, I decided to try to write a generic book which could be understood by all. The terms used are defined in relation to customers arriving at, being served by, and departing from subsystems, from the different viewpoints of the customer and of an outside (sometimes random) observer. The mental image one gets is of humans being served by mechanical objects, while being observed by other human beings.

Another decision to be made was the level at which to present the material, namely, as a first or second course in queueing theory, as a reference book for practitioners, or as a monograph for would-be researchers in the field. Once again, I decided to try to aim for all. There is no reason why this material cannot be taught to mathematically mature college seniors or new graduate students who have already had courses in linear algebra and probability theory, but have not necessarily had any queueing theory. Unfortunately this would have required that the first two chapters be expanded to more than twice their present size, without ever mentioning LAQT. Since there are already many books available that give an excellent introduction to queueing theory, I opted for either a first course, where the student already has had some background in Markov processes and elementary queueing theory, or a second course. For instance, many students in computer science and electrical engineering take a course in applied probability which includes material such as that in Chapters 7 and 8 of Trivedi's book [TRIV82]. Alternately, many courses in performance modeling (e.g., courses using [MOLL89] or [LAZO84]) are adequate to serve as an introduction to this book.

We assume that the reader is already familiar with matrix theory. However, except for such elementary formulas as that defining matrix multiplication, we do not expect the student to have any particular theorem at his or her fingertips. Therefore background information is introduced as needed. There is no special section put aside for reviewing linear algebra. We assume the same about the reader's knowledge of integral and differential calculus (in particular, Taylor's series and l'Hospital's rule) and elementary probability theory. For those whose mathematics is a bit rusty, we recommend that an elementary text in each of these areas be kept handy. But worry not; for all the mathematical content, this is not a rigorous text. It is a *why and how to* book. Whenever we would like a matrix to have a particular property, we assume it is so, whether or not we can prove it.

The material is rather densely packed, so several readings and rereadings may be necessary for the less experienced queueing theorist, particularly since there are numerous definitions in the text, and definitions do not usually stick in one's mind without some effort. This problem is reduced somewhat by the book's layout. We are inclined to introduce an idea in one chapter, and then use it again in a subsequent section, but in a more intricate way. We have done our best to give explicit reference to material previously discussed.

For Instructors and Practitioners

One might say that the "father" of LAQT is Victor Wallace, who in the 1960s introduced the concept of *quasi-birth-death-processes* and proved that there exists a matrix geometric solution for a large class of such systems, including the open GI/G/C queue [WALL69]. His presentation, though motivated by queueing theory [WALL72], was couched in terms of abstract Markov chains, and so, though acknowledged, was never picked up as a practical way of dealing mathematically, conceptually, or computationally with specific problems in elementary or intermediate queueing theory.

The first researcher actually to take this viewpoint in solving problems specific to queueing theory was Marcel Neuts, who in the mid-1970s introduced *phase* distributions [NEUT75] and showed that they had matrix representations which could be manipulated algebraically, while operating on state vectors corresponding to the queue length probabilities (one vector for each value of n, the queue length). He strongly argued that a matrix formulation could more easily be handled by computers than could integration or differentiation [NEUT81]. Also, since so many problems seemed to have a recursive solution, algorithms for their numerical evaluation became straightforward. However, he and his students concentrated most of their efforts attacking hitherto unsolved problems, and thus remained too abstract to be appreciated by the *practical users* (as I was then) of queueing theory. It seemed as though this was just another one of the many techniques one might use to solve a small set of problems.

This researcher became interested in the subject in the late 1970s in studying the problem of what happens to a subnetwork of exponential servers when the number of customers who can be active simultaneously is restricted. My students and I soon realized that if the subsystem was restricted to one active customer, then that subsystem was equivalent to a single server with a nonexponential (Coxian, or Kendall [KEND64], or RLT, or matrix exponential) distribution. Then, after John Carroll reduced the balance equations from second-order to first-order difference equations [CARR79], we independently, and virtually simultaneously with Neuts, found the explicit matrix geometric solution to M/G/1 and GI/M/C queues. The two papers appear back-to-back in the May 1982 issue of *Operations Research* [CARR82], [NEUT82]. I consider this to be the true beginning of LAQT, for then it became clear that many seemingly diverse problems could be solved using one technique and one viewpoint.

It is interesting to realize that the basis for LAQT was established by Erlang himself [ERLA17] when he represented a single server by a series of exponential stages, but linear algebra was not in vogue at the turn of this century, so queueing theory had to be developed entirely within the framework of what is called *modern analysis*. The *method of stages* is really a part of LAQT, distorted so it could fit into the classical view, while D. R. Cox's statement in the 1950s [COX55], that "Every pdf can be approximated arbitrarily closely by a function whose Laplace transform can be written as the ratio of two polynomials (RLT functions)" is really the basis for claiming that there exists a linear algebraic formulation of every problem which can be formulated otherwise.

You might question whether LAQT really is a peer to the *standard* variety of queueing theory. Well, for decades now, it has been standard technique in various areas of *applied mathematics* to replace differential operators on a solution function by an equivalent linear operator on a vector in Hilbert space. In fact, the pair of representations of quantum theory, Werner Heisenberg's *matrix mechanics* and Erwin Schrodinger's *wave mechanics*, is the prime example of this duality. The proof by John von Neumann that they are mathematically equivalent is closely related to Cox's completeness statement in extending A. K. Erlang's *method of stages* to include all functions with rational Laplace transforms [COX55]. Fortunately for physics, linear algebra was a known quantity by the 1920s, so the two viewpoints grew together and have become so intertwined that the typical quantum practitioner switches from one to the other and back again with little difficulty. A similar statement can be made about *linear control theory*. Both of those disciplines deal with functions of complex variables, even though what is actually observed must be real. If physicists can talk about the *charm* of *quarks*, which can never be seen outside their nuclear home, and electrical engineers can have imaginary currents, surely our customers should be allowed to travel with negative probabilities and complex service times from one phase to another, as long as they remain inside one subsystem or another.

The reader should avoid mapping this material onto already familiar techniques, at least until Chapter 4 has been covered. By then you will see the power and elegance of this methodology, as well as its usefulness, and be able to "switch back and forth without difficulty." Furthermore, since most solutions are in terms of matrix operations rather than integrals, or roots of equations, highly efficient algorithms for both single and parallel computer systems can easily be written. In fact, a software package that runs under the *MATLAB*† environment and that can solve all the examples and computational exercises in this book has been written by Dilip Tagare and Ed Bigos [LIPS90] and is available at cost from the author upon request.

† MATLAB is a registered trademark of MathWorks, Inc.

Organization

The book is laid out by chapter in order of increasing complexity of structure. There is more than enough material for a two-semester course, but a one-semester first course or a one-semester second course can easily be fashioned.

In Chapter 1 we make a quick survey of those topics normally connected to Markov chains. Chapter 2 starts out as a continuation of Chapter 1 by using the Chapman-Kolmogorov equations to set up the M/M/1 queue. But we soon switch to the simpler and intuitively more satisfying view associated with steady-state transition diagrams. Every queueing system is made up of two subsystems, each of which contains one exponential server. In Chapter 3 we show that by adding structure to a subsystem we give it a nonexponential (called matrix exponential) service time distribution. In Chapter 4 we combine the ideas of the two previous chapters to study the M/G/1 queue (i.e., one nonexponential and one exponential subsystem). As long as our system is closed (finite population of customers), there is no difference between an M/G/1//*N* loop and a G/M/1//*N* loop. But if the population is increased unboundedly, one or the other server will saturate. So, if the nonexponential server is the faster one, we have the open M/G/1 queue as given in Chapter 4. However, in Chapter 5 we assume that the exponential server is faster, and derive the properties of an open G/M/1 queue. In Chapter 6 two or more customers can independently be active at once in one subsystem, the M/G/*C* system. This increases the complexity of the mathematics required, as well as the computational complexity and size of matrices. But it also enormously increases the range of problems which can be solved, the so-called *generalized* M/G/*C* systems. In Chapter 7 we revert to one active customer per subsystem, but now both subsystems have structure, and we are dealing with a G/G/1//*N* loop. This leads to a different increase in complexity, requiring a *direct product* of vector spaces, which we must first discuss before actually finding the steady-state solution. Finally, in Chapter 8 we try to give a linear algebraic formulation which does not depend upon a physical interpretation of individual states. As such, it acts as a review of the book.

The chapters are all structured in more or less the same way, with obvious deviations because of the material. First we find the closed steady-state solution. Then we *open* the loop by increasing the customer population unboundedly. Then we look at certain specialized topics (e.g., load-dependent servers, renewal theory, comparison with other methods) Finally we explore the transient behavior of the appropriate queue.

A one-semester first course would cover Chapter 1 and the steady-state parts of Chapters 2, 3, 4, and 5. Depending on the background of the students, the instructor might want to add some descriptive material to Chapters 1 and 2.

Assuming that students have already had a course in queueing theory, but not one that covered LAQT, a one-semester second course would skim through Chapter 1 and the first part of Chapter 2. But then Section 2.3 must be covered in earnest, as must the first part of Chapter 3. Except for the material on *residual times*, which must be covered, Section 3.3 can be omitted. Most of

Chapters 4 and 5 should be covered, but the instructor can skip Chapter 6 if desired and go directly to Chapter 7. However, since Chapter 6 is potentially of great practical importance, the instructor may prefer to skip Chapter 7 instead. Chapter 8 can be put in or left out, as per taste.

A two-semester course can be given which combines the two one-semester courses in the order just described, or one can go sequentially from beginning to end, skipping those topics which seem inappropriate. However, one cannot study Section 6.4, for example, without first covering the related material in Chapters 2, 4, and 5.

Acknowledgments

I would like to thank Professor Howard Sholl and the Booth Research Center for their continued support in the technical creation of this book. In particular, Anthony Guzzi has rewritten DITROFF† so it actually does what it is supposed to do (at least on my work station). His devotion to my needs has been beyond the call. Also, I thank John Marshall and Sue Zajac for keeping the system up (most of the time) and the secretaries (Jean, Sue, Ruth, Sandi, and Sherry) for keeping *me* up (most of the time). To my former students, now collaborators, Appie van de Liefvoort (University of Missouri-KC), Aby Tehranipour (Eastern Michigan University), and Yiping Ding (now at BGS Systems, Inc.), I give thanks for technical advice in the various chapters where they are experts. Thanks to Seva nanda Adari, Jinzhu (Jim) Chen, and Houzhong Yan for reading the first draft and pointing out how ideas could be made clearer. I thank the students who were in my class in the fall of 1990 (Somnath Deb, Sharad Garg, Rudi Hackenberg, Chengdong Lu, Jim Moriarty, Carolyn Pe, and Cien Xu), who used the second version of this book and searched for errors of content. Siddhartha Roy and Dilip Tagare meticulously went through the final draft, searching for errors of all kinds (and they found many). Dilip was also responsible for generating the 30 or so graphs and, with Ed Bigos, has created a software package which can be used with this book. To Professors George Nagy (Rensselaer Polytechnic Institute); Sharad Seth (University of Nebraska); Don Towsley (University of Massachusetts); Victor Wallace (University of Kansas); and Arnie Russek (*my* thesis advisor), Jim Galligan, and Krishna Pattipati (all of the University of Connecticut), thanks for useful critical comments. I must also thank Macmillan Publishing Company's Ed Maura for taking the initiative in inviting me to do the book, John Griffin who oversaw the project from cover to cover, the unknown proofreader who went through the text with the devotion of a mother combing her daughter's hair, and especially Leo Malek who was determined that this would be a good-looking book. Thanks to Erikson/Dillon Art Services for creating such fine figures and page layout, and finally, Janet Pecorelli and the American Mathematical Society's printing service in Providence, RI, for producing such high-quality galley proofs at short notice.

† DITROFF is a registered trademark of AT&T.

CONTENTS

CHAPTER 1
INTRODUCTION

The ultimate Markov assumption: "Today is the first day of the rest of your life."

The author is often asked what *queueing theory* is. First we state that *queueing* is the only word in the English language with five successive vowels (there is an alternative spelling that deletes the second e, but it is obviously inferior) and that a queue is a line of customers waiting to be served, such as one sees in banks, supermarkets, and fast-food outlets. More often than not, the questioner will interrupt to say, "You've been doing a bad job, and common sense would tell us how to do things better." Of course we have not even begun to explain where the *theory* comes in (*So, you're in queueing?*). The goal of a mathematical modeler (the class of researchers to which queueing theorists belong) is to describe and understand what is really going on, for only then can someone (not necessarily the modeler) make an informed decision on what should be done to improve things. You, the reader, presumably already know what a queue is and will have the patience to learn some of the theory, particularly that related to our *linear algebraic approach to queueing theory* (LAQT), which doesn't show up until Chapter 3.

1.1. BACKGROUND

Any system in which the available resources are not sufficient to satisfy the demands placed upon them at all times is a candidate for queueing analysis. This is quite a general statement, but we will keep our picture as simple and as explicit as possible. We will be dealing with *subsystems* (or *service centers*), denoted by $S_1, S_2, \cdots, S_m$. *Customers* then wander from one subsystem to another, perhaps forever. We will not be so ambitious as to consider in detail more than two subsystems at a time, but we will allow each subsystem to have one or more *servers* in it, where each server is itself made up of one or more *stages*, or *phases*. But we are getting ahead of ourselves with such detail.

1.1.1. Basic Formulas

What can we say about such systems? What do we know? Well, the single most important rule is *Little's theorem* [LITT61], which we now describe. Consider an arbitrary subsystem, as shown in Figure 1.1.1. Customers come and go and wander from one service center to another. An outside observer can count the

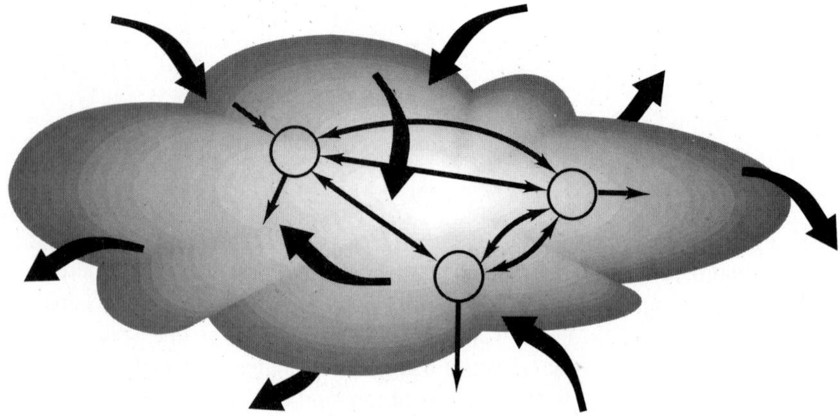

Figure 1.1.1. Arbitrary subsystem of servers. Any number of customers may enter the subsystem (therefore, it is *open*) and travel from one server to another repeatedly before leaving. Customers are not *marked*, therefore if one leaves and reenters, he is counted twice.

number of customers who enter that subsystem over a period of time t, symbolizing that count by $N(t)$. The same customers may have come and gone more than once, but they are counted each time. After a very long period of time, we would suppose that the difference between the number who have arrived and the number who have left is negligible compared to either (no one stays forever). Then the observer would say that the measured *arrival rate*, $N(t)/t$, approaches a constant,

$$\Lambda := \frac{N(t)}{t}$$

after a very long time t.

"How long is *very long*?" is an important question, so it pays to pause for a moment to discuss it. Mathematicians and statisticians have a procedure for dealing with this; they say, "in the limit as t goes to infinity," or

$$\Lambda = \lim_{t \to \infty} \frac{N(t)}{t}. \tag{1.1.1a}$$

Because it actually *is* important, we discuss seriously what is really meant by this limit. The mathematicians' definition of (1.1.1a), as you may recall, is the following.

Definition 1.1.1

For all $\varepsilon > 0$, *there exists a* t_0 such that for all $t > t_0$, the following is true

$$\left| \frac{N(t)}{t} - \Lambda \right| < \varepsilon,$$

where $|X|$ denotes *absolute value of* X. ◊†

On the other hand, when dealing with measurements, even those that could be carried out exactly, we can talk only about the probability that something will happen. Thus the statisticians' definition of (1.1.1a) is the following:

Definition 1.1.2

For all δ *and* $\varepsilon > 0$, *there exists a* t_0 *such that for all* $t > t_0$, the following is true:

$$\Pr\left(\left| \frac{N(t)}{t} - \Lambda \right| > \varepsilon \right) < \delta.$$

The symbol $\Pr(X)$ stands for the phrase "*the probability that the expression represented by* X *is true*." ◊

Depending on the context, we will mean one or the other definition when we write something like (1.1.1a). The reader should spend a few moments reviewing these two ideas.

We now return to our discussion of Little's theorem. As a second measurement, our observer could keep track of how long each customer spends in the subsystem, for each visit, calling it T_i for the ith visitor. Then the average time spent in the subsystem by a *typical* customer is given by

$$\overline{T} = \frac{1}{N(t)} \sum_{i=1}^{N(t)} T_i \tag{1.1.1b}$$

for very large t (as $t \to \infty$).

As a third measurement, or set of measurements, our observer might frequently count how many customers are in the subsystem at any particular time, and call it n_i. If she does this often enough, say, m times, she can claim that the average number of customers in S at any time is given by (as $m \to \infty$)

$$\overline{q} = \frac{1}{m} \sum_{i=1}^{m} n_i . \tag{1.1.1c}$$

Little's theorem relates these three measurements by the simple formula

$$\overline{q} = \Lambda \overline{T}. \tag{1.1.2}$$

† Symbol ◊ designates end of definition.

Although its proof is not very difficult, we will accept it as law, since the proof can be found in many books, for example, [KLEI75] or [MOLL89]. This law tells us that given any two of the performance parameters, the third parameter is uniquely determined by (1.1.2). In other words, the three measurements, in principle, are not independent of each other. In studying real-world systems, cautious experimenters will usually measure all three parameters and then use Little's law to check for self-consistency and/or reliability of data. In mathematical modeling, since the limit as t goes to infinity can be taken correctly, (1.1.2) holds exactly (except for some pathological systems that we ignore here).

The second most important formula, and the first one always derived in any discussion of queueing systems, is the *steady-state solution of the open* M/M/1 *queue*. We derive and discuss this in Chapter 2, but for now we merely look at the result. Suppose that customers arrive randomly and independently of each other to a lone server and that the average rate at which they arrive is given by the parameter λ. Suppose further that the time between arrivals is a random number taken from the exponential distribution, with mean $\bar{x}_2 = 1/\lambda$ and that the arrivals are independent of each other. This is known as a *Poisson arrival process*, which we will run across again and again throughout the book. Let X be the time needed by a customer once he gets to be served. The actual time he needs is a random number, also taken from an exponential distribution, but with mean $\bar{x}_1$. We have thus described the M/M/1 queue. The M stands for *memoryless*, or *Markovian* (nobody seems to know which), and means for us that the process being represented by M comes from an exponential distribution. The first symbol, [A], in the notation

$$\mathrm{A}/\mathrm{B}/C$$

describes the arrival process, the second symbol, [B], describes the service distribution, and the third symbol, [C], tells us how many servers there are in the subsystem. Thus we have a Poisson arrival process [A = M] to a single [C = 1] exponential server [B = M].

We next define the *utilization factor* (or *parameter*) to be

$$\rho := \lambda \bar{x}_1 = \frac{\bar{x}_1}{\bar{x}_2}. \tag{1.1.3}$$

Suppose, for instance, that customers need 9 minutes of service, on average, and that they are arriving at the rate of 6 per hour (or 10 minutes between arrivals, on average); then $\rho = 0.9$, and we would expect our server to be busy 90% of the time. Therefore, it will be idle $1 - \rho = 0.1$, or 10% of the time. In Chapter 2, Equation (2.1.6b), we will show that

$$\bar{q} = \frac{\rho}{1-\rho}, \tag{1.1.4a}$$

and from Little's theorem,

$$\bar{T} = \frac{\bar{q}}{\lambda} = \frac{\bar{x}_1}{1-\rho}. \tag{1.1.4b}$$

According to these formulas, the average customer (remember, a very large number of customers have gone through) will arrive at a queue that already has nine other customers in it (counting the one in service) and will have to wait 90 minutes (give or take 10) from the time he arrives at the subsystem to the time he leaves. This behavior is represented in Figure 1.1.2 by the curve labeled M/M/1.

Before going on, we must clarify what is meant by "*will see, on average*," since in this case an arriving customer will see *more* than nine customers in the queue one-third of the time, and one-third of the time he will see three or fewer. In fact, he will see exactly nine customers less than 4% of the time. Actually we are still being loose with our words. What we really mean is that *a customer will find nine customers in the queue with a probability less than* 0.04. More rigorously, we say the following:

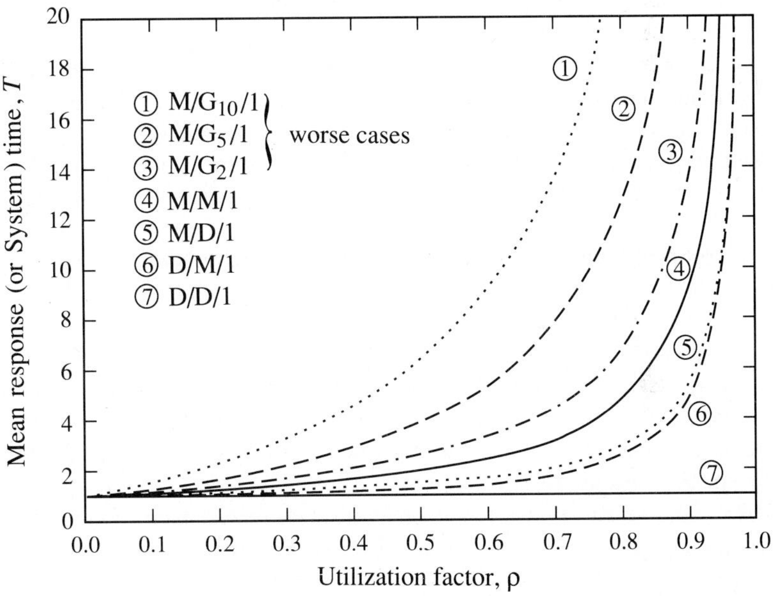

Figure 1.1.2 Steady-state mean response times for various single-server queues. The horizontal axis is the *utilization factor*, ρ, from (1.1.3). The average arrival rate must be smaller than the service rate ($\rho < 1$), otherwise the queue will back up indefinitely. All but two of the curves represent queues with Poisson arrivals, for which (1.1.6) is applicable. Their coefficients of variation are $C^2 = 0$ (M/D/1), $C^2 = 1$ (M/M/1), and $C^2 = 2, 5, 10$ (*worse cases*). The other two curves have *deterministic* arrivals (the time between successive arrivals is constant) corresponding to the D/M/1 and D/D/1 queues, respectively. See the text for further details.

Definition 1.1.3

Let N be a random variable designating the number of customers an arriving customer finds in the queue. Then

$$\Pr(N = 9) < 0.04.$$

The *average* or *mean* number of customers is given by

$$E(n) = \bar{n} := \sum_{n=0}^{\infty} n \Pr(N = n), \tag{1.1.5a}$$

where we read

$$E[f(n)] := \sum_{n=0}^{\infty} f(n) \Pr(N = n) \tag{1.1.5b}$$

as the *expected value* (or *expectation value*) *of* $f(n)$. ◊

Because the precise terminology is so bulky, we will tend to use the vague expressions that we hope we have clarified in this section. The reader should always be prepared to insert the precise wording when necessary.

Returning to the M/M/1 queue, we see an apparent contradiction. From an outside observer's (e.g., manager's) viewpoint, the server is idle (dawdling) 10% of the time, while from a customer's point of view, the queue is *always* very long. The explanation has to do with the unpredictability of arrivals and time for service, for sometimes customers will seem to come in bunches, and sometimes one customer will require far more than the average service time. For instance, the probability that 2 or more customers will arrive in one mean interarrival time, is greater than 26% [$\Pr(n \geq 2) = 2/e$], while 13% ($1/e^2$) of the customers will need more than twice the mean service time. These large fluctuations will cause the queue to back up at times. Once the queue backs up, it will be difficult for it to drain. As an example, suppose that at some time there are 10 customers at S. Then it will take about 90 minutes (on average) to service them. But in that time, approximately 9 new customers will arrive, and it will take another 81 minutes to satisfy them. So on and on it goes. The inverse question, "How long will it take to get 10 customers in the queue in the first place?" is also important. After all, some systems may not exist long enough to reach their steady state. In this case, (2.3.3b) tells us that we can *expect* over 13 hours to elapse before an observer will find 10 customers in the queue, and over 80 customers will have come and gone by then. We will look at this *transient behavior* very closely throughout the book.

This very simplest of systems tells us that if you try to keep your server busy most of the time, you will have to pay for it in vastly degraded service to your customers. This has nothing to do with overworking your servers and thereby making them less efficient or tired or lazy. Nor is it due to the arrival of an unexpectedly large group of customers all at once. It is due *entirely* to the irregularity and unpredictability of arrivals and service demands. More complicated or more sophisticated or more realistic systems share this behavior; that is, they all depend on the term $1/(1-\rho)$. Unfortunately, these explanations are not intuitively satisfying to the typical observer, despite their validity. Somehow, people always say, "If I were in charge, I would do things better." Even people who *are* in charge say it, but they say, "If I *really* were in charge $\cdots$." This only points out all the more strongly that we, the human species, have extremely poor intuition concerning statistically fluctuating phenomena. This is all the more reason to have mathematical models to protect us from our faulty feelings. That is why the M/M/1 queue is so important even though it oversimplifies almost all real systems. However inaccurate it is, it is far more reliable than our intuition, because it contains that ubiquitous denominator $1-\rho$.

What, then, can be done to improve service? Obviously, more servers can be added (i.e., go to an M/M/C queue in Section 2.1.4, where C is the number of customers who can be served simultaneously), if more money is available. Alternatively, the behavior of the customers can be controlled for example, by not permitting them to demand much more service time than the average customer gets. (*Sorry, your time is up.*)† In the early 1930s, F. Pollaczek [POLL30] and A. Y. Khinchin [KHIN32] separately studied the steady-state M/G/1 queue (the G means that the service distribution is *General*; i.e., it can be almost anything) and derived what has come to be known as the P-K formula, which we discuss in detail in Chapter 4. That formula [Equations (4.2.5)] shows that the mean number of customers waiting for service (counting the one being served) depends only on their arrival rate λ, and the mean $[\bar{x} = E(x)]$ and *variance* $[\sigma^2 := E(x^2) - \bar{x}^2]$ of the service time distribution. It is usually expressed in terms of the *coefficient of variation*, $C^2 := \sigma^2/\bar{x}^2$, as given here.

$$\bar{q} = \rho + \frac{\lambda^2}{1-\rho}\frac{E(x^2)}{2} = \frac{\rho}{1-\rho} + \frac{\rho^2}{1-\rho}\cdot\frac{C^2-1}{2}, \tag{1.1.6}$$

This reduces to (1.1.4a) when $C^2 = 1$. It shows that even if every customer were given exactly the same amount of service time (i.e., $C^2 = 0$), the mean queue length would only be reduced to half, and $1-\rho$ is still in its denominator. Furthermore, if no constraint were placed on truly selfish (needy, demanding, important) customers, the mean queue length (and mean waiting time) could become arbitrarily large (i.e., when $C^2 \gg 1$). That is, there is no upper bound on how bad it could get. Equation (1.1.6) for $C^2 = 0$ is plotted on Figure 1.1.2 with the label *best of Poisson arrivals* (M/D/1). The *worse cases* are for various values of C^2 greater than 1. There is no *worst* case, since we can always find a service distribution with with a larger C^2.

Another way to modify the performance of a single steady-state queue is to control the arrival pattern of customers by, for instance, scheduling them to come

Only complete control of both arrivals and service times will yield the desired efficiency of no waiting. This is shown in Figure 1.1.2 by the horizontal line labeled *ideal* (D/D/1). But in that case, will the customers really get what they

† Note, however, that if this is done, not only will the fluctuations be reduced, but the *mean* service time will be reduced as well. Customers will not necessarily get what they came for.

came for? Our job as queueing theorists is to analyze systems with *given*, or *possible*, performance characteristics as described by their arrival and service distributions. Optimally, we would prefer to leave the arrival and service demands (needs) alone and change the inanimate system characteristics. We leave it to the CEO's, politicians, management consultants, and other self-proclaimed *efficiency experts* to modify or control customer and server behavior to suit their goals.

Since the early 1970s, networks of queues have been studied and applied to numerous areas in computer science and engineering, with a high degree of success. The basis for this success was due to Jackson [JACK63], and Gordon and Newell [GORD67], who showed that certain classes of steady-state queueing networks with any number of service centers could be solved using a *product-form solution*. But it was not until Buzen [BUZE73] showed that the ominous-looking formulas the previous researchers had derived were actually computationally manageable that *performance analysis of queueing networks* began to blossom into a research field of its own. The theory has ultimately been extended to include, for instance, multiple classes, other service time distributions if the queueing discipline is not first-come, first-served (FCFS), and state-dependent routing [BASK75]. *Jackson networks*, as they are now called, have been so successful in so many areas that it is hard to see where they do not apply. Their success lies in their ability to fit the measurements of any given queueing network. The reason is that the product-form solution has enough free parameters in it to fit anything (see Section 4.4.3). They also contain, hidden within them, the all-important denominator $1-\rho$. The fact that it is hidden within the complex formalism can be valuable, since questioners are then unlikely to say "I can do better." One is far less likely to argue with the output of a sophisticated computer program which requires an enormous amount of data input, than with an algebraic formula or the verbal arguments of an "expert."

But the ability of Jackson networks to *predict* is an open question. It is important to state that *they do not apply to systems where there are population size constraints*, or to *non steady-state systems*, or *nonexponential servers with FCFS queueing discipline*. This book therefore goes in a direction orthogonal to that covered by Jackson networks. Only in Chapters 4 and 6 do we discuss the connection. We show in Section 4.4.3 in what sense they are not valid for FCFS M/G/1 queues. In Section 6.2.3 we show that a *generalized* M/G/C//N network reduces to a (single-class) Jackson network when $N \le C$. The meaning of all this will become clear as the reader goes through the book. At the moment it is only
at 10-minute intervals, as doctors and dentists do. In Chapter 5 we will look at the G/M/1 queue and see that even if customers come exactly at their appointed times, the waiting time, [from Equation (5.1.7c)], will again only be cut approximately in half (again with a variant of $1-\rho$ in its denominator), because of the uncertainty of how long service will take. This is shown in Figure 1.1.2 with the curve labeled *best of exponential servers* (D/M/1) [COHE69].
important for those already familiar with Jackson networks to realize that there is much in queueing theory that is not covered by Jackson networks.

As you might surmise from this discussion, just about everything that has been done in queueing theory has assumed the steady state. Very little is known about transient behavior, for one of two reasons. We do not know which of these two statements is valid:

1. Transient behavior is unimportant; therefore, it is not studied.

2. Transient behavior is too difficult to measure and analyse; therefore, it is declared to be unimportant.

A growing number of researchers (including this author) have *declared* that transient behavior should be considered important (see, e.g., [NEUT77]); therefore, we will devote a considerable amount of space and effort in each chapter to its analysis. If it should no longer prove too difficult to study, perhaps more researchers will agree that it is important.

1.1.2. Markov Property

The reader is not expected to know anything special about *Markov chains*, the property that Markov introduced and built on in 1907 [MARK07]. It is, however, an important underpinning of the approach expounded here. We must first describe what we mean by a *state*. A complete specification of a system or subsystem is collectively called a state. No two states can have the same complete specification; therefore, they must differ in at least one aspect. For example, for the purposes of coin-flipping, a coin can be in only one of two states, heads or tails. Two coins, collectively, can be in one of three states, *HH, HT, or TT*. So after the flips we could say (assuming it was true) that *the system is in state HT*. We also use the word to describe the probability that the flips will result in one state or another. That is, we could say before the flips (or without having seen the results) that *the system is in state*

$$[\,0.25\,,\,0.50\,,\,0.25\,],$$

corresponding to the probability for each of the three states to occur. If we must make a distinction, we will call the former a *pure state*, and the latter a *composite state*. We will also refer to the triplet of values as a *state vector*. Now you might ask if *HT* is completely specified, since it could have been *ht* or *th*. From the coin flipping game point of view, or if we cannot tell which coin is which, there is only the one *external state, HT*, but it has two *internal states*. The other two external states, *HH* and *TT*, have only one internal state each. The set of internal states corresponding to an external state is referred to as its *state space*. Thus we say that

$$\Xi_{HT} := \{\, ht\,, th \,\}$$

or that *ht is an element of* Ξ_{HT}, written

$$ht \in \Xi_{HT}.$$

If this seems confusing, it will become clearer over time, since we will use these terms regularly. For instance, suppose that you are studying a subsystem, represented by the symbol S. If you could look inside you would see, say, three exponential servers (which we call *phases*). Next suppose that only one customer can be inside the subsystem at a time, and there are five customers there altogether (one inside and four outside). Then we say that the subsystem is in external state [5] and that the system has three internal states. We might write $\Xi_5 = \{ 1, 2, 3 \}$.

In general, the sum of probabilities of being in a set of internal states with the same external state will be the probability of being observed in that external state. This is quite analogous to the terms "sample" and "event" used in many probability texts, where an event is a set of samples, and the probability of an event is the sum (or integral) of probabilities over the sample points [TRIV82]. In our case, the external states are mutually exclusive, and the internal states (sample points) may not be individually measurable or even physically meaningful (we may not always be allowed to *look* inside S). Even so, we will use the rather picturesque description of customers meandering, sometimes with negative probabilities, through networks of exponential servers (phases), whose service times may be complex numbers. Even if this annoys the realist within each of us, it helps us to picture and remember the process being discussed and to distinguish it from similar processes that might also be of interest. But in the end, the mathematical conclusions *must* be correct if the theory is to be meaningful.

Suppose that a system can be completely described as being in one of a countable (either finite or infinite) number of states. Since the set of states is discrete, the system cannot gradually go from one state to another. Therefore at a later time it will *hop* to another of those states. In time, then, the history of the system can be described by a sequence of states. Such a sequence is called a *chain*. The *Markov property* states that the probability that the system will be in a particular state at the next moment of time (i.e., after the next hop) depends only on the state it is in now, *not* where it was previously. A *Markov chain* is a sequence of states generated by a process that satisfies the Markov property. This abstract idea will become meaningful once we look at some simple systems.

Although not all aspects of queueing theory are described by Markov processes, there are few known analytical techniques that go beyond the Markov property. Thus we should say a few words about the so-called "memoryless" property. Only a system with one state is truly memoryless. A system can be extended to include pseudostates that serve the purpose of "remembering" some of the past. It is not uncommon to construct such states even though they are not observable, as long as the formalism is maintained.

The question then is: What is a non-Markovian system? This can be answered in the following way. In general, a system's future behavior depends on its entire past history and thus it must "remember" everything. A Markovian system, on the other hand, can remember only a part of its history and thus must discard old information as new events occur. Two points follow directly from this idea. First, for short amounts of time (depending on the size of the state space), a Markovian model would be an excellent representation of a non-Markovian system.

Over long periods of time, however, a Markovian system will forget its initial state and thus would be a poor approximation for those systems that do depend on their initial state.

1.1.3. Notation, Pronouns, Examples, etc.

The following notational standards will be adhered to as closely as possible. All matrices (two-subscripted arrays) will be represented by boldface capital letters (e.g., **M**), while their components will be noted in either of two ways, M_{ij} or $(\mathbf{M})_{ij}$, depending on the context. Similarly, all vectors (single-subscripted arrays) will be represented by boldface lowercase letters [e.g., **v** has components v_i or $(\mathbf{v})_i$]. Row vectors and column vectors play distinctly different roles in the formalism presented here. As in many books on matrix theory, the symbol " ' ", means *transpose*, but we will always be interested in an object *or* its transpose, never both, so $\mathbf{v}'$ will always denote a column vector. Sometimes we will discuss a set of vectors or matrices, such as $\{\mathbf{v_1}, \mathbf{v_2}, \mathbf{v_3}\}$. In this case, the subscripts are also set in boldface type. So the jth component of the ith vector is $(\mathbf{v_i})_j$.

We also strictly adhere to the following convention on the use of pronouns. We are always talking about *customers*, the *author*, *random* or *outside observers*, *servers*, *service centers* and *subsystems*, and the *reader*. To minimize the ambiguity, we will always refer to the reader as *you*; a customer is *he*; an observer (random or outside) is *she*; and a server, service center, or subsystem is *it*. Thus the following statement has an unambiguous meaning: "*We* point out to *you* that *she* sees *him* enter *it*." Translation: "The reader should note that the observer sees the customer enter the subsystem." However she may not be able to see what he does after he enters (although she might figure out what he is *probably doing*).

All equations, definitions, figures, examples, and exercises are numbered in sequence by chapter and section (but not subsection). Thus "Figure 2.3.4" is the fourth figure in Section 2.3. Also, "(4.1.13d)" is the fourth [d] equation in the thirteenth set of equations in Section 4.1, while "Equations (4.1.13)" refers to all four of (4.1.13a), (4.1.13b), (4.1.13c), and (4.1.13d). Note that an object like "(4.1.13)" without a qualifier *always* refers to an equation. Otherwise we say "Definition 4.5.7," and so on. Since *lemmas* are really *theorems*, and both can have *corollaries*, we have chosen to number them together in a single sequence. Thus we have Lemma 4.2.1, Theorem 4.2.2, and Corollary 4.2.2, but no Theorem 4.2.1. Clearly, Corollary 4.2.2 is a corollary to Theorem 4.2.2.

The symbol ":=," as used in expressions of the form

$$A := B,$$

means *Symbol A is defined by expression B.* In such cases, A is appearing for the first time.

We have given many *examples* throughout the book, most of them involving numerical computation, invariably summarized by a family of curves in a graph. Most of the *exercises* we have asked the student to perform are proofs or other mathematical manipulations. The examples can easily be made into exercises by having the student redo the example using a different distribution function. In a class environment, each student can be assigned a different function. Then a comparison study can be made by the class as a whole to see how the different functions affect the particular phenomenon being studied.

1.2. DISTRIBUTION FUNCTIONS OVER TIME

We use the word *system* in referring to a closed entity, one in which customers neither enter nor leave, while a *subsystem* is one to which customers come and go. The simplest of all subsystems has only one state, with at most one customer. Then that state is either occupied or unoccupied. In the next two sections we will see how such a simple system evolves in time, where we assume that events could occur at any time (continuous) or at equally spaced moments (discrete).

1.2.1. Exponential Distribution (Continuous Time, t)

Let $R(t)$ be the probability that the subsystem is busy at time t, where t is a continuous parameter. Assume that at $t=0$ it definitely was busy $[R(0)=1]$. The probability that it will still be busy at a time $t+\delta$ is, by the memoryless property, equal to the probability that it was busy at time t, $[R(t)]$, times the probability that it is still busy a time δ later, $[S(\delta, t)]$; that is, $R(t+\delta) = R(t) \times S(\delta, t)$. Now, whatever else S is, it must have $S(0, t) = 1$, and if we assume that it is a smooth function, it is expandable in a Maclaurin series, $S(\delta, t) = 1 - \mu\delta + O(\delta^2)$, where μ also depends on t (but not on δ). At this point we make the assumption that S does not depend on its second argument [i.e., $S(x, t_1) = S(x, t_2)$], thus making μ a constant, the meaning of which, will be clear in a moment. Such systems are called *homogeneous*. Then

$$R(t+\delta) = R(t)[1 - \mu\delta + O(\delta^2)].$$

Subtracting $R(t)$ from both sides of this equation and dividing by δ, we get

$$\frac{R(t+\delta) - R(t)}{\delta} = -\mu R(t) + O(\delta).$$

Next let δ go to 0 and get

$$\frac{dR(t)}{dt} = -\mu R(t). \qquad (1.2.1a)$$

It is well known, and can be proven by direct substitution, that the solution of (1.2.1a) is

$$R(t) = e^{-\mu t}, \tag{1.2.1b}$$

an exponential function.

Let us pause here to discuss some notational difficulties and conventions. We enumerate some well-known terms from probability theory. For any distribution, $R(t)$ is the probability that the event being awaited has not yet occurred by time t, and is often called the *reliability function* for the subsystem. The probability that it will have occurred is

$$B(t) = 1 - R(t) \tag{1.2.2a}$$

and is called the *probability distribution function* (*PDF*), with derivative

$$b(t) := \frac{dB(t)}{dt}, \tag{1.2.2b}$$

called the *probability density function* (*pdf*).

A slight terminology problem shows up when we deal with functions. For instance, is $f(x)$ different from $f(t)$? If one is thinking of the function as a whole (i.e., the entire set of points, $\{(x, f(x)) \mid 0 \le x < \infty\}$, or equivalently, $\{(t, f(t)) \mid 0 \le t < \infty\}$), they are the same set. In other words, the graphs of the two expressions are the same curve. Therefore, mathematicians will sometimes use $f(\cdot)$ instead, to mean that *any* symbol could go inside the parentheses. Extending the confusion, we often write

$$\bar{t} := \int_0^\infty t f(t)\, dt,$$

but since t is a *dummy* variable, this integral is the same as

$$\int_0^\infty x f(x)\, dx$$

(or any other symbol), which we would be inclined to represent by $\bar{x}$. We see, then, that the important information is the f, not the variable symbol. The situation gets even more complicated when we are dealing with several functions at the same time. To get around this (and for other reasons), statisticians use the idea of a *random variable* which is associated with a given function. One now says:

Definition 1.2.1

Let T be a random variable, distributed according to $f(t)$ [or $f(x)$, or even $f_T(x)$]. Then we can write

$$\mathrm{E}(T) := \int_0^\infty t\, f(t)\, dt = \int_0^\infty x\, f(x)\, dx. \tag{1.2.3a}$$

(Random variables are *always* capital letters.) In words, we read this as: "the *expected value* (or *expectation value*) of T is equal to $\cdots$." In general, we can write

$$\mathrm{E}(T^n) := \int_0^\infty t^n\, f(t)\, dt. \tag{1.2.3b}$$

We read this as: "The *expected value* of T^n" or "the nth *moment of* $f(t)$." The symbol $E(X)$ is a different object, since X must be a random variable distributed according to a *different* function, perhaps $g(x)$ or $f_X(x)$, or even $f_X(t)$. ◊

We will tend to use the *bar* notation when it is not too confusing. Also, both letters, x and t, will be used as the *time* variable. Therefore, in general, for any function $h(t)$ we write

$$E(h(T)) := \overline{h(t)} := \int_0^\infty h(t) f(t)\, dt, \tag{1.2.3c}$$

where T is distributed according to $f(\cdot)$. If $h(t) = t$, we have the *first moment* of $f(t)$ or *mean value*, or *expected value* of T. Another much used function is the *variance*, symbolized by σ^2 and defined by

$$\sigma^2 := E\left([T - E(T)]^2\right) = \int_0^\infty \left(t - \bar{t}\right)^2 f(t)\, dt \tag{1.2.4a}$$

which can be shown to be equal to

$$\sigma^2 = E(T^2) - [E(T)]^2 = \overline{t^2} - \bar{t}^2. \tag{1.2.4b}$$

The *standard deviation* of $f(t)$ is symbolized by σ, which satisfies the obvious, $\sigma := \sqrt{\sigma^2}$. In words, σ represents the average *spread* about the mean; the smaller σ is, the narrower the distribution. Since we will usually deal with functions that are defined only for positive t, a *relative* width is often useful. Therefore, we have the *coefficient of variation*, defined by

$$C^2 = \frac{\sigma^2}{[E(T)]^2}. \tag{1.2.4c}$$

We hope this discussion has not brought on more confusion than it has allayed. We have found that trivial notational problems such as these often prevent understanding of expressions with which the reader would otherwise have no trouble.

Let us return to where we were before the pause. Continuing from (1.2.1b) and (1.2.2a), the pdf for this function is

$$b(t) = -\frac{dR(t)}{dt} = \mu\, e^{-\mu t}, \tag{1.2.5a}$$

the exponential distribution. The mean lifetime for the process (when we are dealing with time variables, the expected value of a function is usually referred to as the *mean lifetime*, or *mean service time*, or simply, *lifetime*) is

$$E(T) := \int_0^\infty t\, \mu e^{-\mu t}\, dt = \frac{1}{\mu}. \tag{1.2.5b}$$

The reciprocal of the mean lifetime, in this case μ, is interpretable as the *mean service rate*, or the mean rate of leaving. The nth moment for the exponential distribution equals $n!/\mu^n$. The variance is $1/\mu^2$ and the coefficient of variation is equal to 1.

1.2.2. Geometric Distribution (Discrete Time, n)

Suppose that events can occur only at discrete moments of time, such as at the tick of a clock, and suppose that the subsystem stays busy at each tick with probability p. If $R_0 = 1$, then $R_1 = p$ and $R_2 = pR_1 = p^2$. In general, the probability R_n that it will still be busy by the nth step is equal to $p \cdot R_{n-1}$, from which it follows that

$$R_n = p^n. \tag{1.2.6a}$$

R_n is the discrete analog of $R(t)$, so we could call it the *discrete reliability function.* The analog to $b(t)$ (sometimes called the *probability mass function* or *discrete density function*), symbolized by b_n, is the probability that the server will finish in exactly n steps. It is known that b_n is the *geometric distribution*, or the negative binomial distribution of order 1, but we calculate it here by doing the analog of differentiation:

$$b_n = R_{n-1} - R_n = (1-p)\, p^{n-1}. \tag{1.2.6b}$$

The expected number of steps before completion is simply

$$\mathrm{E}(n) := \sum_{n=1}^{\infty} n\, b_n = (1-p) \sum_{n=1}^{\infty} n\, p^{n-1} = \frac{1}{1-p}. \tag{1.2.6c}$$

Equation (1.2.1b) and Equations (1.2.5) are much closer to Equations (1.2.6) than it would seem by superficial examination. Suppose that although time is a continuous parameter, the system of Section 1.2.1 is examined only at regular intervals, as with moving pictures. Let δ be the time between snapshots. Then $t = n\delta$. Using this in (1.2.1b), we get

$$R(t) = e^{-\mu n \delta} = \left(e^{-\mu\delta}\right)^n. \tag{1.2.7a}$$

Let $p = e^{-\mu\delta}$; then $R(t) = R_n$. At least as far as the reliability function is concerned, a discrete time system is indistinguishable from a continuous-time system in which observations are made at regular intervals. Equations (1.2.5b) and (1.2.6c) do not yield identical values, but the following inequality is satisfied:

$$\delta[\mathrm{E}(n) - 1] < \mathrm{E}(T) < \delta\,\mathrm{E}(n). \tag{1.2.7b}$$

The proof follows directly by substituting for p and letting $u = \mu\delta$. Then (1.2.7b) converts [after multiplying all terms by $(e^u - 1)/\delta$] to the inequality

$$1 < \frac{e^u - 1}{u} < e^u, \qquad u > 0.$$

Equation (1.2.7b) says that the uncertainty in E(T), when measured to the nearest (rounded up) multiple of δ, is less than one time unit, which is as close as a discrete and continuous system can come to each other. This is true even for more general systems, where the strict inequality may be replaced by $\leq$.

1.3. CHAPMAN-KOLMOGOROV EQUATIONS

We now consider a system that has many states of possible existence. In Chapter 2, when we deal with queues the states will be explicitly described. For now it is sufficient to consider a state to be one possible complete specification of the system's condition. The system can be in one and only one state at a time, and in the course of time it will change from one state to another. The set of all possible states is called the *state space*. Probability books often identify these with *samples* in a *sample space*. If the space is finite, or at most countably infinite, we have a *discrete state space*. We will be interested exclusively in systems with discrete state spaces.

As our system evolves in time, it must "jump" from one state to the next, since there is no continuum of states in a discrete space to match the continuous time parameter. A sequence of such states is called a *chain*, and if the Markov property holds, we have a *Markov chain*. Of course, time can be continuous or discrete, giving a *continuous* or *discrete Markov chain*. If the state space is uncountable, change *chain* to *process*.

1.3.1. Continuous Time

As with most expositions purporting to start from scratch, the first few sections are overladen with definitions. Let i and j take on positive integer values, corresponding to the possible states of the system. Then

Definition 1.3.1

$\Xi := \{ i \mid i \textit{ is a state of the system} \}$. We read this as: Ξ *is the set of all i, such that i is a state of the system.* We will also call i a *pure state* of the system. If Ξ is a finite set of states with, say, m members (i.e., $m = |\Xi|$), we can write $1 \le i \le m$, or $i \in \Xi$ (i is an element of Ξ). ◊

Next, define the following:

Definition 1.3.2

$\pi_i(t) :=$ *probability that the system will be in state* $i \in \Xi$ *at time* t. $\boldsymbol{\pi}(t)$ is an m-dimensional row vector whose ith component is $\pi_i(t)$, and is called the *state probability vector*. $\boldsymbol{\pi}(0)$ is referred to as *the initial state of the system*. ◊

We will often say that *the system is in state* $\boldsymbol{\pi}$ when we mean that *the system is in state* $i \in \Xi$ *with probability* π_i. If a distinction between the two ideas is necessary, we will say that the system is in *composite state* $\boldsymbol{\pi}$, as opposed to *pure state i*. In this case, $\pi_j = \delta_{ij}$, where

$$\delta_{ij} = \begin{cases} 0 & \text{for } i \neq j \\ 1 & \text{for } i = j \end{cases}$$

is the *Kronecker delta.*

Definition 1.3.3

$S_i(\delta)$:= *probability that the system will do nothing in the interval* δ *when in state* $i \in \Xi$. $S_i(\delta)$ has the same properties as the $S(\delta)$ in Section 1.2, except that its Taylor series coefficients depend on i [i.e., $S_i(\delta) = 1 - \mu_i\,\delta + O(\delta^2)$]. This function is used only as a convenience for deriving some preliminary formulas. It will not be used beyond this chapter. ◊

Also,

Definition 1.3.4

P_{ij} := *probability that the system will jump to* $j \in \Xi$ *upon leaving state* $i \in \Xi$ *The matrix* **P**, defined by $(\mathbf{P})_{ij} = P_{ij}$, *is called a transition probability matrix* if $P_{ij} \geq 0$ *and* $\sum_{j=1}^{m} P_{ij} \leq 1$ *for all i and j. It is also referred to as a Markov matrix* or a *stochastic matrix*, or simply *probability matrix* if $\sum_{j=1}^{m} P_{ij} = 1$. If $\sum_{j=1}^{m} P_{ij} < 1$ for some i, then **P** is called a *sub-stochastic matrix.* ◊

Since our system is closed, the sum of probabilities of all possible jumps must be 1. That is,

$$\sum_{j=1}^{m} P_{ij} = 1. \tag{1.3.1a}$$

By introducing the special row vector,

$$\boldsymbol{\varepsilon} := [\,1, 1, 1, \cdots, 1\,],$$

with $\boldsymbol{\varepsilon}'$ being the transpose (i.e., column vector) of $\boldsymbol{\varepsilon}$, (1.3.1a) can be rewritten in matrix form as

$$\mathbf{P}\boldsymbol{\varepsilon}' = \boldsymbol{\varepsilon}'. \tag{1.3.1b}$$

Many matrices in this book have this property so we give it a special name. We say that any matrix which satisfies (1.3.1b) is an *isometric matrix*. Thus **P** is *isometric*. Using an extended view of the definition, we can say that $\boldsymbol{\varepsilon}'$ itself is isometric, since its row sum (only one term) is 1. In Section 3.2.3 we will give an extended rationale for this nomenclature.

Note that (1.3.1b) is a matrix equation, whereas (1.3.1a) looks explicitly at the components. The reader need not be concerned at the moment with the subtle distinction we are trying to make. However, as the book evolves, we will tend to ignore the properties of the individual matrix elements. It is the matrix as a whole that *operates* on the system's present state vector and changes it to the future state vector. Therefore, we will almost never make use of the property $P_{ij} \geq 0$. However, we will always be concerned to see if a matrix is isometric, for this is an algebraic property of the matrix. When we prove that a square matrix is isometric, the reader is welcome to think of it as being a stochastic matrix, but we will seldom prove it.

We next derive the generalization of (1.2.1a), keeping in mind that the system can go to any state, including the one it is presently in, or one it previously visited. We have

$$\pi_i(t+\delta) = \pi_i(t)S_i(\delta) + \sum_j \pi_j(t)[1-S_j(\delta)]P_{ji} + O(\delta^2).$$

In words, the probability that the system will be in state i at time $t+\delta$, $[\pi_i(t+\delta)]$, is equal to the probability that it was in state i at time t, $[\pi_i(t)]$, and remained there for time δ, $[S_i(\delta)]$, plus the sum of probabilities that it was in some other state, j (including i), at time t, $[\pi_j(t)]$, left that state within the interval δ, $[1-S_j(\delta)]$, and went to i, $[P_{ji}]$, plus multiple transitions, $[O(\delta^2)]$. As with the derivation of (1.2.1a), replace S_i with its Taylor expansion, subtract $\pi_i(t)$ from both sides of the equation, divide by δ and take the limit for δ goes to 0, and get

$$\frac{d\pi_i(t)}{dt} = \sum_j \pi_j(t)\mu_j P_{ji} - \pi_i(t)\mu_i. \tag{1.3.2a}$$

This is one form of the *Chapman-Kolmogorov (C-K) equation.* It can be expressed more elegantly as a matrix equation in the following way. We have already defined the row vector:

$$\boldsymbol{\pi}(t) := [\ \pi_1(t), \pi_2(t), \ \cdots\ ],$$

and now introduce a diagonal matrix:

Definition 1.3.5

$(\mathbf{M})_{ij} = \mu_i \delta_{ij}$, *where* δ_{ij}, *the Kronecker delta, is* 0 *unless* $i = j$, in which case it is 1. In other words, **M** is a diagonal matrix, with diagonal elements $M_{ii} = \mu_i$, where μ_i is the rate of leaving state i. **M** is called the *completion rate matrix.* It is also referred to as the *holding rate matrix*, but we will not use that term here. ◊

We now can rewrite (1.3.2a) as

$$\frac{d\pi(t)}{dt} = \pi(t)\mathbf{MP} - \pi(t)\mathbf{M} = -\pi(t)\boldsymbol{Q}, \tag{1.3.2b}$$

where the *transition rate matrix* (also called the *infinitesimal rate*, or simply *rate matrix*), $\boldsymbol{Q}$, is defined by

$$\boldsymbol{Q} := \mathbf{M}(\mathbf{I} - \mathbf{P}). \tag{1.3.2c}$$

Although equivalent to the usual definition (most researchers define $-\boldsymbol{Q}$ as the transition rate matrix), $\boldsymbol{Q}$ is given in a somewhat different form because we have separated the process of leaving a state, [**M**], from that of deciding which state to go to next, [**P**]. This will be most useful to us in succeeding chapters. **M** governs the time between events and **P** controls what happens when an event occurs. Thus we can look at the behavior of systems conditioned by the occurrence of specific events. For instance, in Chapter 4 we will not only study the steady-state probabilities of finding an M/G/1 queue in a given state, but will also analyze the probabilities of being in a given state after a departure or after an arrival.

Let us define **o** to be the row vector of all 0's. It is clear from (1.3.1b) and (1.3.2c) that $\boldsymbol{Q}\boldsymbol{\varepsilon}' = \mathbf{o}'$, so upon multiplying (1.3.2b) from the right with $\boldsymbol{\varepsilon}'$, it follows that

$$\frac{d}{dt}[\boldsymbol{\pi}(t)\boldsymbol{\varepsilon}'] = 0. \tag{1.3.2d}$$

In other words, $\pi(t)\boldsymbol{\varepsilon}'$ is a constant that we may presume to be 1 for all t, since the sum of the π's is 1 at $t = 0$. This is no more than would be expected in a closed system.

The solution to (1.3.2b), another form of the C-K equation, is the matrix equivalent of (1.2.1b), namely

$$\boldsymbol{\pi}(t) = \boldsymbol{\pi}(0)\boldsymbol{G}(t), \tag{1.3.2e}$$

where

$$\boldsymbol{G}(t) := \exp(-t\boldsymbol{Q}). \tag{1.3.3a}$$

Some explanation is required, however. Only multiplication, addition and subtraction of matrices are defined. Division is replaced by taking the inverse, if it exists. Therefore, a function of a matrix must be defined in terms of these primitives. So, in general, any function of a matrix is formally defined by a Maclaurin series expansion, satisfying:

Theorem 1.3.1: Let $f(t)$ be any function of t whose Maclaurin series converges for all $|t| < r$ (its radius of convergence), and let ξ be the spectral radius of any square matrix, **X**. If **X** is of finite dimension, then

$$\xi := \max_i |\lambda_i|,$$

where the λ's are the eigenvalues of $\mathbf{X}$. The matrix function, $f(t\mathbf{X})$, is well defined by the Maclaurin expansion of $f(t)$, for all $|t| < r/\xi$. Note that $f(\cdot)$ takes on the algebraic structure of its argument. If its argument is a scalar, $[t]$, then $f(t)$ is a scalar. If its argument is a square matrix, $[\mathbf{X}]$, then $f(\mathbf{X})$ is a square matrix, with the same dimension. ■†

For example

$$\exp(-t\mathbf{X}) := \mathbf{I} - t\mathbf{X} + \frac{t^2}{2!}\mathbf{X}^2 - \frac{t^3}{3!}\mathbf{X}^3 + \cdots. \tag{1.3.3b}$$

The radius of convergence for the exponential function is infinite, so (1.3.3b) is valid for all t.

Corollary 1.3.1: Let $f(t)$ be any function of t whose Maclaurin series converges for all $|t| < r$. Then $f(t\mathbf{X})$ commutes with $\mathbf{X}$ whenever $f(t\mathbf{X})$ is defined. That is,

$$f(t\mathbf{X})\mathbf{X} = \mathbf{X}f(t\mathbf{X})$$

for all $|t| < r/\xi$, as defined in Theorem 1.3.1. ■

Exercise 1.3.1: Use (1.3.3b) to show that (1.3.2e) satisfies (1.3.2b).

Since $\boldsymbol{Q}\boldsymbol{\varepsilon}' = \mathbf{o}'$, it is a straightforward step using (1.3.3b), to show that

$$\boldsymbol{G}(t)\boldsymbol{\varepsilon}' = \boldsymbol{\varepsilon}' \text{ for all } t, \tag{1.3.4a}$$

so $\boldsymbol{G}(t)$ is an isometric matrix. It can also be shown that if $\mathbf{P}$ is a stochastic matrix, then $\boldsymbol{G}(t)$ is also, but only for $t \geq 0$. That is, $[\boldsymbol{G}(t)]_{ij} \geq 0$ if $t \geq 0$. (It is dangerous to try to go backward in time.)

One cannot take it for granted that all relations in elementary algebra follow through for matrix algebra. For instance,

Theorem 1.3.2: Let $\mathbf{A}$ and $\mathbf{B}$ be two square matrices of the same dimension, then

$$\exp[t(\mathbf{A}+\mathbf{B})] = \exp(t\mathbf{A})\exp(t\mathbf{B}) \quad \text{for all } t, \quad \text{iff } \mathbf{AB} = \mathbf{BA}.$$

(*iff* stands for *if and only if*). We restate for emphasis: If $\mathbf{A}$ and $\mathbf{B}$ do *not* commute, the equation is not valid. ■

† Symbol ■ designates end of Theorem, Lemma, or Corollary.

Exercise 1.3.2: Prove Theorem 1.3.2 by direct substitution of the appropriate Taylor expansions.

It is clear from (1.3.3a) that $\boldsymbol{G}(t)$ is the operator that translates a system directly from time 0 to time t. Theorem 1.3.2 allows the most familiar form of the C-K equation to be written:

$$\boldsymbol{G}(s+t) = \boldsymbol{G}(s)\boldsymbol{G}(t). \tag{1.3.4b}$$

Remember that $\boldsymbol{G}(t)$ is an isometric matrix (think *transition matrix*) whose elements change with t.

Exercise 1.3.3: Let

$$\mathbf{M} = \begin{bmatrix} 1 & 0 \\ 0 & 2 \end{bmatrix} \quad \text{and} \quad \mathbf{P} = \begin{bmatrix} 0 & 1 \\ 1 & 0 \end{bmatrix}$$

Find $\boldsymbol{G}(t)$. Show that it is a transition matrix. What is $\boldsymbol{G}$ in the limit as t goes to infinity?

1.3.2. Discrete Time

The discrete-time analog of (1.3.2b) is self-evident from the definition of the transition matrix **P**. Let $\pi_{\mathbf{d}}(n)$ be the vector whose ith component is the probability that the system will be in state i at step n. Then

$$\pi_{\mathbf{d}}(n) = \pi_{\mathbf{d}}(n-1)\mathbf{P} = \pi_{\mathbf{d}}(0)\mathbf{P}^n. \tag{1.3.5a}$$

The discrete analog to $\boldsymbol{G}(t)$ is $\mathbf{G_d}(n) := \mathbf{P}^n$. The obvious analog to (1.3.4b) is

$$\mathbf{G_d}(n+m) = \mathbf{G_d}(n)\mathbf{G_d}(m). \tag{1.3.5b}$$

As in Section 1.2, if a continuous-time system is observed only at integral multiples of some time interval δ, that system is indistinguishable from a discrete-time system with transition matrix

$$\mathbf{P_d} := \mathbf{G_d}(1) = \boldsymbol{G}(\delta) = \exp(-\delta\boldsymbol{Q}).$$

Although every $\boldsymbol{Q}$ maps onto some $\mathbf{P_d}$, not every transition matrix can be expressed in this way. In general, all elements of $\mathbf{P_d}$ will be greater than 0. unless:

1. The graph associated with Q is made up of two or more disjoint subgraphs (this would be the case iff 1 is a multiple eigenvalue of Q). We would then say that Q is *reducible*.

2. There exists a state, or set of states, which are *transient*, (i.e., states that cannot be reached from, but can reach, the rest of the network).

If it is possible to get from state i to state j at all, then $(\mathbf{P_d})_{ij} > 0$, and in fact, is of order δ^n, where n is the number of steps it takes to get there. One might say that for every Q, the matrix $\mathbf{P_d} = \exp(-\delta Q)$ exists, but we cannot say the inverse for every $\mathbf{P}$. ["log($\mathbf{P}$)" (whatever that is) does *not* necessarily exist for a given $\mathbf{P}$, since $\log(x)$ does not have a Maclaurin expansion, although $\log(1+x)$ does.]

Exercise 1.3.4: A simple example of both (1) and (2) is given by

$$\mathbf{P} = \begin{bmatrix} 0 & 1 & 0 & 0 \\ 0 & 1 & 0 & 0 \\ 0 & 0 & 0 & 1 \\ 0 & 0 & 1 & 0 \end{bmatrix}.$$

Find the eigenvalues and eigenvectors of $\mathbf{P}$. Clearly, states 1 and 2 are disjoint from 3 and 4, and state 1 is transient.

1.3.3. Time-Dependent and Steady-State Solutions

As you may have seen from Exercise 1.3.3, (1.3.2e) is not as explicitly useful as it seems. More useful solutions of this are covered in depth in the literature. We will discuss it slightly here, enough to see how $\boldsymbol{\pi}(t)$ varies with time, and will do some examples in detail in Chapter 2. First we review a little matrix theory.

Some Properties of Matrices

The *eigenvalues* of a matrix, $\mathbf{X}$, are the roots of its *characteristic equation*,

$$\phi(\lambda) := | \lambda\mathbf{I} - \mathbf{X} | = 0, \tag{1.3.6}$$

where $|\cdot|$ denotes the *determinant* of any square matrix. In other words, λ_i is an eigenvalue of $\mathbf{X}$ if and only if it is a root of $\phi(\lambda)$ [i.e., $\phi(\lambda_i) = 0$]. If $\mathbf{X}$ is of finite dimension, say m, then $\phi(\lambda)$ is a polynomial of degree m, with m roots. If a particular root appears more than once, it is a *multiple root*, and we say there is a *degeneracy* in that eigenvalue. Otherwise, it is a *simple root*.

Corresponding to each λ_i is at least one *left eigenvector* and one *right eigenvector*, satisfying the following:

$$\mathbf{u_i X} = \lambda_i \mathbf{u_i} \quad \text{and} \quad \mathbf{X v'_i} = \lambda_i \mathbf{v'_i}. \tag{1.3.7a}$$

The number of right eigenvectors belonging to each eigenvalue is greater than or equal to 1, and less than or equal to the degree of multiplicity of that root. If the number of eigenvectors belonging to a given eigenvalue is strictly less than the degree of multiplicity of that root, then the matrix is said to be *defective*. There are as many left as there are right eigenvectors, and they satisfy the following *orthogonality condition*:

$$\mathbf{u_i v'_j} = 0 \quad \text{for} \quad \lambda_i \neq \lambda_j. \tag{1.3.7b}$$

The general case can be treated with some difficulty, but for now assume that the λ_i's are distinct. Then it can be assumed that

$$\mathbf{u_i v'_i} = 1. \tag{1.3.7c}$$

Note that each $\mathbf{u_i}$ is a row vector with m components $[(\mathbf{u_i})_k,\ 1 \le k \le m]$ and that each $\mathbf{v'_i}$ is a column vector, also with m components. Consider the $m \times m$ matrices

$$(\mathbf{U})_{ik} := (\mathbf{u_i})_k$$

and

$$(\mathbf{V})_{ki} := (\mathbf{v'_i})_k.$$

Equations (1.3.7) imply that $\mathbf{U}$ and $\mathbf{V}$ are inverses of each other, (i.e., $\mathbf{UV} = \mathbf{VU} = \mathbf{I}$).

The *spectral decomposition theorem* states that (where m is the dimension of $\mathbf{X}$)

$$\mathbf{X} = \sum_{i=1}^{m} \lambda_i \mathbf{v'_i u_i}, \text{ and } \mathbf{I} = \sum_{i=1}^{m} \mathbf{v'_i u_i}. \tag{1.3.8a}$$

Note that whereas

$$\mathbf{u_i v'_j} = \sum_{k=1}^{m} (\mathbf{u_i})_k\, (\mathbf{v'_j})_k$$

(*inner, dot* or *scalar product*) is a scalar, the object $\mathbf{v'_j u_i}$ (*outer product*) is an m-dimensional matrix of rank 1, where all rows are proportional to each other and to $\mathbf{u_i}$. That is, $(\mathbf{v'_j u_i})_{kl} = (\mathbf{v'_j})_k(\mathbf{u_i})_l$. It follows from the orthogonality conditions above that

$$\mathbf{X}^k = \sum_{i=1}^{m} \lambda_i^k\, \mathbf{v'_i u_i} \tag{1.3.8b}$$

and more generally,

$$f(t\mathbf{X}) = \sum_{i=1}^{m} f(t\lambda_i) \mathbf{v'_i u_i}, \tag{1.3.8c}$$

where $f(x)$ is any function expressible in a Maclaurin series. Theorem 1.3.1 follows directly from this.

How System Approaches Its Steady State

Recall that for the transition rate matrix, $\boldsymbol{Q}\boldsymbol{\varepsilon}' = \mathbf{o}' = 0\boldsymbol{\varepsilon}'$, so $\boldsymbol{\varepsilon}'$ is a right eigenvector of $\boldsymbol{Q}$ with eigenvalue 0. Every eigenvalue must have a left eigenvector as well. Since we have assumed that all eigenvalues are distinct, a unique $\boldsymbol{\pi}$ satisfying

$$\boldsymbol{\pi}\boldsymbol{Q} = \mathbf{o} \quad \text{and} \quad \boldsymbol{\pi}\boldsymbol{\varepsilon}' = 1 \tag{1.3.9a}$$

exists and is known as the *steady-state* (*s.s.*) or *equilibrium vector*. Since the order in which eigenvalues and eigenvectors are labeled is arbitrary, let $\lambda_1 = 0$, $\mathbf{u_1} = \boldsymbol{\pi}$, and $\mathbf{v'_1} = \boldsymbol{\varepsilon}'$. Then (1.3.3a) and (1.3.2e) become

$$\boldsymbol{G}(t) = \boldsymbol{\varepsilon}'\boldsymbol{\pi} + \sum_{i=2}^{m} e^{-t\lambda_i}\, \mathbf{v'_i}\, \mathbf{u_i}, \tag{1.3.9b}$$

and [where $\alpha_i := \boldsymbol{\pi}(0)\,\mathbf{v'_i}$]

$$\pi(t) = \pi + \sum_{i=2}^{m} \alpha_i\, e^{-t\lambda_i}\, \mathbf{u_i}. \tag{1.3.9c}$$

Recall from the theory of complex variables that if z is a complex number, then $z = x + iy$, where x and y are real numbers. $x :=$ Re(z) is the *real part* of z, $y :=$ Im(z) is the *imaginary part*, $(-i)^2 = -1$, and $|z|^2 := x^2 + y^2$. Therefore,

$$|e^z| = |e^x e^{iy}| = |e^x|\ |e^{iy}| = |e^x|,$$

since $e^{iy} = \cos y + i \sin y$, and $|e^{iy}| = \sqrt{\cos^2 y + \sin^2 y} = 1$. It follows that if Re($\lambda_j$) > 0 for all $j > 1$ (which is the case for transition rate matrices as defined so far),

$$\lim_{t\to\infty} \boldsymbol{G}(t) = \boldsymbol{\varepsilon}'\boldsymbol{\pi} \tag{1.3.10a}$$

and

$$\lim_{t\to\infty} \boldsymbol{\pi}(t) = \boldsymbol{\pi}. \tag{1.3.10b}$$

Clearly, the asymptotic behavior of $\boldsymbol{\pi}(t)$ is independent of its initial status, $\boldsymbol{\pi}(0)$.

Before going on, we note that if the set of states can be partitioned into two subsets, such that there is no way to get from one subset to the other, then $\lim_{t\to\infty}\boldsymbol{\pi}(t)$ depends on the probability that the system began in one subset or the other. But this also implies that the eigenvalue, 0 is degenerate, and there are at least two left eigenvectors with eigenvalue 0, call them $\boldsymbol{\pi}_1$ and $\boldsymbol{\pi}_2$. In other words, $\lim_{t\to\infty}\boldsymbol{\pi}(t) = a\,\boldsymbol{\pi}_1 + (1-a)\,\boldsymbol{\pi}_2$. It is not hard to see that if such a partition exists, we can treat both subsets independently and solve them separately. Therefore, we can assume that our system is connected, i.e., *irreducible*, so the 0 eigenvector, $\boldsymbol{\pi}$, is unique.

The question "How long will it take to get to the asymptotic region?" is not easy to answer, but one rule of thumb involves the *relaxation time* (RT) [MORS58], (also called the *settling time*) defined by

$$\frac{1}{RT} := \min_{i=2}^{m} [\mathrm{Re}(\lambda_i)]. \tag{1.3.11}$$

In words, list the real parts of all the eigenvalues of $\boldsymbol{Q}$. (They all must be positive, or else we are in trouble.) Pick the smallest one. Then the reciprocal of that number is RT. If t is much greater than RT, we can expect that $\boldsymbol{\pi}(t)$ will be close to $\boldsymbol{\pi}$. For t small enough, the system is said to be in the *transient region* and displays *transient behavior*. But as t gets larger, the difference between $\boldsymbol{\pi}(t)$ and $\boldsymbol{\pi}$ eventually becomes small. Look at the following string of inequalities for the kth component of their difference:

$$|(\boldsymbol{\pi}(t) - \boldsymbol{\pi})_k| = \left|\sum_{i=2}^{m} \alpha_i e^{-t\lambda_i} (\mathbf{u_i})_k\right| \le \sum_{i=2}^{m} |\alpha_i| \, |(\mathbf{u_i})_k| \, |\exp[-t\,\mathrm{Re}(\lambda_i)]|$$

$$\le e^{-t/RT} \sum_{i=2}^{m} |\alpha_i| \, |(\mathbf{u_i})_k| = Ce^{-t/RT}.$$

We see that the upper bound of the difference drops at least by a factor of e for each time unit, RT, *but* C could be enormous, so it could take a long time before the actual difference, $|\boldsymbol{\pi}(t) - \boldsymbol{\pi}|$, shows this behavior.

Equations (1.3.10) can be interpreted in the following way. Set the system of interest going and wait some time longer than RT before looking to see what state the system is in. The probability that it is in state k is close to π_k. But one observation is meaningless, so more data must be taken. After the measurement, the system continues to evolve as though it just started in the measured state. Thus one must wait another long time before measuring it again. This is not a particularly efficient way to validate (1.3.10b). Consider, instead, the conceptual experiment of setting up a large number of identical systems (sometimes called an *ensemble*), let them all run simultaneously, and observe the state each is in after a time t. The fraction of them that are in state k should be close to $[\pi(t)]_k$.

A more practical viewpoint is available. Suppose that one wishes to know the fraction of time a system spends in each state over a long period of time T. This would correspond to the time average of $\boldsymbol{\pi}(t)$,

$$\overline{\boldsymbol{\pi}}(T) := \frac{1}{T}\int_0^T \boldsymbol{\pi}(t)\, dt = \boldsymbol{\pi}(0)\, \overline{\boldsymbol{G}}(T), \tag{1.3.12a}$$

where

$$\overline{\boldsymbol{G}}(T) := \frac{1}{T}\int_0^T \exp(-t\boldsymbol{Q})\, dt = \frac{1}{T}\int_0^T \left[\boldsymbol{\varepsilon}'\boldsymbol{\pi} + \sum_{i=2}^{m} e^{-t\lambda_i} \mathbf{v'_i u_i}\right] dt$$

$$= \boldsymbol{\varepsilon}'\boldsymbol{\pi} + \sum_{i=2}^{m} \mathbf{v'_i u_i} \left[\frac{1 - e^{-T\lambda_i}}{T\lambda_i} \right]. \tag{1.3.12b}$$

Again, as long as $\text{Re}(\lambda_i) > 0$,

$$\lim_{T \to \infty} \overline{G}(T) = \boldsymbol{\varepsilon}'\boldsymbol{\pi} \tag{1.3.13a}$$

and

$$\lim_{T \to \infty} \overline{\boldsymbol{\pi}}(T) = \boldsymbol{\pi}, \tag{1.3.13b}$$

the same as their unbarred counterparts in (1.3.10).

> **Exercise 1.3.5:** Prove that $(\boldsymbol{\varepsilon}'\boldsymbol{\pi})^2 = \boldsymbol{\varepsilon}'\boldsymbol{\pi}$. More generally, show that for any two vectors satisfying $\mathbf{uv}' = 1$, it follows that $(\mathbf{v'u})^2 = \mathbf{v'u}$. Matrices that have this property are said to be *idempotent.*

Note that whereas $G(t)$ and $\boldsymbol{\pi}(t)$ converge exponentially to their asymptotic limits according to Equations (1.3.9) (which may be a very long time if RT is very large), $\overline{G}(T)$ and $\overline{\boldsymbol{\pi}}(T)$ approach their limits much slower, as $1/T$ and Equations (1.3.12). That is, although the initial state has little influence on the long-term behavior of a system, its effect on the time average of system behavior lingers on.

As an aside, it is interesting to note that researchers in discrete simulation methods usually throw away the first 100 or so data points if they want their results to converge much more rapidly to the steady state. Accumulated simulation statistics are equivalent to $\overline{\boldsymbol{\pi}}(T)$. One is led to question the significance of the steady-state $\boldsymbol{\pi}$, for those systems that run only for a time T comparable to, or less than, RT. In the succeeding chapters we discuss other parameters that describe relatively short-term behavior of queueing systems.

Discrete systems behave in a manner similar to continuous systems with one exception, namely those that are *periodic*. Intuitively, these are systems that have at least one state to which the system returns in exactly n, $n > 1$ steps. These correspond to transition matrices that have at least one eigenvalue with modulus 1, other than 1 itself.

We now turn our attention to Equations (1.3.5), and as in the continuous case, let $\{\lambda_i\}$ be the set of eigenvalues of $\mathbf{P}$, while $\{\mathbf{u_i}\}$ and $\{\mathbf{v'_i}\}$ are its left and right eigenvectors, respectively. Note that except for $\mathbf{v'_1} = \boldsymbol{\varepsilon}'$, these objects are different from those for Q in the continuous case. Also, $\lambda_1 = 1$, and $\boldsymbol{\pi}_\mathbf{d} = \mathbf{u_1}$ satisfies

$$\boldsymbol{\pi}_{\mathbf{d}}\mathbf{P} = \boldsymbol{\pi}_{\mathbf{d}} \quad \text{and} \quad \boldsymbol{\pi}_{\mathbf{d}}\boldsymbol{\varepsilon}' = 1 \tag{1.3.14a}$$

and is *not* the same $\boldsymbol{\pi}$ as that defined in (1.3.9a), although they are closely related.

> **Exercise 1.3.6:** Prove that the $\boldsymbol{\pi}$ in (1.3.9a), when right-multiplied by **M**, is a constant times the $\boldsymbol{\pi}_{\mathbf{d}}$ in (1.3.14a).

The limit of (1.3.5a) as n goes to infinity can be evaluated with the aid of the spectral decomposition theorem. Inserting (1.3.8a) (with **P** replacing $\boldsymbol{Q}$) into (1.3.5a) leads to

$$\lim_{n\to\infty} \boldsymbol{\pi}_{\mathbf{d}}(n) = \lim_{n\to\infty} \boldsymbol{\pi}_{\mathbf{d}}(0)\,\mathbf{P}^n = \boldsymbol{\pi}_{\mathbf{d}}(0) \lim_{n\to\infty} \left[\boldsymbol{\varepsilon}'\boldsymbol{\pi}_{\mathbf{d}} + \sum_{i=2}^{m} \lambda_i^{\,n}\, \mathbf{v}'_{\mathbf{i}}\mathbf{u}_{\mathbf{i}}\right]$$

$$= \boldsymbol{\pi}_{\mathbf{d}} + \sum_{i=2}^{m} \left[\boldsymbol{\pi}_{\mathbf{d}}(0)\mathbf{v}'_{\mathbf{i}}\right]\left[\lim_{n\to\infty} \lambda_i^n\right]\mathbf{u}_{\mathbf{i}}. \tag{1.3.14b}$$

Clearly, if $|\lambda_i| < 1$ for $i > 1$, then $\lim_{n\to\infty} \lambda_i^n = 0$ and

$$\lim_{n\to\infty} \boldsymbol{\pi}_{\mathbf{d}}(n) = \boldsymbol{\pi}_{\mathbf{d}}. \tag{1.3.14c}$$

Similarly, again with $|\lambda_i| < 1$,

$$\lim_{n\to\infty} \mathbf{P}^n = \boldsymbol{\varepsilon}'\boldsymbol{\pi}_{\mathbf{d}}. \tag{1.3.14d}$$

As already mentioned, although all irreducible chains have only one eigenvalue equal to 1, they can have other eigenvalues whose modulus is 1. For example, the **P** in Exercise 1.3.3 has eigenvalues 1 and −1. When this is the case, the limit as n goes to infinity of λ_i^n *does not* exist for some i, so $\boldsymbol{\pi}_{\mathbf{d}}(n)$ has no limit [unless $\boldsymbol{\pi}_{\mathbf{d}}(0)\cdot\mathbf{v}'_{\mathbf{i}} = 0$]. In other words, there may be no steady state.

What, then, does $\boldsymbol{\pi}_{\mathbf{d}}$ mean? The answer comes from the discrete-time average equivalent to Equations (1.3.12). Define

$$\overline{\mathbf{G}}_{\mathbf{d}}(N) := \frac{1}{N}\left(\mathbf{I} + \mathbf{P} + \cdots + \mathbf{P}^{N-1}\right) = \boldsymbol{\varepsilon}'\boldsymbol{\pi}_{\mathbf{d}} + \frac{1}{N}\sum_{k=0}^{N-1}\left(\sum_{i=2}^{m}\lambda_i^{\,k}\,\mathbf{v}'_{\mathbf{i}}\mathbf{u}_{\mathbf{i}}\right)$$

$$= \boldsymbol{\varepsilon}'\boldsymbol{\pi}_{\mathbf{d}} + \frac{1}{N}\sum_{i=2}^{m}\mathbf{v}'_{\mathbf{i}}\mathbf{u}_{\mathbf{i}}\left(\sum_{k=0}^{N-1}\lambda_i^{\,k}\right)$$

or

$$\overline{\mathbf{G}}_{\mathbf{d}}(N) = \boldsymbol{\varepsilon}'\boldsymbol{\pi}_{\mathbf{d}} + \frac{1}{N}\sum_{i=2}^{m}\mathbf{v}'_{\mathbf{i}}\mathbf{u}_{\mathbf{i}}\left(\frac{1-\lambda_i^{\,N}}{1-\lambda_i}\right). \tag{1.3.15a}$$

Clearly, as long as $|\lambda_i| \le 1$ (the term corresponding to $\lambda_i = 1$ has already been excluded), we can write

$$\lim_{n\to\infty} \frac{1-\lambda_i{}^n}{n(1-\lambda_i)} = 0,$$

so

$$\lim_{n\to\infty} \overline{\mathbf{G}}_{\mathbf{d}}(n) = \boldsymbol{\varepsilon}'\boldsymbol{\pi}_{\mathbf{d}} \tag{1.3.15b}$$

even for cyclic chains. We see, then, that the "average" interpretation for $\boldsymbol{\pi}_{\mathbf{d}}$ still holds, even though there may be no steady state.

Is it disturbing that discrete chains have at least one property that continuous chains do not have? This dilemma can be resolved by the following argument. Discrete chains assume that exactly n transitions have occurred by time n, while for continuous t, even after a relatively short time, one cannot be sure exactly how many steps have occurred. So even if the system is cyclic in the physical sense, one cannot be sure how many cycles have occurred. This carries over to the discrete chain if one loses track of the exact number of steps. For instance, suppose that a system has been running for 10,000 units of time, take or leave a few. Then the average of $\boldsymbol{\pi}_{\mathbf{d}}(n)$ over those few would be $\boldsymbol{\pi}_{\mathbf{d}}$. Mathematically, suppose that our system has a cycle of length $k > 1$, so that

$$\lim_{n\to\infty} [\boldsymbol{\pi}_{\mathbf{d}}(n+j) - \boldsymbol{\pi}_{\mathbf{d}}(n)] = 0$$

only if j is a multiple of k. Then

$$\boldsymbol{\pi}_{\mathbf{d}} = \lim_{n\to\infty} \frac{1}{k}\sum_{j=1}^{k} \boldsymbol{\pi}_{\mathbf{d}}(n+j).$$

Exercise 1.3.7: Let (where $0 < a < 1$)

$$\mathbf{P} = \begin{bmatrix} 0 & a & 1-a \\ 1 & 0 & 0 \\ 1 & 0 & 0 \end{bmatrix}.$$

Find all the eigenvalues and eigenvectors of $\mathbf{P}$, solve for $\boldsymbol{\pi}_{\mathbf{d}}(n)$, and show that $\frac{1}{2}[\boldsymbol{\pi}_{\mathbf{d}}(n) + \boldsymbol{\pi}_{\mathbf{d}}(n+1)]$ approaches $\boldsymbol{\pi}_{\mathbf{d}}$ for large n.

Despite the fact that there was much matrix theory in this chapter, we have not yet touched upon what is meant by LAQT. That will have to wait until Chapter 3.

From now on we consider only continuous-time systems. It should not be inferred from this that discrete-time systems are less utilitarian. There is some belief, in fact, that they could be more useful. Some day we may try to treat the queueing world as a movie in discrete time.

CHAPTER 2
M/M/1 QUEUE

"I'm sure that I've never been in a queue as slow as this." – Any Customer, Anywhere, Anytime

Since the M/M/1 queue is the simplest and most elementary of all queues, we cover it here in some detail. But what we discuss will differ from that covered in the usual first course in queueing theory, and we use different techniques to accomplish our goals. Our purpose is threefold. First, we want to connect Chapter 1 with queueing theory and familiarize the reader with our terminology. Second, we want to set up points of view and techniques that will be used in later chapters when LAQT is finally introduced. Third, we want to reinforce the view that the behavior of a queueing system in the transient or small time region may be important more often than we have thought heretofore, and that it is possible to study that region realistically and perform calculations relatively easily, in fact, in some cases with the same ease (or difficulty) as with the steady state.

All systems treated in this book will be *closed.* That is, there will always be a fixed number of customers in the system. Each system is made up of two subsystems that interact with each other exclusively by exchanging customers. If N, the fixed number of customers, is large enough, we will show that one of the subsystems must become saturated. It then becomes a steady source of customers to the other subsystem. *Open* systems, then, are those where one of the subsystems is at full capacity almost all the time.

2.1. STEADY-STATE M/M/1//N LOOPS

Consider the system shown in Figure 2.1.1. It is made of two subsystems, called S_1 and S_2. At any time, S_1 has n customers, S_2 has k customers, and the system as a whole has $N = n + k$ customers. In this chapter both S_1 and S_2 are memoryless and thus have exponential service time pdf's of the form $\mu \exp(-\mu x)$ and $\lambda \exp(-\lambda x)$, respectively (which from a formal point of view means that each external state has only one internal state, but more of that in Chapter 3). The system is completely specified at any time if n and k are known. Since N is fixed, k is known if n is known, so the states of the system can be labeled by $n = 0, 1, 2, \cdots N$ (i.e., there are $N + 1$ states).

The notation $M_2/M_1/1//N$ corresponds to Figure 2.1.1 in the following way. First assume that S_1 has a shorter mean service time than S_2. The first symbol, [M_2], indicates that S_2 is *memoryless* or *Markovian* or *exponential*, or

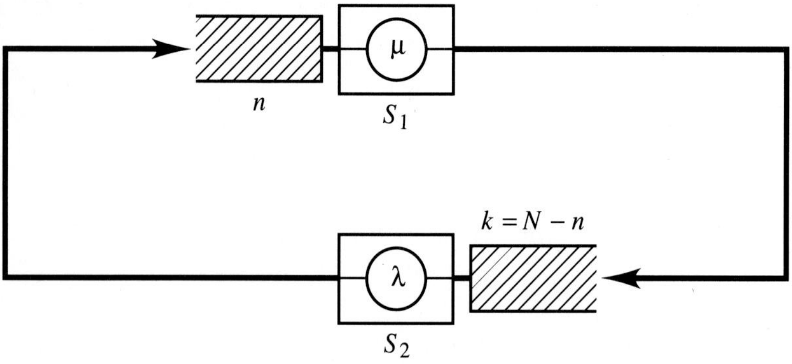

Figure 2.1.1. Closed loop made up of two subsystems, S_1 and S_2. The number of customers at S_1 (counting the one in service) is n, and k is the number at S_2. Their sum $N = k + n$ is fixed, thus the system is *closed.*

equivalently, has only one internal state. M_1 says the same thing about S_1. The third position, containing the number "1," means that S_1 can serve only one customer at a time. The space between the third and fourth slashes tells us that there is no limit as to how many customers can be in the queue at S_1. If there had been a number there, S_1 would have had a *finite waiting room.* We take a brief look at this *slot* in Section 5.3. The last symbol, N, indicates that there are a total of N customers in the system. Some books assume that S_2 has N identical servers, so all customers at S_2 can be served simultaneously, as in the *machine minding problem* (also known as *machine repairman model*) or in *time-sharing systems.* This is discussed in detail in Section 2.1.4.

Recall from Equations (1.3.2) that the completion rate matrix, **M**, is diagonal, where M_{ii} is the rate at which the system leaves state i given that it is in state i. Here i stands for the integer pair $(n, N-n)$, so, for instance, for $n = 0$, all customers are at S_2, and since only one can be served at a time, $M_{00} = \lambda$. Similarly, when all the customers are at S_1 ($n = N$), no customers can be served at S_2, so $M_{NN} = \mu$. However, for n in between, both subsystems are servicing customers, so the total departure rate is the sum of two service rates, namely, $M_{ii} = \mu+\lambda$. We prove this by deriving the density function for the first subsystem to complete service. First let $R_1(x) = \exp(-\mu x)$ be the probability that S_1 will still be unchanged at time x. Similarly, let $R_2(x) = \exp(-\lambda x)$. Then $R_1(x)R_2(x) = \exp[-(\mu+\lambda)x]$ is the probability that both S_1 and S_2 are unchanged at time x. Next define

$$B_<(x) := 1 - R_1(x)R_2(x)$$

as the probability that at least one of the subsystems has done something by time x. Then

$$b_<(x) := \frac{d}{dx}B_<(x) = (\mu+\lambda)\, e^{-(\mu+\lambda)x}$$

is the desired pdf. So the process in which one of two things can happen is exponentially distributed, with service (departure in this case) rate $(\mu+\lambda)$.

In summary, the completion rate matrix looks like

$$\mathbf{M} = \begin{bmatrix} \lambda & 0 & 0 & \cdots & 0 \\ 0 & \mu+\lambda & 0 & \cdots & 0 \\ 0 & 0 & \mu+\lambda & \cdots & 0 \\ : & : & : & \cdots & : \\ 0 & 0 & 0 & \cdots & \mu \end{bmatrix}. \tag{2.1.1a}$$

The transition matrix **P** from Equations (1.3.1) has the following values. For $n = 0$, the only thing that can happen is for a customer to leave S_2 and go to S_1, so $P_{01} = 1$. Similarly, $P_{N,N-1} = 1$. For all other n, one of two things could happen. Either a customer could leave S_2 and go to S_1, or the reverse. In the first case the system would go from state n to $n+1$, and in the other case the system would go from n to $n-1$. The probability that one would happen over the other is proportional to the separate subsystems' (servers') service rates, μ and λ. In other words, $P_{n,n+1} = \lambda/(\mu+\lambda)$. We show this by evaluating the probability that S_2 will finish before S_1. This will occur if S_2 finishes around time t $[b_2(t)\,dt]$ while S_1 is still running $[R_1(t)]$ for any $t > 0$ (integrate over t). This gives us

$$\Pr(S_2 \text{ will finish before } S_1) = \int_0^\infty b_2(t) R_1(t)\,dt = \int_0^\infty \lambda e^{-\lambda t} e^{-\mu t}\,dt$$

$$= \lambda \int_0^\infty e^{-(\mu+\lambda)t}\,dt = \frac{\lambda}{\mu+\lambda}. \tag{2.1.1b}$$

What we have just shown is important enough to be summarized in a theorem.

Theorem 2.1.1: Let X_1 and X_2 be independent random variables, each having exponential distribution functions with rates μ and λ, respectively. Then the PDF for the first one to finish, given that both have already started, but have not finished, by time $x = 0$, is also exponentially distributed, with parameter $\mu + \lambda$. That is, let

$$X = \min[X_1, X_2].$$

Then

$$\Pr(X < x) := B_<(x) = 1 - e^{-(\lambda+\mu)x},$$

and

$$b_<(x) = (\mu+\lambda)\,e^{-(\lambda+\mu)x}.$$

Furthermore, $\Pr(X_2 < X_1)$ is given by (2.1.1b). Since both X_1 and X_2 are exponentially distributed, these results do not depend which server started first. ■

The entire **P** matrix is the following:

$$\mathbf{P} = \begin{bmatrix} 0 & 1 & 0 & \cdots & 0 & 0 \\ \frac{\mu}{\mu+\lambda} & 0 & \frac{\lambda}{\mu+\lambda} & \cdots & 0 & 0 \\ 0 & \frac{\mu}{\mu+\lambda} & 0 & \cdots & 0 & 0 \\ \vdots & \vdots & \vdots & \cdots & \vdots & \vdots \\ 0 & 0 & 0 & \cdots & 0 & \frac{\lambda}{\mu+\lambda} \\ 0 & 0 & 0 & \cdots & 1 & 0 \end{bmatrix}. \qquad (2.1.1c)$$

Finally, $Q = \mathbf{M}(\mathbf{I} - \mathbf{P})$ can easily be calculated to give us

$$Q = \begin{bmatrix} \lambda & -\lambda & 0 & \cdots & 0 & 0 \\ -\mu & \mu+\lambda & -\lambda & \cdots & 0 & 0 \\ 0 & -\mu & \mu+\lambda & \cdots & 0 & 0 \\ \vdots & \vdots & \vdots & \cdots & \vdots & \vdots \\ 0 & 0 & 0 & \cdots & \mu+\lambda & -\lambda \\ 0 & 0 & 0 & \cdots & -\mu & \mu \end{bmatrix}. \qquad (2.1.1d)$$

This procedure of calculating Q in two steps rather than directly, as is usually done, seems cumbersome, but its utility will become clear in later chapters.

2.1.1. Time-Dependent Solution for $N = 2$

The time-dependent solution for $N = 1$ was actually done in Exercise 1.3.3. The next simplest nontrivial case is $N = 2$. Here

$$Q = \begin{bmatrix} \lambda & -\lambda & 0 \\ -\mu & \mu+\lambda & -\lambda \\ 0 & -\mu & \mu \end{bmatrix}. \qquad (2.1.2)$$

Obviously, $\boldsymbol{\varepsilon}'$ ($\boldsymbol{\varepsilon} = [1, 1, 1]$) is a right eigenvector of Q with eigenvalue 0, and it is not hard to find its companion, the left eigenvector with eigenvalue 0 [i.e., $\boldsymbol{\pi}(2)Q = \mathbf{o}$]. One proves by direct substitution that

$$\boldsymbol{\pi}(2) = \frac{1}{1 + \rho + \rho^2}[1, \rho, \rho^2],$$

where $\rho = \lambda/\mu$ and $\boldsymbol{\pi}\boldsymbol{\varepsilon}' = 1$. The components of the total probability vector, $[\boldsymbol{\pi}(2)]_j$, are the steady-state probabilities of finding $(j-1)$ customers at S_1. Put colloquially, after a long time, a random observer who may come along will find $j - 1$ customers at S_1 with probability $[\boldsymbol{\pi}(2)]_j$. The eigenvalues of Q satisfy the polynomial equation coming from Equations (1.3.6)

$$\phi(\beta) = \beta^3 - 2(\mu+\lambda)\,\beta^2 + (\mu^2 + \mu\lambda + \lambda^2)\beta = 0. \qquad (2.1.3a)$$

The roots of this equation are (for convenience we let the indices take on values 0 to $N = 2$ rather than the convention used in Chapter 1)

$$\begin{aligned} \beta_0 &= 0 \\ \beta_1 &= \mu(1+\rho+\sqrt{\rho}) \\ \beta_2 &= \mu(1+\rho-\sqrt{\rho}). \end{aligned} \qquad (2.1.3b)$$

β_0 is the root corresponding to the steady-state solution, while β_1 and β_2 moderate the transient behavior. Now $\beta_2 < \beta_1$, so the relaxation time from Equations (1.3.11) is $1/\beta_2$. Since the time units are arbitrary, we must establish some comparison to learn something from the formula. One convenient time unit to use in this case is the mean time for a single customer to go around the loop once, unimpeded. A simple way to do this is to let $1/\mu + 1/\lambda = 1$; then, from Equations (1.3.11),

$$RT = \frac{\rho}{(1+\rho)(1+\rho-\sqrt{\rho})}.$$

In this case it should be easy to see that RT is maximal when $\rho = 1$ and that $RT(\rho) = RT(1/\rho)$. We examine the general case in Section 2.2, but we note that these results are typical.

Exercise 2.1.1: For a cycle time of 1 ($1/\mu + 1/\lambda = 1$) show that the formula above is true, and draw a graph of RT versus ρ. When is RT a maximum? Prove that $RT(\rho) = RT(1/\rho)$.

Exercise 2.1.2: Find all the left and right eigenvectors of $\boldsymbol{Q}$ and verify that Equations (1.3.8a) are satisfied. Construct $\boldsymbol{G}(t)$ from (1.3.9a), and then $\boldsymbol{\pi}(t; 2)$, where $\boldsymbol{\pi}(0; 2)$ is one of [1 0 0], or [0 1 0], or [0 0 1].

2.1.2. Steady-State Solution for Any N

The steady-state solution for the M/M/1//N queue is, of course, well known and is shown in every book that discusses queueing theory to any extent. We discuss it briefly here to show how one goes from closed to open systems. Our assumption in this section is that S_2 is load independent. That is, the service rate of S_2 is the same irrespective of how many customers are in its queue.

Before going on, we mention that the solution for the M/M/1//N loop is identical with that for the open M/M/1/N queue. You might well ask what the difference is, since it turns out that the M/G/1/N and M/G/1//N queues also have identical solutions. We go into this in detail in Section 5.3, when we compare the G/M/1//N and G/M/1/N queues, (which *do not* have identical solutions) but give a short explanation here. When a customer arrives at an M/M/1/N queue that already has N customers, the arriving customer is turned away. Each subsequent arrival will be turned away until S_1 has a completion. Since the arrival process is a Poisson process, which implies that the interarrival times are exponentially distributed, the time for the next arrival is exponentially distributed, with the same mean, having no memory of the previous arrival. The M/M/1//N loop behaves in the following way. If all N customers are at S_1, there can be no further arrivals until S_1 has a completion. After such a completion, S_2 can service its new arrival, thereby preparing a new arrival for S_1. We see that shutting off the arrival process has the same affect as turning away arrivals, but only if the arrival process is memoryless.

From (1.3.9) and (1.3.10), the steady-state solution of our loop satisfies $\pi \boldsymbol{Q} = \mathbf{o}$, which from (1.3.2c) is the same as $\pi\mathbf{M} = \pi\mathbf{MP}$. These equations are referred to as the steady-state *balance equations*. In the notation of Chapter 1, the left-hand side $(\pi_i \mu_i)$ is interpreted as the probability rate of leaving state i, while the right-hand side is the probability rate of entering state i. And, of course, they are equal when a system reaches its steady state.

At this point it is advantageous for us to change our notation, to be consistent with succeeding chapters, where $\boldsymbol{\pi}$ will take on a different meaning. Since the abstract state i stands for there being $n = i - 1$ customers at S_1, define the following.

Definition 2.1.1

$r(n;N)$:= *steady-state probability that there are n customers at S_1, where N is the (fixed) number of customers in the system overall.* Then $r(n;N)$ replaces $[\pi(N)]_i$ $(n = i - 1)$ everywhere. ◊

For the M/M/1//N queue, these equations become, using (2.1.1d),

$$\lambda r(0;N) = \mu r(1;N),$$

$$(\mu + \lambda) r(n;N) = \lambda r(n-1;N) + \mu r(n+1;N) \quad \text{for } 0 < n < N \qquad (2.1.4a)$$

$$\mu r(N;N) = \lambda r(N-1;N).$$

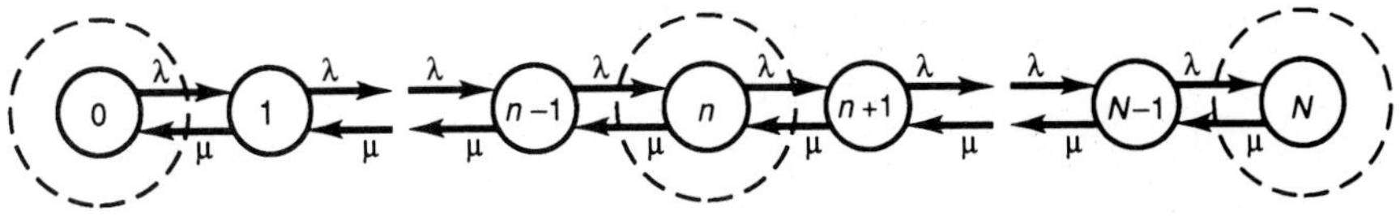

Figure 2.1.2. State transition rate diagram for an M/M/1//N queue, representing the probability rate of going from the tail to the head of each arrow. The three closed, dashed curves correspond to the three equations of (2.1.4a).

It is common to represent these equations graphically by what are called *state transition rate diagrams*, as shown in Figure 2.1.2. Each arrow corresponds to going from the state represented by the circle at the tail to the state represented by the circle at the head, with probability rate equal to the probability of being at the tail times the rate corresponding to the arrow. Every closed curve encompassing part of the graph represents a valid balance equation, where the sum of the rates represented by the arrows going into the loop equals the sum of the rates leaving the loops. In particular, each closed loop enclosing only one state (circle) yields one of the equations in (2.1.4a).

In any case, the solution to (2.1.4a) is well known to be

$$r(n;N) = \frac{\rho^n}{K(N)}, \quad 0 \le n \le N, \tag{2.1.4b}$$

where

$$K(N) := \sum_{n=0}^{N} \rho^n = \frac{1-\rho^{N+1}}{1-\rho} \quad (\rho \ne 1). \tag{2.1.4c}$$

The proof follows by substituting (2.1.4b) into (2.1.4a). Equation (2.1.4c) follows from the requirement that $\sum_{n=0}^{N} r(n;N) = 1$. For future reference, observe that $K(N)$ satisfies the recurrence relation

$$K(N) = 1 + \rho K(N-1). \tag{2.1.4d}$$

When $\rho = 1$, $r(n;N) = 1/(N+1)$ for all n. That is, the steady-state probability for all queue lengths is the same. Yet if the system initially had all its customers at S_1, it would be a long time indeed before a majority of them would be found at S_2. Of course, for very large N, and after a long period of time, we are unlikely to find the system in any particular state. Thus the steady-state solution, if anything, is warning our random observer to be wary of any conclusions concerning the behavior of a system that are based on short-term observations. We look at this again in Section 2.3.

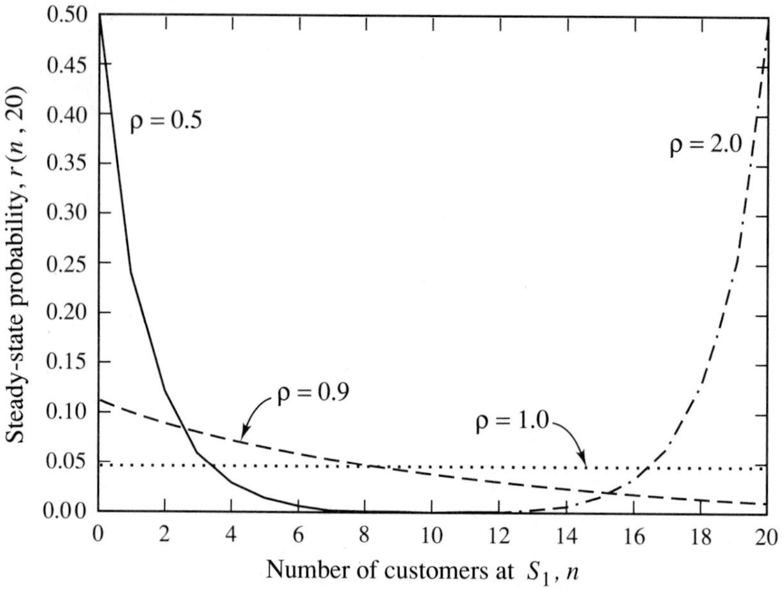

Figure 2.1.3. Steady-state probabilities, $r(n; 20)$, that there will be n customers at or in S_1, for $\rho = 0.5, 0.9, 1$, and 2. The curves for $\rho = 0.5$ and 2 are mirror images of each other. Also, the curve for $\rho = 1$ is a constant; that is, all queue lengths are equally likely. These observations are not necessarily true for more general queues. Equations (2.1.4b) and (2.1.4c or d) are used to compute the values plotted.

Example 2.1.1: In Figure 2.1.3 we have plotted the steady-state queue length probabilities for the M/M/1//20 queue for various values of ρ. Notice that when $\rho < 1$, $r(n; 20)$ is a monotonically decreasing function of n, while when $\rho > 1$, it is a monotonically decreasing function of $N - n$. As you might expect, the curves labeled $\rho = 0.5$ and $\rho = 2$ are mirror images of each other. The most significant feature of these curves is that they are so broad, particularly when ρ is near 1. It is best to think of $r(n; N)$ as being the fraction of time that n customers will be at S_1 over a very very, long period of time. ●†

What is often of interest in closed systems is the activity of each of the servers. The probability that a server is busy is equivalent to the fraction of time it is busy over a long period of time. This, in turn, determines the amount of "work" done per unit time by that server. Now suppose that customers somehow enter our closed loop, travel around until they have received a total of T_i units of service from S_i $(i = 1, 2)$, and then leave, being replaced instantly by a statistical clone. Then, of course, $T_1/T_2 = \rho$. Next define the steady-state probabilities.

† Symbol ● designates end of Example.

Definition 2.1.2

$P_i(N)$:= steady-state probability that S_i, $i = 1, 2$, is busy, given that there are N customers in the loop. ◊

Then

$$\Lambda(N) := \frac{P_i(N)}{T_i} \tag{2.1.5a}$$

is the rate at which customers enter and leave the loop, and is independent of i. $\Lambda(N)$ can be referred to as the *system throughput*. Now $P_1(N)$ is 1 minus the probability that S_1 is idle, so from (2.1.4b) with $n = 0$, and (2.1.4d),

$$P_1(N) = 1 - r(0; N) = 1 - \frac{1}{K(N)} = \frac{K(N)-1}{K(N)} = \rho\frac{K(N-1)}{K(N)}. \tag{2.1.5b}$$

Similarly, from (2.1.4c),

$$P_2(N) = 1 - r(N; N) = \frac{K(N) - \rho^N}{K(N)} = \frac{K(N-1)}{K(N)}. \tag{2.1.5c}$$

Then, since $\rho = T_1/T_2$, we show that the throughput as seen at S_1 is the same as that seen at S_2:

$$\Lambda(N) = \frac{P_1(N)}{T_1} = \frac{1}{T_2}\frac{K(N-1)}{K(N)} = \frac{P_2(N)}{T_2}. \tag{2.1.5d}$$

Example 2.1.2: We can understand the throughput behavior by looking at Figure 2.1.4, which shows $\Lambda(N)$ as a function of N for several values of ρ. Note that $\Lambda(N; \rho) = \Lambda(N; 1/\rho)$. In all cases, $\Lambda(N)$ saturates as N becomes increasingly large, and we see behavior typical of even more complicated queueing systems. That is, $\Lambda(N+1) > \Lambda(N)$ for all N, but

$$[\Lambda(N+2) - \Lambda(N+1)] < [\Lambda(N+1) - \Lambda(N)].$$

This is the law of diminishing returns: *Adding yet one more customer to the system will increase throughput, but the increase will not be as much as it was in adding the previous customer.* Finally,

$$\lim_{N\to\infty} [P_1(N) + P_2(N)] = 1 + \rho \quad \text{for } \rho \le 1.$$

That is, in general, only *one* server will saturate, while the other will be busy only a fraction of the time. Only when $\rho = 1$ will both servers approach full capacity with ever-increasing N. ●

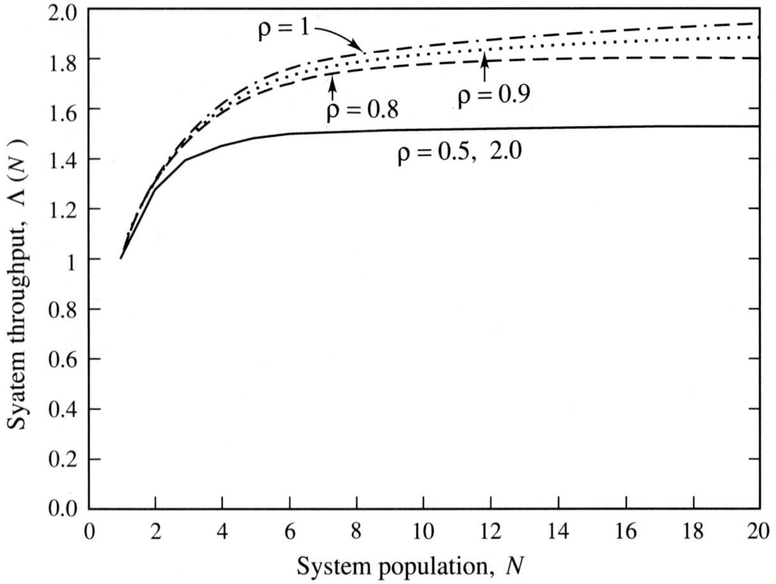

Figure 2.1.4. Throughput for steady-state M/M/1//N queues, where the total resource time needed for a customer to go around once is $T_1 + T_2 = 1$. The curves for $\rho = 0.5$ and $\rho = 2$ are identical because ρ and $1/\rho$ yield the same system with S_1 and S_2 interchanged. All the curves will saturate (become horizontal) if N is made large enough. Use Equations (2.1.5c), (2.1.5d) and (2.1.4d).

Exercise 2.1.3: Prove that the limit given in the preceding equation is indeed true. What is the limit when ρ is greater than 1? Also prove that $\Lambda(N;\rho) = \Lambda(N; 1/\rho)$ when $T_1 + T_2 = 1$.

2.1.3. Open M/M/1 Queue ($N \to \infty$)

We can next find the open system solution by doing the following. When $\rho < 1$, Equations (2.1.4) retain their meaning for large N. In this case,

$$\lim_{N\to\infty} K(N) = \frac{1}{1-\rho},$$

so

$$r(n) := \lim_{N\to\infty} r(n;N) = (1-\rho)\rho^n \tag{2.1.6a}$$

and

$$\lim_{n\to\infty} r(n) = 0.$$

That is, when N is very large, the probability that S_2 will be idle is negligible, so it is continually serving customers whose interdeparture times are exponentially distributed. Each new customer starts up in the same way the previous one did, so S_2 becomes a steady Poisson process of arrivals to S_1. Thus we have the equivalent of an open M/M/1 queue, with a mean queue length of

$$\bar{q}_s := \sum_{n=1}^{\infty} n\, r(n) = (1-\rho) \sum_{n=1}^{\infty} n\, \rho^n = \frac{\rho}{1-\rho}. \tag{2.1.6b}$$

When $\rho > 1$ it follows from (2.1.4c) that $1/(N)$ becomes vanishingly small for very large N, and thus for small n, $r(n;N)$ is essentially zero. Now S_1 is never idle and becomes a Poisson source for S_2. One would expect a certain duality between S_1 and S_2, which indeed is the case. Simply interchange 1 and 2, and thus replace ρ by $1/\rho$.

It is also interesting to evaluate the asymptotic throughput of our loop. We are thus interested in [from (2.1.5d)]

$$\lim_{N\to\infty} \Lambda(N) = \frac{1}{T_2}\frac{K(N-1)}{K(N)}.$$

We have already noted that when $\rho < 1, K(N)$ approaches $1-\rho$, while from (2.1.4c), $K(N)$ grows as ρ^N when ρ is greater than 1. This leads easily to the following limiting values:

$$\lim_{N\to\infty} \Lambda(N) = \frac{1}{T_2} \quad \text{for} \quad \rho \le 1$$

and

$$\lim_{N\to\infty} \Lambda(N) = \frac{1}{T_2}\frac{1}{\rho} = \frac{1}{T_1} \quad \text{for} \quad \rho \ge 1.$$

In other words, we have proven what should be obvious. The throughput of the system is bounded by the maximal throughput of the slower server, the *bottleneck*. The two equations can be summarized by

$$\lim_{N\to\infty} \Lambda(N) = \min\left(\frac{1}{T_1}, \frac{1}{T_2}\right). \tag{2.1.6c}$$

A perhaps more interesting question to answer is: How long will a customer be at S_1, both waiting for and being served? This turns out to be easy to answer once the mean queue length is known. The relevant formula, *Little's law,* or *theorem*, or *result*, which we introduced in (1.1.2), existed for many years before being proven formally by J. D. C. Little in 1961 [LITT61]. Recall that it is valid for any subsystem that has been in operation long enough so that the number of customers who have come and gone is far greater than the number presently there or who were there originally. Restated simply,

$$\bar{q} = \Lambda \bar{T}_s, \tag{2.1.7a}$$

where Λ is the mean arrival rate to (and departure rate from) the subsystem and $\overline{T}_s$ is the mean time spent there by each customer. In our case, $\Lambda = \lambda$ and $\rho = \lambda / \mu$, so from (2.1.6b), we have proven (1.1.4b)

$$\overline{T}_s = \frac{\overline{q}_s}{\lambda} = \frac{\overline{x}}{1-\rho}, \tag{2.1.7b}$$

where $\overline{x} = 1/\mu$ is the mean service time of S_1. Note that if $\rho = 0$ (no customers waiting at all), the mean time a customer remains in the system is the expected $\overline{x}$, while as with the mean queue length, the time a customer must wait grows unboundedly as ρ approaches 1.

It is useful to tighten up our terminology somewhat. Often, one wishes to make a distinction between the time spent waiting for service and the time in service. We shall use the term *system time* or *total time* spent in, say, S_1 as the time spent by a customer from the moment he enters S_1's queue until he leaves that subsystem. In a closed loop, this also corresponds to the time interval from the moment the customer leaves S_2 until he returns. For that reason, this time interval is also called *response time* for S_1. We shall use the three terms interchangeably, tending to prefer the first two when discussing open systems, whereas the latter tends to be used more in dealing with time-sharing systems.

In many applications, the time spent being served is considered useful, and only the time spent waiting in the queue is wasted. This time is called both *queueing time* and *waiting time*. We will try to use the latter term, for there is some ambiguity here when load-dependent servers are considered (see the following section and Section 5.4), or when we consider "generalized M/G/C systems" in Chapter 6, for then it is not always clear when waiting ends and service begins. We will often talk about *queue length*, or the *number of customers in the queue*, and when we do, we will invariably mean *the number of customers at, or in, S_i*, i.e., including those being served.

If only one customer can be served at a time, and the performance of S_1 is the same no matter how many customers are in its queue, the steady-state mean system time $\overline{T}_s$ and mean waiting time $\overline{T}_w$ are related by the simple relation

$$\overline{T}_s = \overline{T}_w + \overline{x}_1. \tag{2.1.7c}$$

From Little's theorem, the number in the queue and the number in S_1 are related by the slightly strange formula

$$\overline{q}_s = \overline{q}_w + \rho. \tag{2.1.7d}$$

The reason ρ appears instead of 1 is that sometimes there is no one waiting when someone is being served. It is pleasant to realize that (2.1.7c) and (2.1.7d) are true for any distribution, *but* the reader should be careful to observe the restrictions as stated in the beginning of this paragraph.

2.1.4. Load-Dependent Servers

The solutions for the M/M/1 queue can be extended without much difficulty to the M/M/C//N, and even somewhat more general, queues. Suppose that there are C identical exponential servers in S_1, each with service rate μ, feeding off a single queue. That is, as long as there are $n \geq C$ customers at S_1, all of the servers will be active, and as long as $n \leq C$, none of the customers will be waiting to be served. As we already know, if several exponential servers are busy, the probability rate for something to happen is the sum of their service rates. Therefore, we can define a service rate for S_1 that depends on the number of customers there. That is, let $\mu(n)$ be the service rate of S_1 when there are n customers there; then

$$\mu(n) = n\,\mu \quad \text{for} \quad n \leq C \tag{2.1.8a}$$

and

$$\mu(n) = C\,\mu \quad \text{for} \quad n \geq C. \tag{2.1.8b}$$

We think of S_1 as a load-dependent server. Actually, the formulas we derive in this section do not depend on the explicit form we have just given the μ's; thus we can immediately generalize, and let $\mu(1)$, $\mu(2)$, and so on, be any positive numbers. The reader may think of S_i as a multiple server subsystem, or as a single server whose service rate changes (not necessarily by integral units) with change of queue length. See the end of this section for further notational discussion.

Another formulation, which we adopt here, is to introduce the *load-dependence factor*, $\alpha_1(n)$, which is the ratio of service rates, $\mu(n)$ and $\mu(1)$. By definition, $\mu(1) := \mu$, $\alpha_1(1)$ always equals 1, and $\alpha_1(n) = \mu_1(n)/\mu$, which for a subsystem with C identical servers gives the following:

$$\alpha_1(n) = n \quad \text{for} \quad n \leq C \tag{2.1.8c}$$

and

$$\alpha_1(n) = C \quad \text{for} \quad n \geq C. \tag{2.1.8d}$$

Clearly, $\mu(n) = \alpha_1(n)\,\mu$. Similarly, we can view S_2 as a load-dependent server, with load-dependence factor, $\alpha_2(n)$. Then $\lambda(n) = \alpha_2(n)\lambda$. Next look at Figure 2.1.2. The arrow going from n to $n-1$ corresponds to the probability rate of going from n to $n-1$, which can happen only if there is a completion at S_1. The rate for this to happen is $\mu(n)$. Similarly, the arrow going from n to $n+1$ corresponds to an arrival from S_2, whose rate must be $\lambda(N-n)$. Then all the arrows pointing to the left should be labeled (reading from right to left)

$$\mu(N),\ \mu(N-1),\ \cdots,\ \mu(n+1),\ \mu(n),\ \mu(n-1),\ \cdots,\ \mu(1),$$

while those pointing to the right are labeled (reading, this time, from left to right)

$$\lambda(N),\ \lambda(N-1),\ \cdots,\ \lambda(N-n+1),\ \lambda(N-n),\ \lambda(N-n-1),\ \cdots,\lambda(1).$$

Before solving for the M/M/C//N loop, let us review the meaning of a *state transition-rate diagram*. If, as in Figure 2.1.2, a single node is encircled, the sum of the probability rates entering the circle minus the sum of those leaving must be zero in the steady state. Suppose, instead, that two adjacent nodes are enclosed together. Then the arrows connecting them would not be included in the balance equations. But this would yield the same as one would get by adding the single equations together. After all, each of the two arrows appears in each equation, once as leaving one node, and once as entering the other, canceling out when the two equations are added. In general, then, we can say that for *any* closed curve, what goes in must equal what goes out for the steady state to occur. Now consider the closed curve that encompasses all nodes from 0 to n. Only one arrow goes in, and one arrow goes out, so we have the simple set of first-order difference equations:

$$\lambda(N-n)\,r(n;N) = \mu(n+1)\,r(n+1;N) \quad \text{for} \quad 0 \le n < N. \tag{2.1.9a}$$

In particular,

$$r(1;N) = \frac{\lambda(N)}{\mu(1)} r(0;N) \tag{2.1.9b}$$

and

$$r(2;N) = \frac{\lambda(N-1)}{\mu(2)} r(1;N) = \frac{\lambda(N)\lambda(N-1)}{\mu(1)\mu(2)} r(0;N). \tag{2.1.9c}$$

Next, following the notation of [GORD67], let $\rho = \lambda/\mu$, $\beta_i(0) := 1$, and for $n > 0$,

$$\beta_i(n) := \alpha_i(n)\,\beta_i(n-1) = \alpha_i(1)\alpha_i(2) \cdots \alpha_i(n). \tag{2.1.10a}$$

For a subsystem with C identical servers, we have

$$\beta_i(n) := n! \quad \text{for } n \le C \tag{2.1.10b}$$

and

$$\beta_i(n) := C!\,C^{n-c} \quad \text{for } n \ge C. \tag{2.1.10c}$$

Then with only a little trickery, the general solution becomes

$$r(n;N) = \frac{1}{K(N)} \frac{\rho^n}{\beta_1(n)\,\beta_2(N-n)}, \tag{2.1.11a}$$

where, owing to the fact that the sum of probabilities must be 1,

$$K(N) := \sum_{n=0}^{N} \frac{\rho^n}{\beta_1(n)\,\beta_2(N-n)}. \tag{2.1.11b}$$

The reader may recognize this as a discrete convolution of the reciprocals of the μ's and λ's.

Next consider a generalization of the throughput as defined in (2.1.5a). The probability that S_1 is busy no longer can yield the throughput, since its service rate depends on n. Therefore, it is somewhat more difficult to express for a load-dependent server, but turns out to be just as simple to compute. The rate at which S_1 serves customers depends on the distribution of the number in the queue. Then $\Lambda(N)$ is a weighted average of the $\mu(n)$'s:

$$\Lambda(N) = \sum_{n=1}^{N} \mu(n) r(n;N) = \sum_{n=1}^{N} \mu \alpha_1(n) r(n;N) = \frac{\mu}{K(N)} \sum_{n=1}^{N} \frac{\alpha_1(n)\rho^n}{\beta_1(n)\beta_2(N-n)}.$$

But $\alpha_1(n)/\beta_1(n) = 1/\beta_1(n-1)$ and $\mu\rho = \lambda$, so (change the summation variable from n to $n-1$)

$$\Lambda(N) = \frac{\lambda}{K(N)} \sum_{n=1}^{N} \frac{\rho^{n-1}}{\beta_1(n-1)\beta_2(N-n)} = \lambda \frac{K(N-1)}{K(N)}. \tag{2.1.12}$$

This is identical to the throughput for the load-independent system described in (2.1.5d) with $\lambda = 1/T_2$, except that now $K(N)$ does not satisfy (2.1.4d). There is no simple recursive relationship among the $K(N)$'s for arbitrary β's.

There are three different ways to "open up" our load-dependent system, two of which yield equivalent results. For the first way, merely let $\beta_2(n) = 1$ for all n. Then, if $\lambda/\mu(N)$ is less than 1 for large N, S_2 is a Poisson source to S_1 and we have the standard M/M/C queue when $\beta_1(n)$ satisfies (2.1.10b and 10c). That is, from Equations (2.1.11),

$$K := \lim_{N\to\infty} K(N) = \sum_{n=0}^{\infty} \frac{\rho^n}{\beta_1(n)} \tag{2.1.13a}$$

and

$$r(n) := \lim_{N\to\infty} r(n;N) = \frac{1}{K}\frac{\rho^n}{\beta_1(n)}. \tag{2.1.13b}$$

Actually, one can make a somewhat more general statement. If

$$\lambda_\infty := \lim_{N\to\infty} \lambda(N)$$

exists and $\lambda_\infty/\mu(N)$ is less than 1 for large N, everything still holds except that now $\rho = \lambda_\infty/\mu$.

A second approach is to argue that $\lambda(n)$ is really a function of N and n by way of their difference, $N-n$. That is, let

$$\bar{\lambda}(n) := \lim_{N\to\infty} \lambda(N-n)$$

and

$$K = \sum_{n=0}^{\infty} \frac{\rho^n}{\beta_1(n)\bar{\beta}_2(n)}$$

where $\bar{\alpha}_2(n) := \bar{\lambda}(n) / \bar{\lambda}(1)$, $\bar{\beta}_2(0) := 1$, and

$$\bar{\beta}_2(n) := \bar{\alpha}_2(n)\bar{\beta}_2(n-1).$$

The $\bar{\alpha}_2$'s can be interpreted as a slowdown of the arrival process because of the increasing queue length, so this is referred to as an M/M/C queue with *discouraged arrivals*. This may be a misnomer in some countries where consumer goods are scarce. In those places, we are told, arrival rates to queues actually increase with queue length. Mathematically, since K in this case is not a convolution, β_1 and β_2 can be combined into a single load-dependent factor. However, for more general queues (e.g., M/G/C and G/M/C) the two must still be kept separate.

The third view, which ends up being the same as the first, considers all customers, while they are at S_2, to act independently. That is, each customer spends a random amount of time at S_2, with mean Z, and then, independently of the other customers, goes to S_1. The completion rate is exactly $(N-n)/Z$. Z is called the *think time*, or *delay time*, and S_2 is called a *thinking stage* or *time-sharing stage* or *delay stage*, as well as some other names. Clearly, as N goes to infinity, the arrival rate grows unboundedly, thereby swamping S_1. In reality, there never are an infinite number of potential customers, but there may be so many and they stay at S_2 so long, that n (the number at S_1) is always small compared to N, so the departure rate from S_2 is more-or-less constant. In mathematical terms, let Z grow unboundedly with N, and let

$$\lambda_\infty = \lim_{N\to\infty} \frac{N}{Z}.$$

This yields the same solution as case 1.

In all these cases we can make a statement that generalizes (2.1.6c). Let μ_∞ be the limiting value of $\mu(N)$; then

$$\lim_{N\to\infty} \Lambda(N) = \min\left(\mu_\infty, \lambda_\infty\right). \tag{2.1.14}$$

Once again, the throughput of the system is bounded by the maximal capacity of its slowest server.

Finally, let us consider our open M/M/C queue, and let C go to infinity. Then S_1 is a place where customers arrive randomly,"hang around" for a while, $[1/\mu]$, and then leave. The number present at any time is distributed according to the Poisson distribution. Since $\beta_1(n) = n!$,

$$K = \sum_{n=0}^{\infty} \frac{\rho^n}{n!} = e^\rho,$$

leading to

$$r(n) = \frac{\rho^n}{n!} e^{-\rho}. \tag{2.1.15}$$

This is just one of the many nonequivalent derivations of the Poisson distribution.

As a final comment in this section, we point out that all the formulas are valid whether or not the α's and μ's satisfy Equations (2.1.8). If they do, we will retain the notation "M/M/C//N *loop*," including the system with a time-sharing subsystem, for which we will use the notation "M/M/∞//N" or "M/M/C//C." If we wish to look at systems in which the α's are *not* integers but satisfy a weakened version of Equations (2.1.8), namely, for $n \le C$, $\alpha_1(n) =$ anything > 0, but

$$\alpha_1(n) = \alpha_1(C) \quad \text{for} \quad n \ge C,$$

then we would refer to it as an "M/M/C – *type loop.*" If the α's can be anything whatsoever, we will use the notation, "M/M/X//N *loop.*" To maintain a connection with the outside literature, we will refer to all of these generically as "M/M/C-*type systems,*" or, "*systems with load-dependent servers.*" We will also adhere to this notation in dealing with more general distributions in Sections 4.4.3 and 5.4, and Chapter 6 (e.g., G/M/X and M/G/C queues). In Chapter 6, we will also introduce the *generalized* M/G/C *system*.

2.1.5. Departure Process

Let us now consider one last steady-state process before moving on to the transient behavior of the M/M/1 queue. Suppose that an observer is sitting just downstream from S_1, measuring the time between departures, without knowing the state of the system. What would she expect to see? In other words, given that a customer has just left, what is the time until the next one leaves S_1? We are asking for the distribution of *interdeparture times.* This question was originally considered by P. J. Burke [BURK56]. This is easy enough to find out once we accept a theorem about M/M/1 queues that will be proven in Section 4.1.3, Theorem 4.1.4. This theorem states that for both open and closed M/M/1 queues (more generally, M/G/1 – but *not* G/M/1 – queues), the steady-state probability that a departing customer will leave n fellow customers behind at S_1 is the same as the steady-state probability of finding n there, except that he will never leave N customers behind, since he, at least, must be at S_2. Let $d(n;N)$ be this probability; then from (2.1.4b) we can write

$$d(n;N) = \frac{\rho^n}{c(N)}, \tag{2.1.16}$$

where $c(N)$ is found by summing over n, from 0 to $N-1$. Thus $c(N) = K(N-1)$ from (2.1.4c).

Now, as long as S_1 is busy, the density function for the departure of the next customer is simply the same as the pdf of S_1 (i.e., $\mu e^{-\mu t}$). But if S_1 is idle, our downstream observer must wait first for a customer to finish being served at S_2 and then be processed by S_1. This is the *convolution* of the two pdfs:

$$[b_1 \times b_2](t) := \int_0^t b_1(s)\, b_2(t-s)\, ds = \int_0^t b_1(t-s)\, b_2(s)\, ds,$$

which for two exponential distributions yields

$$[b_1 \times b_2](t) := \int_0^t \mu e^{-\mu s} \lambda e^{-\lambda(t-s)} ds = \mu\lambda e^{-\lambda t} \int_0^t e^{-(\mu-\lambda)s} ds = \frac{\mu\lambda}{\mu-\lambda}\left(e^{-\lambda t} - e^{-\mu t}\right).$$

The overall distribution is the weighted average of the two possibilities. Let $b_d(t;N)$ be the pdf for the interdeparture time from S_1 of an M/M/1//N queue, and recall that $\rho = \lambda/\mu$; then

$$b_d(t;N) = d(0;N)[b_1 \times b_2](t) + [1 - d(0;N)]\mu e^{-\mu t}$$

$$= \frac{1-\rho}{1-\rho^N}\frac{\lambda}{1-\rho}\left(e^{-\lambda t} - e^{-\mu t}\right) + \left(1 - \frac{1-\rho}{1-\rho^N}\right)\mu e^{-\mu t}.$$

We can regroup the terms to get the following simple form:

$$b_d(t;N) = \frac{1}{1-\rho^N}\lambda e^{-\lambda t} - \frac{\rho^N}{1-\rho^N}\mu e^{-\mu t}. \tag{2.1.17a}$$

For the closed loop, the departure process is *not* a Poisson process, since the interdeparture times are not exponentially distributed. For the open queue, where $\rho < 1$ and $N \to \infty$, $b_d(t)$ *is* exponential. The mean time between departures is easy enough to get:

$$\bar{t}_d(N) := \int_0^\infty t\, b_d(t;N)dt = \frac{1-\rho^{N+1}}{1-\rho^N}\frac{1}{\lambda}. \tag{2.1.17b}$$

We leave it to the following exercise to show that $\bar{t}_d$ is the reciprocal of the mean throughput given by (2.1.5d).

Exercise 2.1.4: Verify that (2.1.17b) is true, and show that $\bar{t}_d(N) = 1/\Lambda(N)$.

Either from (2.1.17b) or from (2.1.6c), we have

$$\lim_{N\to\infty} \bar{t}_d(N) = \max\left(\frac{1}{\lambda}, \frac{1}{\mu}\right). \tag{2.1.18a}$$

For the open queue, if ρ is less than 1, ($\lambda < \mu$), the mean departure rate from S_1 is the same as the mean arrival rate. But if ρ is greater than 1, the mean departure rate is governed by the service rate of S_1. We can now prove the well-known result, first given by P. J. Burke in 1956, that the departures from an open M/M/1 queue are exponentially distributed. Simply let N go to infinity on (2.1.17a),

$$b_d(t) := \lim_{N\to\infty} \bar{t}_d(N) = \lambda e^{-\lambda t} \quad \text{for } \rho < 1, \tag{2.1.18b}$$

$$= \mu e^{-\mu x} \quad \text{for } \rho > 1.$$

As long as ρ is less than 1, it is as though S_1 did not exist (exponential in $\to$ exponential out). We also see once again that S_2, with its unbounded number of customers, is a Poisson source for S_1. But if ρ is greater than 1, S_1 releases customers at its service rate and becomes a Poisson source for S_2. The symmetry of our loop would require this, anyway.

We must emphasize that this result (exponential in $\to$ exponential out, for an open, unsaturated M/M/1 queue) is indeed extraordinary. It is also valid for load-dependent (i.e., M/M/C) queues, *but it is not true* for first-come first-served M/G/1 queues or even G/M/1 queues. It is not even true for closed M/M/1//N loops. We must be careful not to generalize too quickly from what we learn about the M/M/1 queue.

2.2. RELAXATION TIME FOR M/M/1//N LOOPS

For the rest of this chapter we examine systems for which not enough time has elapsed to declare that our system is in its steady state. We call this time range the *transient region*. In principle we would like to solve the Chapman-Kolmogorov equations (1.3.2b), but in practice, if N is large, this is not an easy task. Aside from the M/M/1 queue, there are very few known analytic solutions to this equation. A rather ingenious solution for the open M/M/1 queue, where N is infinite, is given in [TAKA62]. That may well be the only explicit solution for an infinite state-space, time-dependent queueing system in existence. But even the existence of that solution does not help much, since it is so difficult to evaluate or interpret.† Therefore we must find some simpler ways of parameterizing transient behavior. For our initial view, we remind the reader of the discussion about relaxation times in Section 1.3.3 and (1.3.11).

In general, finding the eigenvalues of a matrix is not a trivial task, particularly if one wants to express them in terms of unspecified parameters rather than numerically. If the dimension of the matrix is small enough, as with (2.1.2) and (2.1.3), the eigenvalues can be found by straightforward, if tedious, methods. In the case of our $\boldsymbol{Q}$, since one of the eigenvalues is zero, the characteristic equation can be written as degree N rather than $N+1$, the size of $\boldsymbol{Q}$. It is well known that no general formula (such as the quadratic equation) exists for the roots of polynomials of degree greater than four, nor can one ever be found. (If you have ever used the cubic or quartic formulas to get analytic expressions, you might be inclined to say that even four is too big.) Therefore, unless one is "lucky" (as with the zero eigenvalue), the task is hopeless for $N > 4$.

By a fortuitous stroke of good fortune, because the $\boldsymbol{Q}$ of (2.1.1c) is so repetitive, $\phi_N(\beta) = |\boldsymbol{Q} - \beta \mathbf{I}|$ satisfies a recurrence relation in N which turns out to be similar to that satisfied by Chebyshev polynomials of the second kind, from which all the eigenvalues can be obtained. The details can be found in [MORS58]. As always, $\beta_0 = 0$, and

$$\beta_k = \mu + \lambda + 2\sqrt{\mu\lambda}\cos\frac{k\pi}{N+1} \qquad \text{for } k = 1, 2, 3, \cdots, N. \qquad (2.2.1a)$$

† Takacs actually supplies two different forms for the solution, neither of which is easy to evaluate. Most texts list the second form, which involves an infinite sum of Bessel functions, but the first form turns out to be more useful (particularly in the region where the time parameter is neither very small nor very large) if one is comfortable with numerical integration.

The smallest β is β_N, which therefore must be $1/RT$. As in Exercise 2.1.1, it is convenient to express the relaxation time in units of the time it takes a lone customer to make one cycle, $(1/\mu + 1/\lambda)$. Then, recalling that $\rho = \lambda/\mu$, and $\cos[\pi N/(N+1)] = -\cos[\pi/(N+1)]$, we get the following expression for the *normalized relaxation time*:

$$T(\rho, N) := \frac{\mu\lambda}{\mu+\lambda} RT = \frac{\rho}{(1+\rho)}\left(1 + \rho - 2\sqrt{\rho}\cos\frac{\pi}{N+1}\right)^{-1}. \tag{2.2.1b}$$

T is invariant to the replacement of ρ with $1/\rho$, that is, $T(\rho, N) = T(1/\rho, N)$. Next, we look at $T(\rho, N)$ when N is very large. For $\rho \neq 1$, $T(\rho, N)$ has a finite limit as N goes to infinity. Thus the relaxation time for an open system (normalized so that $1/\mu + 1/\lambda = 1$), is

$$T(\rho) := \lim_{N\to\infty} T(\rho, N) = \frac{\rho}{(1+\rho)(1-\sqrt{\rho}\,)^2} = T(1/\rho). \tag{2.2.2a}$$

It is not hard to show that $T(\rho)$ approaches $0.5/(1-\rho)^2$ when ρ is close to 1 [LIPS83]. As so often happens, $\rho = 1$ must be treated as a special case. We can either set $\rho = 1$, or let $N\to\infty$, but not both at the same time. Here $T(1, N)$ goes to infinity as $O(N^2)$. We show this by setting ρ equal to 1 in (2.2.1b) to get

$$T(1, N) = \frac{1}{2}\left(2 - 2\cos\frac{\pi}{N+1}\right)^{-1} = \frac{1}{4}\left(1 - \cos\frac{\pi}{N+1}\right)^{-1},$$

and then use Maclaurin's expansion for $\cos x$ $[\cos x = 1 - x^2/2 + O(x^4)]$:

$$T(1, N) = \frac{1}{4}\left[\frac{1}{2}\left(\frac{\pi}{N+1}\right)^2 + O\left(\frac{1}{N^4}\right)\right]^{-1}$$

$$= \frac{1}{2}\left(\frac{N+1}{\pi}\right)^2\left[1 + O\left(\frac{1}{N^2}\right)\right]. \tag{2.2.2b}$$

Naturally, the relaxation time for an open system ($N = \infty$) is infinite when $\rho = 1$. That is, the system *never* reaches a steady state.

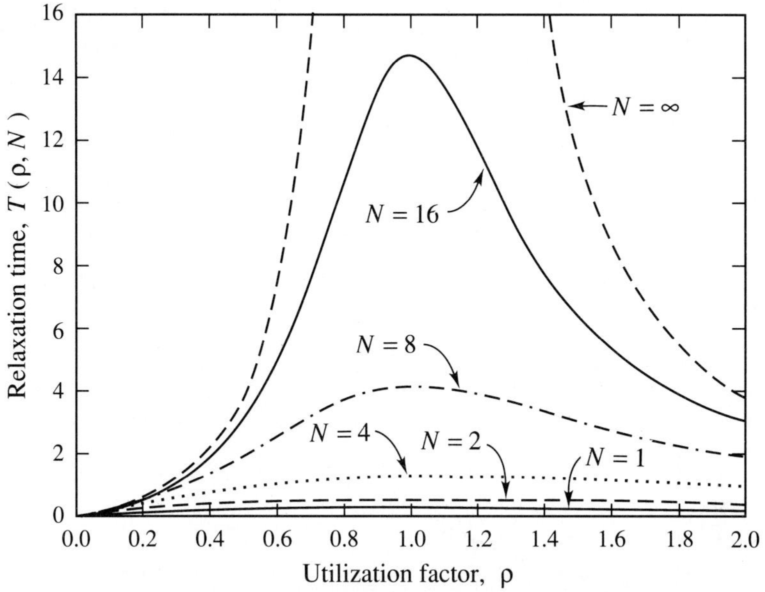

Figure 2.2.1. Relaxation time as a function of ρ for M/M/1//N queues, as given by (2.2.1b). $T(\rho, N)$ is in units of cycle time for one customer. All curves peak at $\rho = 1$, while at $\rho = 1$, $T(1, N)$ goes to infinity as N becomes increasingly large. For all values of ρ, the relaxation time increases with N.

Example 2.2.1: Figure 2.2.1 summarizes what we have said about relaxation times. What is most important is to observe that as systems get bigger (in this case, N larger) and more saturated (ρ close to 1), the time it takes to approach the steady-state solution grows as well. This puts into question the steady-state solution as a description of systems that are in existence for relatively short times. ●

Example 2.2.2: Figure 2.2.2 presents the same information in a different way. Now N varies for fixed $\rho = 0.5, 0.9, 0.95$, and 1. As with the throughput curves, $T(\rho, N) = T(1/\rho, N)$, so $\rho = 2$ yields the same curve as $\rho = 1/2$. As $N \to \infty$, each curve approaches its limit as given by (2.2.2a), except, of course, for $\rho = 1$, which has no limit. ●

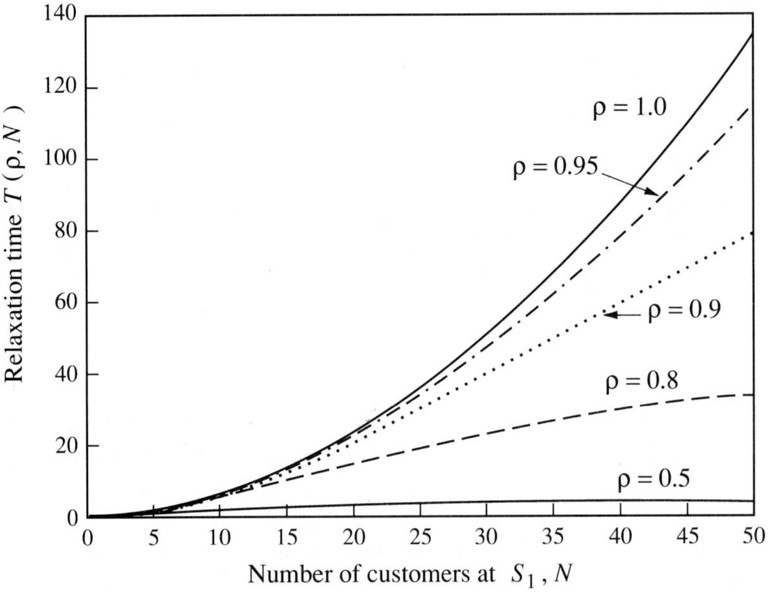

Figure 2.2.2. Relaxation times as a function of system population N for M/M/1//N loops. The RT's for ρ and $1/\rho$ are identical; therefore, we only show curves for $\rho \le 1$. As $N \to \infty$, all curves except that for $\rho = 1$ will saturate.

Clearly, if ρ is close to 1, the relaxation time can be very large. However, if ρ is very small (or very large), $T(\rho, N)$ is small. This may be an underestimate of how long it takes a system to come close to its steady state. If all customers are initially at the slower server, very few completions would have to occur to approach the steady state, since very few customers are ever likely to be at the faster server at any one time. Even so, the mean time for one slow server completion (in units of the cycle time) is $1/(1+\rho)$, which is $1/\rho$ (for small ρ) times larger than $T(\rho, N)$. On the other hand, if all the customers are initially at the faster server, the steady state cannot be approached until almost all of them have been served at least once. The mean time for this is of the order of $\rho N/(1+\rho)$. The two conditions together imply that

$$0 \le RT \le \frac{N}{\rho} T(\rho, N), \quad \rho < 1. \tag{2.2.2c}$$

RT could be 0 if the system was initially in its steady state, which means that all queue lengths are possible from the beginning (i.e., we do not know anything).

2.3. OTHER TRANSIENT PARAMETERS

In this section we introduce alternative ways (other than *RT*) of examining the transient region. We will be pleased to find that some of the objects we needed for the steady-state solution will also be used here. As with every Markov chain, only one thing at a time can happen in a queueing network; the evolution of the system in time is marked by a discrete sequence of events. Such sequences can be represented by *time-dependent state transition diagrams*. The technique described here is easily generalized to include nonexponential and even more general service centers, and that will be done in succeeding chapters.

2.3.1. Mean First-Passage Time for Queue Growth

As a first application, we will examine the time it takes for a queue to grow from 0 to some integer n. Such processes are referred to as *first passages*, and the average time for such events to take place are called *first-passage times*. The points at which a Markov chain reaches each length for the first time are called *ladder points*. Looking at Figure 2.1.1, suppose that initially all the customers are at S_2; then in mean time $1/\lambda$ the first event occurs, corresponding to an arrival to S_1. After that, one of two events can occur: either the customer at S_1 returns to S_2, or another customer from S_2 goes to S_1. The sequence of possible events grows factorially after that, and it becomes thoroughly impractical to enumerate all of them. However, if in any sequence the system returns to a state it was in previously, a recursive relation can be set up that may be solvable. This is known as a *regenerative process* [KING72]. We shall see how this works in this section and will use it frequently in subsequent chapters.

To apply this method, one must start with single jumps. So we define

Definition 2.3.1

$\tau_u(n)$ = *mean first-passage time for the queue at* S_1 *to go from* n *to* $n+1$. The process begins with n customers at S_1. Customers may then leave and arrive in arbitrary order, but eventually there will be $n+1$ customers at S_1 for the first time. The mean time for this to happen is $\tau_u(n)$. (In subsequent sections we shall have occasion to use d for *down*, and m for *max*, as well as u for *up*.) ◊

Consider Figure 2.3.1. The circles on the lowest horizontal line correspond to the set of states the system can be in initially, which in the present case is labeled by the number of customers at S_1. The second horizontal line represents the state the system is in after one transition. The average time elapsed between the two lines depends on the initial state. Thus if the system started with all customers at S_2 [$n = 0$], the mean time for the first transition would be $1/\lambda$. Similarly, if all customers were initially at S_1 [$n = N$], the average time elapsed would be $1/\mu$. For all other initial states, the time would be $1/(\mu + \lambda)$. A straight arrow corresponds to a single, direct transition, with the probability that

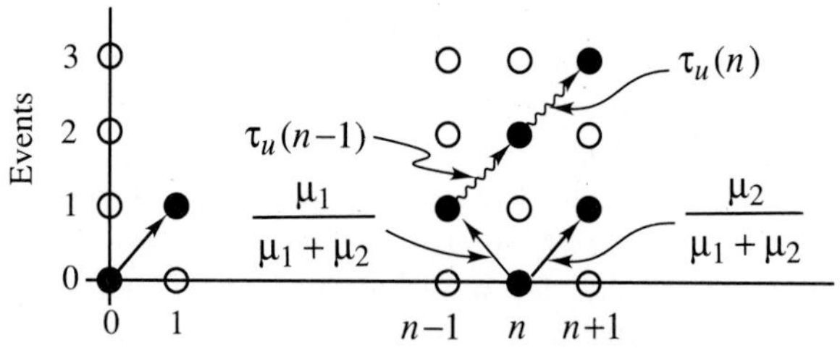

Figure 2.3.1: Time-dependent state transition diagram for a closed M/M/1//*N* loop, describing the mean time $[\tau_u(n)]$ for a queue to grow by one customer.

it will occur written near it. For instance, the system can go from $n \rightarrow n+1$ in one step, with probability $\lambda/(\mu+\lambda)$, with a mean time delay of $1/\mu+\lambda)$. A wavy arrow corresponds to the sum of all possible ways the system can get from the tail to the head for the first time, irrespective of the number of transitions taken. Thus the arrow labeled "$\tau_u(n)$" includes not only the direct transition $(n \rightarrow n+1)$, but also $(n \rightarrow n-1 \rightarrow n \rightarrow n+1)$, and $(n \rightarrow n-1 \rightarrow n-2 \rightarrow n-1 \rightarrow n \rightarrow n-1 \rightarrow n \rightarrow n+1)$, and the infinite number of other sequences that eventually lead to $n+1$.

Our ability to represent an infinite number of sequences by a single symbol is the key to setting up a soluble set of recursive relations. If the system starts with n at S_1, an event will occur in mean time $1/(\mu+\lambda)$. That event can be one of two things. Either the queue will go directly to $n+1$, or it will drop to $n-1$, in which case it will take time $\tau_u(n-1)$ to get back to n, and a further $\tau_u(n)$ to finally get to $n+1$. Mathematically we can write

$$\tau_u(n) = \frac{\lambda}{\mu+\lambda} \cdot \frac{1}{\mu+\lambda} + \frac{\mu}{\mu+\lambda}\left[\frac{1}{\mu+\lambda} + \tau_u(n-1) + \tau_u(n)\right],$$

where $\tau_u(0) = 1/\lambda$. For convenience, drop the subscript u when no confusion is likely to arise. The two terms without a τ in them combine to yield the following:

$$\tau(n) = \frac{1}{\mu+\lambda} + \frac{\mu}{\mu+\lambda}[\tau(n-1) + \tau(n)] . \tag{2.3.1a}$$

We interpret this as follows. It takes a mean time of $1/(\mu + \lambda)$ for something to happen. If the event was an arrival, we are done. The probability that it was not an arrival is $\mu/(\mu+\lambda)$, in which case the queue will have dropped back to $n-1$ and take a mean time of $[\tau(n-1) + \tau(n)]$ to first get back to n and then to $n+1$.

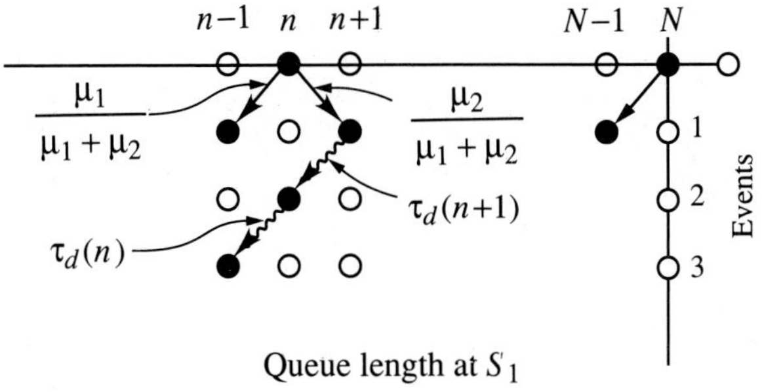

Figure 2.3.3: Time-dependent state transition diagram for a closed M/M/1//N loop describing the mean time $[\tau_d(n)]$ for a queue to decrease by 1 customer. $\tau_d(1)$ is the mean busy period.

$$\tau(n;N) = \frac{1}{\mu+\lambda} + \frac{\lambda}{\mu+\lambda}[\tau(n+1;N) + \tau(n;N)], \tag{2.3.5a}$$

where $\tau(N;N) = 1/\mu$. Making the substitution $\rho = \lambda/\mu$ and the usual rearrangements, we get

$$\mu\tau(n;N) = 1 + \rho\mu\tau(n+1;N). \tag{2.3.5b}$$

Directly substituting into (2.3.5b) for $n = N-1$ and $N-2$, it follows that $\mu\tau(N-1;N) = 1+\rho$, and $\mu\tau(N-2;N) = 1+\rho+\rho^2$. One can easily guess, and prove by induction, that

$$\tau(N-k;N) = \frac{1}{\mu}\sum_{i=0}^{k}\rho^i = \frac{1-\rho^{k+1}}{\mu(1-\rho)} \quad \text{for } \rho \neq 1 \tag{2.3.6a}$$

and

$$\tau(N-k;N) = \frac{k+1}{\mu} \quad \text{for } \rho = 1, \tag{2.3.6b}$$

where $k = N-n$. It is clear that when $\rho \geq 0$, τ_d grows unboundedly with N (and k), but when $\rho < 1$, then

$$\tau_d(n) := \lim_{N\to\infty}\tau_d(n;N) = \frac{1}{\mu(1-\rho)}. \tag{2.3.6c}$$

With some thought, it is not hard to see that for an open system, the mean time for a queue to drop by 1 should be the same for all n, but it is an interesting result nonetheless.

By definition, the mean time for a busy (1-busy) period is the same as the mean time to eventually go from $n = 1$ to $n = 0$. The *k-busy time* is defined as follows:

Definition 2.3.4

$t_d(k \to 0; N)$:= *the mean time for the k-busy period of an* M/M/1//*N loop*. The process begins with k customers at S_1, and ends when there are 0 customers there for the first time. ◊

First we have

$$t_d(1 \to 0; N) = \tau(1; N) = \frac{1-\rho^N}{\mu(1-\rho)} \quad \text{for } \rho \neq 1 \tag{2.3.7a}$$

and

$$t_d(1 \to 0; N) = \frac{N}{\mu} \quad \text{for} \quad \rho = 1. \tag{2.3.7b}$$

As with the τ_d's, when $\rho \geq 1$, the mean extent of the busy period grows unboundedly to infinity with N, while when $\rho < 1$, the limit for $t_d(1 \to 0; N)$ exists and approaches $t_d(1 \to 0) = 1/[\mu(1-\rho)]$ for large N. The expression for $t_d(1 \to 0)$ looks familiar and tells us that the mean busy period for an open M/M/1 queue is the same as its mean system time given by (2.1.7b). This equality is purely coincidental. We show in Chapter 4 that this formula even gives the mean time for the busy period of *all* open M/G/1 queues (but not G/M/1 queues), whereas the expression for the mean system time for M/G/1 queues, the well-known P-K formula, is more complicated.

In direct analogy with Equations (2.3.3) we see that the mean time for the k-busy period is

$$t_d(k \to 0; N) = \sum_{j=1}^{k} \tau_d(j; N) = \frac{1/\mu}{1-\rho} \sum_{j=1}^{k} (1 - \rho^{N-k+1}),$$

which after some straightforward manipulation yields

$$\mu\, t_d(k \to 0; N) = \frac{k}{1-\rho} - \frac{\rho^{N-k+1}}{(1-\rho)^2} + \frac{\rho^{N+1}}{(1-\rho)^2} \quad \text{for } \rho \neq 1 \tag{2.3.8a}$$

and

$$\mu\, t_d(k \to 0; N) = kN - \frac{k(k-1)}{2} \quad \text{for } \rho = 1. \tag{2.3.8b}$$

As with the τ_d's for open systems, the k-busy period is infinite when $\rho \geq 1$, while when $\rho < 1$,

$$\mu\, t_d(k \to 0) = \frac{k}{1-\rho}. \tag{2.3.8c}$$

This makes sense, since it takes a time $1/[\mu(1-\rho)]$ [or what is the same thing, $\lambda\rho/(1-\rho)$] for an open queue to drop by 1, so if there were k customers to start with, it should take k times $\lambda\rho/(1-\rho)$ to drop to 0.

Probability That Queue Will Reach at Least Length *k*

Although the time for a busy period may be important, it is by no means the only parameter worth looking at. From an experimental point of view, it is easy to measure, for instance, the number of busy periods in which a given queue length was reached or the maximum queue length reached. It is desirable, therefore, to be able to compute these quantities as well.

By now we should be getting pretty good at working with time-dependent state transition diagrams. The procedure for calculating probabilities for queue changes is very similar to that for calculating the mean time for the change to occur. First we must calculate the probabilities for one step at a time, and then take the product of the probabilities (note that we take the *sum* of the times) for the complete process. First define:

Definition 2.3.5

$W_u(n)$:= *probability that the queue at* S_1 *will go from* n *to* $n+1$ *during a busy period (i.e., without going to* 0). The process begins with n customers at S_1, and ends when the queue (including the active customer) either reaches $n+1$ or 0. The queue can fall and rise any number of times before the process ends. ◊

The queue either goes up [with probability $\lambda/(\mu+\lambda)$], or goes down [$\mu/(\mu+\lambda)$], in which case it must eventually get back to n without first going to 0 [$W_u(n-1)$], and then get to $n+1$, [$W_u(n)$, another regenerative process]. The equation describing this is

$$W_u(n) = \frac{\lambda}{\mu+\lambda} + \frac{\mu}{\mu+\lambda}[W_u(n-1)W_u(n)]. \tag{2.3.9a}$$

This reorganizes to

$$W_u(n) = \rho[1+\rho - W_u(n-1)]^{-1}, \tag{2.3.9b}$$

where $W_u(1) = \lambda/(\mu+\lambda) = \rho/(1+\rho)$. Our "great" experience with these things allows us to guess and prove by induction, with $K(0) = 1$, that

$$W_u(n) = \rho\frac{K(n-1)}{K(n)}, \tag{2.3.9c}$$

where $K(n)$ was defined in (2.1.4c) and satisfies the recursive and explicit formulas

$$K(n) = \sum_{j=0}^{n}\rho^j = 1 + \rho K(n-1) = \frac{1-\rho^{n+1}}{1-\rho} \quad \text{for } \rho \neq 1 \tag{2.3.10a}$$

and

$$K(n) = n+1 \quad \text{for } \rho = 1. \tag{2.3.10b}$$

We will not always be so fortunate to find explicit expressions for more complicated queues.

As the final effort of this section, we calculate the probability that the queue will get at least to k during a busy period. This is the same as

Definition 2.3.6

$W_u(1 \rightarrow k)$:= *probability that the queue at* S_1 *will go from* 1 *to* k *before going to* 0. The process begins with one customer at S_1 and ends when the queue (including the active customer) reaches either k or 0. ◊

Then $W_u(1 \rightarrow 1) = 1$, and for $k > 1$,

$$W_u(1 \rightarrow k) = \prod_{n=1}^{k-1} W_u(n) := W_u(1) W_u(2) \cdots W_u(k-1), \qquad (2.3.11a)$$

which due to (2.3.9c) gives us

$$W_u(1 \rightarrow k) = \frac{\rho}{K(1)} \rho \frac{K(1)}{K(2)} \rho \frac{K(2)}{K(3)} \cdots \rho \frac{K(k-2)}{K(k-1)}.$$

As long as ρ does not equal 1, this conveniently simplifies to

$$W_u(1 \rightarrow k) = \frac{\rho^{k-1}}{K(k-1)} = \frac{(1-\rho)\rho^{k-1}}{1-\rho^k}. \qquad (2.3.11b)$$

For $\rho = 1$ we get the much simpler expression

$$W_u(1 \rightarrow k) = \frac{1}{k} \qquad (\rho = 1). \qquad (2.3.11c)$$

Note that (2.3.11a, b, and c) are valid for any customer population as long as $k \leq N$. Thus they are valid for open systems as well. Observe that as might be expected if $\rho \leq 1$, then $W_u(1 \rightarrow k)$ approaches 0 as n gets increasingly large. However, if $\rho > 1$, then

$$\lim_{k \to \infty} W_u(1 \rightarrow k) = \lim_{k \to \infty} \frac{(1-\rho)\rho^{k-1}}{1-\rho^k} = 1 - \frac{1}{\rho}. \qquad (2.3.11d)$$

In other words, for an open system with $\rho > 1$, the probability that the queue will grow to infinity without the busy period ever ending is $1 - 1/\rho$. That is, the probability that a busy period will end is $1/\rho$. Processes which are not guaranteed to end are sometimes referred to as *defective* [FELL71]. When $\rho = 1$, we have the interesting apparent contradiction that each busy period will surely end $[1 - W_u(1 \rightarrow \infty) = 1]$, but on the average it will take an infinite amount of time to do so.

Maximum Queue Length During a Busy Period

The last property that we study in this chapter is the probability that S_1's maximum queue length in a busy period will be k. Call this $W_m(k;N)$, where N is the total number of customers in the system. To evaluate this, we not only use the W_u's of the preceding section, but we also evaluate the probabilities of coming down without ever exceeding $k < N$. So, define

Definition 2.3.7

$W_d(n,k;N)$ = *probability that the queue at S_1 will go from n to $n-1$ without exceeding k, where $N \geq k \geq n > 0$.* The process begins with n customers at S_1 and ends when the queue either reaches $n-1$ or $k+1$. Put differently, $W_d(n,k;N)$ is also the probability that the queue will reach $n-1$ before going to $k+1$. For $k=N$, then, $W_d(n,N;N)=1$, since it is certain that the queue will eventually drop by 1 from any n. ◊

Next we recognize that for $k<N$,

$$W_d(k,k;N) = \frac{\mu}{\mu+\lambda} = \frac{1}{1+\rho}. \tag{2.3.12a}$$

For $n<k$, the recursive formulas are exactly analogous to (2.3.9), namely

$$W_d(n,k;N) = \frac{\mu}{\mu+\lambda} + \frac{\lambda}{\mu+\lambda}[W_d(n+1,k;N)\, W_d(n,k;N)],$$

which leads to

$$W_d(n,k;N) = [1+\rho-\rho\, W_d(n+1,k;N)]^{-1}. \tag{2.3.12b}$$

The usual guess and proof by induction gives us an explicit expression for $W_d(n,k;N)$:

$$W_d(n,k;N) = \frac{K(k-n)}{K(k-n+1)} \qquad \text{for } k<N. \tag{2.3.12c}$$

Notice that this expression is independent of N, as long as $k<N$. For $k=N$ it is clear that $W_d(N,N;N)=1$, since the queue cannot grow beyond N. It follows from (2.3.12b) that if $W_d(n+1,N;N)=1$, then $W_d(n,N;N)$ must also equal 1. Therefore,

$$W_d(n,N;N)=1 \qquad \text{for } 1 \leq n \leq N. \tag{2.3.12d}$$

This merely states the obvious, that a closed system will experience every queue length with certainty (not once, but over and over), and of course, irrespective of what ρ is. It is nice to know that our mathematics sometimes produces the expected. Remember, though, that (2.3.12d) is not necessarily true of open systems.

> **Exercise 2.3.2:** Given Equations (2.3.10) and (2.3.12a), prove by induction that (2.3.12c) is the unique solution of (2.3.12b).

Our next task is to calculate the object in the following definition.

Definition 2.3.8

$W_d(k \to 0; N)$:= *probability that the queue at* S_1 *will drop from* k *to* 0 *without ever exceeding* k*, in an* M/M/1//N *loop.* The process begins with k customers at S_1, and ends when it reaches either $k+1$ or 0. ◊

This must be the product of the probabilities of cascading downward one step at a time. Therefore, since $K(0) = 1$, this is

$$W_d(k \to 0; N) = \prod_{n=1}^{k} W_d(n, k; N) = \frac{K(k-1)}{K(k)} \frac{K(k-2)}{K(k-1)} \cdots \frac{K(1)}{K(2)} \frac{K(0)}{K(1)}.$$

All but one of the terms cancel, leaving us with the simple formula

$$W_d(k \to 0; N) = \frac{1}{K(k)} = \frac{1-\rho}{1-\rho^{k+1}}, \qquad (2.3.13a)$$

for $k = 1, 2, 3, \cdots, N-1$, while

$$W_d(N \to 0; N) = 1. \qquad (2.3.13b)$$

This last equation must be true. Since it is impossible for the queue to exceed N, it must drain eventually.

Our final exercise is to calculate the probability described in this section's title. Clearly, this is equal to the probability that the queue at S_1 will reach k $[W_u(1 \to k)]$ and then drop to 0 without ever exceeding k $[W_d(k \to 0; N)]$. Therefore, we define for the M/M/1//N queue:

Definition 2.3.9

$W_m(k; N)$:= *probability that the queue at* S_1 *will reach a maximum of* k *during a busy period for an* M/M/1//N *queue.* The process begins with 1 customer at S_1, and ends when there are either $k+1$ or 0 customers there. The process is a *success* only if it ends with 0 customers, and the queue reaches k at least once during the interval. ◊

This turns out to be

$$W_m(k;N) = W_u(1 \to k)\, W_d(k \to 0; N)$$

$$= \frac{\rho^{k-1}}{K(k-1)} \frac{1}{K(k)} \quad \text{for } 1 \le k < N \qquad (2.3.14a)$$

and

$$W_m(N;N) = W_u(1 \to N) = \frac{\rho^{N-1}}{K(N-1)}. \qquad (2.3.14b)$$

Note that $W_m(k,N)$ does not depend on N as long as $k < N$; thus we can write that

$$W_m(k;N) = W_m(k;\infty) \quad \text{for } k < N.$$

Since the queue at S_1 must grow to some maximum length during a busy period, it must follow that

$$\sum_{k=1}^{N} W_m(k;N) = 1. \qquad (2.3.15)$$

This is shown to be true by recognizing that since $K(n) = 1 + \rho K(n-1)$,

$$W_m(k;N) = \frac{\rho^{k-1}}{K(k-1)K(k)} = \frac{\rho^{k-1}}{K(k-1)} - \frac{\rho^k}{K(k)}. \qquad (2.3.16a)$$

Clearly, in validating (2.3.15), the negative term of $W_m(k;N)$ exactly cancels the positive term of $W_m(k{+}1;N)$, and since $W_m(N;N)$ has only a positive term, all terms cancel except the positive part of $W_m(1;N)$, which is $\rho^0/K(0) = 1$.

Equation (2.3.16a) tells us something else, which we should have suspected in the first place. Notice from (2.3.11b) that

$$W_m(k;N) = W_u(1 \to k) - W_u(1 \to k{+}1), \qquad (2.3.16b)$$

but still, it is nice to know that we have derived it.

Example 2.3.2: As our truly final example for this chapter, we observe how $W_m(k;N)$ behaves when both k and N are very large. This is shown in Figure 2.3.4 for $N = 10$ and various values of ρ. Clearly, when $\rho < 1$, W_m goes to 0 as ρ^k. That is, the probability of reaching long queues becomes highly unlikely. Now, if $\rho = 1$, then $W_m(k;N) = 1/k(k+1)$ for $k < N$ and $W_m(N;N) = 1/N$. Thus very large queue lengths can be expected during a busy period, in fact, so large that it may take forever for some busy periods to end. ●

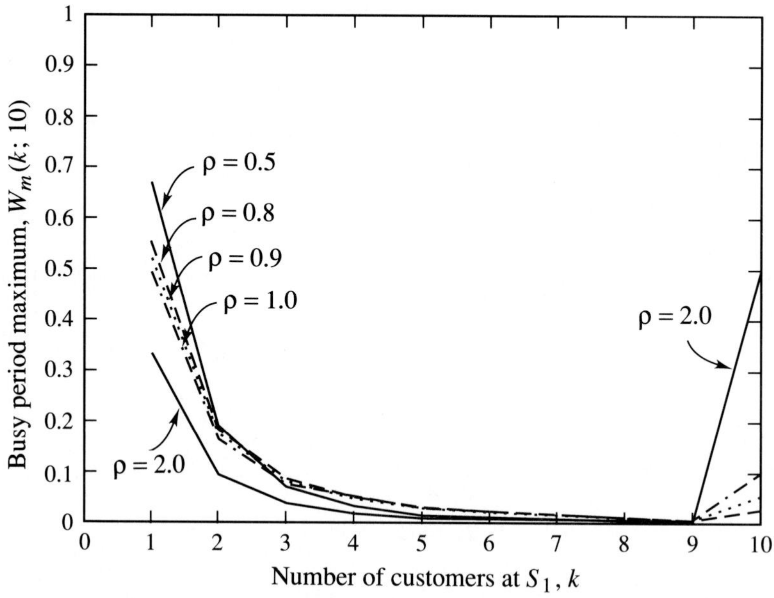

Figure 2.3.4: Probability, $W_m(k;\, 10)$, that the queue at S_1 will reach a maximum of k during a busy period of an M/M/1//10 loop. Curves for $\rho = 0.5,\ 0.8,\ 0.9,\ 1.0$, and 2.0 are displayed. All the curves decrease for increasing k, except at $k = 10$. Since $W_m(k;\, 10) = W_m(k;\, \infty)$ for all $k < 10$, $W_m(10;\, 10)$ corresponds to the probability that the open queue will exceed a length 9 during a busy period. At $k = 1$, $W_m(1;\, 10)$ decreases with ρ, but at $k = 10$, the reverse is true.

Exercise 2.3.3: Evaluate $W_m(k;\, \infty)$ and $W_m(N;\, N)$ for all k for N equals 5 and 20, and $\rho = 0.1, 0.5, 0.9, 1, 1.1$, and 2. Make sure that your numbers satisfy (2.3.15). How do your numbers compare with Figure 2.3.4?

Perhaps the most interesting results for maximum queue length occur for $\rho > 1$. In this case $W_m(k;\, N)$ goes to 0 as $1/\rho^k$, just as it does for $\rho < 1$. But $W_m(N;\, N)$ approaches the finite limit, $1 - 1/\rho$. This, of course, is the probability that the busy period will never end in an open system. For those busy periods that do end (the probability of which is $1/\rho$), $W_m(k;\, \infty)$ is still the correct probability that k will be the maximum queue length.

CHAPTER 3

MATRIX EXPONENTIAL FUNCTIONS

"I shall never believe that God plays dice with the universe" – Albert Einstein

"God not only plays dice, He also sometimes throws the dice where they cannot be seen." – Stephen Hawking

We are now ready to give structure to the subsystems S_1 and S_2. In Chapter 2 we assumed that each subsystem had only one internal state, which was equivalent to assuming that they were exponential servers. Now we assume that S_1 has m states, but defer consideration of S_2 until Chapter 4. Without loss of generality, a subsystem with m states can be viewed as a network of exponential *phases*, or *stages*, which can be accessed by only one customer at a time; the rest of the customers wait *outside* until the active one leaves. We will show that such a subsystem is in turn equivalent to a single server whose pdf is certainly *not* exponential. In fact, every pdf that can be written as a finite sum of terms of the form $x^k \exp(-\mu x)$ (any number of terms with any nonnegative integer k, with any number of different μ's whose real part is positive) is equivalent to a subsystem of this form. We know that functions of this type can approximate every pdf arbitrarily closely in some sense. Therefore, we can say that the *closure* of this set (infinite sums) contains all (well, maybe *almost* all) pdf's. We also know that every one of these functions has a Laplace transform that can be written as a ratio of two polynomials. Such functions are said to have *rational Laplace transforms* (*RLT* s).

3.1. PROPERTIES OF A SUBSYSTEM, *S*

Once again, we must start with a series of definitions. Let our subsystem, S, be made up of a collection of *phases* as shown in Figure 3.1.1. The term *stage* is often used instead of *phase*, and if we are thinking of a subsystem made up of real components, each of these phases, or stages, would be an exponential server in its own right. The reader is welcome to think of them in this light, and indeed we will talk of them as though they *are* real. However, in the long run they are merely meant to be mathematical building blocks for constructing the matrix operators we need for LAQT. Therefore, we (almost always) adhere to Neuts' convention and call them *phases* [NEUT75], since that word is as far from the real thing as we can get.

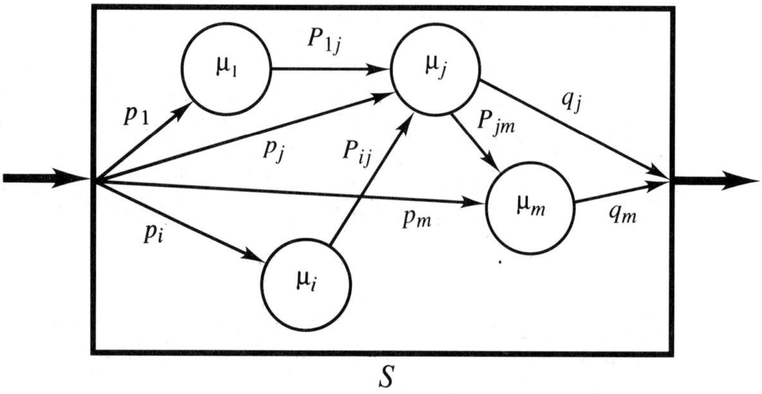

Figure 3.1.1. Typical subsystem, S, with m phases, and where only one customer can be active at a time. **p** is the *entrance vector*, whose ith component is the probability that a customer, upon entering S, will go to phase i. **q'** is the *exit vector*, whose ith component is the probability that a customer, upon completing service at phase i, will leave S. **P** is the sub-stochastic transition matrix, whose ijth component is the probability that a customer who has just finished service at i will go to j. Each phase has exponentially distributed completion time, with mean completion rate $\mu_i = (\mathrm{M})_{ii}$.

As in Section 1.3.1, **M** is the completion rate matrix whose diagonal elements are the completion rates of the individual phases in S. **P** is again the transition matrix where P_{ij} is the probability that a customer will go from phase i to phase j when service is completed at i. However, now **P** is not isometric, since it *does not* satisfy (1.3.1b). For now it is possible for a customer to leave. We define an *exit vector* **q'**, where q_i is the probability of leaving S when service is completed at i. It then follows that

$$\mathbf{P}\boldsymbol{\varepsilon}' + \mathbf{q}' = \boldsymbol{\varepsilon}'. \tag{3.1.1a}$$

If $\mathbf{q}' \neq \mathbf{o}'$ (with no negative components) and $P_{ij} \geq 0$, then **P** is said to be *substochastic*.

Assume that for each i there exists a path to some j for which $q_j \neq 0$. This is equivalent to saying that no matter where a customer starts in S, he will eventually leave. It is also equivalent to the statement that $(\mathbf{I} - \mathbf{P})$ has an inverse. We now show that if $(\mathbf{I} - \mathbf{P})$ has an inverse, the customer can always get out (eventually). Let x_i be the probability that a customer who started at phase i will eventually leave, and let **x'** be the column vector whose ith component is x_i. Then we can say that the probability of leaving eventually is equal to the probability of leaving immediately, $[q_i]$, plus the probability of going instead to some other phase j, $[P_{ij}]$, and eventually leaving from there, $[x_j]$. Mathematically, this is

We will have occasion to describe processes other than the time a customer spends in S. Therefore we provide the following generic definition.

Definition 3.1.3

Let X be the random variable for some process (e.g., *system time*, or *interdeparture time*) whose pdf is $b_X(t)$. Then, $<\mathbf{p}_X, \mathbf{B}_X>$ is a generator of process X if the equations of Theorem 3.1.1 are satisfied. $\mathbf{p}_X$ is the *startup* or *initial* vector for the process (or *startup process vector*), and $\mathbf{B}_X$ *is the process rate matrix*, or the *rate matrix* for the process X. Only when we are dealing with the service time distribution of S will we use the terms *entrance vector* and *service rate matrix*. ◊

3.1.4. Numerical Algorithm for Evaluating *b(x)* and *R(x)*

The formulas given in Theorem 3.1.1 are not merely formal connections between functions and matrices. They can actually be used to calculate, efficiently and accurately, the values of $b(x)$, $R(x)$, and therefore $B(x)$, over a set of equally spaced values of x. First note that because of Theorem 1.3.2, and Equation (3.1.6b),

$$\mathbf{R}(x+y) = \exp[-(x+y)\mathbf{B}] = \mathbf{R}(x)\,\mathbf{R}(y),$$

for any x and y. Now pick some *small*, positive δ and some positive integer k (bigger than 1, but not *too big* – more about this later), and evaluate

$$\mathbf{R}(\delta) \approx \mathbf{I} - \delta\mathbf{B} + \frac{1}{2}\delta^2\mathbf{B}^2 - \frac{1}{6}\delta^3\mathbf{B}^3 + \cdots + \frac{1}{k!}(-\delta)^k\,\mathbf{B}^k. \qquad (3.1.11)$$

If this expression is sufficiently accurate (it certainly can be if δ and k have been chosen wisely), then we have for $x = n\delta$:

$$\mathbf{R}(x) = [\mathbf{R}(\delta)]^n = \mathbf{R}\Big((n-1)\delta\Big)\,\mathbf{R}(\delta),$$

where n can be as large as one needs to get sufficiently large $x = n\delta$. If it is desired that $\mathbf{R}(x)$ be evaluated on N equally spaced points, then [using *Horner's rule*, a nested multiplication algorithm, to evaluate (3.1.11)] $N + k$ matrix–matrix multiplications and k matrix additions are required. The computational complexity is linear in the number of points (one multiplication for each successive point) and of order m^3 in the dimension of the matrix. That is, the computational complexity is of order

$$\mathrm{O}\Big((N+k)\,m^3\Big).$$

We can do much better if we are interested only in the vector $\mathbf{r}(x)$ and the scalars $b(x)$, $R(x)$, and $B(x)$. We can compute them in the following way. Given a matrix representation, $<\mathbf{p}, \mathbf{B}>$:

1. Calculate $\mathbf{b}' := \mathbf{B}\boldsymbol{\varepsilon}'$; $b(0) = \mathbf{p}\mathbf{b}'$; $R(0) = 1$.
2. Calculate $\mathbf{R}(\delta)$ from (3.1.11) using Horner's rule.
3. Calculate $\mathbf{r}(0) = \mathbf{p}$.
4. Then,

BEGIN FOR $n = 1$ to $n = N$, calculate

$$\mathbf{r}(n\delta) = \mathbf{p}\mathbf{R}(n\delta) = \mathbf{r}\left((n-1)\delta\right)\mathbf{R}(\delta)$$

$$R(n\delta) = \mathbf{r}(n\delta)\boldsymbol{\varepsilon}'$$

$$b(n\delta) = \mathbf{r}(n\delta)\mathbf{b}'.$$

END FOR

This involves only k matrix multiplications and additions, N matrix on vector multiplications, and $2N$ vector on vector multiplications (*dot* products). This means, then, that the computational complexity is of order

$$\mathrm{O}\left(N\,m^2\right) + \mathrm{O}\left(k\,m^3\right).$$

Since the term with N is sure to be the larger by far, we see that this algorithm saves a factor of m in computational time over the brute-force procedure we started with. Throughout this book, judicious selection of procedures can make many computations feasible that were previously impossible by other methods.

We point out to our numerical analysis experts that the problem of selecting appropriate δ and k is the same as the problem one has in trying to solve (3.1.6a) as m coupled differential equations, using kth order *ordinary differential equation (ODE)* formulas. In fact, the method we gave above is related to a method gaining favor in some quarters for more general ODEs, namely the *Taylor series expansion method.* We can even claim that our method is very stable, because all the eigenvalues of $\mathbf{B}$ have positive real parts (so there is no exponential blowup, the primary cause of instability). There may be a *stiffness* problem if the largest eigenvalue is very large compared to the desired distance between points, we must make δ small enough to accommodate this.

Our last point has to do with accuracy. From Taylor's remainder theorem, we know that the error in $\mathbf{R}(\delta)$ is of order

$$\mathrm{O}\left(\delta^{k+1}\right).$$

Since the method is stable, the roundoff error accumulates linearly with n; therefore, the roundoff error at x is

$$\mathrm{Err}(x) = n\,\mathrm{O}\left(\delta^{k+1}\right) = x\,\mathrm{O}\left(\delta^{k}\right).$$

This expression can actually be used to estimate the error by evaluating for two different δ's, and performing an *extrapolation* procedure.

Exercise 3.1.3: Evaluate $R(t)$, $B(t)$, $b(t)$, $b^{(k)}(0)$, $\mathrm{E}(t^k)$, and $B^*(s)$ for an Erlangian-2 distribution, using the formulas of this section.

Exercise 3.1.4: Repeat Exercise 3.1.3 for a 2-phase hyperexponential distribution.

3.2. MATRIX EXPONENTIAL DISTRIBUTIONS

Up to now we have been vague about the constraints for **p** and **P**, and so on. As long as $M_{ii} > 0$, $P_{ij} \geq 0$, $(\mathbf{P}\varepsilon')_i \leq 1$ for all i, j, and $(\mathbf{I}-\mathbf{P})^{-1}$ exists, proofs abound that guarantee good behavior. Such distributions are called *phase (PH) distributions* by Marcel F. Neuts [NEUT75], [NEUT81], who has studied them extensively. On the other hand, there is a larger class of pdf's for which the conditions above may not hold, yet still have a matrix representation. In fact, any pdf that has a rational Laplace transform (RLT) also has a matrix representation [LIPS85b]. Such functions are sometimes called *Kendall distributions* [KEND64], and their representations are often referred to as *Coxian servers* [COX.55], [LIPS86].

An interesting and mathematically important point of view is to start with a representation and see if it corresponds to a true pdf. This and related questions are discussed in great detail in the literature, so we will only summarize here. By a *matrix representation* of some distribution function, we mean a vector-matrix pair $\langle \mathbf{p}, \mathbf{B} \rangle$ (or equivalently, $\langle \mathbf{p}, \mathbf{V} \rangle$, since $\mathbf{B} = \mathbf{V}^{-1}$) which can be used in (3.1.7) to (3.1.10), and thus *generates* that function. This much we know. As long as **B** is finite-dimensional (as will always be the case here unless we say otherwise), $B^*(s)$ as defined in (3.1.10) will always be a ratio of two polynomials, and $b(t)$, from (3.1.7d), will always be a sum of polynomials times exponentials. Furthermore, if all the eigenvalues of **B** have positive real parts, then $b(t)$ is integrable and integrates to 1. The critical question remains as to whether $b(t)$ is a pdf [i.e., is it true that $b(t) \geq 0$ for all real $t > 0$?]. At present, there is no way one can look at $\langle \mathbf{p}, \mathbf{B} \rangle$, or $B^*(s)$ to answer this. The only sure way that it can be done is to examine $b(t)$ for all relevant t.

We shall first describe the simple and well-known Erlangian [ERLA17] and hyperexponential distributions. Then we discuss matrix representations from a purely algebraic point of view, avoiding any probability assumptions. In doing this, we will introduce the idea of *isometric transformations*, a special class of *similarity transformations* to which all formulas are invariant, in the sense that they are of identical form and meaning before and after the transformation.

3.2.1. Erlangian and Hyperexponential Distributions

Before we look at the general classes of matrix exponential (ME) distributions, we discuss the two most commonly used, over-used, and abused types. The reader was already introduced to their simplest nontrivial representatives in the exercises, but it pays to discuss them in some depth. The Erlangian-m distribution [for which we use the symbol $E_m(t;\,\mu)$] describes the time it takes for a customer to be served by m identical exponential servers, one at a time (or one server exactly m times). Formally, let X_i be the random variable representing the time it takes for a customer to be served by the ith server, with pdf, $\mu e^{-\mu t}$ (same μ for each server). Let Y_i be the total time it takes for the customer to be served by i servers (i.e., $Y_i = X_1 + X_2 + \cdots + X_i$). Then since $Y_1 = X_1$, its pdf is also $E_1(t;\,\mu) := \mu e^{-\mu t}$. The pdf for Y_2 is the *convolution* of X_1 with X_2. That is,

$$E_2(t;\,\mu) := \int_0^t b_{X_1}(s)\, b_{X_2}(t-s)\, ds = \int_0^t \mu e^{-\mu s}\, \mu e^{-\mu(t-s)}\, ds$$
$$= \mu^2 e^{-\mu t} \int_0^t ds = \mu\,(\mu t)\, e^{-\mu t}.$$

We deliberately introduced the notation $b_{X_i}(t)$ to represent the pdf's of the X_i, even though in this case they are the same exponential function. Then by the definition of the Erlangians, we can say that $b_{Y_i}(t) = E_i(t;\,\mu)$. It is well known in general that

$$b_{Y_m}(t) = \int_0^t b_{Y_{m-1}}(s)\, b_{X_m}(t-s)\, ds,$$

which gives us, for exponentials (provable by induction),

$$E_m(t;\,\mu) := \int_0^t E_{m-1}(s;\,\mu)\, \mu e^{-\mu(t-s)}\, ds = \mu \frac{(\mu t)^{m-1}}{(m-1)!} e^{-\mu t}. \tag{3.2.1a}$$

The nth moment for the Erlangian-m† is known from elementary calculus to be

$$\mathrm{E}\left(Y_m^n\right) = \int_0^\infty t^n E_m(t;\,\mu)\, dt = \frac{(n+m-1)!}{\mu^n (m-1)!}. \tag{3.2.1b}$$

† Observe that we use the *italic E* for the *Erlangian* pdf and roman E for the E*xpected value* symbol.

In particular, the first two moments are

$$E(Y_m) = \frac{m}{\mu} \quad \text{and} \quad E(Y_m^2) = \frac{m^2+m}{\mu^2},$$

with a variance of

$$\sigma_m^2 := E(Y_m^2) - [E(Y_m)]^2 = \frac{m}{\mu^2},$$

giving a coefficient of variation of

$$C_m^2 = \frac{\sigma_m^2}{[E(Y_m)]^2} = \frac{1}{m}. \tag{3.2.1c}$$

We make three trivial observations before moving on. The Erlangian-1 is an exponential distribution; only two parameters (μ and m) need be specified; and all other Erlangians have $C^2 < 1$. There is a natural extension of this family, called γ *distributions*, where m can take on noninteger as well as integer values. Although the nonintegral Gamma distributions have awkward mathematical properties at $t = 0$ (try to find their derivatives there), they are used by statisticians because any coefficient of variation less than 1 can be fit by these. We will not use them, since they cannot be represented by finite-dimensional matrices. Besides, we have other ways to fit any C^2.

From our own description, this function should be generated by the subsystem, which looks like Figure 3.2.1. Since this is merely a *representation* of a distribution, we change our terminology from *server* to *phase*. Remember, a phase is always exponentially distributed, with a *completion rate* that may be a complex number, but its real part is always positive. For Erlangian-m distributions, all the phases have the same μ, so the completion rate matrix is the m-dimensional matrix satisfying $\mathbf{M} = \mu\mathbf{I}$. The transition matrix is given by the following:

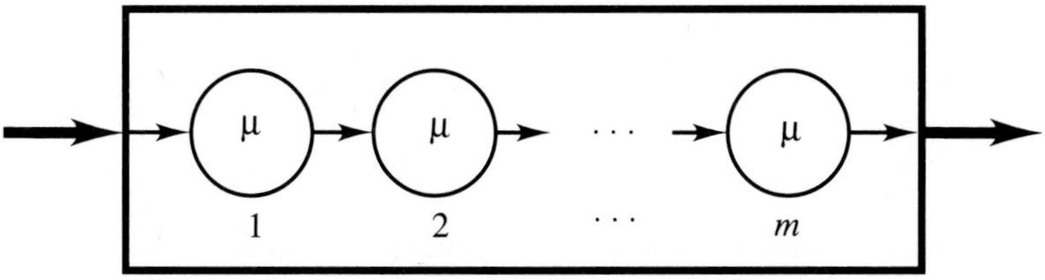

Figure 3.2.1: Subsystem containing m phases arranged into a string. Each phase has completion rate μ. A customer, upon entering, always goes to phase 1. When finished there he goes to phase 2, and so on until phase m, after which he leaves. The density function for this excursion is the Erlangian–m $E_m(t;\mu)$, as given in (3.2.1a).

$$\mathbf{P} = \begin{bmatrix} 0 & 1 & 0 & \cdots & 0 & 0 \\ 0 & 0 & 1 & \cdots & 0 & 0 \\ : & : & : & \cdots & : & : \\ 0 & 0 & 0 & \cdots & 0 & 1 \\ 0 & 0 & 0 & \cdots & 0 & 0 \end{bmatrix} \Bigg\} \; m \text{ rows and columns.} \tag{3.2.2a}$$

Next define the auxiliary matrix

$$\mathbf{L} := \mathbf{I} - \mathbf{P} = \begin{bmatrix} 1 & -1 & 0 & \cdots & 0 & 0 \\ 0 & 1 & -1 & \cdots & 0 & 0 \\ : & : & : & \cdots & : & : \\ 0 & 0 & 0 & \cdots & 1 & -1 \\ 0 & 0 & 0 & \cdots & 0 & 1 \end{bmatrix}. \tag{3.2.2b}$$

Then the completion rate matrix for the process is

$$\mathbf{B} = \mathbf{M}(\mathbf{I} - \mathbf{P}) = \mu \mathbf{I} \mathbf{L} = \mu \mathbf{L},$$

with service time matrix

$$\mathbf{V} = \mathbf{B}^{-1} = \frac{1}{\mu} \mathbf{L}^{-1}.$$

and m-dimensional entrance vector

$$\mathbf{p} = [1 \;\; 0 \;\; 0 \;\; \cdots \;\; 0].$$

One can verify directly that the inverse of $\mathbf{L}$ is given by

$$\mathbf{L}^{-1} = \begin{bmatrix} 1 & 1 & 1 & \cdots & 1 & 1 \\ 0 & 1 & 1 & \cdots & 1 & 1 \\ : & : & : & \cdots & : & : \\ 0 & 0 & 0 & \cdots & 1 & 1 \\ 0 & 0 & 0 & \cdots & 0 & 1 \end{bmatrix}. \tag{3.2.2c}$$

From the well-known summation rule of binomial coefficients,

$$\binom{n+m+1}{m} = \sum_{j=0}^{m} \binom{n+j}{j}, \tag{3.2.3a}$$

it follows that

$$\left[\mathbf{L}^{-n}\right]_{ij} = \binom{n+j-i-1}{j-i} \quad \text{for} \quad i \le j. \tag{3.2.3b}$$

For instance (if you are concerned, just try a few matrix multiplications),

$$\mathbf{L}^{-3} = \begin{bmatrix} 1 & 3 & 6 & 10 & 15 & \cdots \\ 0 & 1 & 3 & 6 & 10 & \cdots \\ 0 & 0 & 1 & 3 & 6 & \cdots \\ 0 & 0 & 0 & 1 & 3 & \cdots \\ 0 & 0 & 0 & 0 & 1 & \cdots \\ \vdots & \vdots & \vdots & \vdots & \vdots & \vdots \end{bmatrix}.$$

Note that all these matrices are *triangular*, in that every element below the diagonal is 0 (e.g., $L_{ij} = 0$ if $i > j$). If **P** (or **B** or **V**) is of this form, this is referred to as a *feed-forward network*. In any case, it can be shown that these matrices reproduce Equations (3.2.1) by purely algebraic manipulation of the equations in Theorem 3.1.1. Indeed, $<\mathbf{p}, \mathbf{B}>$ as given here is a faithful representation of the Erlangian-*m* pdf.

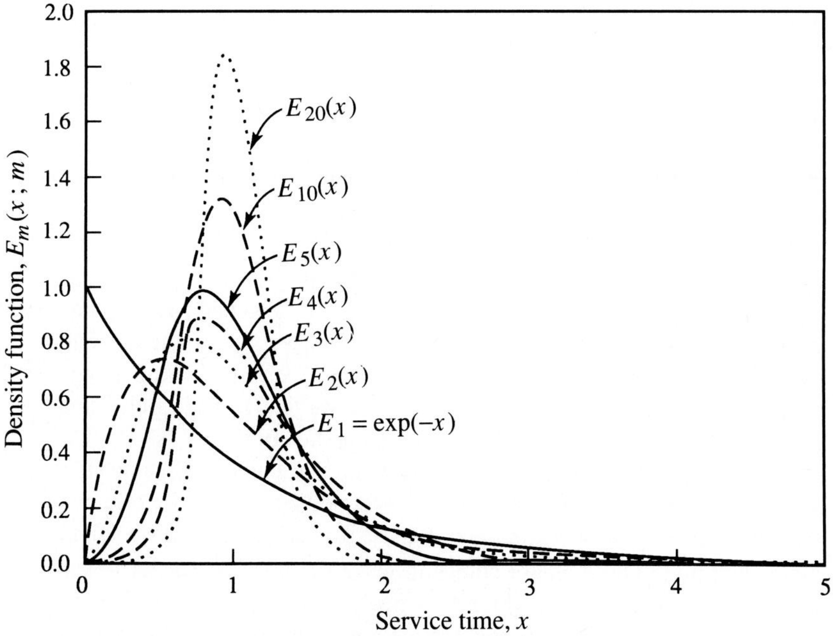

Figure 3.2.2: The pdf's for the Erlangians with parameter, $m = 1, 2, 3, 4, 5, 10$, and 20. All have a mean of 1. They all peak at a value less than their means, namely $(m-1)/m$, and get narrower with increasing m, agreeing with the fact that $C^2 = 1/m$ also gets smaller.

Example 3.2.1: The values of Erlangians for several values of m have been calculated and plotted in Figure 3.2.2. These all have a mean of 1 (we set $\mu = m$) and, except for the exponential, are 0 at $t = 0$. Consistent with their values for C^2, these functions get narrower and narrower with increasing m ($\sigma = 1/\sqrt{m}$). In fact, the function defined by the limit

$$\delta(t-1) := \lim_{m \to \infty} E_m(t;\, m)$$

[the *Dirac*$-\delta$ function - see (5.1.12a) and following] can be used as a representation of the *deterministic distribution*, the one that always gives a service time of 1. ●

The other abused class of functions is the family of *hyperexponential distributions* of the form

$$H_m(t) := p_1 \mu_1 e^{-\mu_1 t} + p_2 \mu_2 e^{-\mu_2 t} + \cdots + p_m \mu_m e^{-\mu_m t}$$

$$= \sum_{j=1}^{m} p_j \mu_j e^{-\mu_j t}, \tag{3.2.4a}$$

where μ_j and $p_j \geq 0$ are real, and the sum of the p's is 1. We can assume without loss of generality that the μ's are all distinct (otherwise, we could combine two equal ones together), and all the p's are strictly greater than 0 (otherwise, we just make m smaller). With these conditions we can say that for $m > 1$, C^2 is always greater than 1. The limit when all the μ's become equal is the exponential distribution, and, of course, so is $H_1(t)$. Let Z_m be the random variable described by the distribution, $H_m(t)$; then the moments are

$$\mathrm{E}\left(Z_m^n\right) = n! \sum_{i=1}^{m} \frac{p_i}{\mu_i^n}. \tag{3.2.4b}$$

The H_m's have an obvious representation. A customer enters a subsystem, and with probability p_i, goes to phase i, which has a completion rate of μ_i. When finished, he leaves. Then $M_{ii} = \mu_i$, $\mathbf{P} = \mathbf{O}$, and $\mathbf{p}$ is the entrance vector whose ith component is p_i. From this it follows that $\mathbf{B} = \mathbf{M}$ (pretty simple). It is trivial to show that this is a faithful representation of the H_m's.

This family is so rich in parameters that one is usually left in a quandary as to what values to give them. Even the H_2 function has three free parameters (p_1, μ_1, and μ_2), and after we have chosen a specific E(Z_2) and σ_2^2, we still need one more condition. Often, abuse comes in when the third parameter is picked for mathematical convenience, which may badly distort physical reality. We shall be guilty of that in the following example, but since we are not looking at any particular system at present, there is nothing much else we can do.

Example 3.2.2: The hyperexponential distribution, $H_2(t)$, with mean of 1, has been calculated and plotted in Figure 3.2.3 for several different values for C^2. For mathematical convenience we have let $\alpha^2 - 2\alpha C^2 + 1 = 0$, where $\alpha = \mu_2/\mu_1$. The pdf's themselves are as innocent looking as their representations, but as we shall see in Chapters 4 and 5, since they can have any value for the coefficient of variation (as long as it is greater than 1), they can disastrously affect mean system times. All true hyperexponentials are strictly greater than 0 at $t = 0$ and decay smoothly thereafter. Note that with this family, the larger C^2, the bigger $H(0)$ is. But as t gets larger, they all cross (not necessarily at the same place), and in the intermediate region, they are in reverse order. The most important aspect of these curves, which is not shown clearly here, is that they cross over once more, so for very large t they are in the same order in which they began. The higher moments of these distributions are completely dominated by this *tail* behavior. ●

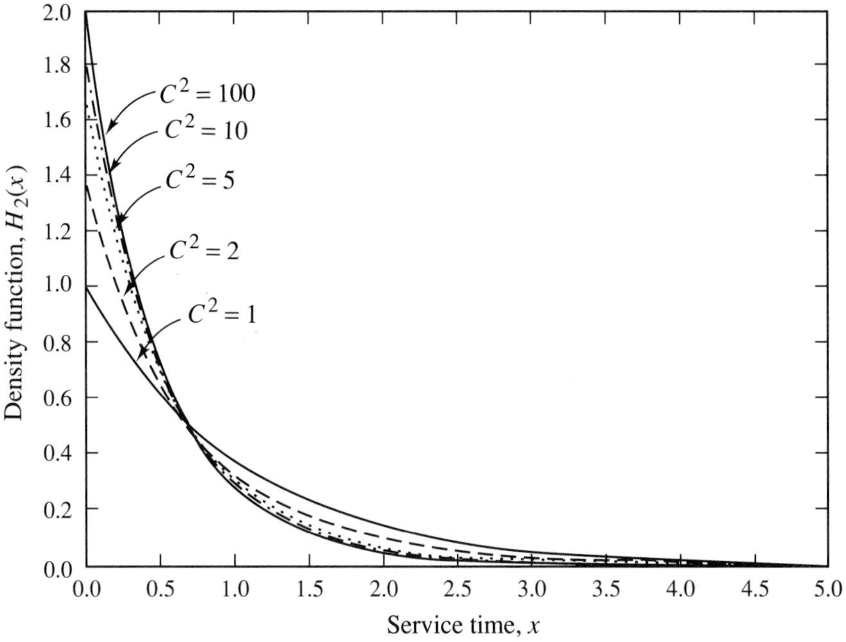

Figure 3.2.3: Family of hyperexponential-2 distributions, all with a mean value of 1, with $C^2 = 1$, 2, 5, 10, and 100, respectively. The curve for $C^2 = 1$ is the exponential distribution. The third condition chosen for this three-parameter family was $2\alpha C^2 = 1 + \alpha^2$ ($\alpha = \mu_2/\mu_1$) for no good reason, except that we had to do something. Although this graph does not show it, all the curves cross *twice*, so they asymptotically are in the same sequence as they were at $t = 0$, ordered according to their value of C^2.

We are reluctant to make general claims about the behavior of functions beyond the significance of their second moments.

3.2.2. Sums of Erlangian Functions

Consider functions of the form

$$b(t) = \sum_{k=1}^{K} f_k(t) e^{-\mu_k t} \quad \text{with} \quad \mathrm{Re}(\mu_k) > 0, \tag{3.2.5a}$$

where $f_k(t)$ is a polynomial of degree $m_k - 1$, and m_k can be any positive integer (sums of polynomials times exponentials). That is,

$$f_k(t) = \sum_{j=0}^{m_k} a_{jk}\, t^j .$$

We give a different look to the equation by introducing the Erlangian functions of order j as given in (3.2.1a). Then the expression for $b(t)$ can be rewritten in the form

$$b(t) = \sum_{k=1}^{K} a_k \left(\sum_{j=1}^{m_k} p_j^{(k)} E_j(t;\, \mu_k) \right), \tag{3.2.5b}$$

where we have arbitrarily split a_{jk} into two terms such that

$$\sum_{j=1}^{m_k} p_j^{(k)} = 1, \quad \text{for } all\ k.$$

Furthermore, since

$$\int_0^\infty b(t)\, dt = 1,$$

we must also have $\sum_{k=1}^{K} a_k = 1$. The number of terms all told is $m = \sum_{k=1}^{K} m_k$. As you might expect, m turns out to be the dimension of the representation we will be constructing. In general, from (3.2.1b) we can write down the nth moments of $b(t)$ in terms of the binomial coefficients. If T is the random variable described by this process,

$$\mathrm{E}(T^n) := \int_0^\infty t^n b(t) dt = n! \sum_{k=1}^{K} \frac{a_k}{\mu_k^n} \left[\sum_{j=1}^{m_k} \binom{n+j-1}{j-1} p_j^{(k)} \right]. \tag{3.2.6}$$

The only requirement for these integrals to exist is that the real part of each μ be positive, which we have already assumed. We can even let the a_k's and $p_j^{(k)}$'s be complex (negative or positive) numbers, as long as they appear in complex conjugate pairs to guarantee that $b(t)$ and its moments are real (this is a subsidiary requirement, which we will more or less ignore).

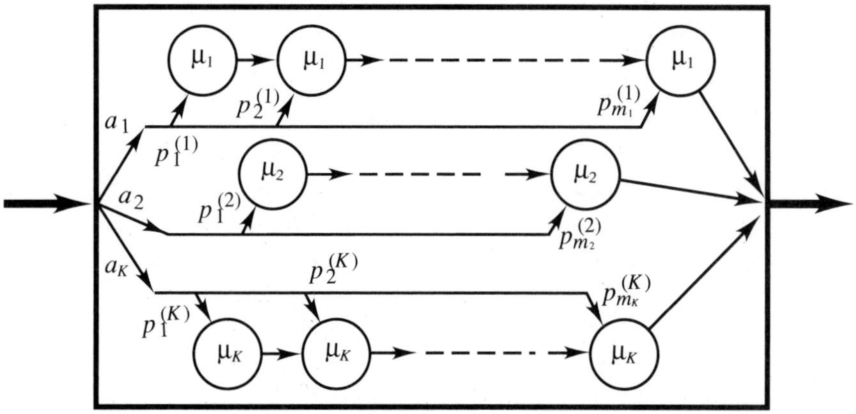

Figure 3.2.4: Subsystem containing m exponential phases. The phases are arranged into K strings, where the kth string has m_k identical phases in tandem, each with completion rate μ_k. A customer can go to any phase (with probability $a_k p_j^{(k)}$), but then must proceed along that string until the end, and then leave. The probability density function for this excursion is $b(t)$ as given in (3.2.5b). a_k and $p_{m_k}^{(k)}$ are assumed to be non-zero, otherwise the corresponding phases can be eliminated, yielding a smaller representation. Every representation is equivalent to one and only one of these, where the a's, μ's, and p's are allowed to be complex, if necessary. No equivalent representation can have fewer dimensions.

Next we give a *pseudophysical* interpretation to $b(t)$. Look at Figure 3.2.4. This exactly describes the expression for $b(t)$. A customer enters a subsystem and goes to string k (made up of m_k identical phases with completion rate μ_k) with probability a_k. He then goes directly to the jth phase (counting from the departing end) of that string with probability $p_j^{(k)}$, and proceeds to the end, being served en route by each of the j phases. The customer then leaves. We will now construct a matrix representation for this process. The transition matrix for a single line, $\mathbf{P}_k$, is an m_k–dimensional matrix of form given in (3.2.2a): The *completion rate* matrix for the kth string is also of dimension m_k, and since all the completion rates for that string are equal, it is simply $\mathbf{M_k} := \mu_k \mathbf{I_k}$. Next let $\mathbf{L_k}$ to be the square matrix of dimension m_k of the form in (3.2.2b). The m-dimensional completion and transition matrices for the entire process can be written as

$$\mathbf{P} := \begin{bmatrix} \mathbf{P_1} & \mathbf{O} & \mathbf{O} & \cdots & \mathbf{O} \\ \mathbf{O} & \mathbf{P_2} & \mathbf{O} & \cdots & \mathbf{O} \\ \vdots & \vdots & \vdots & \cdots & \vdots \\ \mathbf{O} & \mathbf{O} & \mathbf{O} & \cdots & \mathbf{O} \\ \mathbf{O} & \mathbf{O} & \mathbf{O} & \cdots & \mathbf{P_K} \end{bmatrix}; \quad \mathbf{M} := \begin{bmatrix} \mathbf{M_1} & \mathbf{O} & \mathbf{O} & \cdots & \mathbf{O} \\ \mathbf{O} & \mathbf{M_2} & \mathbf{O} & \cdots & \mathbf{O} \\ \vdots & \vdots & \vdots & \cdots & \vdots \\ \mathbf{O} & \mathbf{O} & \mathbf{O} & \cdots & \mathbf{O} \\ \mathbf{O} & \mathbf{O} & \mathbf{O} & \cdots & \mathbf{M_K} \end{bmatrix},$$

where each entry is an appropriately sized and valued matrix (e.g., **O** is a matrix of all 0's). The *entrance* or *initial vector* for the process is given by the *direct sum* (the components of the vectors are concatenated together to produce a single vector with as many components as all the others put together):

$$\mathbf{p} := a_1 \mathbf{p}^{(1)} \oplus a_2 \mathbf{p}^{(2)} \oplus \cdots \oplus a_K \mathbf{p}^{(\mathbf{K})}, \tag{3.2.7a}$$

where $\mathbf{p}^{(\mathbf{k})}$ is the m_k–dimensional vector whose components are $p_1^{(k)}$, $p_2^{(k)}$, $\cdots, p_{m_k}^{(k)}$, and $p_{m_k}^{(k)} \neq 0$.

We next construct the m–dimensional *service rate* matrix for this subsystem:

$$\mathbf{B} := \mathbf{M}(\mathbf{I}-\mathbf{P}) = \begin{bmatrix} \mu_1\mathbf{L}_1 & \mathbf{O} & \mathbf{O} & \cdots & \mathbf{O} \\ \mathbf{O} & \mu_2\mathbf{L}_2 & \mathbf{O} & \cdots & \mathbf{O} \\ \vdots & \vdots & \vdots & \cdots & \vdots \\ \mathbf{O} & \mathbf{O} & \mathbf{O} & \cdots & \mathbf{O} \\ \mathbf{O} & \mathbf{O} & \mathbf{O} & \cdots & \mu_K\mathbf{L}_\mathbf{K} \end{bmatrix}. \tag{3.2.7b}$$

Since all the elements below the diagonal are zero (remember, the **L**'s are triangular), we see that μ_k is an eigenvalue of **B**, with multiplicity m_k. We know this because there is an obvious theorem in matrix theory which states that the eigenvalues of a triangular matrix are its diagonal elements. If none of the μ's are equal to 0, we can let

$$\mathbf{V} = \mathbf{B}^{-1}, \tag{3.2.7c}$$

where **V** looks just like **B** with each $\mu_k \mathbf{L}_\mathbf{k}$ replaced by $(1/\mu_k)\mathbf{L}_\mathbf{k}^{-1}$.

Our purpose now is to show that the important properties of matrix representations are valid for purely algebraic reasons. To do this, first recall (3.2.3). Then look at the scalar reduction of the matrices by multiplying both sides of $\mathbf{L}_\mathbf{k}^{-n}$ with the vectors $\mathbf{p}^{(\mathbf{k})}$ and $\varepsilon'_\mathbf{k}$ to get

$$\mathbf{p}^{(\mathbf{k})}\mathbf{L}_\mathbf{k}^{-n}\varepsilon'_\mathbf{k} = \sum_{j=1}^{m_k} p_j^{(k)} \binom{n+j-1}{j-1}. \tag{3.2.8}$$

This expression is identical to the term in the brackets in (3.2.6).

Now when we put (3.2.6) to (3.2.8) together and recall the definition of $\Psi[\,\cdot\,]$; we get

$$\mathrm{E}(T^n) = n!\,\Psi[\mathbf{V}^n].$$

Note that this relation is valid even if $b(t)$ is not a pdf. It only requires that the moments exist (i.e., the moments must be finite). Since we are dealing with finite sums of terms, the moments exist if and only if $\mathrm{Re}(\mu_k) > 0$. No probability assumptions are required for the component parts.

By algebraic manipulations and arguments similar to the preceding paragraph, we can show that the Laplace transform of $b(t)$ satisfies the following:

$$B^*(s) := \Psi[(\mathbf{I}+s\mathbf{V})^{-1}] = \frac{q_2(s)}{q_1(s)},$$

where $q_1(s)$ is a polynomial of degree m, and $q_2(s)$ is of lesser degree. The matrix, $(\mathbf{I} + s\,\mathbf{V})^{-1}$, appears often, and is called the *resolvent matrix* by researchers in control theory. All the roots of q_2 are distinct from those of q_1. For finite-dimensional representations where $\mathbf{V}$ exists, this equation is valid even if some μ_k have negative real part. In other words, this equation is an algebraic relation between $\mathbf{V}$, $\mathbf{p}$, and the polynomials, $q_i(s)$ which reduces to the Laplace transform when all the μ_k's have positive real parts (but that is not important here). Furthermore, it follows that

$$\phi(y) := y^m q_1(1/y) \tag{3.2.9a}$$

is the characteristic polynomial for $\mathbf{V}$ and has K distinct roots, each with multiplicity m_k:

$$\phi(y) = \mathrm{Det}\,[y\mathbf{I} - \mathbf{V}] = (y - \mu_1)^{m_1}(y - \mu_2)^{m_2} \cdots (y - \mu_K)^{m_K}. \tag{3.2.9b}$$

If $K \neq m$, or equivalently, if $m_k > 1$ for at least one k, then V is *defective.*

The third relationship that can be proven by direct algebraic manipulation is the following:

$$b(t) = \Psi\big[\,\mathbf{B}\exp(-t\mathbf{B})\,\big] := \sum_{n=0}^{\infty} \frac{(-t)^n}{n!} \Psi\big[\,\mathbf{B}^{n+1}\big],$$

whereas $(\mathbf{I} + s\,\mathbf{V})^{-1}$ has direct matrix meaning, $\exp(-t\mathbf{B})$ is only defined in terms of its Maclaurin series expansion (as given in rightmost expression of the equation above). However, since the exponential function has an infinite radius of convergence, this formula is also valid for all $\mathbf{B}$. That is why it is so tempting to call this class of functions *matrix exponential*.

So our wonderful formulas from Theorem 3.1.1, namely,

$$b(t) = \Psi\big[\,\mathbf{B}\exp(-t\mathbf{B})\,\big], \tag{3.2.10a}$$

$$B^*(s) := \Psi\big[\,(\mathbf{I} + s\,\mathbf{V})^{-1}\,\big]\frac{q_2(s)}{q_1(s)}, \tag{3.2.10b}$$

and

$$\mathrm{E}\Big(T^n\Big) = n!\,\Psi\big[\,\mathbf{V}^n\,\big]. \tag{3.2.10c}$$

are purely matrix identities, having no dependence on probability laws. We define the vector-matrix pair $<\mathbf{p}\,,\mathbf{B}>$ (or $<\mathbf{p}\,,\mathbf{V}>$) to be a (*faithful*) *representation* of $b(t)$ if these equations hold.

3.2.3 Isometric Transformations: Representation Theorem

One might ask if there are more general representations than those discussed in the previous section. The answer is *no*. We can show that for every finite-dimensional square matrix (no other restrictions whatever), call it $\mathbf{B}$, there always exists a nonsingular isometric matrix $\mathbf{S}$, (i.e., $\mathbf{S}\boldsymbol{\varepsilon}' = \boldsymbol{\varepsilon}'$, and $\mathbf{S}^{-1}\boldsymbol{\varepsilon}' = \boldsymbol{\varepsilon}'$), such that

$$\mathbf{SBS}^{-1} = \begin{bmatrix} \mu_1\mathbf{L}_1 & \mathbf{O} & \mathbf{O} & \cdots & \mathbf{O} & \mathbf{E}_1 \\ \mathbf{O} & \mu_2\mathbf{L}_2 & \mathbf{O} & \cdots & \mathbf{O} & \mathbf{E}_2 \\ \vdots & \vdots & \vdots & \cdots & \vdots & \vdots \\ \mathbf{O} & \mathbf{O} & \mathbf{O} & \cdots & \mu_K\mathbf{L}_K & \mathbf{E}_K \\ \mathbf{O} & \mathbf{O} & \mathbf{O} & \cdots & \mathbf{O} & \mathbf{E} \end{bmatrix}, \qquad (3.2.11)$$

where the μ's are distinct, the **E** matrices are of column dimension m_e (possibly equal to 0), but otherwise irrelevant to Equations (3.2.10). The reader should compare this with (3.2.7b) before going on.

It is well known that $\mathbf{SB}^n\mathbf{S}^{-1} = (\mathbf{SBS}^{-1})^n$ for all n. For instance,

$$\mathbf{SB}^2\mathbf{S}^{-1} = \mathbf{SBS}^{-1}\mathbf{SBS}^{-1} = (\mathbf{SBS}^{-1})^2.$$

Therefore,

$$\mathbf{S}\exp(-t\mathbf{B})\mathbf{S}^{-1} = \mathbf{S}\sum_{n=0}^{\infty}\frac{(-t)^n}{n!}\mathbf{B}^n\mathbf{S}^{-1} = \sum_{n=0}^{\infty}\frac{(-t)^n}{n!}\mathbf{SB}^n\mathbf{S}^{-1}$$
$$= \sum_{n=0}^{\infty}\frac{(-t)^n}{n!}(\mathbf{SBS}^{-1})^n = \exp(-t\mathbf{SBS}^{-1}).$$

Thus we can write :

$$b(t) = \Psi[\mathbf{B}\exp(-t\mathbf{B})] = \Psi[\mathbf{S}^{-1}\mathbf{SB}\exp(-t\mathbf{B})\mathbf{S}^{-1}\mathbf{S}]$$
$$= (\mathbf{pS}^{-1})\mathbf{SBS}^{-1}\exp(-t\mathbf{SBS}^{-1})(\mathbf{S}\boldsymbol{\varepsilon}').$$

Now we define $\tilde{\mathbf{B}}$ to be the square submatrix of $\mathbf{SBS}^{-1}$ with the last m_e rows and columns thrown away, and similarly, let $\tilde{\mathbf{p}}$ be equal to the vector $\mathbf{pS}^{-1}$ with the last m_e components deleted. In the notation of (3.2.7a), if $p_{m_k}^{(k)} = 0$ for any k, then that element can also be eliminated from $\tilde{\mathbf{p}}$, together with the corresponding row and column for $\tilde{\mathbf{B}}$. What remains satisfies

$$\tilde{\mathbf{p}}\boldsymbol{\varepsilon}' = 1 \qquad (3.2.12a)$$

($\boldsymbol{\varepsilon}'$ is of appropriate reduced dimension), and

$$b(t) = \tilde{\mathbf{p}}\,\tilde{\mathbf{B}}\exp(-t\tilde{\mathbf{B}})\boldsymbol{\varepsilon}'. \qquad (3.2.12b)$$

That is, $<\tilde{\mathbf{p}}, \tilde{\mathbf{B}}>$ generates exactly the same function as $<\mathbf{p}, \mathbf{B}>$. One gets similar results for the other two of Equations (3.2.10). As a matter of fact, Equations (3.2.10) are invariant under *any* similarity transformation where **S** is isometric, not merely to the special similarity transformation which satisfies (3.2.11). Any transformation of this class is called an *isometric similarity transformation*, or simply an *isometric transformation*.

Note that if we consider the sum of the components of a vector, **r** to be its "*length*", then an isometric transformation preserves the "*length*" of every row vector, i.e., $\mathbf{r}\boldsymbol{\varepsilon}'$ is invariant. However, we must be careful, since the sum can be negative and thus cannot be used as a "*metric*" in the mathematical *metric space* sense. In any case, a matrix which does not change that length is *iso–metric* (*iso* means *same*, and *metric* means *length*). Let $\mathbf{r}$ be any row vector, and $\mathbf{S}$ be a square matrix. Then

$$\mathbf{r}\boldsymbol{\varepsilon}' = \mathbf{r}\mathbf{S}\boldsymbol{\varepsilon}' \quad \text{for all } \mathbf{r} \quad \Leftrightarrow \quad \mathbf{S}\boldsymbol{\varepsilon}' = \boldsymbol{\varepsilon}'$$

The proof follows from the fact that the only column vector which is *orthogonal* to every row vector is the one with all zeros, therefore, $\mathbf{v}' := \mathbf{S}\boldsymbol{\varepsilon}' - \boldsymbol{\varepsilon}' = \mathbf{o}'$, since $\mathbf{r}\mathbf{v}' = 0$ for all $\mathbf{r}$. See [LIPS85b] for details.

We have now seen that all functions of exponential type [i.e., of the form given in Equations (3.2.5)] have rational Laplace transforms (RLTs) and can be represented by a vector-matrix pair, $\langle \mathbf{p}, \mathbf{B} \rangle$ of the form (3.2.7a) and (3.2.7b). Conversely, for every $\langle \mathbf{p}, \mathbf{B} \rangle$ there exists an equivalent vector-matrix pair, $\langle \tilde{\mathbf{p}}, \tilde{\mathbf{B}} \rangle$, of equal or *lesser* dimension which is of the form (3.2.7b) and represents the same function as given by Equations (3.2.10). There is no representation that has a smaller dimension. In general, one can say that if $\langle \mathbf{p}, \mathbf{B} \rangle$ is a representation of $b(t)$, so is $\langle \mathbf{p}\mathbf{S}^{-1}, \mathbf{S}\mathbf{B}\mathbf{S}^{-1} \rangle$, where $\mathbf{S}$ is *any* nonsingular isometric matrix of appropriate dimension. Clearly, there are an infinite number of equivalent representations of every pdf (including, interestingly enough, the exponential distribution, which has an infinite number of equivalent representations, but of dimension > 1). Thus it would seem useless to try to give real physical meaning to the individual components of $\mathbf{p}$ or $\mathbf{B}$.

As a final thought on this, we note that for *any* $\mathbf{V}$, the polynomial

$$f(y) := y^m q_1(1/y)$$

[q_1 was defined in (3.2.10b)] divides the characteristic polynomial of $\mathbf{V}$. We made a similar-sounding statement in Equations (3.2.9), but that referred only to a $\mathbf{V}$ of the form already given in Equations (3.2.7). This statement refers to *any* matrix, including those *not* of canonical form. That is, every root of $f(y)$ is an eigenvalue of $\mathbf{V}$, and the multiplicity of this root in $f(y)$ is less than or equal to the degeneracy in $\mathbf{V}$. The difference between the dimension of $\mathbf{V}$ and the degree of f is equal to m_e.

It is important to note that if the a_i's, $p_j^{(k)}$'s, and μ_i's are real and positive, we are dealing with a *phase distribution* [NEUT81], but they do not have to be for $b(t)$ to be a proper pdf. Neuts has defined a phase distribution to be a distribution for which there exists a representation where $\mathbf{p}$, $\mathbf{P}$, and $\mathbf{M}$ have only real, non-negative components. Such a representation is of *phase type*. There are numerous examples where the original representation is of phase type, but the canonical one is not (e.g., Erlangian distributions with feedback). A detailed classification of matrix exponential functions is given in [LIPS86].

We summarize the above with a definition and theorem that we have not quite proven.

Definition 3.2.1

Let $<\mathbf{p}_1 , \mathbf{B}_1>$ *and* $<\mathbf{p}_2 , \mathbf{B}_2>$ *be two vector-matrix pairs. Then they are equivalent if and only if they have the same moments according to (3.2.10c), or have the same Laplace transform according to (3.2.10b), or represent the same function according to (3.2.10a).* Any one of the three can prove the other two if the **B**'s are invertible. If they are equivalent, we write

$$<\mathbf{p}_1 , \mathbf{B}_1> = <\mathbf{p}_2 , \mathbf{B}_2>.$$

They do not have to be of the same dimension to be equivalent. ◊

Now we state the representation theorem.

Theorem 3.2.1: Consider *any* finite vector-matrix pair, $<\mathbf{p} , \mathbf{B}>$ with the following properties. **p** is isometric; **B** is invertible and has no eigenvalues with nonpositive real part. Then:

1. The three equations in Theorem 3.1.1 [or (3.2.10)] are algebraically correct.

2. For any isometric, invertible matrix, **S**, the following is true:

$$<\mathbf{p} , \mathbf{B}> \equiv <\mathbf{p}\mathbf{S}^{-1} , \mathbf{S}\mathbf{B}\mathbf{S}^{-1}>.$$

3. There exists a special **S** such that $\mathbf{S}\mathbf{B}\mathbf{S}^{-1}$ is of the form (3.2.11). If the last m_e rows and columns are discarded, together with those rows and columns (if any) corresponding to $p_{m_k}^{(k)} = 0$, for some k, then the reduced vector-matrix pair $<\tilde{\mathbf{p}} , \tilde{\mathbf{B}}>$ is *the canonical representation* with the following properties:

   (a) Is unique to within an exchange of blocks.
   (b) Is equivalent to $<\mathbf{p} , \mathbf{B}>$.
   (c) No other representation is of smaller dimension.
   (d) The characteristic equation for $\tilde{\mathbf{V}}$ satisfies Equations (3.2.9).

The various components may not be physically realizable even if the components of the original representation are, but diagrammatically it looks like Figure 3.2.4. If the reduced $m_k > 1$ for any k, then the canonical representation is defective. It follows then, that if the canonical representation is defective, then no diagonal representation of $b(x)$ exists. ■

Remember, even if individual components of **p** or **B** are complex, $b(x)$ is unchanged by an isometric transformation, so the physical consequences are unchanged. D. R. Cox was the first one to consider complex probabilities [COX.55] in this context. In the next chapters we derive numerous equations, *all* of which are invariant to this class of isometric transformations.

3.2.4. Several Specific Representations

Despite the powerful theorem we stated in the preceding section about minimal and unique representations, a feeling persists that somehow one can construct a set of phases that can do better (sort of like looking for a perpetual motion machine). A common example, which some people have called a *generalized Erlangian*, has the historically older name of *hypoexponential distribution*. The simplest example of this consists of two phases in tandem that do *not* have equal completion rates. The straightforward representation of this is

$$\mathbf{p} = [\,1\ \ 0\,]; \quad \mathbf{M} = \begin{bmatrix} \mu_1 & 0 \\ 0 & \mu_2 \end{bmatrix}; \quad \mathbf{P} = \begin{bmatrix} 0 & 1 \\ 0 & 0 \end{bmatrix},$$

and thus,

$$\mathbf{B} = \begin{bmatrix} \mu_1 & -\mu_1 \\ 0 & \mu_2 \end{bmatrix}.$$

Since this is a triangular matrix, the eigenvalues of **B** are equal to μ_1 and μ_2. If the μ's are not equal, **B** can be diagonalized by the matrix made up of its eigenvectors. Look at

$$\mathbf{S}^{-1} = \frac{1}{\mu_2 - \mu_1} \begin{bmatrix} \mu_2 & -\mu_1 \\ 0 & \mu_2 - \mu_1 \end{bmatrix}, \quad \mathbf{S} = \frac{1}{\mu_2} \begin{bmatrix} \mu_2 - \mu_1 & \mu_1 \\ 0 & \mu_2 \end{bmatrix}.$$

First note that $\mathbf{S}\boldsymbol{\varepsilon}' = \boldsymbol{\varepsilon}'$. Then

$$\tilde{\mathbf{p}} = \mathbf{p}\mathbf{S}^{-1} = \left[\frac{\mu_2}{\mu_2 - \mu_1} \quad \frac{\mu_1}{\mu_1 - \mu_2} \right], \quad \tilde{\mathbf{B}} = \mathbf{S}\mathbf{B}\mathbf{S}^{-1} = \begin{bmatrix} \mu_1 & 0 \\ 0 & \mu_2 \end{bmatrix}.$$

So a hypoexponential distribution is just another hyperexponential distribution, *but* with a difference. Suppose that $\mu_2 > \mu_1$. Then $(\tilde{\mathbf{p}})_2 < 0$ and $(\tilde{\mathbf{p}})_1 > 1$. This is not very physical, but it gives the right pdf, namely,

$$b(t) = \frac{\mu_2}{\mu_2 - \mu_1}\left(\mu_1 e^{-\mu_1 t}\right) + \frac{\mu_1}{\mu_1 - \mu_2}\left(\mu_2 e^{-\mu_2 t}\right) = \frac{\mu_1 \mu_2}{\mu_1 - \mu_2}\left(e^{-\mu_2 t} - e^{-\mu_1 t}\right).$$

You can check this out by taking the direct convolution of two nonequivalent exponentials.

This equation, by the way, gives us a second difference. The value of $b(0)$ is 0, whereas every *true* hyperexponential must be greater than 0 at the origin. In any case, which representation would you want to use? Which is easier to handle? In either case there is a problem. If μ_1 and μ_2 get arbitrarily close to each other, then *poof*. But if they are exactly equal, we have an *Erlangian*–2 distribution. In fact, if we take the limit very carefully, we will get $E_2(t)$. The problem is somewhat complicated algebraically. This is what happens in general. Take any two-dimensional representation. As long as the two eigenvalues are different, we have two pairs of eigenvectors, and the canonical representation looks like a hyperexponential. If we then let the two eigenvalues approach each other, suddenly the two left (and right) eigenvectors become equal to each other, and we are left with a degenerate eigenvalue with only one pair of eigenvectors. This means that there is no isometric (or any other) transformation that will diagonalize **B**, so we are stuck with the E_2 representation (not so terrible), but the entrance vector has two components. Sometimes (not here), the first component could be 0. In such cases we can throw the first component away and just have a single phase. What remains is the exponential distribution.

An alternative possibility is for there to be two independent eigenvectors with the single eigenvalue (i.e., a true H_2 function where the two completion rates are the same). In this case we can again throw away one of the phases and permit our customer to go to the other phase with the sum of the probabilities†, again yielding the exponential distribution. Confusing enough? Well, let us try another example.

Consider once again two phases in tandem, but now the customer can leave after finishing phase 1 with probability θ. Then the **P** matrix is

$$\mathbf{P} = \begin{bmatrix} 0 & 1-\theta \\ 0 & 0 \end{bmatrix}.$$

This leads to a completion rate matrix of the form

$$\mathbf{B} = \begin{bmatrix} \mu_1 & \mu_1(\theta-1) \\ 0 & \mu_2 \end{bmatrix}.$$

This too looks like an H_2 function, but if $\mu_2 = \theta\mu_1$, then suddenly, $\boldsymbol{\varepsilon}'$ is a right eigenvector of **B**, with eigenvalue μ_2. This means that this subsystem is only an exponential server, even though nothing special seems to have happened. We can show this easily. Suppose that $\mu_2 = \theta\mu_1$; then

$$\mathbf{B}\boldsymbol{\varepsilon}' = \mu_2\boldsymbol{\varepsilon}'.$$

Going on, we have

$$\exp(-t\mathbf{B})\boldsymbol{\varepsilon}' = \sum_{j=0}^{\infty} \frac{(-t)^j}{j!}\mathbf{B}^j\boldsymbol{\varepsilon}' = \sum_{j=0}^{\infty} \frac{(-t)^j}{j!}(\mu_2)^j\,\boldsymbol{\varepsilon}' = \sum_{j=0}^{\infty} \frac{(-\mu_2 t)^j}{j!}\boldsymbol{\varepsilon}' = e^{-\mu_2 t}\,\boldsymbol{\varepsilon}'.$$

Then $\Psi[\exp(-t\mathbf{B})] = \exp(-t\mu_2)$ for *any* entrance vector.

This is a special case of a theorem that goes as follows. Consider the set of column vectors:

$$\boldsymbol{\varepsilon}',\ \mathbf{B}\boldsymbol{\varepsilon}',\ \mathbf{B}^2\boldsymbol{\varepsilon}',\ \cdots,\ \mathbf{B}^{m-1}\boldsymbol{\varepsilon}',$$

† This property is quite different from the apparently similar situation in control theory. There, if a degenerate eigenvalue has two eigenvectors, that implies a feedback loop, which causes instability.

where m is the dimension of **B**. The job is to find that smallest integer, j, such that $\mathbf{B}^j\,\boldsymbol{\varepsilon}'$ can be written as a linear combination of the vectors with a lower power of **B**. We know that $j \le m$, since it is well known that the characteristic polynomial (an mth degree polynomial) always allows us to write $\mathbf{B}^m\,\boldsymbol{\varepsilon}'$ in terms of the others. If j is strictly less than m, we can actually find a representation of lesser dimension that is equivalent to the one given. In the case above, $j = 1$.

Perhaps we have tried to say too much in too little space. Rest assured that there are an infinity of examples that can force us into confusing interpretations of the various components. Therefore, we reiterate that it is the matrix as a whole that describes any subsystem, or nonexponential server, not its components.

3.2.5. On the Completeness of ME Functions

Despite the richness of possibilities we have just seen, not every pdf has an exact representation. On the other hand, it is well known that the set of polynomials times exponentials forms a complete set in that (almost?) every integrable function can be approximated arbitrarily closely by a sum of members of that set. This approximation concept is discussed in detail elsewhere and is becoming an increasingly important area of research, with few clear answers at present.

By *completeness* we mean that for every pdf of interest there exists a sequence of finite dimensional vector-matrix pairs (perhaps of ever increasing dimension) whose properties converge to those of that pdf. Suppose that $\{<\mathbf{p_n}\,,\mathbf{B_n}>\,|\; n = 1, 2, \cdots\}$ is such a sequence, then in some meaningful sense,

$$\lim_{N\to\infty} \Psi[\exp(-t\mathbf{B_n})] = R(t) \tag{3.2.13a}$$

and

$$\lim_{N\to\infty} \Psi[\mathbf{V_n}^k] = \int_0^\infty t^k b(t)\,dt, \qquad k \ge 0, \tag{3.2.13b}$$

with equivalent limits for other properties of $b(t)$. We can now state a formal definition.

Definition 3.2.2

Matrix exponential (ME) function: Any function that is faithfully represented by a member of the closure set of finite-dimensional vector-matrix pairs, $<\mathbf{p}\,,\mathbf{V}>$ *is an ME function.* The technical meaning of *closure* is that the limit of every sequence is in the closure of the set. ◊

In a practical sense this means dealing with infinite-dimensional matrices, or permitting the $f_k(t)$'s in (3.2.5a) to be infinite power series functions, not merely polynomials. It is not clear at the moment just how big the *closure set* is compared with the set of physical reliability functions, $\{R(t)\}$, that is, the class of nonincreasing functions of t which are bounded from above by 1, and go to 0 as t goes to infinity.

Before moving on, we discuss the meaning of a representation for which **B** or **V** is singular (i.e., either matrix has no inverse). First we consider finite representations. If **B** is singular, exactly one of the parallel paths of its minimal representation (Figure 3.2.4) has a single phase with a 0 completion rate, or equivalently, an infinite completion time (multiple phases in tandem with infinite completion times would be redundant). This corresponds to one of two possibilities. Either (at least) one of the phases in the original representation is broken, or there is a possibility that a customer can be trapped in an infinite loop. In either case, there is a greater-than-zero probability that a customer can take an infinite time to complete service. In other words, the mean service time for this distribution is infinite. This is consistent with (3.2.10c) since **V** does not exist if **B** is singular.

Next, suppose that **V** is singular. Then it has at least one 0 eigenvalue, and exactly one parallel path of its minimal representation has a single phase with zero completion time. This is physically equivalent to the possibility that a customer may bypass service altogether. In other words, there is a probability greater than 0 that a customer will have 0 service time, or the PDF, $B(t)$ at $t = 0$, is greater than 0. Since

$$B(t) = \int_0^t b(x)\,dx,$$

$b(x)$ (and all its derivatives) must be singular at $x = 0$. This is consistent with (3.1.8b) and the fact that **B** does not exist. We will get around this without much difficulty later.

Infinite matrices have a much greater variety of singularities, thus making them more difficult to analyze mathematically (aside from the technical problem of dealing with an infinite set of numbers). A detailed discussion of this must wait for further research. It is sufficient for our purposes to deal only with finite matrices, and to represent distribution functions that have special properties (e.g., functions with nonexponential tails, which have infinite moments) by a sequence of finite representations of ever-increasing dimension, as described above in our definition of ME functions.

3.2.6. Setting Up Matrix Representations

It is one thing to talk abstractly about a pdf that describes the behavior of a server or process. It is another thing to say explicitly what the pdf is, based on real-world examination. "What is the pdf?" is definitely *not* a trivial question to answer, nor is "What is its matrix representation?" We discuss the two questions briefly here.

There are various ways in which the behavior of subsystems show up in the course of examining queueing systems. Several of them, together with how they can be represented in LAQT, follow.

1. If a Markov chain, or transition graph description is given, **p, P**, and **M** are included as part of the description. In effect, this gives us Figure 3.1.1.

2. If a density function is given, and one of the following is true:

   (a) $b(t)$ only has terms that are exponentials times powers of t, then it can be rewritten in the form of (3.2.1), from which the appropriate parameters corresponding to Figure 3.2.1 can be found;

   (b) $b(t)$ is not as in part (a), then an approximation must be found that obeys (a).

3. If the Laplace transform for the pdf of the subsystem is given, and:

   (a) $B^*(s)$ is RLT, then expand it in terms of its partial fractions, which requires finding the roots of its denominator polynomial. Each term will be of the form $[a_i p_j^{(i)}]/(1+s\mu_i)^j$, from which Figure 3.2.1 can be drawn; or

   (b) $B^*(s)$ is not RLT, then an approximation must be found that is RLT.

4. Only a finite set of data is available that reflects the performance of the subsystem. Then a suitable function must be constructed that reflects this performance (i.e., another approximation).

It is not at all clear what a *good* approximation might mean in a queueing theory context. Its goodness depends very much on what use will be made of the function and in what context. Even if two functions seem to look alike, they may yield radically different results in any given application. The commonly accepted procedure of picking approximate functions that have the correct first two (or more) moments (i.e., $\bar{t}$ and σ^2, etc.) has been shown to be inadequate and even very misleading in solving various problems. The value of $b(t)$ and its derivatives at 0 may be more important at times. This is discussed in Chapter 5. Until then, we assume in all topics we cover here that a representation, or a series of approximate representations converging to $b(t)$, has already been selected.

3.3. RENEWAL PROCESSES AND RESIDUAL TIMES

Consider a sequence of random variables (e.g., service times), $\{X_1, X_2, \cdots X_k, \cdots\}$ where all the X_k's, except perhaps X_1, are taken from the same distribution. Its relationship with the material in this chapter is straightforward. Let our subsystem, S, start with an infinite number of customers waiting to enter while c_1 is being served. Call them $c_2, c_3, c_4, \cdots$. When c_1 is finished, he leaves, and c_2 immediately enters, and so on (see Figure 3.3.1). The time c_k spends in S is X_k. It is clear that the X_k's are random numbers taken from the same pdf, (3.1.7d). There is one possible exception, namely if c_1 had entered S at some indeterminate time before our (almost) omnipresent observer started her clock.

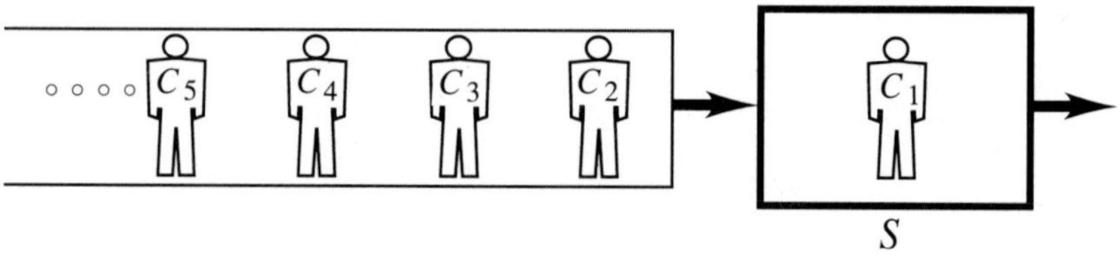

Figure 3.3.1: Renewal process viewed as a sequence of departures from a single subsystem S, able to serve one customer at a time. There is an infinite queue of customers waiting to enter S, and at time = 0, c_1 is already being served. He leaves at time $Y_1\,(=X_1)$, after which c_2 immediately enters. At time $Y_2\,(=Y_1+X_2)$ he leaves and c_3 enters. This goes on indefinitely, generating the renewal epochs, $Y_1, Y_2, Y_3, Y_4, \cdots, Y_k \cdots$.

Then X_1 would come from the PDF [see (3.1.7c)]

$$\Pr(X_1 < x) := B_1(x) = 1 - \pi[\exp(-x\mathbf{B})]\varepsilon',$$

where π_i is the probability that c_1 was at phase i when she started looking, and x is the time elapsed thereafter. This subtle change of π for $\mathbf{p}$ in (3.1.7a) is quite a powerful technique in analyzing a sequence of events. In general, one can take the vector that describes what happened up to the present [π] and premultiply it to the vector of probabilities of future events (recall from Definition 3.1.1 that $[\exp(-x\mathbf{B})\varepsilon']_i$ is the probability that c_1 will still be in S at time x, given that it was in state i at time 0) to get the total probability for any given event to occur. If observation began at the moment c_1 entered S, then $\pi = \mathbf{p}$, and X_1 has the same distribution as the other X_k's.

An interesting event to examine is the time for the nth customer to complete service. That time is the random variable, Y_n, called a *renewal epoch*, and defined by the obvious relation

$$Y_n := \sum_{k=1}^{n} X_k = Y_{n-1} + X_n, \quad n > 1,$$

where $Y_1 = X_1$. The sequence $\{Y_n\}$ constitutes a *renewal process* if $\pi = \mathbf{p}$ and is a *delayed* [FELL71] or *generalized renewal process* if $\pi \neq \mathbf{p}$. We already know quite a bit about this variable from elementary probability theory [TRIV82]. The classic survey is given in [COX.62].

3.3.1. Matrix Representations for the pdf of Y_n

We know that the mean time for c_n to finish is the sum of the mean service times for all customers up to and including n. Since all customers, except perhaps c_1, share the same distribution, let

$$E(X_k{}^j) := \overline{x^j}, \qquad k > 1,\ j \geq 0.$$

The variance of X_k is

$$\mathrm{Var}(X_k) := \sigma^2 := \overline{x^2} - \bar{x}^2.$$

Then [where for c_1 only, $E(X_1{}^j) = \overline{x_1{}^j}$ and $\sigma_1^2 = \overline{x_1^2} - \bar{x}_1{}^2$]

$$E(Y_n) = \bar{x}_1 + (n-1)\bar{x} \quad \text{and} \quad \mathrm{Var}(Y_n) = \sigma_1{}^2 + (n-1)\sigma^2.$$

These properties follow from the fact that the pdf for Y_n [call it $b_n(x)$] is the convolution of the pdfs of the X_k's. For instance,

$$b_2(x) = \int_0^x b_1(s)\, b(x-s)\, ds,$$

and generally,

$$b_{k+1}(x) = \int_0^x b_k(s)\, b(x-s)\, ds \quad k > 1, \tag{3.3.1a}$$

where $b(x)$ comes from (3.1.7d), and $b_1(x)$ is similar, with $\boldsymbol{\pi}$ replacing $\mathbf{p}$. Similarly, the PDFs $B_k(x)$ satisfy

$$B_{k+1}(x) = \int_0^x B_k(s)\, b(x-s)\, ds. \tag{3.3.1b}$$

Note that $b_1(x)$ is the pdf for both X_1 and Y_1, and $b(x)$ is the pdf for every other X_k.

Equations (3.3.1) are not in a form conducive to producing a useful matrix representation, but let us try anyway. After all, it is often as important to know what cannot be done as it is to know what can. For $k = 2$,

$$b_2(x) = \int_0^x \pi \mathbf{B} \exp(-s\mathbf{B})\, \boldsymbol{\varepsilon}'\, \mathbf{p} \exp[-(x-s)\mathbf{B}]\, \boldsymbol{\varepsilon}'\, ds$$

$$= \pi\mathbf{B}\left[\int_0^x \exp[-s\mathbf{B}]\, \mathbf{Q} \exp[-(x-s)\mathbf{B}]\, ds\right] \varepsilon',$$

where† $\mathbf{Q} := \boldsymbol{\varepsilon}'\mathbf{p}$ is an idempotent matrix (see Exercise 1.5) and *does not* commute with $\mathbf{B}$ or $\mathbf{V}$. This matrix will appear over and over and has many useful properties, some of which we summarize here as a lemma.

† We emphasize to the reader that this $\mathbf{Q}$ has nothing whatever to do with the transition rate matrix $\boldsymbol{Q}$, of Chapter 1.

Lemma 3.3.1: Let $\mathbf{Q} = \boldsymbol{\varepsilon}'\mathbf{p}$; then $\mathbf{Q}$ has one eigenvalue equal to 1, and $m-1$ eigenvalues equal to 0, where m is the dimension of $\mathbf{Q}$. In other words, $\mathbf{Q}$ is of rank 1. $\boldsymbol{\varepsilon}'$ and $\mathbf{p}$ are its right and left eigenvectors belonging to eigenvalue 1. That is, they satisfy

$$\mathbf{Q}\boldsymbol{\varepsilon}' = \boldsymbol{\varepsilon}' \quad \text{and} \quad \mathbf{pQ} = \mathbf{p}.$$

$\mathbf{Q}$ is *idempotent*, since $\mathbf{Q} = \mathbf{Q}^2$. Also, for any square matrix, $\mathbf{D}$, the following are true:

$$\mathbf{QDQ} = \Psi[\mathbf{D}]\mathbf{Q}, \quad \mathbf{QD}\boldsymbol{\varepsilon}' = \Psi[\mathbf{D}]\boldsymbol{\varepsilon}', \quad \mathbf{pDQ} = \Psi[\mathbf{D}]\mathbf{p}.$$

The proofs are straightforward by substituting for $\mathbf{Q}$. Recall that $\Psi[\mathbf{D}] = \mathbf{pD}\boldsymbol{\varepsilon}'$ is a scalar and can be brought outside any matrix algebraic expression. For instance, for some arbitrary multiplication string of matrices with at least two appearances of $\mathbf{Q}$, we can write

$$\mathbf{AQBQC} = \mathbf{A}\ \mathbf{QBQ}\ \mathbf{C} = \Psi[\mathbf{B}]\ \mathbf{AQC}.$$

Therefore, every string can be reduced to a scalar times a matrix string with only one $\mathbf{Q}$ in it. ■

We return now to the last equation preceding Lemma 3.3.1. Because $\mathbf{Q}$ is in the middle of this expression, no simplification can be made in this form. Expanding both exponentials in power series will allow the integrals to be performed. There are two paths that one can take. We will do both in turn. First,

$$b_2(x) = \pi\mathbf{B}\int_0^x \sum_{k=0}^{\infty} \frac{(-\mathbf{B})^k}{k!}\mathbf{Q}\sum_{j=0}^{\infty}\frac{(-\mathbf{B})^j}{j!} s^k (x-s)^j\, ds\, \mathbf{B}\boldsymbol{\varepsilon}'$$

$$= \pi\mathbf{B}\sum_{k=0}^{\infty}\sum_{j=0}^{\infty}\frac{(-\mathbf{B})^k\,\mathbf{Q}\,(-\mathbf{B})^j}{k!\,j!}\left[\int_0^x s^k (x-s)^j\, ds\right]\mathbf{B}\boldsymbol{\varepsilon}'.$$

The expression in brackets is the *beta function* as defined in [ABRA64] and is equal to

$$B(k+1, j+1) := \int_0^x s^k (x-s)^j\, ds = \frac{k!\,j!}{(k+j+1)!}x^{k+j+1}.$$

Therefore,

$$b_2(x) = x\pi\mathbf{B}\left[\sum_{k=0}^{\infty}\sum_{j=0}^{\infty}\frac{(-x\mathbf{B})^k\,\mathbf{Q}\,(-x\mathbf{B})^j}{(k+j+1)!}\right]\mathbf{B}\boldsymbol{\varepsilon}'. \tag{3.3.2a}$$

As far as we know, there is no closed-form expression for this. However, see Exercises 3.3.5 and 3.3.6 for a meaning of these terms.

Exercise 3.3.1: If S is one-dimensional (i.e., exponential), then $\mathbf{Q} = 1$, $\pi = 1$, and $\mathbf{B} = \mu$. Show directly from the expression above that $b_2(x) = \mu^2 x\, e^{-\mu x}$.

Alternatively, we can write

$$b_2(x) = \pi\mathbf{B}\left[\int_0^x \exp(-s\mathbf{B})\,\mathbf{Q}\exp(+s\mathbf{B})\,ds\right]\exp(-x\mathbf{B})\,\mathbf{B}\varepsilon'$$

$$= \pi\mathbf{B}\sum_{k=0}^{\infty}\sum_{j=0}^{\infty}\frac{(-\mathbf{B})^k\,\mathbf{Q}\,(\mathbf{B})^j}{k!\,j!}\left[\int_0^x s^{k+j}\,ds\right]\exp(-x\mathbf{B})\,\mathbf{B}\varepsilon',$$

and finally,

$$b_2(x) = x\,\pi\mathbf{B}\left[\sum_{k=0}^{\infty}\sum_{j=0}^{\infty}\frac{(-x\mathbf{B})^k\,\mathbf{Q}\,(x\mathbf{B})^j}{k!\,j!\,(k+j+1)}\right]\exp(-x\mathbf{B})\,\mathbf{B}\varepsilon'. \tag{3.3.2b}$$

This does not seem to be much better, if at all, and we can expect $b_3(x)$ to yield even messier expressions for either form. Our purpose in going through this at all was to warn the reader that matrix functions are *not always* as easy to manipulate as their scalar counterparts. We must look elsewhere for useful expressions.

Exercise 3.3.2: Show that (3.3.2b) also reduces to $b_2(x) = \mu^2 x\, e^{-\mu x}$ when S is one-dimensional.

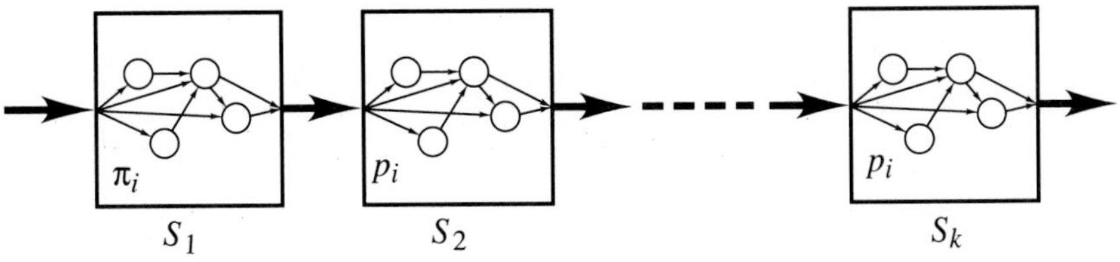

Figure 3.3.2: Representation of the distribution of Y_k, the kth convolution of S with itself. All the S's are identical, except that the starting vector for S_1 may be different.

Rather than trying to convert a convolution into a matrix expression, let us, instead, look at Y_k as a single process. Consider Figure 3.3.2 where we have k

identical subsystems in tandem, each described by the same pair $<\mathbf{p}, \mathbf{V}>$, except for S_1, which has π instead of $\mathbf{p}$. A customer starts at the ith phase of S_1 with probability π_i. After meandering for a while, [$\mathbf{P}$], he leaves, [$\mathbf{q}'$], and immediately goes to S_2, entering there and going to phase i with probability p_i. Instead of having a convolution of k m-dimensional objects, our process is described by the $(k \times m)$-dimensional arrays $\{\mathbf{p_k}, \varepsilon'_\mathbf{k}, \mathbf{P_k}, \mathbf{M_k}, \text{etc.}\}$. $\varepsilon'_\mathbf{k}$ is a $k \times m$ vector of all 1's. The process must start in one of the first m states, so

$$\mathbf{p_k} = [\, \pi, \mathbf{o}, \mathbf{o}, \cdots, \mathbf{o}] \tag{3.3.3a}$$

(each $\mathbf{o}$ is an m-vector of all 0's) and will go from i to j in S_1 with probability P_{ij}, or go to the jth phase in S_2 with probability $q_i p_j = (\mathbf{q}'\mathbf{p})_{ij}$. For $k = 3$, for instance,

$$\mathbf{P_3} = \begin{bmatrix} \mathbf{P} & \mathbf{q'p} & \mathbf{O} \\ \mathbf{O} & \mathbf{P} & \mathbf{q'p} \\ \mathbf{O} & \mathbf{O} & \mathbf{P} \end{bmatrix} \quad \text{and} \quad \mathbf{M_3} = \begin{bmatrix} \mathbf{M} & \mathbf{O} & \mathbf{O} \\ \mathbf{O} & \mathbf{M} & \mathbf{O} \\ \mathbf{O} & \mathbf{O} & \mathbf{M} \end{bmatrix}. \tag{3.3.3b}$$

The rate matrix for the process is (remember that $\mathbf{Bq'p} = \mathbf{BQ}$)

$$\mathbf{B_3} = \mathbf{M_3}(\mathbf{I_3} - \mathbf{P_3}) = \begin{bmatrix} \mathbf{B} & -\mathbf{BQ} & \mathbf{O} \\ \mathbf{O} & \mathbf{B} & -\mathbf{BQ} \\ \mathbf{O} & \mathbf{O} & \mathbf{B} \end{bmatrix} = \mathbf{B} \begin{bmatrix} \mathbf{I} & -\mathbf{Q} & \mathbf{O} \\ \mathbf{O} & \mathbf{I} & -\mathbf{Q} \\ \mathbf{O} & \mathbf{O} & \mathbf{I} \end{bmatrix} \tag{3.3.3c}$$

with process time matrix

$$\mathbf{V_3} = \mathbf{B_3}^{-1} = \begin{bmatrix} \mathbf{I} & \mathbf{Q} & \mathbf{Q} \\ \mathbf{O} & \mathbf{I} & \mathbf{Q} \\ \mathbf{O} & \mathbf{O} & \mathbf{I} \end{bmatrix} \mathbf{V}. \tag{3.3.3d}$$

The generalization to any k should be clear. We can now write down the pdf for this process:

$$b_k(x) = \mathbf{p_k} \left[\mathbf{B_k} \exp(-x\mathbf{B_k}) \right] \varepsilon'_\mathbf{k} \quad \text{and} \quad \mathrm{E}_k(x^j) = \frac{1}{j!} \mathbf{p_k} \left[\mathbf{V_k}^j \right] \varepsilon'_\mathbf{k}. \tag{3.3.3e}$$

Exercise 3.3.3: Prove by direct calculation that (3.3.3d) is the inverse of $\mathbf{B_3}$. Also, give the general expression for $\mathbf{V_k}$ and $\mathbf{B_k}$, and find the mean and variance of $b_k(x)$ using only these formulas.

3.3.2. Renewal Function and Transient Renewal Processes

If all that renewal theory had to offer was another view of convolutions, the topic would not have arisen at all. Its importance comes in studying the number of events that occur in a given interval of time. Suppose that we observe S for a time 0 to T. What is the probability that exactly n customers will depart in that time? Let that probability be P_n; then

$$P_n(T) = \Pr(Y_n \leq T \leq Y_{n+1}). \tag{3.3.4a}$$

Now, the PDF corresponding to (3.3.3e), $B_n(T)$ is the probability that Y_n is less than T, but does not exclude the possibility that Y_m, $m > n$, is also less than T. Therefore,

$$B_n(T) = P_n(T) + P_{n+1}(T) + P_{n+2}(T) + \cdots$$

$$= \sum_{m=n}^{\infty} P_m(T) = P_n(T) + B_{n+1}(T).$$

It then follows that

$$P_n(T) = B_n(T) - B_{n+1}(T). \tag{3.3.4b}$$

We are already familiar with the well-known example for exponential distributions, namely the *Poisson distribution*, for which

$$P_n(T) = \frac{(\mu T)^n}{n!} e^{-\mu T}. \tag{3.3.4c}$$

We now derive this formula with the aid of (3.3.4b). The nth convolution of an exponential with itself is the Erlangian density function, already defined in (3.2.1a) and satisfying (3.3.1a),

$$E_n(x) = \mu \frac{(\mu x)^{n-1}}{(n-1)!} e^{-\mu x},$$

whose PDF is

$$B_n(T) = \int_0^T E_n(x)\,dx = 1 - \left[\sum_{k=0}^{n-1} \frac{(\mu T)^k}{k!}\right] e^{-\mu T} = \sum_{k=n}^{\infty} \frac{(\mu T)^k}{k!} e^{-\mu T}.$$

The desired result follows directly.

A useful function in renewal theory is the average number of departures in the interval $(0, T]^\dagger$. The initial time, $t = 0$, is not included in the interval because we do not wish to count the departure of c_0, if it existed. Define the *renewal function* to be the expected number of departures in this interval. Then

† We follow standard mathematical practice, where $(a, b]$ stands for: "all the points between a and b, not including a, but including b." Put another way, '(' and ')' mean *open* (does not include), while '[' and ']' mean *closed* (does include).

$$M(T) := \sum_{n=0}^{\infty} nP_n(T) = \sum_{n=0}^{\infty} nB_n(T) - \sum_{n=0}^{\infty} nB_{n+1}(T)$$

$$= \sum_{n=1}^{\infty} nB_n(T) - \sum_{n=1}^{\infty} (n-1)B_n(T).$$

Note that two terms cancel, leaving the well-known formula

$$M(T) = \sum_{n=1}^{\infty} B_n(T). \tag{3.3.5}$$

Following [TRIV82], from (3.3.1b),

$$M(T) = B_1(T) + \int_0^T B_1(s)b(T-s)\,ds + \int_0^T B_2(s)b(T-s)\,ds + \cdots$$

$$= B_1(T) + \int_0^T [B_1(s) + B_2(s) + B_3(s) + \cdots]b(T-s)\,ds,$$

yielding an integral equation (the *renewal equation*) for the renewal function

$$M(T) = B_1(T) + \int_0^T M(s)b(T-s)\,ds = B_1(T) + \int_0^T M(T-s)b(s)\,ds. \tag{3.3.6}$$

Actually, it has been found to be easier to study the derivative of $M(T)$, since $M(T)$ goes to infinity as T goes to infinity, but its derivative does not. Therefore, define the *renewal density*

$$m(x) := \frac{dM(x)}{dx} = \sum_{n=1}^{\infty} b_n(x) = b_1(x) + \int_0^x m(s)b(x-s)\,ds. \tag{3.3.7}$$

$m(x)$ can be interpreted as the instantaneous completion (or service, or arrival) rate of the renewal process. A solution of this equation is not easy to come by, although its Laplace transform is, giving us little insight into what is going on. It is known, however, that if $b(x)$ is exponential, then $m(T) = \mu$ is constant for all T, which is what one would expect for all distributions. The expected number of completions in the interval $(0, T]$ is $M(T) = \mu T$, no matter how big or small T is. It is somewhat surprising that this *is not* true for general servers. We will find $m(x)$ by looking at a different problem that turns out to simultaneously solve (3.3.7).

Consider our same S with only one customer, who, after visiting S leaves, and with probability α immediately returns to S, as shown in Figure 3.3.3. Feller calls this a *transient renewal process*. We will use the symbol S_r for the subsystem that generates this. For the rest of this chapter, r will be used as a subscript for objects that are properties of S_r, as distinct from the use of any other symbols, such as i, j, k, l, and n, which are numerical subscripts. So, for instance, S_k is the kth subsystem in tandem in Figure 3.3.2.

We can construct the pdf for S_r by the following argument. Our customer will visit S exactly k times with probability $\alpha^{k-1}(1-\alpha)$. Since the pdf for visiting k times is $b_k(x)$, the pdf for S_r is

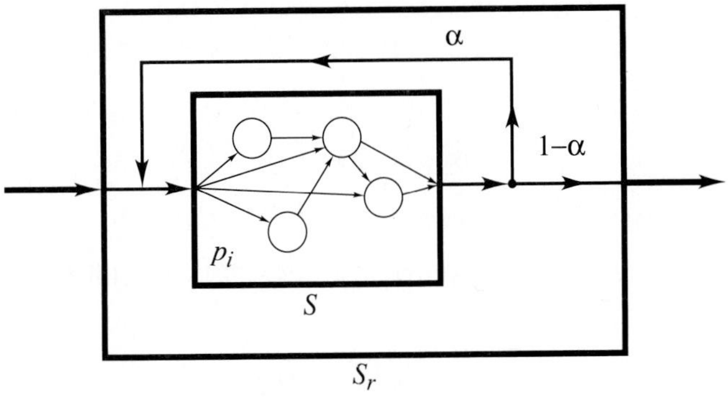

Figure 3.3.3: Representation of the pdf of $f(x;\alpha)$. After leaving S, a customer can either leave S_r or return to S with probability α.

$$f(x;\alpha) := (1-\alpha)\sum_{k=1}^{\infty}\alpha^{k-1}b_k(x) := (1-\alpha)m(x;\alpha). \qquad (3.3.8a)$$

From previous discussion, it follows that $m(x;\alpha)$ satisfies the integral equation

$$m(x;\alpha) = b_1(x) + \alpha\int_0^x m(s;\alpha)b(x-s)\,ds, \qquad (3.3.8b)$$

from which we get

$$m(x) = m(x;1) = \lim_{\alpha\to 1}\frac{f(x;\alpha)}{1-\alpha}. \qquad (3.3.8c)$$

By integrating (3.3.8a) and (3.3.8b) one can also get a formal expression for $M(x;\alpha)$ with the following properties:

$$M(x;\alpha) := \int_0^x m(s;\alpha)\,ds = B_1(x) + \alpha\int_0^x M(s;\alpha)B(x-s)\,ds$$
$$= \frac{F(x;\alpha)}{1-\alpha}. \qquad (3.3.8d)$$

Parallel with (3.3.8c), the renewal function satisfies

$$M(x) = M(x;1) = \lim_{\alpha\to 1}\frac{F(x;\alpha)}{1-\alpha}, \qquad (3.3.8e)$$

where $F(x;\alpha)$ is the PDF of S_r.

This process can be viewed directly as an m-dimensional subnetwork. Let $(\mathbf{P_r})_{ij}$ be the probability of going from i to j, either directly, $[P_{ij}]$, or by leaving, $[q_i]$, and then immediately returning, $[\alpha]$, and going to j, $[p_j]$. Then

$$\mathbf{P_r} = \mathbf{P} + \alpha\mathbf{q}'\mathbf{p}. \qquad (3.3.9a)$$

$\mathbf{M_r}$ is the same as $\mathbf{M}$, so the service rate matrix for S_r is

$$\mathbf{B_r}(\alpha) = \mathbf{M_r}(\mathbf{I} - \mathbf{P_r}) = \mathbf{M}(\mathbf{I} - \mathbf{P} - \alpha\mathbf{q}'\mathbf{p}) = \mathbf{B} - \alpha\mathbf{M}\mathbf{q}'\mathbf{p}.$$

Since $\mathbf{q}' = \mathbf{B}\boldsymbol{\varepsilon}'$, we have

$$\mathbf{B_r}(\alpha) = \mathbf{B}(\mathbf{I} - \alpha\mathbf{Q}), \tag{3.3.9b}$$

and since $f(x;\, \alpha)$ is a density function, generated by $<\pi\,,\, \mathbf{B_r}(\alpha)>$, we can write

$$f(x;\, \alpha) = \pi \exp[-x\mathbf{B_r}(\alpha)]\ \mathbf{B_r}(\alpha)\varepsilon' = \pi \exp[-x\mathbf{B}(\mathbf{I} - \alpha\mathbf{Q})]\ \mathbf{B}(\mathbf{I} - \alpha\mathbf{Q})\boldsymbol{\varepsilon}'.$$

But $(\mathbf{I}-\alpha\mathbf{Q})\boldsymbol{\varepsilon}' = (1-\alpha)\boldsymbol{\varepsilon}'$, so

$$f(x;\, \alpha) = (1-\alpha)\,\pi \exp[-x\mathbf{B}(\mathbf{I}-\alpha\mathbf{Q})]\varepsilon' \tag{3.3.9c}$$

and

$$m(x;\, \alpha) = \pi \exp[-x\mathbf{B}(\mathbf{I}-\alpha\mathbf{Q})]\mathbf{B}\ \boldsymbol{\varepsilon}'. \tag{3.3.9d}$$

A similar expression can be found for its integral,

$$M(x;\, \alpha) = \frac{1 - \pi \exp[-x\mathbf{B_r}(\alpha)]\ \boldsymbol{\varepsilon}'}{1-\alpha}. \tag{3.3.9e}$$

Let us digress for a moment to observe that $m(x;\, \alpha)$ is actually the generating function of the $b_k(x)$'s. Thus $m(x;\, 0) = b_1(x)$, and in general the kth derivative of $m(x;\, \alpha)$ with respect to α evaluated at $\alpha = 0$ is $(k)!\, b_{k+1}(x)$, where $b_k(x)$ is the kth convolution of $b(x)$ with itself. That is,

$$b_k(x) = \frac{1}{(k-1)!}\left[\left(\frac{\partial}{\partial\alpha}\right)^{(k-1)} m(x;\, \alpha)\right]_{\alpha=0}. \tag{3.3.9f}$$

This must reduce to the familiar when S is a pure exponential, where as usual, $\mathbf{Q}$, $\mathbf{p}$, and $\boldsymbol{\varepsilon}'$ all equal 1, and $\mathbf{B} = \mu$, so

$$m(x;\, \alpha) = \mu e^{-\mu x(1-\alpha)} = \mu e^{-\mu x}\, e^{-\mu x \alpha}.$$

For $\alpha = 0$, we get as expected, $b_1(x) = \mu e^{-\mu x}$. There is more to be said about this result. As we pointed out previously in (3.3.8a), the density function of a subsystem with feedback, $f(x;\, \alpha)$, is equal to $(1-\alpha)m(x;\, \alpha)$. So if S is exponential, with mean service rate μ, then $f(x;\, \alpha)$ is also exponentially distributed, with mean service rate $(1-\alpha)\mu$. But this is a special case of the following rather interesting lemma, which we state in two apparently unrelated ways.

Lemma 3.3.2: Let $b(x)$ be any pdf with m-dimensional representation $<\mathbf{p}\,,\mathbf{B}>$.

(a) Then the representation of $b(x)$ with feedback [i.e., $<\mathbf{p}\,,\, \mathbf{B_r}(\alpha)> \rightarrow f(x;\, \alpha)$] is also m-dimensional.

(b) The average of $b(x)$ with its convolutions of all order, weighted over a geometric distribution $[(1-\alpha)\alpha^{k-1}]$, has a representation which has the same dimension as $b(x)$ itself [(3.3.8a)]. ■

What makes this interesting is the following. We know from the discussion in Section 3.3.1 that the representation of $b_k(x)$ is of dimension $m \times k$. Yet an appropriately weighted average of all these functions has a representation that is no more complicated than the one for $b(x)$. The example of an exponential distribution shows this clearly. The exponential distribution has a one-dimensional representation. Its kth convolution, the Erlangian-k distribution, has a smallest representation that is of dimension k. Yet the weighted sum is again exponential (i.e., its representation is one-dimensional).

Exercise 3.3.4: Show, using (3.3.9f), that the expression above for $m(x;\alpha)$ does indeed generate the Erlangian-k distributions defined in (3.2.1).

It is not as easy to take the derivative of a matrix function as it would seem. First, since **B** and **Q** do not commute (see Theorem 1.3.2) we *cannot* write $[\exp(-x\mathbf{B})][\exp(x\alpha\mathbf{BQ}]$ for $\exp[-x\mathbf{B}(\mathbf{I}-\alpha\mathbf{Q})]$. Second, the function must be replaced by its Taylor series expansion, and then each term differentiated separately. For instance (here, ' stands for the derivative with respect to α),

$$\frac{d}{d\alpha}\mathbf{B_r}^3 = \frac{d}{d\alpha}(\mathbf{B_r}\mathbf{B_r}\mathbf{B_r}) = \mathbf{B_r}'\mathbf{B_r}^2 + \mathbf{B_r}\mathbf{B_r}'\mathbf{B_r} + \mathbf{B_r}^2\mathbf{B_r}'.$$

Now from (3.3.9b), $\mathbf{B_r} = \mathbf{B}$ for $\alpha = 0$, and

$$\frac{d}{d\alpha}\mathbf{B}(\mathbf{I}-\alpha\mathbf{Q}) = -\mathbf{BQ},$$

so

$$\left(\frac{d\mathbf{B_r}^3}{d\alpha}\right)_{\alpha=0} = -\left(\mathbf{BQB}^2 + \mathbf{B}^2\mathbf{QB} + \mathbf{B}^3\mathbf{Q}\right).$$

This expression clearly *is not* the same as $-3\mathbf{B}(\mathbf{BQ})^2$, and in fact is typical of the terms appearing in Equations (3.3.2).

Exercise 3.3.5: Show in general that

$$\left(\frac{d\mathbf{B_r}^n}{d\alpha}\right)_{\alpha=0} = -\mathbf{B}\left(\sum_{k=0}^{n-1}\mathbf{B}^k\,\mathbf{Q}\,\mathbf{B}^{n-k-1}\right).$$

Exercise 3.3.6: Use Exercise 3.3.5 to show that the expression for $b_2(x)$ in (3.3.2a) is actually

$$-\left(\frac{\partial}{\partial\alpha}\Psi\big[\exp(-x\mathbf{B_r})\big]\right)_{\alpha=0},$$

which is the derivative of $B_r(x)$.

Returning to (3.3.9d), we can take the limit as α goes to 1 directly and get

$$m(x) = \pi \exp[-x\mathbf{B}(\mathbf{I}-\mathbf{Q})]\mathbf{B}\boldsymbol{\varepsilon}'. \tag{3.3.9g}$$

It is not so easy to find $M(x)$ from $M(x;\,\alpha)$ in (3.3.9e) because both numerator and denominator approach 0 as $\alpha \to 1$. For exponential distributions, $\mathbf{B} \Rightarrow \mu$, $\mathbf{Q} \Rightarrow 1$, and $\pi \Rightarrow 1$, so $m(x) = \mu$. Clearly, for all other distributions $m(x)$ varies with x. However, we have the first of three versions of the *renewal theorem*:

Theorem 3.3.3a: Let S represent an ME distribution generating a renewal process; then

$$\lim_{x\to\infty} m(x) = \frac{1}{\bar{x}}. \tag{3.3.10a}$$

■

Proof: First observe that since $\mathbf{V} = \mathbf{B}^{-1}$, we can state that

$$\pi_{\mathbf{r}} := \frac{\mathbf{pV}}{\Psi[\mathbf{V}]} \tag{3.3.10b}$$

is the left eigenvector of $\mathbf{B}(\mathbf{I}-\mathbf{Q})$ with eigenvalue 0, and corresponding right eigenvector, $\boldsymbol{\varepsilon}'$, with length $\pi_{\mathbf{r}}\boldsymbol{\varepsilon}' = 1$. (We take another look at the *mean residual vector*, $\pi_{\mathbf{r}}$, in the next section.) Then as we showed in Chapter 1 [Equations (1.3.3a) and (1.3.10a)], for large x,

$$\exp[-x\mathbf{B}(\mathbf{I}-\mathbf{Q})] \to \boldsymbol{\varepsilon}'\pi_{\mathbf{r}}.$$

Recall from (3.1.4b) that $\Psi[\mathbf{V}] = \bar{x}$ and that $\pi\boldsymbol{\varepsilon}' = 1$, so we have

$$\lim_{x\to\infty} m(x) = (\pi\boldsymbol{\varepsilon}')\pi_{\mathbf{r}}\mathbf{B}\boldsymbol{\varepsilon}' = \frac{\mathbf{pV}}{\bar{x}}\mathbf{B}\boldsymbol{\varepsilon}' = \frac{\mathbf{pVB}\boldsymbol{\varepsilon}'}{\bar{x}} = \frac{1}{\bar{x}}.$$

QED†

† These letters stand for the time-honored Latin phrase *Quod Erat Demonstrandum*, whose translation is "which was to be demonstrated." QED designates the end of a proof.

Since there are no convergence difficulties in talking about an infinite-dimensional $\mathbf{B}_\infty$, we delete the argument, n, in $P_k(x, n)$ and $\mathbf{r}(x, n)$, and define the set of m-dimensional vectors, $\pi(x, k)$, as

$$\pi(x, k) := \frac{1}{P_k(x)} [\, r_{km+1}(x), r_{km+2}(x), \cdots , r_{k(m+1)}(x) \,], \qquad (3.3.15a)$$

where

$$\pi(x, k)\, \boldsymbol{\varepsilon}' = 1.$$

Put differently,

$$\mathbf{r}(x, \infty) = [P_0(x)\pi(x, 0),\ P_1(x)\pi(x, 1),\ P_2(x)\pi(x, 2),\ \cdots\ ] \quad (3.3.15b)$$

[the P's are scalars and the $\pi(x, k)$'s are m-vectors]. We are coming down the home stretch now.

It should be clear that $[\pi(x, k)]_i$ is the conditional probability that c_{k+1} is at phase i at time x, given that c_k has finished. Therefore, these π's can be used in the same way that the initial vector, π, is used, except that we now start measuring at time x. For instance, the renewal density for the interval $(x, x+T]$, given that exactly k customers were served from 0 to x, is

$$m(T; x, k) := \pi(x, k) \exp(-T\mathbf{B_r})\boldsymbol{\varepsilon}'. \qquad (3.3.15c)$$

The service time remaining for c_{n+1} given that it was in service at time x (a conditional residual time) is

$$\bar{x}(x, n) := \pi(x, n)\, \mathbf{V}\boldsymbol{\varepsilon}'. \qquad (3.3.15d)$$

The number of sequences of events that can be analyzed in this way is unlimited. For instance, one can analyze the renewal process starting at some time x_2, given that k_1 customers were served in the interval before x_1, and k_2 customers were served in the interval $(x_1, x_2]$, etc, etc. Of course, the longer the sequence of conditions, the less interesting the results, for they must ultimately converge to the results using $\pi_{\mathbf{r}}$. Well, maybe.

3.3.4. Two Illustrations of Renewal Processes

In discussing renewal theory, we have introduced three views, corresponding to Figures 3.3.1 to 3.3.3, none of which actually correspond to the standard description in terms of arrivals. There should be no problem of changing our view from arrivals to departures, but the formulas derived from the three distinct viewpoints given in the previous sections is bound to be at least somewhat confusing. In this subsection we illustrate the various formulas for two distributions. The first, as always, assumes that S has only one internal state, and thus represents an exponential server. This will lead us to yet another derivation of the Poisson process. In the second example S represents the Erlangian-2 distribution.

The Poisson Process

As always, for exponential distributions, $\mathbf{B} \Rightarrow \mu$, while $\mathbf{p}$, $\mathbf{Q}$, and $\boldsymbol{\varepsilon}'$ all equal 1. Many formulas have already been reduced to their exponential results (or have been left to the exercises). We finish the job here. First consider (3.3.9c). The pdf $f(x;\alpha)$ is the density function for a subsystem with external feedback, as shown in Figure 3.3.3. If S itself is exponential, so is S_r, for, as we showed for $m(x;\alpha)$ following (3.3.9f),

$$f(x;\alpha) = (1-\alpha)\,\mu\, e^{-(1-\alpha)\mu x}.$$

This is an exponential distribution with mean service rate $\mu' = (1-\alpha)\mu$. We discussed the underlying significance of this in Lemma 3.3.2. But this tells us something else as well, which we state as another lemma.

Lemma 3.3.5: If any diagonal element of a transition matrix is greater than 0, it can be replaced by 0, with a commensurate change in its service rate and the other elements of $\mathbf{P}$ in that row. That is, suppose that $P_{ii} > 0$. Then let $\alpha = P_{ii}$ and

$$\text{new } P_{ii} = 0; \quad \text{new } P_{ij} = \frac{P_{ij}}{1-\alpha} \quad \text{for } j \neq i; \quad \text{new } M_{ii} = M_{ii}(1-\alpha).$$

The new $\mathbf{P}$ and new $\mathbf{M}$ will yield the same results as the original ones. Thus one can assume (if convenient) that the diagonal elements of a transition matrix are all 0, without loss of generality. ■

The discussion on residual and delayed times has no significance when discussing exponential servers, since $\pi_{\mathbf{r}}(x)$ as defined in (3.3.11b) is always 1 because $\mathbf{B}_{\mathbf{r}} = \mathbf{O}$. Everything is memoryless, and remains the same as it was at the beginning, until the customer leaves.

We then go to (3.3.14), for $n = \infty$. Here [compare with (3.3.3c) and (3.2.2b)]

$$\mathbf{B}_\infty = \mu \begin{bmatrix} 1 & -1 & 0 & 0 & 0 & \cdots \\ 0 & 1 & -1 & 0 & 0 & \cdots \\ 0 & 0 & 1 & -1 & 0 & \cdots \\ 0 & 0 & 0 & 1 & -1 & \cdots \\ \vdots & \vdots & \vdots & \vdots & \vdots & \cdots \end{bmatrix}.$$

To evaluate $\exp(-x\mathbf{B}_\infty)$, one needs $(\mathbf{B}_\infty)^k$ for all k. It can be proven by induction that

$$(\mathbf{B}_\infty{}^k)_{ij} = \mu^k\,(-1)^{j-i}\binom{k}{j-i} \quad \text{for } j \geq i, \tag{3.3.16a}$$

and 0 otherwise. Therefore, without too much mathematical difficulty, we get the expression (using $y = \mu x$)

$$\exp(-x\mathbf{B}_\infty) = e^{-y}\begin{bmatrix} 1 & y & y^2/2! & y^3/3! & y^4/4! & \cdots \\ 0 & 1 & y & y^2/2! & y^3/3! & \cdots \\ 0 & 0 & 1 & y & y^2/2! & \cdots \\ 0 & 0 & 0 & 1 & y & \cdots \\ \vdots & \vdots & \vdots & \vdots & 1 & \cdots \\ \vdots & \vdots & \vdots & \vdots & \vdots & \cdots \end{bmatrix}.$$

From its definition in (3.3.3a),

$$\mathbf{p}_\infty = [\,1, 0, 0, 0, 0, \cdots\,],$$

so $\mathbf{r}(x, \infty)$ is the top row of $\exp(-x\mathbf{B}_\infty)$, or

$$\mathbf{r}(x,\infty) = \left[\, e^{-y},\, ye^{-y},\, \frac{y^2 e^{-y}}{2!},\, \frac{y^3 e^{-y}}{3!},\, \cdots \right].$$

Since $m = 1$ from Section 3.3.3, we get (as no surprise to anyone)

$$P_k(x) = \frac{(\mu x)^k e^{-\mu x}}{k!}, \tag{3.3.16b}$$

the Poisson probabilities of finding k departures in time interval x [compare with (2.1.15) and (3.3.4c)].

Renewal Process with E_2 Interdeparture Times

One of the great strengths of the method in this book is that the expressions can easily be numerically evaluated automatically by computer. However, it is not easy to get physical insight unless one carries out many parametric studies, presenting the results graphically. As it happens, if m (the dimensionality of S) is small enough, we can find explicit expressions from the matrix formulas. The smallest nontrivial case is then $m = 2$. We now consider one such example.

The Erlangian distribution was discussed in Section 3.2.1 and Equation (3.2.1a). Recall that $E_k(x)$ corresponds to k identical exponential phases in tandem, each with service rate μ. Then for $k = 2$,

$$\mathbf{B} = \mu\begin{bmatrix} 1 & -1 \\ 0 & 1 \end{bmatrix} \quad \text{and} \quad \mathbf{Q} = \begin{bmatrix} 1 & 0 \\ 1 & 0 \end{bmatrix}.$$

From (3.3.9b),

$$\mathbf{B_r}(\alpha) = \mathbf{B}(\mathbf{I} - \alpha\mathbf{Q}) = \mu\begin{bmatrix} 1 & -1 \\ -\alpha & 1 \end{bmatrix}.$$

To get explicit expressions for $f(x;\alpha)$, $m(x;\alpha)$, and whatever else might be interesting, we must first get an explicit form for $\exp[-x\mathbf{B_r}(\alpha)]$. It is not hard to show that the eigenvalues for $\mathbf{B_r}(\alpha)$ are $(1 \pm \sqrt{\alpha})$, with eigenvectors (for convenience let $\beta = \sqrt{\alpha}$):

$$\mathbf{u}_{\pm} = [\,-1, \pm\beta] \quad \text{and} \quad \mathbf{v}'_{\pm} = \frac{1}{2}\begin{pmatrix} -1 \\ \pm 1/\beta \end{pmatrix}.$$

Since the eigenvalues are distinct, we can use (1.3.8c) to get (where $y = \mu x \beta$)

$$\exp[-x\mathbf{B_r}(\alpha)] = e^{-\mu x}\begin{bmatrix} \cosh y & \dfrac{\sinh y}{\beta} \\ \beta \sinh y & \cosh y \end{bmatrix}. \tag{3.3.17}$$

We use this to find $f(x;\,\alpha)$ from (3.3.9c). Let π have components π_1 and π_2, whose sum is 1; then

$$f(x;\,\alpha) = (1-\alpha)\,\mu e^{-\mu x}\left(\pi_1 \frac{\sinh y}{\beta} + \pi_2 \cosh y\right).$$

This certainly is not a simple expression even if $\pi_2 = 0$, that is, when $\pi = \mathbf{p}$. In this case

$$f(x;\,\alpha) = \frac{1-\alpha}{2\beta}\mu\left(e^{-\mu x(1-\beta)} - e^{-\mu x(1+\beta)}\right).$$

It is not clear what the generalization for Lemma 3.3.5 is when S is not exponential. We have already noted (Lemma 3.3.2) that a subsystem with external feedback, as in Figure 3.3.3, has the same dimensionality as the subsystem without feedback (what else could it be?). The last equation shows that an Erlangian-2 with feedback is equivalent to a subsystem of two unequal phases in tandem, with *no* feedback. The service rates of the two phases are the eigenvalues of $\mathbf{B_r}(\alpha)$. But, of course, this should have been clear from Section 3.2.1. As might be expected, when $\alpha = 1$, $f(x;\,1)$ is identically 0, corresponding to the fact that our looping customer is forever imprisoned in S_r.

We know from (3.3.8a) that $m(x;\,\alpha) = f(x;\,\alpha)/(1-\alpha)$, therefore

$$m(x;\,\alpha) = \frac{\mu}{2}e^{-\mu x}\left(\frac{\pi_1}{\beta}(e^{\mu\beta x} - e^{-\mu\beta x}) + \pi_2(e^{\mu\beta x} + e^{-\mu\beta x})\right).$$

Recall from (3.3.9f) that $m(x;\,\alpha)$ is the generator of the convolutions of $b(x)$. In this case is

$$b(x) = b_1(x) = m(x;\,0) = \pi_1 E_2(x;\,\mu) + \pi_2 \mu e^{-\mu x}.$$

This makes sense, since if our customer starts at the second phase (π_2), it will leave in exponential time. But if it starts at the first phase (π_1), it must go through both phases, taking Erlangian-2 time to leave.

Exercise 3.3.8: From (3.3.9f), find the kth convolutions of $b(x)$. In particular, show that if $\pi_1 = 1$, then the kth convolution is the Erlangian-$2k$, while if $\pi_2 = 1$, then the kth convolution is the Erlangian of order $2k-1$.

From (3.3.8c) the renewal density for our example is

$$m(x) = m(x;\, 1) = \frac{\mu}{2} + \mu(\pi_2 - \pi_1)e^{-2\mu x}. \tag{3.3.18a}$$

Observe that as x goes to infinity $m(x)$ approaches $\mu/2$, which is 1 over $\bar{x}$, consistent with the first form of the renewal theorem (Theorem 3.3.3a). Also note that if $\pi_1 = \pi_2$, then $m(x)$ is always $1/\bar{x}$. This is consistent with the third form of the renewal theorem (Theorem 3.3.3c), since the mean residual vector [from (3.3.10b)] is $\boldsymbol{\pi}_\mathbf{r} = [0.5, 0.5]$. In words, since both phases have equal service times and we do not know where our customer started, it will be at either one with equal probability.

The renewal function can be found from $m(x)$ by simple integration,

$$M(T) = \int_0^T m(x)\,dx = \frac{T}{\bar{x}} + \frac{\pi_2 - \pi_1}{2\mu\bar{x}}\left(1 - e^{-2\mu T}\right). \tag{3.3.18b}$$

Since $M(T)$ is the mean number of departures in interval T, $M(T)/T$ is the mean number of departures per unit time in that interval. This has a finite limit as T goes to infinity, and should be compared with $m(T)$, which is the departure rate at the end of the interval. Note that

$$\frac{M(T)}{T} = \frac{1}{\bar{x}} + \frac{\pi_2 - \pi_1}{\bar{x}}\,\frac{1 - e^{-2\mu T}}{2\mu T}. \tag{3.3.18c}$$

We see that $M(T)/T$ approaches the same limit as $m(T)$, *but* much more slowly. Even when $\exp(-\mu T)$ is negligible, a term in $1/T$ persists (unless the system started in the mean residual state). This is analogous to the average system behavior described in Chapter 1 [see (1.3.12b) and the discussion following it], and in fact, the dependence on T is identical.

We next move on to the residual vector defined in (3.3.11b). Instead of starting with $\mathbf{p}$, we start with the more general π, and get with the aid of (3.3.17) for $\alpha = \beta = 1$,

$$\begin{aligned}\boldsymbol{\pi}_\mathbf{r}(x) &= \pi \exp(-x\mathbf{B}_\mathbf{r}) \\ &= e^{-\mu x}[\,\pi_1 \cosh \mu x + \pi_2 \sinh \mu x\,,\ \pi_1 \sinh \mu x + \pi_2 \cosh \mu x\,],\end{aligned}$$

which in this case rearranges to

$$\boldsymbol{\pi}_\mathbf{r}(x) = [0.5, 0.5] + \frac{\pi_2 - \pi_1}{2} e^{-2\mu x}\,[-1, 1]. \tag{3.3.19}$$

This can be used, for instance, to calculate the time remaining for our trapped customer to complete his present service, given that he has been going in circles for time x. That is,

$$\overline{x_r}(x) := \boldsymbol{\pi}_{\mathbf{r}}(x)\mathbf{V}\boldsymbol{\varepsilon}' = \frac{1}{2\mu}[3 - (\pi_2 - \pi_1)e^{-2\mu x}] .$$

As x goes to infinity, we get the mean residual time, $\overline{x_r}$, which is $3/2\mu$, or $3\bar{x}/4$, irrespective of the initial state.

Exercise 3.3.9: Prove the formula above. Show that $\overline{x_r}(0) = \bar{x}$ for $\pi_1 = 1$.

We next evaluate the delayed renewal density, either using the material preceding Theorem 3.3.3b or by taking (3.3.19) as the starting vector for $m(x)$. In either case we get

$$m_r(T; x) = \frac{\mu}{2} + \mu(\pi_2 - \pi_1)e^{-2\mu(x+T)}. \tag{3.3.20a}$$

For any finite T, as x grows large, $m_r(T; x)$ approaches $1/\bar{x}$, as was described in the second form of Theorem 3.3.3.

The delayed renewal function also follows easily. As above,

$$M_r(T; x) := \int_0^T m_r(s; x)ds = \frac{\mu T}{2} + \frac{\pi_2 - \pi_1}{2} e^{-2\mu x}(1 - e^{-2\mu T}). \tag{3.3.20b}$$

As with (3.3.18c) the behavior as x goes to infinity can be examined best by looking at M/T. Then for any T,

$$\frac{M(T; x)}{T} = \frac{1}{\bar{x}}\left[1 + (\pi_2 - \pi_1)e^{-2\mu x}\frac{1 - e^{-2\mu T}}{2\mu T}\right]. \tag{3.3.20c}$$

Note that $M(T; x)$ approaches the expected limit $(1/\bar{x})$ much more rapidly than does $M(T)$ [which is really $M(T; 0)$]. Thus if one waits some time, x, before beginning measurements, successive intervals of T will yield the same average number of completions.

Exercise 3.3.10: Let $2\mu = 1$ and $\pi_1 = 0$; then compare Equations (3.3.19) and (3.3.20) for $T = 2$ and increasing x.

In dealing with residual vectors, we have given the impression that all information about the internal state of the subsystem is gradually lost as time goes on. This is true only because observations concerning past behavior have not been included in estimating the future. In the discussion following (1.3.15b) it was pointed out that in a discrete Markov chain, time and the counting of events were synonymous, whereas a continuous chain soon loses track of the number of events. In the second part of Section 3.3.3 it was shown that knowledge of the number of past departures can be incorporated into estimations of future behavior. We will show presently that such information can affect appreciably what is likely to happen.

First we must determine $\mathbf{r}(x,\infty)$ from (3.3.14) for our present example. To do this, besides the matrices already evaluated at the beginning of this section, we need **BQ**, which is easily shown to be

$$\mathbf{BQ} = \mu\begin{bmatrix} 1 & -1 \\ 0 & 1 \end{bmatrix}\begin{bmatrix} 1 & 0 \\ 1 & 0 \end{bmatrix} = \mu\begin{bmatrix} 0 & 0 \\ 1 & 0 \end{bmatrix}.$$

We must also have $\mathbf{p}_\infty$, which is the same as (3.3.3a), where each element is a two-vector, with $\pi = [\pi_1, \pi_2]$.

We next set up $\mathbf{B}_\infty$ and find that it is identical with the $\mathbf{B}_\infty$ we had for the exponential distribution, except that all rows and columns are taken two at a time. Observe that each 2 by 2 block on the diagonal of $\mathbf{B}_\infty$ in the first part of Section 3.3.4 is precisely **B**, while the 2 by 2 blocks above and to the right of the diagonal blocks are all $-\mathbf{BQ}$. We are indeed fortunate, since $\exp(-x\mathbf{B}_\infty)$ is the same as that in the preceding section. Equations (3.3.15) imply that (again $y = \mu x$)

$$\pi(x, k) = \left[\pi_1 \frac{y^{2k}}{(2k)!} + \pi_2 \frac{y^{2k-1}}{(2k-1)!}\ ,\ \pi_1 \frac{y^{2k+1}}{(2k+1)!} + \pi_2 \frac{y^{2k}}{(2k)!}\right] \frac{e^{-y}}{P_k(x)}$$

and

$$P_k(x) = \frac{y^{2k-1}e^{-y}}{(2k+1)!}[(2k+1)y + \pi_1 y^2 + \pi_2(2k)(2k+1)]. \tag{3.3.21a}$$

Therefore,

$$\pi(x, k) = \frac{[\pi_1(2k+1)y + \pi_2(2k)(2k+1)\ ,\ \pi_1 y^2 + \pi_2(2k+1)y\,]}{(2k+1)y + \pi_1 y^2 + \pi_2(2k)(2k+1)}. \tag{3.3.21b}$$

Observe that when $x = 0_+$, $\pi(0_+, 0) = [\pi_1, \pi_2]$, and for $k > 0$, $\pi(0_+, k) = [1, 0]$.

Ordinarily, not too much credence should be placed in physical interpretations of the components of the internal states of a subsystem, since there may be many equivalent representations of S. In this case, however, there is some insight to be gained. When x is very small, one should expect c_1 to still be in S, and in his starting state. This is exactly the case, since $P_k(0_+)$ is essentially 0 except for $k = 0$. Given the highly unlikely event that $k-1$ customers have already left in 0_+ time, c_k would almost surely have not progressed much beyond just entering. This is indeed the case, since $\pi(0_+, k) = \mathbf{p}$, the entrance vector.

departing in successive intervals. In a different approach related to the technique

As y increases, the second component of $\pi(x, k)$ also increases, and when y is approximately equal to $2k$, the two components are comparable. As y increases further, the second component becomes much larger than the first, and approaches 1. This has a direct physical interpretation. One would expect approximately k customers to be served in time $2k/\mu$, so if x is much larger than that, c_k has surely been in service a long time and must be at the second phase by now. Again the reader is warned that such interpretations are risky, and is referred to the discussion in Section 3.2.2. The important point to note is that depending on k and x, the internal state of S could be vastly different from $\mathbf{p}$ or $\pi_{\mathbf{r}}$, the mean residual vector (which in this case is [0.5, 0.5]).

There is a useful statement that can be said in general. If many more customers have actually been served than one would expect in the time interval under measurement, the internal state vector of S will be close to the entrance vector. If the number that have been served is comparable to the expected number, the internal state will be closer to $\pi_{\mathbf{r}}$. Finally, if far fewer customers have been served than might be expected, S will be described by a completely different state vector. In any case, the initial vector (in this case, $\pi = [\pi_1, \pi_2]$) will be washed out.

Whatever might or might not be said about the internal state of S, many different predictions can be made. First, we can calculate the mean time to the next departure, given the number that have already departed, from Equations (3.3.21).

Exercise 3.3.11: Let $\mu = 2$ per minute, and suppose that measurement began at the moment a customer began service. In the interval $(0, 2]$, $0 \le k \le 10$ customers have been served. What is the mean time for the next customer to depart $[\bar{x}(2, k)]$? Make a table for k versus $\bar{x}$. Suppose the interval is $(0, 4]$. What are the $\bar{x}$'s now?

Exercise 3.3.12: Do the same as in Exercise 3.3.11, except that now you have no idea when the first customer you counted began service. Compare and discuss the two pairs of results.

Another interesting number to look at is the renewal function conditioned on k departures in the previous interval of time. Equations (3.3.18) can be used for this purpose. But instead of using the initial vector, π, we use $\pi(x, k)$, which is, after all, the initial vector starting at x.

Exercise 3.3.13: Suppose that $0 \leq k \leq 10$ customers have finished in the first 2 minutes, as in Exercise 3.3.11. What is the expected number of departures in the next 2 minutes? In the next 4 minutes? Summarize your answers in a table. Also calculate the number you would have expected in the first 2 minutes.

The marginal probabilities of having n departures in the interval $(x, x+T]$, conditioned on having had k completions up to then can be calculated using (3.3.21a) (where n replaces k and t replaces x), again using the appropriate components from (3.3.21b) instead of the initial vector $\boldsymbol{\pi}$. The number of necessary parameters is growing steadily now, we have $n \times k \times x \times T$ possibilities. A formal presentation of even more complex formulas becomes increasingly difficult, since one loses track of everything that is going on. But still, let us have one more exercise.

Exercise 3.3.14: Compare the $P_n(x)$'s as defined in (3.3.4a) for a Poisson process, and a renewal process where the interdeparture times are distributed according to an Erlangian-2 distribution. Assume that measurement begins when c_1 enters S, and that the mean interdeparture time in all cases is 1 minute. Calculate the E_2 process for four different conditions:

1. The interval for counting the number of arrivals is $(0, 2]$ [Equation (3.3.21a)].
2. The interval for counting arrivals is $(2, 4]$, and no customer completed service previously [(3.3.21a) conditioned by (3.3.21b)].
3. The same as condition 2 except that two customers had completed service in the interval $(0, 2]$.
4. The same as condition 3, but for four customers.

Construct a single table of numbers that has the Poisson and all four Erlangian cases for $0 \leq n \leq 10$, and discuss their similarities and differences.

We have one more extension to discuss before giving up on this chapter. This will be done by example, although it should be clear how one can generalize to any subsystem. Although it may be difficult (and often impossible) to know what is going on inside S, it is easy to keep track of the number of customers of *embedded Markov chains*, one waits until a customer begins service before taking measurements. In that case, the period always begins with $\pi = \mathbf{p}$. When the interval is over, one waits for the next completion before measuring again. But then the mean number of departures is not $T/\bar{x}$, even for large time, since we are always starting over. In Chapter 6, S will be generalized so that several customers can be served at once. In that case, when a customer leaves, the internal state of the residual subsystem is not known, so one cannot start over until S is completely empty. Such behavior is called a *semi-Markov chain*. The technique described herein does generalize to multiple customer service without any conceptual complications.

Example 3.3.1: Consider an example such as Exercise 3.3.13. Initially, S is in state $a[1, 0]$. Suppose that in the first minute c_1 and c_2 have both finished, but c_3 is still in S. At that moment, c_3 will be at phase 1 with probability $[\pi(1, 2)]_1$. Using (3.3.21b) (with $\pi_1 = 1$), this probability is 0.7143. Now suppose that c_3 is still busy by the end of the second minute; then at that moment it is at phase 1 with probability 0.2941. One gets this by using (3.3.21b) again, but this time $\pi_1 = 0.7143$, and of course, $\pi_2 = 0.2857$. If one measures the number of completions in the interval (0, 2] without noting how many finished in the first minute, the probability that c_3 will be at phase 1 is 0.5556.

Interestingly enough, if no customers finish in the first minute, and two finish in the second minute, the sequence for phase 1 to be busy is $1.0000 \rightarrow 0.3333 \rightarrow 0.6757$, while if one customer finishes in each of the two minutes, the sequence is $1.0000 \rightarrow 0.6000 \rightarrow 0.5556$. This happens to be exactly the same as when going to 2 without considering the number at 1, but this is not always the case. Anyway, we see that the three different ways of having two completions in two minutes, keeping track of how many completed in the first minute, yield different results. Now, for instance, in calculating the mean time for c_3 to finish using the data above and (3.3.15d), we get three different answers, 0.837838, 0.777778, and 0.647059 minute, respectively, for 2, then 0; 1, then 1; and 0, then 2. ●

Exercise 3.3.15: Extend the discussion above to three 1-minute intervals where a total of three customers finished service. What is the mean time until c_4 finishes in each case?

It is hoped that the reader can now extend this procedure to any example. Any information that one has concerning past behavior of a system should be usable in calculating conditional events in the future.

CHAPTER 4
M/G/1 QUEUE

"The shortest path between two truths in the real domain passes through the complex domain."
– J. Hadamard

We are finally ready to look at nonexponential queues in earnest. In Chapter 2 we looked at closed loops in which both subsystems were single servers with exponential service time distributions. We showed how to transform a closed system into an open one, and how certain types of non-steady-state behavior should be analyzed. In Chapter 3, we showed how a large class of nonexponential servers (ME distributions) can be treated exactly, using a matrix representation, and applied it to examining various aspects of renewal processes, as well as the specific behavior of a single general server, including residual times. We now combine those two chapters in studying the M/ME/1 queue, first looking at steady-state closed systems, then "opening" them, and finally, extending the transient results of Chapter 2. In those cases where a particular result does not depend on the specific properties of a matrix, the result becomes applicable to M/G/1 queues as well. Much of this material is an outgrowth of the Ph.D. Thesis by John L. Carroll [CARR79], and the associated papers, [CARR82], and [TEHR89].

4.1. STEADY-STATE M/ME/1//N (AND M/ME/1/N) LOOP

We start, as always, by making some new definitions. In Chapter 2 each state could be described uniquely by n, the number at S_1, whereas in Chapter 3 the states were described by identifying the phase in S_1 where the active customer was. Here both must be specified to describe uniquely a state of the system shown in Figure 4.1.1. This figure is itself a combination of Figures 2.1.1 and 3.1.1, where the single server, S_1 of Figure 2.1.1 is replaced by the m-phase subsystem, S, of Figure 3.1.1.

All the objects in the following list are the same as defined in Chapter 3: $\mathbf{p}$, $\mathbf{P}$, $\mathbf{q}'$,$\boldsymbol{\varepsilon}'$, $\mathbf{M}$, $\mathbf{B}$,$\mathbf{V}$, $\mathbf{Q}$, and the linear operator, $\Psi[\cdot]$. For a closed system with N customers, define the following.

Definition 4.1.1

$[\boldsymbol{\pi}(n;N)]_i$:= *steady-state probability that there are n customers in the queue at* S_1*, and the one being served is at phase i.* n includes the customer being served,

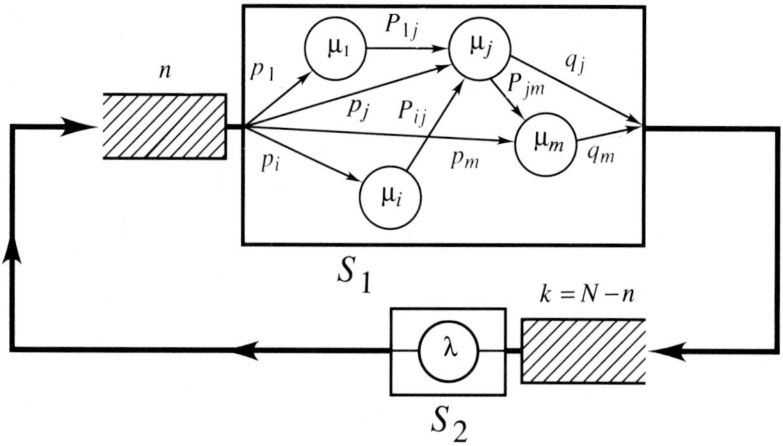

Figure 4.1.1: Closed loop made up of two subsystems. S_2 is purely exponential, with service rate λ, while S_1 is of the type described in Figure 3.1.1 and thus represents a matrix exponential distribution. There are n customers at S_1, with the active customer being at phase i (or in state i), while the other k $(= N - n)$ customers are at S_2.

and $\boldsymbol{\pi}(n; N)$ is a row vector with m components. The associated scalar probability, $r(n; N)$ is the same as Definition 2.1.1. ◊

Although $\pi_i(0; N)$ has no meaning (if no customers are at S_1, no phase can be busy), it will be useful to define the vector

$$\pi(0; N) := r(0; N)\mathbf{p}. \tag{4.1.1}$$

Then for all n, $0 \le n \le N$,

$$r(n; N) = \sum_{i=1}^{m} \pi_i(n; N) = \boldsymbol{\pi}(n; N)\boldsymbol{\varepsilon}'. \tag{4.1.2}$$

From these definitions, we see that there are $mN + 1$ states describing the closed system [or $m(N+1)$ states if we make believe that $\pi_i(0; N)$ has meaning], but they are grouped together as $(N+1)$ m-vectors. This way of grouping is the basis of LAQT and was first used to advantage by Victor Wallace [WALL69] in formalizing *quasi birth-death (QBD) processes*. He recognized that for many systems, the transition rate matrices, $\boldsymbol{Q}$, as defined in Chapter 1, are block tridiagonal. That is, the $\boldsymbol{Q}$ of Chapter 1 can be considered as a matrix whose elements are themselves matrices. Only the diagonal, superdiagonal, and subdiagonal elements are nonzero matrices. He chose the term QBD because birth-death processes are also tridiagonal, but with scalar components. He also speculated about an algebraic theory for Markovian networks [WALL72]. All queueing systems considered here are special cases of QBD processes, but we will not look at them in that context.

4.1.1. Balance Equations

Let us first introduce some new notation.

Definition 4.1.2

$\{i; n; N\}$ is an integer triplet that corresponds to one possible state of an M/ME/1//N loop. N is the total number of customers in the system, n is the number of customers at S_1, including the one in service, and i is the phase in S_1 that is busy. We can say that the system is in state $\{i; n; N\}$. If we are dealing with an open system ($N = \infty$), we use the notation $\{i; n\}$.

$\Xi := \{ i \mid 1 \le i \le m, i \textit{ is a phase in } S_1 \}$. Since only one customer can be active at a time in S_1, Ξ is the set of all *internal states* of S_1. We can say that the system is in internal state $i \in \Xi$, or that the active customer is at phase i in S_1. ◊

Remember, too, we are assuming that S_1 and S_2 operate independently. This means that only *one thing* happens at a time. The term "one thing" means whatever we wish. Thus a customer leaving S_i and being replaced immediately by the next customer in the queue is "one thing." Also, the process whereby a customer leaves one subsystem (and is replaced by a successor), goes to the other, and finding it empty immediately enters into service is "one thing." However, if two customers are active at the same time (e.g., one in each subsystem), only one can change state. In general, those processes that take 0 time (moving from one subsystem to the other, entering S_i, moving from one phase to another) are considered to be part of the previous process.

As a direct generalization of Section 2.1.2., and (2.1.4a), in order for the system to be in a steady state, the probability rate of leaving a state, $\{i; n; N\}$, must be equal to the probability rate of entering that state. Thus for state $\{\cdot; 0; N\}$ we have

$$\lambda r(0; N) = \sum_j \pi_j(1; N) M_{jj} q_j = \pi(1; N)\mathbf{M}\mathbf{q}'.$$

In words, the probability rate of leaving the state where no one is at S_1 is equal to the probability of there being no one there $[r(0; N)]$ times the probability rate of a customer finishing at S_2 $[\lambda]$. The middle term of the equation above is the probability rate of entering state $\{\cdot; 0; N\}$. This is equal to the sum of probability rates of having one customer in S_1 being served by j, $[\pi_j(1; N)]$, who then finishes there, $[\mu_j = M_{jj}]$, and leaves, $[q_j]$. The rightmost expression of the equation is the matrix equivalent of the middle expression.

From (3.1.1b) and (3.1.3) it follows directly that

$$\mathbf{M}\mathbf{q}' = \mathbf{B}\boldsymbol{\varepsilon}'. \tag{4.1.3a}$$

Thus if both sides of the preceding equation are multiplied on the right by $\mathbf{p}$, and we use (4.1.1), we get the vector balance equation:

$$\lambda\pi(0; N) = \pi(1; N)\mathbf{Mq'p} = \pi(1; N)\mathbf{BQ}, \tag{4.1.3b}$$

where $\mathbf{Q} = \boldsymbol{\varepsilon}'\mathbf{p}$ is the idempotent matrix defined in Section 3.3.1 and has nothing to do with the transition rate matrix. Except when direct reference is made to Chapter 1, $\mathbf{Q}$ always has this meaning.

The balance equation for state $\{i; N; N\}$ is derived as follows. Since in this case there is no one at S_2, there can be no arrivals to S_1, but instead, the customer who is active in S_1 can complete service at i, $[\pi_i(N; N)M_{ii}]$, thereby causing the system to leave that state. The probability rate of entering state $\{i; N; N\}$ is made up of two parts. Either the system could be in state $\{i; N-1; N\}$, $[\pi_i(N-1; N)]$, and have the lone customer at S_2 finish, $[\lambda]$, or all N customers could already be at S_1, but the active customer is at some other phase, j, $[\pi_j(N; N)]$, finishes there, $[M_{jj}]$, and goes to i, $[P_{ji}]$. Note that a completion at S_2 changes the external state of the system (n goes from $N-1$ to N) but not the internal state (i remains the same, since the active customer at S_1 does not move merely because a new customer has arrived at the queue). So this equation is

$$\pi_i(N; N)M_{ii} = \lambda\pi_i(N-1; N) + \sum_j \pi_j(N; N)M_{jj}P_{ji},$$

or in matrix form,

$$\pi(N; N)\mathbf{M} = \pi(N-1; N)\lambda\mathbf{I} + \pi(N; N)\mathbf{MP}.$$

Remembering that $\mathbf{B} = \mathbf{M}(\mathbf{I} - \mathbf{P})$, the equation above can be rearranged to

$$\pi(N; N)\mathbf{B} = \lambda\pi(N-1; N),$$

or since $\mathbf{B}^{-1} = \mathbf{V}$,

$$\pi(N; N) = \lambda\pi(N-1; N)\mathbf{V}. \tag{4.1.3c}$$

The balance equations for states where n is greater than 0 but less than N combine all the features of (4.1.3b) and (4.1.3c). It will be useful to describe what happens in these cases with the help of the state transition diagram in Figure 4.1.2. As usual, the sum of the weights of the arrows going to $\{i; n; N\}$ equals the sum of those leaving. So for $i \in \Xi$, we have

$$\pi_i(n; N)(M_{ii} + \lambda) = \sum_j \pi_j(n; N)M_{jj}P_{ji} + \sum_j \pi_j(n+1; N)M_{jj}q_j p_i + \pi_i(n-1; N)\lambda\mathbf{I}.$$

These m equations can be summarized by the vector equation

$$\pi(n; N)(\mathbf{M} + \lambda\mathbf{I}) = \pi(n; N)\mathbf{MP} + \pi(n+1; N)\mathbf{Mq'p} + \pi(n-1; N)\lambda\mathbf{I}.$$

This, in turn can be rearranged, as with the previous equations, to yield the rest of the balance equations. For $0 < n < N$,

$$\pi(n; N)(\mathbf{B} + \lambda\mathbf{I}) = \pi(n+1; N)\mathbf{BQ} + \pi(n-1; N)\lambda\mathbf{I} \tag{4.1.3d}$$

We mention here that (4.1.3d) is valid for $n = 1$ by virtue of Definition (4.1.1).

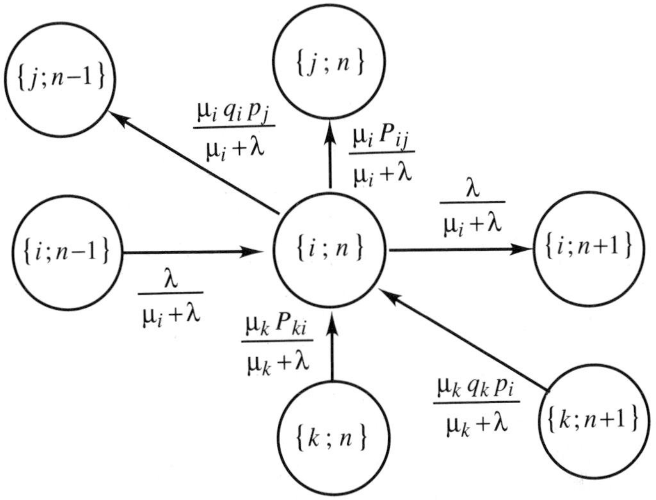

Figure 4.1.2: Steady-state transition diagram for the $\{i; n; N\}$th state of an M/ME/1//N closed loop. An arrow pointing to the left (either horizontally or diagonally) represents a customer finishing at phase i, $\{[(\mathbf{M}+\lambda\mathbf{I})^{-1}\mathbf{M}]_{ii}\}$, and leaving S_1, $[q_i]$, followed by another customer entering and going to j, $[p_j]$. A vertical arrow corresponds to a customer finishing at phase i, $[(\mathbf{M}+\lambda\mathbf{I})^{-1}\mathbf{M}]$, and going to phase j, $[P_{ij}]$. An arrow to the right (no diagonal arrows allowed) corresponds to a customer finishing at S_2, $[(\mathbf{M}+\lambda\mathbf{I})^{-1}\lambda]$, and immediately going to S_1, without changing the internal state.

The set of Equations (4.1.3) fall in the class of "second-order finite-difference vector equations," not a particularly informative name for our purposes. They are similar in appearance to the balance equations for the M/M/1 queue (2.1.4a), and naturally, reduce to them when S_1 is exponential. In the next section we prove that they reduce to first-order equations and then give an explicit expression for the general solution.

4.1.2. Steady-State Solution

First consider (4.1.3d) for $n = N-1$:

$$\pi(N-1; N)(\mathbf{B} + \lambda\mathbf{I}) = \pi(N; N)\mathbf{BQ} + \lambda\pi(N-2; N).$$

Next replace $\pi(N; N)$ with (4.1.3c), divide by λ, and regroup terms to get (recalling that $\mathbf{VB} = \mathbf{I}$)

$$\pi(N-1; N)\left(\mathbf{I} + \frac{1}{\lambda}\mathbf{B} - \mathbf{Q}\right) = \pi(N-2; N).$$

Now define the important pair of matrices

$$\mathbf{A} := \mathbf{I} + \frac{1}{\lambda}\mathbf{B} - \mathbf{Q} \tag{4.1.4a}$$

and (assuming that it exists)

$$\mathbf{U} := \mathbf{A}^{-1}. \tag{4.1.4b}$$

It then follows that

$$\pi(N-1;\,N) = \pi(N-2;\,N)\mathbf{U}. \tag{4.1.5a}$$

Before proving by induction that this equation is true for all $n < N$, we enumerate a collection of simple relations (stated in the form of the following lemma) that will prove useful throughout the rest of the book.

Lemma 4.1.1: For **A** and **U**, defined by Equations (4.1.4), the following are matrix identities:

$$\lambda\mathbf{A}\,\boldsymbol{\varepsilon}' = \mathbf{B}\boldsymbol{\varepsilon}', \quad \lambda\mathbf{V}\mathbf{A}\,\boldsymbol{\varepsilon}' = \boldsymbol{\varepsilon}', \quad \mathbf{U}\mathbf{B}\boldsymbol{\varepsilon}' = \lambda\boldsymbol{\varepsilon}',$$

or since $\mathbf{Q} = \boldsymbol{\varepsilon}'\mathbf{p}$,

$$\mathbf{UBQ} = \lambda\mathbf{Q} \quad \text{and} \quad \lambda\mathbf{AQ} = \mathbf{BQ}.$$

Similarly,

$$\lambda\mathbf{pA} = \mathbf{pB}, \quad \lambda\mathbf{pAV} = \mathbf{p}, \quad \mathbf{pBU} = \lambda\mathbf{p}.$$

Also,

$$\lambda\mathbf{QA} = \mathbf{QB} \quad \text{and} \quad \lambda\mathbf{QAV} = \mathbf{Q}.$$

■

Proof: Since $\mathbf{I}\boldsymbol{\varepsilon}' = \mathbf{Q}\boldsymbol{\varepsilon}' = \boldsymbol{\varepsilon}'$ it follows that

$$\mathbf{A}\boldsymbol{\varepsilon}' = \left(\mathbf{I} + \frac{1}{\lambda}\mathbf{B} - \mathbf{Q}\right)\boldsymbol{\varepsilon}' = \boldsymbol{\varepsilon}' + \frac{1}{\lambda}\mathbf{B}\boldsymbol{\varepsilon}' - \boldsymbol{\varepsilon}' = \frac{1}{\lambda}\mathbf{B}\boldsymbol{\varepsilon}'.$$

All else follows trivially. QED

Now assume that for all k from $N-2$ down to n [by virtue of (4.1.5a) it is true for $k = N-2$],

$$\pi(k+1;\,N) = \pi(k;\,N)\mathbf{U}.$$

Insert this (with $k = n$) into (4.1.3d) and get

$$\pi(n;\,N)(\mathbf{B} + \lambda\mathbf{I}) = \pi(n;\,N)\mathbf{UBQ} + \lambda\pi(n-1;\,N).$$

After using Lemma 4.1.1, and rearranging somewhat, we get what is needed for the proof by induction, namely the following first-order matrix difference equation promised previously:

$$\pi(n;\,N) = \pi(n-1;\,N)\mathbf{U}. \tag{4.1.5b}$$

Now since (4.1.3d) is true for all n from 1 to $N-1$, so must (4.1.5b). In particular,

$$\pi(1; N) = \pi(0; N)\mathbf{U}$$

and

$$\pi(2; N) = \pi(1; N)\mathbf{U} = \pi(0; N)\mathbf{U}^2.$$

In general, we have [using (4.1.1)]

$$\pi(n; N) = \pi(0; N)\mathbf{U}^n = r(0; N)\mathbf{p}\mathbf{U}^n \quad \text{for } 0 \le n < N, \tag{4.1.6a}$$

and with the help of (4.1.3c),

$$\pi(N; N) = \lambda r(0; N)\mathbf{p}\mathbf{U}^{N-1}\mathbf{V}. \tag{4.1.6b}$$

All the π's are conveniently expressed in terms of $\pi(0; N)$, which by virtue of (4.1.1) depends on only one scalar parameter, $r(0; N)$. This, in turn, can be evaluated by the usual requirement that the probabilities add up to 1:

$$\sum_{n=0}^{N} r(n; N) = \sum_{n=0}^{N} \pi(n; N)\boldsymbol{\varepsilon}' = r(0; N)\mathbf{p}\left(\sum_{n=0}^{N-1} \mathbf{U}^n + \lambda \mathbf{U}^{N-1}\mathbf{V}\right)\boldsymbol{\varepsilon}' = 1. \tag{4.1.6c}$$

This equation can be simplified both visually and computationally by defining the matrix in the large brackets to be the *normalization matrix*, $\mathbf{K}(N)$, and observing that $\mathbf{K}(N)$ satisfies a simple recursive formula:

$$\mathbf{K}(N) := \mathbf{I} + \mathbf{U} + \mathbf{U}^2 + \cdots + \mathbf{U}^{N-1} + \lambda \mathbf{U}^{N-1}\mathbf{V}$$

$$= \mathbf{I} + \mathbf{U}(\mathbf{I} + \mathbf{U} + \mathbf{U}^2 + \cdots + \mathbf{U}^{N-2} + \lambda \mathbf{U}^{N-2}\mathbf{V}). \tag{4.1.6d}$$

The expression inside the parentheses is $\mathbf{K}(N-1)$, so, for $N > 1$,

$$\mathbf{K}(N) = \mathbf{I} + \mathbf{U}\mathbf{K}(N-1), \quad \text{where } \mathbf{K}(1) = \mathbf{I} + \lambda \mathbf{V}. \tag{4.1.6e}$$

By virtue of the fact that for any $\mathbf{F}$ that has no unit eigenvalues,

$$\sum_{n=0}^{N-1} \mathbf{F}^n = [\mathbf{I} - \mathbf{F}^N][\mathbf{I} - \mathbf{F}]^{-1},$$

(4.1.6d) can also be written as

$$\mathbf{K}(N) = [\mathbf{I} - \mathbf{U}^N][\mathbf{I} - \mathbf{U}]^{-1} + \lambda \mathbf{U}^{N-1}\mathbf{V}. \tag{4.1.6f}$$

Finally recall that $\Psi[\,\cdot\,] = \mathbf{p}[\,\cdot\,]\boldsymbol{\varepsilon}'$, so (4.1.6c) leads to

$$r(0; N) = \frac{1}{\Psi[\mathbf{K}(N)]}. \tag{4.1.6g}$$

Equations (4.1.6) are very interesting in that they give us an explicit, closed-form expression for the M/ME/1 queue, which retains the simple form the solution of the M/M/1 queue has, as given in (2.1.4) and (2.1.5). In particular, compare $\mathbf{K}(N)$ with $K(N)$ in (2.1.4c) and (2.1.4d). Furthermore, these equations are ideally suited for numerical computation, as well as algebraic manipulation, as will become clear presently. Efficient computational algorithms can be written to compute the steady-state and other properties, and it should become apparent to the reader just how to do this, but these formulas are important enough to be summarized by

Theorem 4.1.2: For any closed loop made of one exponential server with service rate, λ, and one general server that has a matrix exponential representation, $<\mathbf{p}, \mathbf{B}>$, the steady-state queue-length probabilities are given by

$$r(n;N) = r(0;N)\Psi[\mathbf{U}^n] \quad \text{for} \quad 0 \le n < N, \tag{4.1.7a}$$

$$r(N;N) = \lambda r(0;N)\Psi[\mathbf{U}^{N-1}\,\mathbf{V}]. \tag{4.1.7b}$$

The matrix, $\mathbf{U}$, is given by (4.1.4b), and $r(0;N)$ is given by (4.1.6e) or (4.1.6f), and (4.1.6g). The vector probabilities are given by (4.1.6a) and (4.1.6b):

$$\pi(n;N) = r(0;N)\,\mathbf{p}\mathbf{U}^n \quad \text{for} \quad 0 \le n < N, \tag{4.1.7c}$$

$$\pi(N;N) = \lambda r(0;N)\mathbf{p}\mathbf{U}^{N-1}\,\mathbf{V}. \tag{4.1.7d}$$

These equations give us a *matrix geometric* solution of the M/ME/1//N queue analogous to the geometric solution for the M/M/1//N queue given by (2.1.6). ■

The term *geometric* refers to any series where the ratio of successive terms is a constant factor. Thus $1 + x + x^2 + x^3 + \cdots$ is the *geometric series*, where x is the ratio of terms. In our theorem, $\mathbf{U}$ is the ratio of terms, and since it is a matrix, we call this *matrix geometric*.

Example 4.1.1: Figure 4.1.3 shows the steady-state queue-length probabilities of an M/E_2/1//20 loop, for various values of ρ. At first glance, this figure looks similar to Figure 2.1.3 for the M/M/1 queue, but there are several significant differences. First note that since $r(N;N)$ comes from a different formula than $r(n<N;N)$ [Equations (4.1.7)], we can expect the curve to deviate in going from 19 to 20. This does indeed show itself for $\rho = 1$, but it is not so clear for other values of ρ, either because the curve is growing too big (as with $\rho = 2$), or the values are too small to be seen ($\rho \le 0.9$). A second feature is that the two curves corresponding to $\rho = 2$ and $\rho = 0.5$ are not mirror images of each other. For $\rho = 0.5$, $N = 20$ is sufficiently large so that S_2 is saturated, so $r(n;20) \approx r(n)$ (i.e., the M/E_2/1 queue). On the other hand, the curve corresponding to $\rho = 2.0$ is very close to the E_2/M/1 queue, since now S_1 is saturated. As we shall see in detail in Chapter 5, the two queues are distinctly different in their performance.

What we just described might be expected, but for the third distinctive feature observe that unlike the M/M/1 queue, $r(n;20)$ is not a monotonically decreasing function of n, even when $\rho < 1$. Note that for $\rho = 0.9$ and 1, we have $r(1;N)/r(0;N) > 1$. But from (4.1.7a), as long as $N \ge 2$, the ratio of those two probabilities is $\Psi[\mathbf{U}]$ which for the M/E_2/1//N loop can be shown to be (see the end of Section 4.4.3)

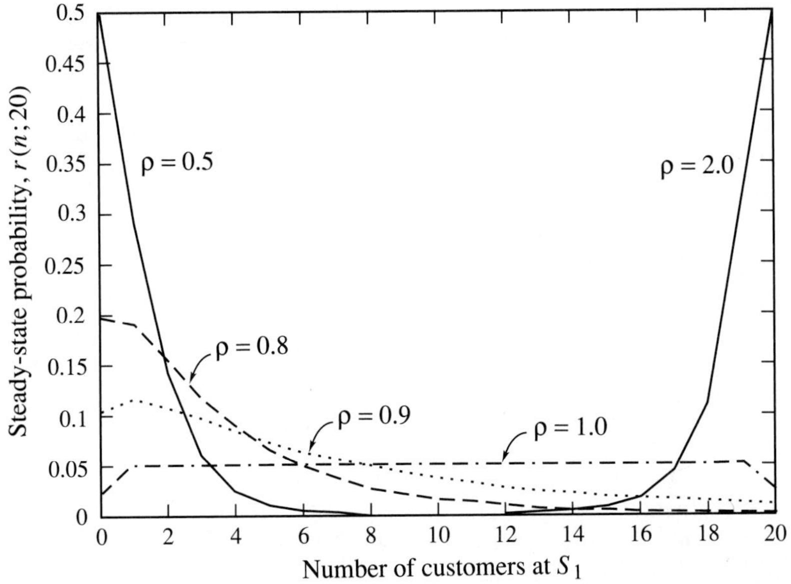

Figure 4.1.3: Steady-state queue-length probabilities for an $M/E_2/1//20$ loop, for $\rho = 0.5, 0.8, 0.9, 1.0$, and 2.0. For $\rho < 1$, the curves tend to decrease with n, but not universally so. Note that the curves for $\rho = 0.5$ and $\rho = 2.0$ are *not* mirror images of each other, nor is the curve for $\rho = 1$ horizontal. (See Figure 2.1.3)

$$\Psi[U] = \frac{\rho}{2}\left(2 + \frac{\rho}{2}\right).$$

This expression is greater than 1 as long as $\rho > 2\sqrt{2} - 2 \approx 0.8284$. Therefore, $r(n; N)$ will first rise and then decrease to zero with increasing n when $0.8284 < \rho \le 1$. ●

Exercise 4.1.1: Calculate the steady-state queue-length probabilities of an $M/E_2/1//N$ loop, for $N = 20$ and $\rho = 0.5$, and 0.9. Compare your answers with those for an M/M/1//N queue by plotting both sets on the same graph, $r(n; N)$ versus n. By what percent do the two sets of numbers differ? Under what conditions could $r(3; N)$ be greater than $r(2; N)$?

Exercise 4.1.2: Do the same calculations as in Exercise 4.1.1, except here let S_1 be equivalent to an H_2 distribution (see Exercise 3.1.2). Let $\alpha = 0.1$ and $\mu_2/\mu_1 = 10$. Are there values of ρ for which $\Psi[U] > 1$ and $r(n; N)$ is not be a monotonic function of n?

The queue-length probabilities are not often used in practice to analyze the performance of closed systems, partly because there is too much to measure. One important performance parameter that *is* measured is the rate at which customers make a full circuit of the loop. Aside from a multiplicative constant, we called this the *system throughput* in Section 2.1.2. The number of customers who leave a single server per unit time is equal to the fraction of time that server is busy (or the probability that the server is busy), divided by the time it takes to service one customer. One can look at this in a different way. S_i releases customers at a constant rate $(1/\bar{x}_i)$ as long as it is busy, but does nothing when it is idle (i.e., when no customers are there). Any way one looks at it [see Equations (2.1.5)], the system throughput is (remember that $\lambda \bar{x}_1 = \rho$)

$$\Lambda(N) = \frac{1 - r(0; N)}{\bar{x}_1} = \frac{\Psi[\mathbf{K}(N)] - 1}{\bar{x}_1 \Psi[\mathbf{K}(N)]} = \frac{\lambda}{\rho} \frac{\Psi[\mathbf{UK}(N-1)]}{\Psi[\mathbf{K}(N)]}. \quad (4.1.8a)$$

The last form for $\Lambda(N)$ is interesting for comparing with (2.1.5b), but computationally, the first two forms are probably better. According to the discussion, and to (2.1.5d), one should be able to calculate the throughput based on the flow through any server. Therefore ($\bar{x}_2 = 1/\lambda$),

$$\Lambda(N) = \frac{1 - r(N; N)}{\bar{x}_2} = \lambda \frac{\Psi[\mathbf{K}(N)] - \Psi[\mathbf{U}^{N-1}\lambda \mathbf{V}]}{\Psi[\mathbf{K}(N)]}. \quad (4.1.8b)$$

Although the two equations for $\Lambda(N)$ must be equal, it takes some effort to prove that they are the same algebraically, which we leave for the following exercise.

Exercise 4.1.3: Prove by direct algebraic manipulation that

$$r(0; N) - \rho r(N; N) = 1 - \rho,$$

or equivalently,

$$(1-\rho)\Psi[\mathbf{K}(N)] = 1 - \rho \Psi[\mathbf{U}^{N-1}\lambda \mathbf{V}].$$

Exercise 4.1.4: Calculate $\Lambda(N)$ for the closed $M/E_2/1//N$ loop, where $\bar{x} + 1/\lambda = 1$ [i.e., $\lambda = 1 + \rho$, and $\bar{x} = \rho/(1+\rho)$] and $\rho = 0.5$ for $N = 1$ through 20. Repeat the calculations for $\rho = 0.9$, 1.0, and 2.0. Draw the four curves on the same graph and compare with the graphs in Exercise 2.1.3.

Exercise 4.1.5: Do the same as in Exercise 4.1.4, but now S_1 is the hyperexponential described in Exercise 4.1.2.

4.1.3. Departure and Arrival Queue-Length Probabilities

In the preceding section we derived the steady-state probabilities of what a random observer would see over a long period of time. There are two special sets of moments in time that deserve separate treatment. These time points are referred to as embedded Markov chains and are used to consider the following questions:

1. What will a customer see upon arriving at S_1?

2. What will a customer leave behind upon exiting S_1?

Since it takes no time for a customer to go from one server to another, these questions are the same as asking the equivalent questions of S_2. In the case of the M/G/1 queue, the two questions turn out to have the same answer, and almost the same as $r(n; N)$, but this is not the case for other systems. We prove the equality here and at the same time demonstrate the method that can be used in the other cases.

First we define a set of steady-state vectors.

Definition 4.1.3

$[\mathbf{w}(n; N)]_i$:= *probability that between events, n customers are at* S_1 *and phase i is busy (or, the system is in internal state,* $i \in \Xi$*).* As with $\boldsymbol{\pi}(0; N)$, $\mathbf{w}(0; N)$ is defined to be proportional to $\mathbf{p}$. ◊

Since nothing happens between events, we can argue that this is the same as the probability of being in state $\{i; n; N\}$ just before, or just after, an event. How then, you may ask, does this differ from $\pi_i(n; N)$? If the time interval between events was always the same (or is taken from some distribution that is independent of the state the system was in), the two would be identical (e.g., $\mathbf{M} = \mu\mathbf{I}$). However, a random observer is less likely to find the system in a state that is relatively short-lived than one that has a long mean time, $[(\lambda\mathbf{I} + \mathbf{M})^{-1}]$. On the other hand, internal and external transitions mark the moments of events and take no notice of the time between them, so you can say (in fact, we will say) that the π's are related to the $\mathbf{w}$'s by time-weighting (not waiting).

The $\mathbf{w}$'s satisfy the following balance equations, which are similar to those for the $\pi(n; N)$'s, except that the elapsed times between events are ignored. So we have the equivalent of a discrete Markov chain. For $n = 0$,

$$\mathbf{w}(0; N)\boldsymbol{\varepsilon}' = \sum_{i=1}^{m} w_i(1; N)\mu_i(\lambda + \mu_i)^{-1} q_i.$$

In words, if there are no customers at S_1, $[\mathbf{w}(0; N)\boldsymbol{\varepsilon}']$, then at the next event the system will certainly leave that state. On the other hand, the system can enter that state by first being in some internal state with one customer at S_1,

$[w_i(1;N)]$, have the next event occur in S_1, $[\mu_i/(\mu_i+\lambda)]$, and have that event be a departure, $[q_i]$. The probability that the next event will occur in S_1, namely $\mu_i/(\mu_i+\lambda)$, comes from Equation (2.1.1b). In vector form,

$$\mathbf{w}(0;N)=\mathbf{w}(1;N)(\lambda\mathbf{I}+\mathbf{M})^{-1}\mathbf{Mq}'\,\mathbf{p}\,.$$

The procedure should be sufficiently clear so that we can write the vector equations for $n>0$ directly. Note that if S_1 and S_2 each have at least one customer, the probability that an event, when it occurs, will be in S_2 is $\lambda[(\lambda\mathbf{I}+\mathbf{M})^{-1}]_{ii}$, while if either is empty, the event will certainly occur in the other. Therefore,

$$\mathbf{w}(n;N)=\mathbf{w}(n;N)(\lambda\mathbf{I}+\mathbf{M})^{-1}\mathbf{MP}+\mathbf{w}(n-1;N)(\lambda\mathbf{I}+\mathbf{M})^{-1}\lambda\mathbf{I}$$
$$+\,\mathbf{w}(n+1;N)(\lambda\mathbf{I}+\mathbf{M})^{-1}\mathbf{Mq}'\,\mathbf{p}\quad 0<n<N,$$

and

$$\mathbf{w}(N;N)=\mathbf{w}(N;N)\mathbf{P}+\mathbf{w}(N-1;N)(\lambda\mathbf{I}+\mathbf{M})^{-1}\lambda\mathbf{I}.$$

We next regroup terms, recall that $\mathbf{B}=\mathbf{M}(\mathbf{I}-\mathbf{P})$ and that $\mathbf{Mq}'=\mathbf{B}\boldsymbol{\varepsilon}'$, to get

$$\mathbf{w}(0;N)=\mathbf{w}(1;N)(\lambda\mathbf{I}+\mathbf{M})^{-1}\mathbf{BQ},\tag{4.1.9a}$$

$$\mathbf{w}(n;N)(\lambda\mathbf{I}+\mathbf{M})^{-1}(\lambda\mathbf{I}+\mathbf{B})$$
$$=\lambda\mathbf{w}(n-1;N)(\lambda\mathbf{I}+\mathbf{M})^{-1}+\mathbf{w}(n+1;N)(\lambda\mathbf{I}+\mathbf{M})^{-1}\mathbf{BQ},\tag{4.1.9b}$$

and

$$\mathbf{w}(N;N)\mathbf{M}^{-1}\mathbf{B}=\lambda\mathbf{w}(N-1;N)(\lambda\mathbf{I}+\mathbf{M})^{-1}.\tag{4.1.9c}$$

These equations look similar to (4.1.3). In fact, we can guess at their solution in

Theorem 4.1.3: The steady-state vector probabilities of finding n customers at S_1 and $N-n$ customers at S_2 between events are

$$\mathbf{w}(0;N)=\lambda C(N)\boldsymbol{\pi}(0;N),\tag{4.1.10a}$$

$$\mathbf{w}(n;N)=C(N)\boldsymbol{\pi}(n;N)(\lambda\mathbf{I}+\mathbf{M}),\tag{4.1.10b}$$

$$\mathbf{w}(N;N)=C(N)\boldsymbol{\pi}(N;N)\mathbf{M}.\tag{4.1.10c}$$

The π's were defined by (4.1.6), and $C(N)$ is the normalizing constant chosen so that the sum of the w's is 1,

$$\frac{1}{C(N)}=\lambda[1-r(N;N)]+\sum_{n=1}^{N}\boldsymbol{\pi}(n;N)\mathbf{M}\boldsymbol{\varepsilon}'.\tag{4.1.10d}$$

■

Proof: Substitute (4.1.10) into (4.1.9) and get (4.1.3). Proof of (4.1.10d) is left as an exercise.

QED

Notice that the $\mathbf{w}$'s are indeed related to the π's by the state-dependent time it takes for an event to occur, $[(\lambda\mathbf{I}+\mathbf{M})^{-1}]$.

Exercise 4.1.6: Prove that Equation (4.1.10d) is correct.

Having found expressions for the **w**'s, we are now prepared to define and then find the following vector probabilities.

Definition 4.1.4

$\mathbf{a}(n; N)$:= *probability vector that a customer arriving at S_1 will find n customers there already. The ith component is the probability that the active customer is at phase $i \in \Xi$. The associated scalar probability is $a(n; N) = \mathbf{a}(n; N)\boldsymbol{\varepsilon}'$.* Note that there are $n+1$ customers at S_1 *after* the arrival. ◊

Definition 4.1.5

$\mathbf{d}(n; N)$:= *probability vector, whose ith component is the probability that a customer departing S_1 will leave n customers behind, with the system in $i \in \Xi$ (immediately after the next customer enters). The associated scalar probability is $d(n; N) = \mathbf{d}(n; N)\boldsymbol{\varepsilon}'$.* Note that there were $n+1$ customers at S_1 *before* the departure. ◊

Of course, $a(n; N)$ is also the probability that the customer will leave $N-n-1$ customers behind at S_2, and $d(n; N)$ is the probability that the customer will find $N-n-1$ other customers already waiting or being served at S_2. We will look at our loop from this point of view in Chapter 5. It is not hard to see that $a(N; N) = d(N; N) = 0$, since the arriving or departing customer cannot count himself. The other probabilities can be evaluated using the following argument.

We know the steady-state vector probabilities, [$\mathbf{w}(n; N)$], of the system's state between events. There are two types of events. Either something happens in S_1, or something happens in S_2. The probability that the event will occur in S_1 is $(\lambda\mathbf{I} + \mathbf{M})^{-1}\mathbf{M}$ if n is not 0 or N, while it is 0 for $n = 0$, and 1 if $n = N$. If the event is in S_2, it will result in an arrival to S_1, while if the event is in S_1, one of two things can happen. Either the active customer will leave, [$\mathbf{q}'$], with another (if available) taking his place, [$\mathbf{p}$], or he will just go to another phase, [$\mathbf{P}$]. All together, there are six different kinds of terms, which we now list with their probabilities. In the following set of equations we use the notation "Pr[$X \rightarrow Y$]" to mean *the probability that the system will go to state Y at the next event, given that it is in state X at present.*

$$
\begin{aligned}
&(1)\ \Pr[\{\,\cdot\,; 0; N\} \to \{i; 1; N\}] = \mathbf{w}(0; N);\\
&(2)\ \Pr[\{i; n; N\} \to \{i; n+1; N\}] = \mathbf{w}(n; N)(\lambda\mathbf{I}+\mathbf{M})^{-1}\lambda\mathbf{I};\\
&(3)\ \Pr[\{j; n; N\} \to \{i; n; N\}] = \mathbf{w}(n; N)(\lambda\mathbf{I}+\mathbf{M})^{-1}\mathbf{MP};\\
&(4)\ \Pr[\{j; n; N\} \to \{i; n-1; N\}] = \mathbf{w}(n; N)(\lambda\mathbf{I}+\mathbf{M})^{-1}\mathbf{Mq'}\,\mathbf{p};\\
&(5)\ \Pr[\{j; N; N\} \to \{i; N; N\}] = \mathbf{w}(N; N)\mathbf{P};\\
&(6)\ \Pr[\{j; N; N\} \to \{i; N-1; N\}] = \mathbf{w}(N; N)\mathbf{q'}\,\mathbf{p}.
\end{aligned}
\tag{4.1.11}
$$

Of course, the sum of these terms is 1, and if rearranged would yield the balance equations we just used to get the **w**'s in the first place. We have enumerated them with a different purpose in mind. First, consider only those transactions that result in an arrival to S_1: namely, (1) and (2). Their sum is the probability of an arrival to S_1 irrespective of n. The sum of the two terms in (4.1.11), whose reciprocal we call $G(N)$, is [use Equations (4.1.10)]

$$
\begin{aligned}
\frac{1}{G(N)} &:= \mathbf{w}(0; N)\boldsymbol{\varepsilon}' + \lambda \sum_{n=1}^{N-1} \mathbf{w}(n; N)(\lambda\mathbf{I}+\mathbf{M})^{-1}\boldsymbol{\varepsilon}'\\
&= \lambda C(N)\left[r(0; N) + \sum_{n=1}^{N-1} r(n; N) \right].
\end{aligned}
$$

Since the sum of the r's must be 1, and the expression in brackets has all but one of them, we get the following:

$$\frac{1}{G(N)} = \lambda C(N)[1 - r(N; N)]. \tag{4.1.12a}$$

By the rule of conditional probabilities, [P(B | A) = P(B ∩ A)/P(A)], the marginal arrival probabilities are $G(N)$ times the appropriate terms above, so after some substitutions and cancellations, the following emerges. For $0 \le n < N$,

$$\mathbf{a}(n; N) = G(N)\mathbf{w}(n; N)(\lambda\mathbf{I}+\mathbf{M})^{-1} = \frac{1}{1-r(N; N)}\boldsymbol{\pi}(n; N)$$

and

$$a(n; N) = \frac{r(n; N)}{1-r(N; N)}.$$

A similar argument holds for $\mathbf{d}(n; N)$. Now, though, we must start in state $\{n+1; N\}$, so that the departing customer leaves n others behind; thus

$$\mathbf{d}(n; N) = G(N)\mathbf{w}(n+1; N)(\lambda\mathbf{I}+\mathbf{M})^{-1}\mathbf{BQ}.$$

As before, we get the probability of a departure from S_1 irrespective of n by adding the contributions from processes (4) and (6).

$$
\begin{aligned}
\frac{1}{G(N)} &= \sum_{n=1}^{N-1} \mathbf{w}(n; N)(\lambda\mathbf{I}+\mathbf{M})^{-1}\mathbf{Mq'}\mathbf{p}\boldsymbol{\varepsilon}' + \mathbf{w}(N; N)\mathbf{q'}\mathbf{p}\boldsymbol{\varepsilon}'\\
&= C(N)\sum_{n=1}^{N-1} \boldsymbol{\pi}(n; N)(\lambda\mathbf{I}+\mathbf{M})(\lambda\mathbf{I}+\mathbf{M})^{-1}\mathbf{B}\boldsymbol{\varepsilon}' + C(N)\boldsymbol{\pi}(N; N)\mathbf{B}\boldsymbol{\varepsilon}',
\end{aligned}
$$

where we have used $\mathbf{p}\boldsymbol{\varepsilon}' = 1$, and $\mathbf{Mq}' = \mathbf{B}\boldsymbol{\varepsilon}'$. Next, recall that $\mathbf{B}\boldsymbol{\varepsilon}' = \lambda\mathbf{A}\boldsymbol{\varepsilon}'$

(Lemma 4.1.1), use Theorem 4.1.2, and get

$$\frac{1}{G(N)} = \lambda C(N) r(0; N) \left(\sum_{n=1}^{N-1} \mathbf{p}\mathbf{U}^n \mathbf{A}\boldsymbol{\varepsilon}' + \mathbf{p}\mathbf{U}^{N-1}\mathbf{V}\mathbf{B}\boldsymbol{\varepsilon}' \right)$$

$$= \lambda C(N) r(0; N) \left(\sum_{n=1}^{N} \mathbf{p}\mathbf{U}^{n-1} \boldsymbol{\varepsilon}' \right),$$

which finally yields what we might have expected,

$$\frac{1}{G(N)} = \lambda C(N)[1 - r(N; N)], \tag{4.1.12b}$$

the same as we got for sum of the arrivals (which is just as well, since we used the same symbol for the two sums). The equality tells us that in the steady state, the overall probability of an arrival to a subsystem is equal to the probability of a departure from that subsystem. We would expect no less. The process of getting the $\mathbf{d}(n; N)$'s and $d(n; N)$'s is the same as that for evaluating $G(N)$, and gives the same results for $d(n; N)$ as for $a(n; N)$. These are summarized by the following theorem.

Theorem 4.1.4: The steady-state vector and scalar probabilities of finding n customers in an M/ME/1//N queue, $[\boldsymbol{\pi}(n; N), r(n; N)]$, of an arriving customer finding n already in the queue $[\mathbf{a}(n; N), a(n; N)]$, and a departing customer leaving n in the queue, $[\mathbf{d}(n; N), d(n; N)]$, are related by the following [from (4.1.7) and (4.1.12)]:

For $0 \le n < N$,

$$\mathbf{a}(n; N) = \frac{1}{1 - r(N; N)} \boldsymbol{\pi}(n; N), \tag{4.1.13a}$$

$$\mathbf{d}(n; N) = \frac{r(n; N)}{1 - r(N; N)} \mathbf{p}, \tag{4.1.13b}$$

$$a(n; N) = d(n; N) = \frac{r(n; N)}{1 - r(N; N)}, \tag{4.1.13c}$$

and finally,

$$a(N; N) = d(N; N) = 0.^{\dagger} \tag{4.1.13d}$$

By virtue of the completeness of ME functions, as described in Section 3.2.1, the scalar equations are true for classes of service time distributions more general than ME. Thus (4.1.13c) and (4.1.13d) are valid for all M/G/1//N queues. ■

You may be wondering what terms (3) and (5) from Equations (4.1.11) contribute to the behavior of an M/ME/1 queue, since no customers are exchanged between S_1 and S_2 during these events. Their role is to give S_1 its nonexponential character, as seen by an outside observer.

---

† Note that although $a(n; N)$ and $d(n; N)$ are equal, their vector counterparts are not. We will show how these results carry over to the open queue in succeeding sections.

4.2. OPEN M/ME/1 QUEUE

In Section 2.1.2 we showed how an M/M/1//N loop becomes an open M/M/1 queue when N becomes unboundedly large. In Section 3.3 we showed that a server with an unboundedly long queue generates a renewal process, and in particular, if the server is exponential, its departures are Poisson distributed. Recall from (3.1.4b) that since $\Psi[\mathbf{V}]$ is the mean service time for S_1 and λ is the mean service rate for S_2,

$$\rho := \textit{utilization factor} = \lambda\Psi[\mathbf{V}] = \lambda\bar{x}.$$

Therefore, if the mean service time of our general server, S_1, is less than the mean service time of S_2 ($\rho < 1$), the M/ME/1//N loop approaches an M/ME/1 open queue for very large N.

4.2.1. Steady-State M/ME/1 Queue

The open queue is described by the following parameters:

$$\mathbf{K} := \lim_{N\to\infty} \mathbf{K}(N), \tag{4.2.1a}$$

$$r(n) := \lim_{N\to\infty} r(n;N), \tag{4.2.1b}$$

$$\pi(n) := \lim_{N\to\infty} \pi(n;N). \tag{4.2.1c}$$

Before going on, we need the following lemma, which will be of use to us in later chapters as well. Its proof is evident by direct substitution.

Lemma 4.2.1: Let $\mathbf{F}$ be any m-dimensional square matrix for which $\Psi[\mathbf{F}] \neq 1$. Then (recall that $\mathbf{Q} = \boldsymbol{\varepsilon}'\mathbf{p}$, and thus $\mathbf{QFQ} = \Psi[\mathbf{F}]\mathbf{Q}$), $(\mathbf{I} - \mathbf{QF})$ is nonsingular and

$$(\mathbf{I} - \mathbf{QF})^{-1} = \mathbf{I} + \frac{1}{1-\Psi[\mathbf{F}]}\mathbf{QF}. \tag{4.2.2a}$$

Similarly,

$$(\mathbf{I} - \mathbf{FQ})^{-1} = \mathbf{I} + \frac{1}{1-\Psi[\mathbf{F}]}\mathbf{FQ}. \tag{4.2.2b}$$

This lemma is valid for any two vectors $\mathbf{x}$ and $\mathbf{y}$ for which $\mathbf{xy}' = 1$ and the scalar $\mathbf{xFy}' \neq 1$, but is used in this book only for $\mathbf{x} = \mathbf{p}$ and $\mathbf{y}' = \boldsymbol{\varepsilon}'$. ■

Assuming that the limit exists, $\mathbf{K}$ from (4.2.1a) can actually be evaluated using (4.1.6e), for we have

$$\mathbf{K} = \lim_{N\to\infty} \mathbf{K}(N+1) = \lim_{N\to\infty} [\mathbf{I} + \mathbf{UK}(N)] = \mathbf{I} + \mathbf{UK},$$

implying that

$$\mathbf{K} = (\mathbf{I} - \mathbf{U})^{-1}. \tag{4.2.3a}$$

The fact that $\mathbf{I} - \mathbf{U}$ is invertible is central to the development of this section and is demonstrated below. From (4.1.4a) we have $\mathbf{A} = \mathbf{I} + (1/\lambda)\mathbf{B} - \mathbf{Q}$ and $\mathbf{U} = \mathbf{A}^{-1}$. Therefore it follows that

$$\mathbf{I} - \mathbf{U} = \mathbf{U}(\mathbf{A} - \mathbf{I}) = \mathbf{U}\frac{1}{\lambda}(\mathbf{B} - \lambda\mathbf{Q}) = \frac{1}{\lambda}\mathbf{U}\mathbf{B}(\mathbf{I} - \lambda\mathbf{V}\mathbf{Q}).$$

We next take the inverse of both sides to get

$$\mathbf{K} = (\mathbf{I} - \mathbf{U})^{-1} = \lambda(\mathbf{I} - \lambda\mathbf{V}\mathbf{Q})^{-1}\mathbf{V}\mathbf{A}.$$

Finally, using the last two equations and (4.2.2b) with $\mathbf{F} = \lambda\mathbf{V}$, $\mathbf{K}$ becomes

$$\mathbf{K} = \lambda\left[\mathbf{I} + \frac{\lambda}{1-\rho}\mathbf{V}\mathbf{Q}\right]\mathbf{V}\mathbf{A} = \lambda\mathbf{A}\mathbf{V}\left[\mathbf{I} + \frac{\lambda}{1-\rho}\mathbf{Q}\mathbf{V}\right]. \tag{4.2.3b}$$

Note that **A**, **V**, and **Q** do not commute with each other, so the order in which these matrices appear is important. This equation for **K** is explicit in terms of known quantities, and therefore exists as long as $\rho \neq 1$, so we have proven that $\mathbf{I} - \mathbf{U}$ has an inverse. This also proves that if $\rho = 1$, then **U** has an eigenvalue equal to 1, and conversely.

An expression that will prove useful in Chapter 5 follows from (4.2.3b) by multiplying it on the right with $\boldsymbol{\varepsilon}'$ to get

$$\lambda\mathbf{A}\mathbf{V}\boldsymbol{\varepsilon}' = (1 - \rho)\mathbf{K}\boldsymbol{\varepsilon}', \tag{4.2.3c}$$

or by multiplying it on the left with **p** to get

$$\lambda\mathbf{p}\mathbf{V}\mathbf{A} = (1 - \rho)\mathbf{p}\mathbf{K}. \tag{4.2.3d}$$

An equally useful form for **K** can be found by substituting for **A** in (4.2.3b) to get

$$\mathbf{K} = \mathbf{I} + \lambda\mathbf{V} + \frac{\lambda^2}{1-\rho}\mathbf{V}\mathbf{Q}\mathbf{V}. \tag{4.2.3e}$$

It follows directly that

$$\mathbf{p}\mathbf{K} = \mathbf{p}\left[\mathbf{I} + \frac{\lambda}{1-\rho}\mathbf{V}\right] \tag{4.2.3f}$$

and

$$\mathbf{K}\boldsymbol{\varepsilon}' = \left[\mathbf{I} + \frac{\lambda}{1-\rho}\mathbf{V}\right]\boldsymbol{\varepsilon}'. \tag{4.2.3g}$$

The last step to the solution is to take $\Psi[\mathbf{K}]$ and use (4.1.6g), yielding

$$[r(0)]^{-1} = \Psi[\mathbf{K}] = 1 + \rho + \frac{\rho^2}{1-\rho} = \frac{1}{1-\rho}. \tag{4.2.3h}$$

Equation (4.2.1), Theorem 4.1.2, and this last result together lead to

Theorem 4.2.2: The steady-state vector and scalar probabilities of finding n customers in an M/ME/1 queue are

$$\pi(n) = (1 - \rho)\mathbf{p}\mathbf{U}^n \tag{4.2.4a}$$

and

$$r(n) = (1 - \rho)\Psi[\mathbf{U}^n]. \tag{4.2.4b}$$

$1 - r(0) = \rho$ is the probability that S_1 is busy. ■

We mention at this point that $\mathbf{K}$, as the limit of $\mathbf{K}(N)$, exists whenever all the eigenvalues of $\mathbf{U}$ are less than 1 in magnitude. We assume without proof that this occurs whenever $\rho < 1$. In any case, Equations (4.2.3) are valid as long as $\mathbf{U}$ has no unit eigenvalues, and thus when $\rho \neq 1$. In Chapter 5 we look at this problem more closely, and in Chapter 7 we will see that for more complicated queues, $\mathbf{U}$ does have a unit eigenvalue (in fact, 1 is a multiple eigenvalue). Having said this, it is easy to argue that $r(N; N)$ [from (4.1.7b)] goes to 0 as N goes to infinity whenever ρ is less than 1, (i.e., S_2 is always busy). We can then extend Theorem 4.1.4 to the open M/G/1 queue, expressed by the following:

Theorem 4.2.3: Let $\mathbf{a}(n)$ and $\mathbf{d}(n)$ be the open M/G/1 queue equivalents to $\mathbf{a}(n; N)$ and $\mathbf{d}(n; N)$ $(\rho < 1)$. Then

$$\mathbf{a}(n) = \lim_{N\to\infty} \mathbf{a}(n; N) \quad \text{and} \quad \mathbf{d}(n) = \lim_{N\to\infty} \mathbf{d}(n; N).$$

Therefore,

$$a(n) = d(n) = r(n)$$

and the vectors,

$$\mathbf{a}(n) = \pi(n).$$

However, $\mathbf{d}(n) = r(n)\mathbf{p} \neq \pi(n)$. ■

This well-known result is discussed, for instance, in [COOP81]. Cooper refers to our random observer as the *outside observer*. Note that $a(n)$, $d(n)$, and $r(n)$ are *not* equal to each other for the open G/M/1 queue, for then $\rho > 1$, so $\lim_{N\to\infty} r(N; N) > 0$. This is discussed fully in Chapter 5.

4.2.2. System Times: Pollaczek–Khinchin Formulas

The prototypical question asked in relation to queueing theory is: How long can a customer expect to wait for service from a busy server? The M/G/1 queue provides an unusually simple answer. This result is amazingly simple, considering that all attempts to find similar answers for somewhat more complicated systems have failed in the 60 or more years since Pollaczek [POLL30] and Khinchin

[KHIN32] separately found that the mean queue length and mean system time for the steady-state open M/G/1 queue depend only on ρ and the first and second moments of S_1's pdf. Even the closed M/G/1//N loop does not share the simple result. These formulas are derived here.

Mean Queue Length

The mean queue length of a general server with Poisson arrivals can be calculated directly from (4.2.4):

$$\bar{q} := \sum_{n=1}^{\infty} n\, r(n) = (1-\rho)\Psi\left[\sum_{n=1}^{\infty} n\mathbf{U}^n\right].$$

We know that

$$\sum_{n=1}^{\infty} n\mathbf{U}^n = (\mathbf{I}-\mathbf{U})^{-1}(\mathbf{I}-\mathbf{U})^{-1}\mathbf{U} = \mathbf{K}\,\mathbf{K}\,\mathbf{U},$$

and also that $\mathbf{KU} = \mathbf{UK} = \mathbf{K} - \mathbf{I}$, so (4.2.3f) and (4.2.3g) can be used to reduce the mean queue-length formula to

$$\bar{q} = (1-\rho)\mathbf{pK}(\mathbf{K}-\mathbf{I})\boldsymbol{\varepsilon}' = (1-\rho)\Psi\left[\left(\mathbf{I} + \frac{\lambda}{1-\rho}\mathbf{V}\right)\left(\frac{\lambda}{1-\rho}\mathbf{V}\right)\right]$$

$$= \lambda\Psi\left[\mathbf{V} + \frac{\lambda}{1-\rho}\mathbf{V}^2\right]. \tag{4.2.5a}$$

But $\Psi[\lambda\mathbf{V}] = \rho$, and $\Psi[\mathbf{V}^2] = \overline{x^2}/2$, [from (3.1.9)], so

$$\bar{q} = \rho + \frac{\lambda^2}{1-\rho}\frac{\mathrm{E}(x^2)}{2}. \tag{4.2.5b}$$

Another form for the P-K formula, which is perhaps more enlightening, can be written by recalling the definition of variance $[\sigma^2 = \mathrm{E}(x^2) - \bar{x}^2]$ and the coefficient of variation $[C^2 = \sigma^2/\bar{x}^2]$. Then

$$\bar{q} = \frac{\rho}{1-\rho} + \frac{\rho^2}{1-\rho}\frac{C^2-1}{2}. \tag{4.2.5c}$$

The mean time a customer spends in S_1 is given by Little's law [Equation (2.1.7)], namely,

$$\bar{T} = \frac{\bar{q}}{\lambda} = \frac{\bar{x}}{1-\rho} + \frac{\bar{x}\rho}{1-\rho}\frac{C^2-1}{2}. \tag{4.2.5d}$$

In this form, for a given ρ and $\bar{x}$, it is clear that if C^2 is greater than 1, the mean queue length and the mean time in the subsystem will be longer than that for an M/M/1 queue (for which $C^2 = 1$), while if C^2 is less than 1, $\bar{q}$ and $\bar{T}$ will be shorter. For a given ρ, the shortest queue length and system time occur for the deterministic distribution, where $C^2 = 0$, but there is no longest mean queue length, since one can always find a distribution whose coefficient of variation exceeds any number. Some examples were shown in Figure 1.1.2. We give one cautionary reminder that (4.2.5d) is true only for the *steady-state* M/G/1 *open queue*.

> **Exercise 4.2.1:** Evaluate the mean system times, $\bar{T}$, for queueing systems with mean service time of 1, and with the following values of the coefficient of variation: $C^2 = 0$, 0.25, 0.50, 1.0, 2.0, 5.0, and 10.0. Use enough values of ρ between 0 and 1 to draw curves for all of them that appear visually smooth. Make sure that they all have the same value at $\rho = 0$, namely, $\bar{x}$, which in this case equals 1.

We have developed the mathematical properties of **K** sufficiently to be able to find the limit of $\Lambda(N)$ in (4.1.8a) without any effort. We also point out in passing, that although **U** and **K**(N) do not commute, **U** and **K** do. When ρ is less than 1, we have [from (4.1.8a), (4.2.3a) and (4.2.3h)]

$$\lim_{N\to\infty} \Lambda(N) = \frac{\lambda}{\rho}\frac{\Psi[\mathbf{UK}]}{\Psi[\mathbf{K}]} = \frac{\lambda}{\rho}\frac{\Psi[\mathbf{K}-\mathbf{I}]}{\Psi[\mathbf{K}]} = \frac{\lambda}{\rho}(1-\rho)\left(\frac{1}{1-\rho} - 1\right) = \lambda.$$

So, as in the M/M/1//N queue [Equation (2.1.6c)], the throughput of the system is limited by the capacity of the slower server, S_2. When ρ is greater than 1, the problem is more difficult, so we will wait until Chapter 5 to deal with it.

Z-Transform

Pollaczek and Khinchin also derived an expression for the *z–transform*, also known as the *generating function*, of the set of queue-length probabilities. We will first derive the vector z-transform and then reproduce the P-K formula. First define

$$\mathbf{q}(z) := \sum_{n=0}^{\infty} \pi(n) z^n = (1-\rho)\mathbf{p} \sum_{n=0}^{\infty} z^n \mathbf{U}^n = (1-\rho)\mathbf{p}(\mathbf{I} - z\mathbf{U})^{-1}. \quad (4.2.6a)$$

This expression can be manipulated into a form that makes use of the properties of **V**, as given in (3.1.9) and (3.1.10). First note that

$$(\mathbf{I} - z\mathbf{U}) = \frac{1}{\lambda}\mathbf{UB}[\mathbf{I} + \lambda(1-z)\mathbf{V} - \lambda\mathbf{VQ}].$$

Next, let $s = \lambda(1-z)$, and define

$$\mathbf{D}(s) := [\mathbf{I} + s\mathbf{V}]^{-1}. \quad (4.2.6b)$$

This matrix shows up often, and from (3.1.10), is related to the Laplace transform by the following:

$$d(s) := \Psi[\mathbf{D}(s)] = B^*[\lambda(1-z)]. \tag{4.2.6c}$$

We manipulate the equation before (4.2.6b) to a form in which Lemma 4.1.2 can be applied, so it follows that

$$\mathbf{I} - z\mathbf{U} = \frac{1}{\lambda}\mathbf{UBD}^{-1}(\mathbf{I} - \lambda\mathbf{DVQ}).$$

Before going on, note that $\lambda\,\mathbf{pDVQ} = \Psi[\lambda\mathbf{DV}]\mathbf{p}$, and from (4.2.6b) that $\lambda\mathbf{DV} = (\mathbf{I} - \mathbf{D})/(1-z)$. We must now take the inverse of $(\mathbf{I} - z\mathbf{U})$ to get

$$(\mathbf{I} - z\mathbf{U})^{-1} = (\mathbf{I} - \lambda\mathbf{DVQ})^{-1}\lambda\mathbf{DVA}.$$

Then as long as $\Psi[\lambda\mathbf{DV}] \neq 1$, Lemma 4.2.1 applies, so (4.2.6a) yields

$$\mathbf{q}(z) = (1-\rho)\mathbf{p}\left(\mathbf{I} + \frac{1}{1-\Psi[\lambda\mathbf{DV}]}\lambda\mathbf{DVQ}\right)\lambda\mathbf{DVA}$$

$$= (1-\rho)\left(1 + \frac{\Psi[\lambda\mathbf{DV}]}{1-\Psi[\lambda\mathbf{DV}]}\right)\lambda\mathbf{pDVA}.$$

By virtue of the fact that $\Psi[\lambda\mathbf{DV}] = (1-d)/(1-z)$, we finally get (after simplifying the expression in the large parentheses)

$$\mathbf{q}(z) = \frac{1-\rho}{1-\Psi[\lambda\mathbf{DV}]}\lambda\mathbf{pDVA} = \frac{(1-\rho)(1-z)}{d-z}\lambda\mathbf{pDVA}. \tag{4.2.6d}$$

This vector z-transform contains the information concerning the internal states of S_1. The sum of its components correspond to the P-K transform formula. Using Lemma 4.1.1, it easily follows that

$$Q(z) := \mathbf{q}(z)\boldsymbol{\varepsilon}' = \frac{(1-\rho)(1-z)}{d(s)-z}d(s) = \frac{(1-\rho)(1-z)B^*(s)}{B^*(s)-z} \tag{4.2.6e}$$

[where again, $s = \lambda(1-z)$]. The rightmost term is the expression normally referred to as the P-K formula. It is not easy to use, since it is indeterminate at $z = 1$, for then $s = 0$ and $B^*(0) = 1$. In fact, (4.2.6a) is surely the easiest form to use for evaluation. From this equation,

$$Q(z) = (1-\rho)\Psi[(\mathbf{I} - z\mathbf{U})^{-1}] \tag{4.2.6f}$$

(the matrix equivalent of the z-transform of the geometric distribution), and from (4.2.3a) and (4.2.3h) it is obvious that $Q(1) = 1$. This must necessarily be true, since by the definitions (4.2.6a) and (4.2.6e) $Q(1)$ is the sum of the queuelength probabilities, which must be 1. The usefulness of the z-transform comes from the ability to get the mean queue length and higher moments without evaluating an infinite sum. Since in our case, the infinite sums are geometric in form and are evaluatable, the potential advantage is questionable†. However, we shall do that here. It is well known that the derivative of $Q(z)$ evaluated at $z = 1$ is $\bar{q}$, while the variance of the queue length is $\sigma_q^2 = Q''(1) - \bar{q}(\bar{q} - 1)$. Now from (4.2.6f)

$$\bar{q} = \left(\frac{dQ(z)}{dz}\right)_{z=1} = (1-\rho)\Psi[(\mathbf{I}-\mathbf{U})^{-2}\mathbf{U}] = (1-\rho)\Psi[\mathbf{K}\mathbf{K}\mathbf{U}],$$

the same expression that led to the derivation of (4.2.5a). It is also straightforward to find the expression for Q'', which is

$$\left(\frac{d^2Q(z)}{dz^2}\right)_{z=1} = 2(1-\rho)\Psi[\mathbf{K}\mathbf{K}^2\mathbf{U}^2] = 2(1-\rho)\Psi[\mathbf{K}(\mathbf{K}-\mathbf{I})^2]$$

$$= 2(1-\rho)\Psi[\mathbf{K}^3 - 2\mathbf{K}^2 + \mathbf{K}].$$

The last two terms in the rightmost Ψ brackets are simple enough to evaluate, while the first one can be evaluated by doing the following:

$$\Psi[\mathbf{K}^3] = \mathbf{p}\mathbf{K}\mathbf{K}\mathbf{K}\boldsymbol{\varepsilon}'$$

$$= \mathbf{p}\left[\mathbf{I} + \frac{\lambda}{1-\rho}\mathbf{V}\right]\left[\mathbf{I} + \lambda\mathbf{V} + \frac{\lambda^2}{1-\rho}\mathbf{V}\mathbf{Q}\mathbf{V}\right]\left[\mathbf{I} + \frac{\lambda}{1-\rho}\mathbf{V}\right]\boldsymbol{\varepsilon}'.$$

It is straightforward, if a bit tedious, to multiply out all terms and regroup them to get

$$Q''(1) = \frac{2}{1-\rho}\Psi[(\lambda\mathbf{V})^2] + \frac{2}{1-\rho}\Psi[(\lambda\mathbf{V})^3] + 2\left(\frac{\Psi[(\lambda\mathbf{V})^2]}{1-\rho}\right)^2.$$

Further manipulation yields

$$\sigma_q^2 = \rho(1-\rho) + \frac{3-2\rho}{1-\rho}\Psi[(\lambda\mathbf{V})^2] + \frac{2}{1-\rho}\Psi[(\lambda\mathbf{V})^3] + \left(\frac{\Psi[(\lambda\mathbf{V})^2]}{1-\rho}\right)^2.$$

We next make use of the fact that $\Psi[(\lambda\mathbf{V})^2] = \rho^2 + \rho^2(C^2-1)/2$; then

$$\sigma_q^2 = \frac{\rho(1-2\rho^2+2\rho^3)}{(1-\rho)^2} + \frac{\rho^2(3-5\rho+3\rho^2)}{(1-\rho)^2}\left(\frac{C^2-1}{2}\right)$$

$$+ \frac{\rho^4}{(1-\rho)^2}\left(\frac{C^2-1}{2}\right)^2 + \frac{2}{1-\rho}\Psi[(\lambda\mathbf{V})^3].$$

† We should point out, however, that since $Q(z) = \sum_{n=0}^{\infty} r(n) z^n$ and $r(n) \geq 0$ for all n, it follows that $Q(z) \geq 0$ for $z \geq 0$. Then (4.2.6f) might be used in the future to find those properties $\mathbf{U}$ must have to guarantee this.

For exponential servers, $C^2 = 1$, and $\Psi[(\lambda \mathbf{V})^n] = \rho^n$, so

$$\sigma_q^2 = \frac{\rho}{(1-\rho)^2}.$$

For the deterministic distribution, $C^2 = 0$, and $\Psi[(\lambda \mathbf{V})^n] = \rho^n / n!$, so

$$\sigma_q^2 = \frac{\rho(12-18\rho+10\rho^2-\rho^3)}{12(1-\rho)^2}.$$

We have seen that (4.2.6f) is easy enough to use, although a bit tedious, to get the moments of the queue length. Use of (4.2.6e) is considerably harder and more tedious to use. In either case, even the second moment is not particularly informative for general analysis, so we leave it for now. However, in the following section we will surprisingly find a better use of (4.2.6f).

4.2.3. System Time Distribution

The P-K transform formulas (4.2.6) turn out to have more significance than that implied in the preceding section. Following standard texts, we will now show that $Q(z)$ is also the Laplace transform, $B_s^*(s)$, of the system time pdf, $b_s(x)$, where $s = \lambda(1-z)$. Then we will go even further (thanks to Appie van de Liefvoort, who first recognized it [LIEF90]) and find the matrix generator, $<\mathbf{p_s}, \mathbf{B_s}>$, of the distribution time itself.

Recall the definition of system time (or total, or response time) from the end of Section 2.1.3, and define the steady-state distribution.

Definition 4.2.1

$B_s(x) :=$ *probability that a customer will leave* S_1 *by time* x *after entering its queue*. That is, $B_s(x)$ is the PDF for system time, $b_s(x)$ is its derivative, and $R_s(x) = 1 - B_s(x)$ is the probability that the customer will still be in the subsystem at time x. ◊

From Corollary 4.2.2 we know that the steady-state probability of finding n customers at S_1, $[r(n)]$, is the same as the probability that a departing customer will leave n customers behind, $[d(n)]$. Now, since the arrival process to S_1 is Poisson, the probability that n customers will arrive in the time interval, x (the time spent there by our now-departing customer), is given by (3.3.16), so the probability that he will leave n customers behind, irrespective of how long he was at S_1, is

$$d(n) = r(n) = \int_0^\infty \frac{(x\lambda)^n}{n!} e^{-x\lambda} b_s(x)\,dx.$$

Next, insert this into the expression for $Q(z)$ [Equations (4.2.6a) and (4.2.6e)], to get

$$Q(z) = \sum_{n=0}^{\infty} z^n r(n) = \sum_{n=0}^{\infty} \int_0^{\infty} \frac{(\lambda xz)^n}{n!} e^{-x\lambda} b_s(x)\, dx$$

$$= \int_0^{\infty} \sum_{n=0}^{\infty} \frac{(\lambda xz)^n}{n!} e^{-x\lambda} b_s(x)\, dx = \int_0^{\infty} e^{\lambda xz} e^{-x\lambda} b_s(x)\, dx.$$

Finally, we identify the Laplace transform in the following theorem.

Theorem 4.2.4: The Laplace transform for the steady-state system-time distribution in an M/G/1 queue is given by

$$B_s^*[\lambda(1-z)] = Q(z) = \int_0^{\infty} e^{-\lambda(1-z)x} b_s(x)\, dx, \tag{4.2.7a}$$

while from (4.2.6f), we have for M/ME/1 queues,

$$B_s^*(s) = (1-\rho)\Psi[(\mathbf{I} - z\mathbf{U})^{-1}], \tag{4.2.7b}$$

where $s = \lambda(1-z)$. ■

This is a most interesting result, but remember that this simple expression occurred for two special reasons. First, $d(n)$ and $r(n)$ are equal, and second, the Poisson arrival process and the Laplace transform are both generated by the exponential function. We cannot expect such simple results for the G/G/1 queue. In attempting an alternative derivation using the arrival probabilities (which also satisfy Corollary 4.2.2) we get a result that so far has not been shown equal to (4.2.7b). We will postpone this derivation until the end of the next section, after we have discussed residual times.

Equation (4.2.7b) can be used to find a vector-matrix pair that generates the moments of $B_s(x)$, and thus, by (3.1.7c), (3.1.8b), (3.1.9) and (3.1.10), the same pair will be a faithful representation of $b_s(x)$ itself, as well as $B_s^*(s)$. First recall that the Laplace transform is also known as the moment generating function, in that its nth derivative evaluated at $s = 0$ is $(-1)^n$ times the nth moment. That is,

$$\overline{x_s^n} = (-1)^n \left(\frac{d^{(n)} B_s^*(s)}{ds^n} \right)_{s=0} = \int_0^{\infty} x^n b_s(x)\, dx.$$

Next, since $s = \lambda(1-z)$, note that

$$\frac{d}{ds} = \frac{dz}{ds}\frac{d}{dz} = -\frac{1}{\lambda}\frac{d}{dz},$$

so (clearly, $z = 1$ when $s = 0$), using (4.2.7b), and recalling that $\mathbf{K} = (\mathbf{I} - \mathbf{U})^{-1}$ [see Equations (4.2.2)], we have

$$\left(\frac{dB_s^*(s)}{ds} \right)_{s=0} = -\frac{1}{\lambda}(1-\rho)(-1)\Psi[(\mathbf{I} - z\mathbf{U})^{-2}(-\mathbf{U})]_{z=1}$$

$$= -\frac{1}{\lambda}(1-\rho)\Psi\left[(\mathbf{I}-\mathbf{U})^{-2}\mathbf{U}\right] = -(1-\rho)\Psi\left[\mathbf{K}\,\frac{\mathbf{KU}}{\lambda}\right].$$

Clearly, the nth differentiation with respect to s introduces two minus signs that cancel, an additional factor of n/λ, and another power of $-\mathbf{UK}$ inside the Ψ brackets. Thus in general we get

$$\left(\frac{d^{(n)}B_s^*(s)}{ds^n}\right)_{s=0} = (-1)^n(1-\rho)n!\,\Psi\left[\mathbf{K}\left(\frac{\mathbf{KU}}{\lambda}\right)^n\right]$$

$$= (-1)^n n!\,[(1-\rho)\mathbf{pK}]\left(\frac{\mathbf{KU}}{\lambda}\right)^n \boldsymbol{\varepsilon}'.$$

Now define

$$\mathbf{p_S} := (1-\rho)\mathbf{pK} \tag{4.2.8a}$$

[which from (4.2.3d) can also be written as $\mathbf{p_S} := \lambda\mathbf{pVA}$] and

$$\mathbf{V_S} := \frac{1}{\lambda}\mathbf{KU}. \tag{4.2.8b}$$

Then the equations preceding (4.2.8a) lead to the familiar-looking expression

$$\overline{x_S^n} = n!\,\Psi_S\left[\mathbf{V_S}^n\right] := n!\,\mathbf{p_S}\mathbf{V_S}^n\,\boldsymbol{\varepsilon}'. \tag{4.2.9a}$$

The resemblance of this equation with (3.1.9) is not superficial. Since $\mathbf{p_S}\boldsymbol{\varepsilon}' = 1$ from (4.2.2g), we can say that $<\mathbf{p_S}, \mathbf{V_S}>$ is a matrix representation of, or generates, the waiting time distribution. By virtue of Theorem 3.1.1 we have

Theorem 4.2.5: Let $\mathbf{p_S}$ and $\mathbf{V_S}$ and $\Psi_S[\cdot]$ be defined by (4.2.8a), (4.2.8b), and (4.2.9a) respectively, then (where $\mathbf{B_S} = \mathbf{V_S}^{-1}$)

$$\overline{x_S^n} = n!\,\Psi_S\left[\mathbf{V_S}^n\right],$$

$$b_S(x) = \Psi_S\left[\mathbf{B_S}\exp(-x\mathbf{B_S})\right], \tag{4.2.9b}$$

$$B_S^*(s) = \Psi_S\left[(\mathbf{I}+s\mathbf{V_S})^{-1}\right]. \tag{4.2.9c}$$

Thus the vector-matrix pair, $<\mathbf{p_S}, \mathbf{B_S}>$ generates a faithful representation of the distribution of system times in a steady-state M/ME/1 open queue. ■

It should be clear from these discussions that the mean system time, $\overline{x_S}$ is the same as $\overline{T}$ of (4.2.5b).

We now find explicitly simple forms for $\mathbf{B_S}$ and $\mathbf{V_S}$. From (4.2.9b),

$$\mathbf{B_S} = \mathbf{V_S}^{-1} = \lambda\mathbf{A}(\mathbf{I}-\mathbf{U}) = \lambda(\mathbf{A}-\mathbf{I}) = \mathbf{B} - \lambda\mathbf{Q}. \tag{4.2.10a}$$

Also, from (4.2.2d), and noting once again that $\mathbf{KU} = \mathbf{K} - \mathbf{I}$,

$$\mathbf{V_s} = \frac{1}{\lambda}(\mathbf{K} - \mathbf{I}) = \mathbf{V} + \frac{\lambda}{1-\rho}\mathbf{VQV}. \tag{4.2.10b}$$

One can also solve for $\mathbf{K}$ in terms of $\mathbf{V_s}$ to get

$$\mathbf{K} = (\mathbf{I} + \lambda\mathbf{V_s}). \tag{4.2.10c}$$

This, by the way, together with (4.2.9c), shows that $B_s^*(\lambda) = 1 - \rho$.

Note that (4.2.3f) yields an expression for $\mathbf{p_s}$ that has a clear physical meaning. Using (3.3.11a), and $\rho = \lambda\Psi[\mathbf{V}]$, we get

$$\mathbf{p_s} = (1-\rho)\mathbf{p} + \lambda\mathbf{pV} = (1-\rho)\mathbf{p} + \rho\boldsymbol{\pi}_\mathbf{r}. \tag{4.2.10d}$$

$(1-\rho)$ is the probability that S_1 will be empty when a customer arrives, ρ is the probability that it will not be empty, and $(\boldsymbol{\pi}_\mathbf{r})_i$ from (3.3.11a) is the probability that phase i will be busy upon the customer's arrival, given that at least one customer is already there. Therefore, $(\mathbf{p_s})_i$ is the probability that phase i will be busy immediately after an arrival, irrespective of S_1's condition before the arrival. It would be nice to find such a simple interpretation of $\mathbf{B_s}$.

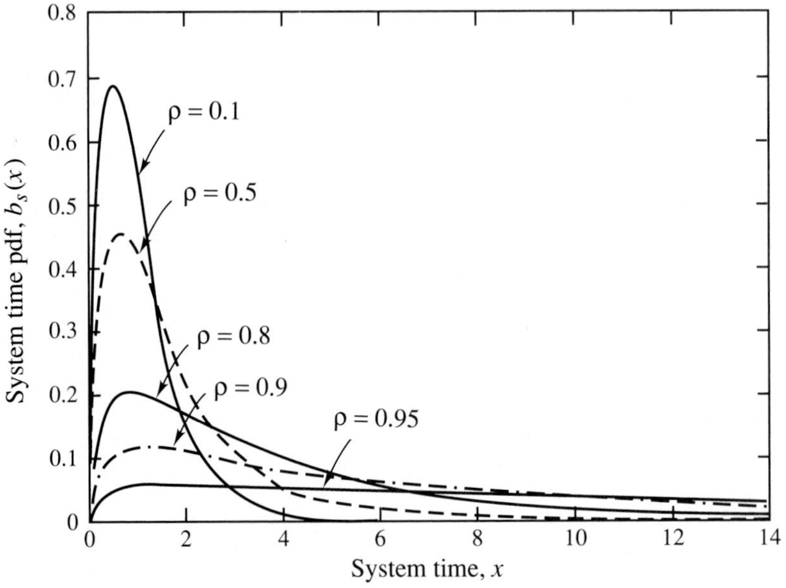

Figure 4.2.1: System time density function, $b_s(x)$ for the M/E_2/1 queue, with $\rho = 0.1$, 0.5, 0.8, 0.9, and 0.95. For small ρ, $b_s(x)$ tends to look like the service-time density function, $4x\exp(-2x)$, while for ρ close to 1, it looks very much like the interarrival time density function, $\lambda\exp(-\lambda x)$, except near $x = 0$.

Example 4.2.1: We have used $<\mathbf{p_s}, \mathbf{B_s}>$ from (4.2.8) to generate $b_s(x)$ for the open M/E_2/1 queue ($N = \infty$) by directly evaluating (4.2.9b) for many values of x, using the algorithm described in Section 3.1.4. We set $\bar{x} = 1$, and selected various values for ρ. The results are shown in Figure 4.2.1. When ρ is very small, then $b_s(x)$ is very peaked, just as is $E_2(x)$. In fact, the curve labeled $\rho = 0.1$ is extremely close to the Erlangian-2 distribution,

$$E_2(x) = 4xe^{-2x},$$

which peaks at $x = 0.5$. When ρ is close to 1, $b_s(x)$ looks more like the interarrival distribution, $\lambda \exp(-\lambda x)$. The curve labeled $\rho = 0.95$ does not seem to support this. Bear in mind, however, that in general, $b_s(0) = (1-\rho)b(0)$, which in this case is 0, while the exponential has a value of λ at the origin. Note that the curve rises rapidly from 0 and then gently decays close to the exponential curve. Another interesting feature of this figure is that all the curves peak at approximately the same place ($x = 0.5$). It seems that for small x, $b_s(x)$ retains the shape of $b(x)$ for all ρ. ●

It should be interesting to study $b_s(x)$ further, using other pdf's. The reader must not be too quick to generalize from what you learn from the exponential and Erlangian-2 distributions.

Exercise 4.2.2: Using the definitions given by Equations (4.2.8) and $s = \lambda(1-z)$, manipulate (4.2.7b) directly to get (4.2.9c). Also, show that $b_s(0) = (1-\rho)b(0)$. Furthermore, prove by direct algebraic manipulation that $\overline{x_s} = \underline{T}$, i.e., show that (4.2.9a) for $n=1$ and (4.2.5d) yield the same result. [Hint: Use (4.2.10b) and (4.2.10d) in (4.2.9a).]

4.2.4. Distribution of Interdeparture Times

We have developed enough results to be able to look once again at departures from S_1. As in Section 2.1.5, we place our observer just outside the exit of S_1 and have her measure the time between departures. The problem is more complicated only because S_1 now represents some general server. It is useful, then, to review Section 2.1.5 before going on.

We ask the following question: Given that a customer, call him c_1, has just left S_1, how long will it be before the next one, call him c_2, leaves? We can assume that our observer has been sitting for a long time, so the system is in its steady state. Also, she has no idea how many customers are at S_1, but if the system is closed, she knows what N is.

Definition 4.2.2

$B_d(x;N)$:= *probability distribution function for the interdeparture times of a steady-state* M/ME/1//N *loop. The process begins immediately after customer* c_1 *leaves* S_1*, and ends as soon as customer* c_2 *leaves.* Customer c_2 may not yet have arrived at S_1 when c_1 left. (In that case, c_1 left behind an empty queue.) $b_d(x; N)$ is the derivative of $B_d(x; N)$, and $R_d(x; N) = 1 - B_d(x; N)$ is the probability that the second customer is still in the subsystem or has not yet arrived at time x. The subscript d reminds us that this is a $d$eparture process. ◊

Only two things are possible: Either S_1 is busy and c_1 must be served from the beginning, or S_1 is idle and our patient observer must wait for c_1 to arrive before being completely served. The probability that the latter will happen is $d(0; N)$ [Equation (4.1.11b)], while the vector probability for the former to happen is $\mathbf{p} - \mathbf{d}(0; N)$. Following this description, the pdf for the process can be found by taking the convolution of the pdf's of S_1 and S_2, but instead, we will give a matrix representation of the process that is more useful and more picturesque.

Look at Figure 4.2.2. Consider S_1 and S_2 together as one subsystem. Since S_2 is only an exponential server with service rate λ, we can assume that service begins there at the moment of the previous departure with probability $d(0; N)$.

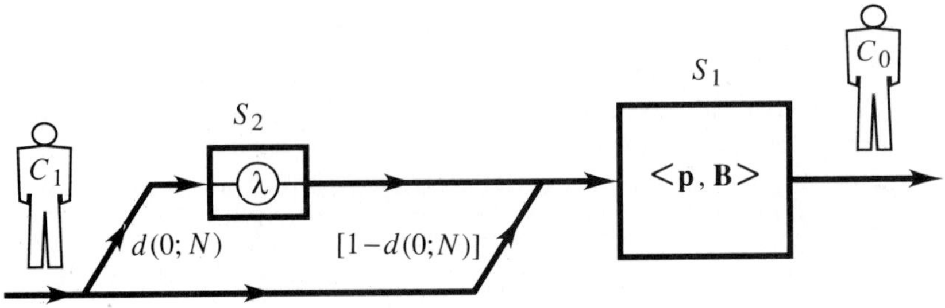

Figure 4.2.2: Pictorial representation of the departure process from S_1, in an M/G/1//N loop. Dependence on the number of customers is implicitly given through the steady-state probabilities at departure times. Given that customer c_0 has just left, c_1 must first enter, [**p**], and travel through S_1 before leaving, [**B**], or if S_1 is empty, c_1 must finish being served by S_2 and then go to S_1 to be served. The probability that no one is at S_1 at the moment of a departure, $d(0; N)$, is given by (4.1.11b).

The dimension of this composite subsystem is $m+1$, corresponding to the possibility that c_1 can either be at one of m phases in S_1 or at the one phase in S_2. This is a *sum-space representation* [the two subspaces, of dimension 1 and m, are concatenated together to produce one $(m+1)$-dimensional space]. In Chapter 7, we will be forced to use a *product space representation* to describe the status of customers at two general subsystems. In this representation, the first component refers to the one phase in S_2 and the next m components (which we replace by an m-vector) refer to the m phases in S_1. The initial vector for the composite subsystem is given by

$$\mathbf{p_d}(N) := [\, d(0; N)\,,\, \{\, 1 - d(0; N)\,\}\, \mathbf{p}\,]. \tag{4.2.11a}$$

If the queue at S_1 is not empty the moment after the departure, then c_1 enters according to $\mathbf{p}$.

The transition matrix, $\mathbf{P_d}$, is easy enough to write down once we recognize that a customer goes from 0 to i with probability p_i, and goes from $i > 0$ to j with probability P_{ij}, where $\mathbf{p}$ and $\mathbf{P}$ are the same objects we used previously to represent S_1. Therefore,

$$\mathbf{P_d} = \begin{bmatrix} 0 & \mathbf{p} \\ \mathbf{o}' & \mathbf{P} \end{bmatrix} \quad \text{and} \quad \mathbf{M_d} = \begin{bmatrix} \lambda & \mathbf{o} \\ \mathbf{o}' & \mathbf{M} \end{bmatrix}.$$

These formulas are to be interpreted in the following way. 0 is a 1×1 matrix filling the element (1, 1). $\mathbf{p}$ is a $1 \times m$ matrix filling elements (1, 2) to $(1, m+1)$. $\mathbf{o}'$ is an $m \times 1$ matrix of 0's, filling elements (2, 1) to $(m+1, 1)$. Finally, $\mathbf{P}$ is an $m \times m$ matrix filling the rest of $\mathbf{P_d}$. We follow the discussion and procedure described in Sections 3.1.2 and 3.1.3 to get the process-rate and process-time matrices, $\mathbf{B_d}$ and $\mathbf{V_d}$. Let $\mathbf{I_d}$ be the identity matrix of dimension $m+1$; then

$$\mathbf{B_d} = \mathbf{M_d}(\mathbf{I_d} - \mathbf{P_d}) = \begin{bmatrix} \lambda & -\lambda\mathbf{p} \\ \mathbf{o}' & \mathbf{B} \end{bmatrix}. \tag{4.2.11b}$$

One can easily prove by direct matrix multiplication that its inverse is

$$\mathbf{V_d} = \mathbf{B_d}^{-1} = \begin{bmatrix} \frac{1}{\lambda} & \mathbf{pV} \\ \mathbf{o}' & \mathbf{V} \end{bmatrix}. \tag{4.2.11c}$$

Now that we have a matrix representation of the departure time distribution, generated by $\langle \mathbf{p_d}, \mathbf{B_d} \rangle$ (or $\langle \mathbf{p_d}, \mathbf{V_d} \rangle$), we can find its moments, and even the pdf itself. First, let us find the mean interdeparture time when there are N customers in the loop. From Theorem 3.1.1, Equation (3.1.9),

$$\bar{x}_d(N) := \int_0^\infty x\, b_d(x; N)\, dx = \mathbf{p_d}(N)[\mathbf{V_d}]\boldsymbol{\varepsilon}'_\mathbf{d}$$

$$= [\, d(0; N)\ ,\ \{1 - d(0; N)\}\,\mathbf{p}\,] \begin{bmatrix} \frac{1}{\lambda} & \mathbf{pV} \\ \mathbf{o}' & \mathbf{V} \end{bmatrix} \boldsymbol{\varepsilon}'_\mathbf{d}$$

$$= \left[\frac{d(0; N)}{\lambda}\ ,\ d(0; N)\mathbf{pV} + \{1 - d(0; N)\}\,\mathbf{pV} \right] \boldsymbol{\varepsilon}'_\mathbf{d}$$

$$= \left[\frac{d(0; N)}{\lambda}\ ,\ \mathbf{pV} \right] \boldsymbol{\varepsilon}'_\mathbf{d}.$$

Now, since $\rho = \lambda\bar{x}$, the mean time reduces to the following simple expression:

$$\bar{x}_d(N) = \frac{1}{\lambda}[d(0; N) + \rho]. \tag{4.2.12a}$$

$d(0; N)$ can be calculated from (4.1.11c), at the same time that the other properties of the steady-state M/G/1//N queue are computed, which as usual, we leave as an exercise. An interesting aspect of this representation is that the departure time's dependence on N appears only in $\mathbf{p_d}(N)$.

Before finding the equation for the pdf, we find the mean interdeparture time for the open system. We already know from (4.2.4b) and Theorem 4.2.3 that $\lim_{N\to\infty} d(0; N) = 1-\rho$ as long as $\rho < 1$, so

$$\bar{x}_d(\rho<1) := \lim_{N\to\infty} \bar{x}_d(N) = \frac{1}{\lambda}(1-\rho+\rho) = \frac{1}{\lambda}. \tag{4.2.12b}$$

Actually, (4.2.12a) is valid for all ρ. If ρ is greater than 1, then $d(0; N)$ goes to 0 as N grows larger, so in this case,

$$\bar{x}_d(\rho>1) := \lim_{N\to\infty} \bar{x}_d(N) = \frac{1}{\lambda}(0+\rho) = \bar{x}. \tag{4.2.12c}$$

Surprised? Of course not. After all, $\bar{x}_d$ is the reciprocal of the mean departure rate, and as long as ρ is less than 1, what goes in must come out, so the arrival rate equals the departure rate (in the steady state, of course). We already saw this for the M/M/1 queue in (2.1.18a). Note that the departure and arrival rates are equal to each other for all N, but they only equal λ for the open queue. In any closed network, even the busiest server will be idle some of the time, so the throughput will be less than maximum in proportion to the time it is not busy.

We can find the second moment in a similar fashion. First observe that

$$\mathbf{V_d}\,\boldsymbol{\varepsilon}'_\mathbf{d} = \begin{bmatrix} \dfrac{1}{\lambda} + \bar{x} \\ \mathbf{V}\boldsymbol{\varepsilon}' \end{bmatrix} = \frac{1}{\lambda}\begin{bmatrix} 1+\rho \\ \lambda\mathbf{V}\boldsymbol{\varepsilon}' \end{bmatrix}.$$

Then, making use of the fact that $\mathbf{p_d}(N)\mathbf{V_d}^2\,\boldsymbol{\varepsilon}'_\mathbf{d} = [\mathbf{p_d}(N)\mathbf{V_d}][\mathbf{V_d}\,\boldsymbol{\varepsilon}'_\mathbf{d}]$, and using the expression preceding (4.2.12a), we get

$$\mathbf{p_d}(N)\mathbf{V_d}^2\,\boldsymbol{\varepsilon}'_\mathbf{d} = \frac{1}{\lambda^2}[\, d(0; N) \quad , \; \lambda\mathbf{pV}\,]\begin{bmatrix} 1+\rho \\ \lambda\mathbf{V}\boldsymbol{\varepsilon}' \end{bmatrix}$$

$$= \frac{1}{\lambda^2}\left(d(0; N)(1+\rho) + \lambda^2\Psi[\,\mathbf{V}^2\,] \right).$$

Next recall that $E(x^2) = 2\Psi[\,\mathbf{V}^2\,]$. The equivalent formula must be true for the departure process, so

$$E_d(x^2; N) = \frac{1}{\lambda^2}\left(2d(0; N)(1+\rho) + \lambda^2 E(x^2) \right).$$

The variance is easy to get now.

$$\sigma_d^2(N) = E_d(x^2; N) - [\bar{x}_d(N)]^2$$

$$= \frac{1}{\lambda^2}\left(2d(0; N)(1+\rho) + \lambda^2[E(x^2) - \bar{x}^2 + \bar{x}^2] - [d(0; N)+\rho]^2 \right).$$

Further trivial manipulation yields the next expression, where σ^2 is the variance for S_1.

$$\sigma_d^2(N) = \frac{1}{\lambda^2}\left(1 - [1-d(0;N)]^2 + \lambda^2\sigma^2\right). \tag{4.2.13a}$$

The open system limit is straightforward, since $d(0;N)$ approaches $1-\rho$ as N goes to infinity. So

$$\sigma_d^2 := \lim_{N\to\infty} \sigma_d^2(N) = \frac{1}{\lambda^2}\left(1 - \rho^2 + \lambda^2\sigma^2\right). \tag{4.2.13b}$$

Recall that the coefficient of variation for any process is defined to be the ratio of variance and mean squared. Thus since $\bar{x}_d = 1/\lambda$ from (4.2.12b),

$$C_d^2 := 1 - \rho^2 + \rho^2 C^2 = 1 + \rho^2(C^2 - 1). \tag{4.2.13c}$$

In this form we can see that for all $\rho < 1$, C_d^2 is less (greater) than 1 whenever C^2 is less (greater) than 1. This expression can be manipulated into the following form:

$$C_d^2 = C^2 - (1-\rho^2)(C^2 - 1), \tag{4.2.13d}$$

which implies that if C^2 is greater (less) than 1, C_d^2 is less (greater) than C^2. Both sets of inequalities can be summarized by the single statement: *For all* $\rho < 1$, C_d^2 *lies between* C^2 *and* 1. The coefficient of variation for the departure process is some sort of average of the coefficients of variation for the interarrival distribution ($C^2 = 1$ for exponential distributions) and the service distribution, C^2.

We are almost ready to find the density function itself. Recall from Theorem 3.1.1 that since $\langle \mathbf{p_d}(N), \mathbf{B_d} \rangle$ generates $b_d(x;N)$, they are related by (3.1.7d), or

$$b_d(x;N) = \mathbf{p_d}(N)[\mathbf{B_d}\exp(-x\mathbf{B_d})]\boldsymbol{\varepsilon}'_\mathbf{d}. \tag{4.2.14}$$

We can make use of this formula by either finding a similarity transformation matrix that diagonalizes $\mathbf{B_d}$, or by replacing $\exp(\cdot)$ with its Taylor expansion and substituting a general expression for $\mathbf{B_d}^n$ (assuming that we can find one). We shall do the latter here. First, from (4.2.11b) let us look at the square of $\mathbf{B_d}$:

$$\mathbf{B_d}^2 = \begin{bmatrix} \lambda^2 & -\lambda^2\mathbf{p}\left(\mathbf{I} + \dfrac{1}{\lambda}\mathbf{B}\right) \\ \mathbf{o}' & \mathbf{B}^2 \end{bmatrix}.$$

If the reader cannot guess at a general expression for the nth power of $\mathbf{B_d}$, then calculating and examining $\mathbf{B_d}^3$ should give sufficient hint. We will leave that step out and write the expression directly. Before we do that, we are beginning to see that the matrix expressions can become rather large and cumbersome, so for convenience, we define the matrix, **X**, for this section only.

$$\mathbf{X} := \left(\mathbf{I} - \frac{1}{\lambda}\mathbf{B}\right)^{-1}.$$

Then the nth power of $\mathbf{B_d}$ is

$$\mathbf{B_d}^n = \begin{bmatrix} \lambda^n & -\lambda^n \mathbf{p} \sum_{k=0}^{n-1} \left(\frac{1}{\lambda}\mathbf{B}\right)^k \\ \mathbf{o}' & \mathbf{B}^n \end{bmatrix} = \begin{bmatrix} \lambda^n & -\lambda^n \mathbf{pX} \left(\mathbf{I} - \left(\frac{1}{\lambda}\mathbf{B}\right)^n\right) \\ \mathbf{o}' & \mathbf{B}^n \end{bmatrix}. \quad (4.2.15a)$$

The proof is by induction and is left as an exercise.

> **Exercise 4.2.3:** Prove by induction that (4.2.15a) is true for all $n > 0$. That is, multiply either of the two matrix expressions by $\mathbf{B_d}$ and show that the resulting expression is of the same form, with the index n increased by 1.

The process of summing all the terms of the form $1/n!(-x\mathbf{B_d})^n$ is not difficult, since it can be done element by element, or block by block. First, define the $(m+1) \times (m+1)$ matrix, $\mathbf{R_d}(x) := \exp(-x\mathbf{B_d})$ [recall the reliability matrix function of Equations (3.1.6)], then

$$[\mathbf{R_d}(x)]_{ij} = [\exp(-x\mathbf{B_d})]_{ij} = \sum_{n=0}^{\infty} \frac{(-x)^n}{n!} \left[(\mathbf{B_d})^n\right]_{ij}.$$

For instance,

$$[\mathbf{R_d}(x)]_{11} = \sum_{n=0}^{\infty} \frac{(-x)^n}{n!} \lambda^n = e^{-x\lambda}$$

and

$$[\mathbf{R_d}(x)]_{j1} = 0, \qquad \text{for } 1 < j \le m+1.$$

The elements $(1, 2)$ to $(1, m+1)$ are best treated as a block, call it $\mathbf{g}$. Then

$$\mathbf{g} = -\sum_{n=0}^{\infty} \frac{(-x\lambda)^n}{n!} \mathbf{pX} \left(\mathbf{I} - \left(\frac{1}{\lambda}\mathbf{B}\right)^n\right)$$

$$= \mathbf{pX} \sum_{n=0}^{\infty} \left(\frac{(-x\mathbf{B})^n}{n!} - \frac{(-x\lambda)^n}{n!}\mathbf{I}\right) = \mathbf{pX}\left(\exp(-x\mathbf{B}) - e^{-x\lambda}\mathbf{I}\right).$$

The block of all elements for which both i and j are greater than 1 is $\exp(-x\mathbf{B})$. We put these all together in the following expression:

$$\mathbf{R_d}(x) = \begin{bmatrix} e^{-x\lambda} & \mathbf{pX}\left(\exp(-x\mathbf{B}) - e^{-x\lambda}\mathbf{I}\right) \\ \mathbf{o}' & \exp(-x\mathbf{B}) \end{bmatrix}. \tag{4.2.15b}$$

We next calculate $\mathbf{B_d}\exp(-x\mathbf{B_d})$ as the last step before evaluating (4.2.14). This is not particularly hard to do, and comes out to be

$$\mathbf{B_d}\mathbf{R_d}(x) = \begin{bmatrix} \lambda e^{-x\lambda} & \mathbf{pX}\left(\mathbf{B}\exp(-x\mathbf{B}) - \lambda e^{-x\lambda}\mathbf{I}\right) \\ \mathbf{o}' & \mathbf{B}\exp(-x\mathbf{B}) \end{bmatrix}.$$

We multiply on the right with $\varepsilon'_\mathbf{d}$ to get the following column vector:

$$\mathbf{B_d}\mathbf{R_d}(x)\varepsilon'_\mathbf{d} = \begin{bmatrix} \lambda e^{-x\lambda} + \Psi[\mathbf{XB}\exp(-x\mathbf{B})] - \Psi[\mathbf{X}]\lambda e^{-x\lambda} \\ \mathbf{B}\exp(-x\mathbf{B})\varepsilon' \end{bmatrix}.$$

Notice that up to now, N does not appear at all, so this expression is good for all N, even in the limit. We are now ready to evaluate (4.2.14) using (4.2.11a).

$$\begin{aligned} b_d(x;N) &= \mathbf{p_d}(N)[\mathbf{B_d}\mathbf{R_d}(x)]\varepsilon'_\mathbf{d} \\ &= d(0;N)\left(\lambda e^{-x\lambda} + \Psi[\mathbf{XB}\exp(-x\mathbf{B})] - \Psi[\mathbf{X}]\lambda e^{-x\lambda}\right) \\ &\quad + [1-d(0;N)]\Psi[\mathbf{B}\exp(-x\mathbf{B})] \\ &= b(x) + d(0;N)\left(\Psi[\mathbf{I}-\mathbf{X}]\lambda e^{-x\lambda} + \Psi[(\mathbf{X}-\mathbf{I})\mathbf{B}\exp(-x\mathbf{B})]\right). \end{aligned}$$

But from the definition of $\mathbf{X}$,

$$\begin{aligned} \mathbf{I}-\mathbf{X} = \mathbf{I} - \left(\mathbf{I}-\frac{1}{\lambda}\mathbf{B}\right)^{-1} &= \left(\mathbf{I}-\frac{1}{\lambda}\mathbf{B}\right)^{-1}\left(\mathbf{I}-\frac{1}{\lambda}\mathbf{B}-\mathbf{I}\right) \\ = \left(\mathbf{I}-\frac{1}{\lambda}\mathbf{B}\right)^{-1}\left(-\frac{1}{\lambda}\mathbf{B}\right) &= -\left(\lambda\mathbf{V}\left(\mathbf{I}-\frac{1}{\lambda}\mathbf{B}\right)\right)^{-1} = (\mathbf{I}-\lambda\mathbf{V})^{-1}. \end{aligned}$$

Therefore, recalling that $d(0;N)$ is given by Theorem 4.1.4,

$$\begin{aligned} b_d(x;N) &= b(x) + d(0;N) \\ &\times\left(\Psi[(\mathbf{I}-\lambda\mathbf{V})^{-1}]\lambda e^{-x\lambda} - \Psi[(\mathbf{I}-\lambda\mathbf{V})^{-1}\mathbf{B}\exp(-x\mathbf{B})]\right). \end{aligned} \tag{4.2.16a}$$

In particular, for $x = 0$,

$$b_d(0;N) = [1-d(0;N)]b(0). \tag{4.2.16b}$$

This formula is as simple as it can get in terms of its dependence on the customer population, so there is no real gain in writing down the limit as N goes to infinity. We point out, though, that [as with the mean interdeparture time (4.2.12)] when ρ is less than 1, $d(0; N)$ is replaced by $1-\rho$, but that does not simplify (4.2.16) any, except when $x = 0$, for then $b_d(0) = \rho b(0)$. If ρ is greater than 1, then $d(0; N)$ goes to 0 for large N, so $b_d(x) = b(x)$, as expected. Also, note that since $b(x)$ and $b_d(x)$ are both density functions, the integral from 0 to infinity of each function is 1. Therefore, the integral of the term multiplying $d(0; N)$ must be 0. In other words, the two terms inside the large parentheses contribute opposing changes to $b(x)$ which exactly cancel out upon integration. This can be shown directly by first recognizing that

$$\int_0^\infty \mathbf{B} \exp(-x\mathbf{B})dx = \mathbf{I}.$$

There is one other limit that is interesting. Under very light loads (i.e., when ρ is very small), $\lambda\mathbf{V}$ is also very small. In this case, $(\mathbf{I} - \lambda\mathbf{V})$ drops out, $d(0; N)$ can be replaced by 1, and we end up with the reasonable result that $b_d(x; N) \to \lambda e^{-x\lambda}$. We see, then, that as ρ increases from 0 to 1, the interdeparture distribution *gradually* changes from the arrival distribution to the service distribution. "Exponential in $\to$ exponential out (EIEO)" is valid only under light loads.

Example 4.2.2: We have used $<\mathbf{p_d}, \mathbf{B_d}>$ from (4.2.11) to generate $b_d(x)$ for the open M/E_2/1 queue ($N = \infty$) by directly evaluating (3.1.7d) for many values of x, using the algorithm described in Section 3.1.4. Just as with Figure 4.2.1, we set $\bar{x} = 1$ and selected various values for ρ. The results are shown in Figure 4.2.3. This figure looks similar to Figure 4.2.1; however, note that their dependence on ρ is completely inverted relative to each other, although they are extremely close for $\rho = 0.5$. When ρ is very close to 1, $b_d(x)$ is very peaked, just as is $E_2(x)$. In fact the curve labeled $\rho = 0.95$ is virtually indistinguishable from the same Erlangian-2 distribution, given above, which peaks at $x = 0.5$. When ρ is very small, $b_d(x)$ will look more like the interarrival distribution, $\lambda\exp(-\lambda x)$. The curve labeled $\rho = 0.1$ does not seem to support this. Bear in mind, analogous to the system-time distribution, that in general, $b_d(0) = \rho b(0)$, which in this case is 0, while the exponential has a value of λ at the origin. Note that the curve rises rapidly from 0 and then gently decays close to the exponential curve. Another interesting feature that this figure shares with Figure 4.2.1 is that all the curves peak at approximately the same place ($x = 0.5$). Again, it seems that for small x, $b_d(x)$ retains the shape of $b(x)$ for all ρ. ●

It would be interesting to find out if this "peaking" property is of typical interdeparture distributions for all M/G/1 queues.

We have one more point to make before moving on. Why should EIEO be true for an open system *even* if S_1 is exponential, as was proven in Chapter 2, Equation (2.1.18b) (it was *not* true for the closed system)? After all, our representation of the departure process has dimensions equal to the sum of the dimensions of S_1 and S_2, which in the case of the M/M/1 queue should be 2. Of

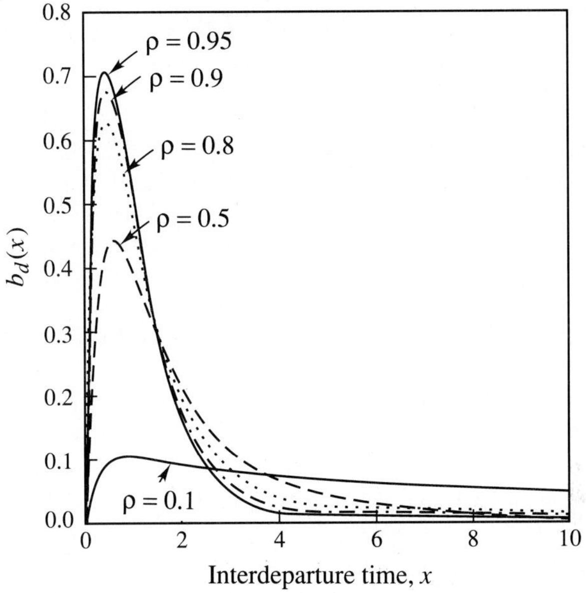

Figure 4.2.3: Interdeparture time density function, $b_d(x)$ for the $M/E_2/1$ queue, with $\rho = 0.1$, 0.5, 0.8, 0.9, and 0.95. For small ρ, except near $x = 0$, $b_d(x)$ tends to look like the interarrival distribution, $\lambda \exp(-\lambda x)$, while for ρ close to 1, it looks very much like the service-time density function, $4x\exp(-2x)$.

course, we would expect (4.2.16) to duplicate (2.1.18b) for the open system, which it does if S_1 is 1-dimensional. In that case, **B** goes to μ, $\lambda\mathbf{V}$ goes to ρ, and b(x) becomes $\mu e^{-x\mu}$. Put this all together with the fact that $d(0; N)$ goes to $1-\rho$ and the negative term in the large brackets of (4.2.16b) exactly cancels $b(x)$, leaving $b_d(x) = \lambda e^{-x\lambda}$. But this argument does not give us much insight. Another view is to look at the matrix representation of b_d. Note that the initial vector for the open system, $[\, 1-\rho\, ,\, \rho\,]$, is a left eigenvector of $\mathbf{B_d}$, with eigenvalue λ. That is,

$$[\, 1-\rho\ ,\ \rho\,]\begin{bmatrix} \lambda & -\lambda \\ 0 & \mu \end{bmatrix} = \lambda[\, 1-\rho\, ,\, \rho\,].$$

We discussed minimal representations in Section 3.2.4, where we showed by example that the dimension of the invariant subspaces of **p** and ε' determine the dimension of the minimal representation. In this case, since the equation above is true, from Theorem 3.1.1, we have for the M/M/1 queue,

$$b_d(x) = \mathbf{p_d}\mathbf{B_d}\exp(-x\mathbf{B_d})\varepsilon'_\mathbf{d} = \lambda\mathbf{p_d}\mathbf{I_d}\exp(-x\lambda\mathbf{I_d})\varepsilon'_\mathbf{d} = \lambda e^{-x\lambda}\mathbf{p_d}\varepsilon'_\mathbf{d} = \lambda e^{-x\lambda}.$$

Whenever either ε' or the entrance vector is an eigenvector of the generating matrix, **B** or **V**, the resulting pdf is exponential.

4.3. DEPENDENCE OF SYSTEM TIME ON n

In Chapter 3 we discussed the idea of residual times, where what can be predicted about the future is contained in what is known about the system now, and is summarized by the residual vector [Equations (3.3.10) to (3.3.13)]. In particular, if nothing is known about the internal state of S_1 (except that it is busy), the mean time until a customer leaves is given by (3.3.12b), with pdf given by (3.3.13). We can extend this to the M/ME/1//N loop and the M/ME/1 queue in the following way.

4.3.1. Residual Time as Seen by a Random Observer

Suppose that a random observer comes to view S_1 without knowing anything about its past history except that n customers are there at present. The probability that she will find n customers there is $r(n; N)$, but we can actually give an expression for the internal state of S_1 at the moment she arrives.

Definition 4.3.1

$\boldsymbol{\pi}_\mathbf{r}(n; N)$:= *residual probability vector of the state* S_1 *is in when a random observer first arrives, given that there are* n *customers in a steady-state* M/ME/1//N *queue.* $\boldsymbol{\pi}_\mathbf{r}(n; N)\boldsymbol{\varepsilon}' = 1$. $[\boldsymbol{\pi}_\mathbf{r}(n; N)]_i$ is the probability that the customer in service in S_1 will be at phase i when the observer comes. There is no internal state if $n = 0$, but for convenience we let $\boldsymbol{\pi}_\mathbf{r}(0; N) = \mathbf{p}$. ◊

From (4.1.6a) and (4.1.6b), we have

$$\boldsymbol{\pi}_\mathbf{r}(n; N) = \frac{\boldsymbol{\pi}(n; N)}{r(n; N)} = \frac{\mathbf{p}\mathbf{U}^n}{\Psi[\mathbf{U}^n]} \qquad \text{for } 0 \le n < N \tag{4.3.1a}$$

and

$$\boldsymbol{\pi}_\mathbf{r}(N; N) = \frac{\mathbf{p}\mathbf{U}^{N-1}\mathbf{V}}{\Psi[\mathbf{U}^{N-1}\mathbf{V}]}. \tag{4.3.1b}$$

These vectors serve as the initial vectors for the process of the active customer completing service. Thus

$$<\boldsymbol{\pi}_\mathbf{r}(n; N), \mathbf{B}>$$

is the generator of the distribution function of the time remaining for the one in service, given that the random observer has found n customers at S_1. For instance, the density function for this process is given by the expression

$$b_r(x; n; N) := \boldsymbol{\pi}_\mathbf{r}(n; N)\mathbf{B}\exp(-x\mathbf{B})\boldsymbol{\varepsilon}' = \frac{\Psi[\mathbf{U}^n\mathbf{B}\exp(-x\mathbf{B})]}{\Psi[\mathbf{U}^n]} \tag{4.3.2a}$$

for $0 < n < N$, and

$$b_r(x; N; N) := \frac{\Psi[\mathbf{U}^{N-1}\mathbf{V}\mathbf{B}\exp(-x\mathbf{B})]}{\Psi[\mathbf{U}^{N-1}\mathbf{V}]} = \frac{\Psi[\mathbf{U}^{N-1}\exp(-x\mathbf{B})]}{\Psi[\mathbf{U}^{N-1}\mathbf{V}]}. \quad (4.3.2b)$$

These formulas are not as hard to compute as they look. First, the vectors $\mathbf{pU}^n$ can be calculated recursively, and in any case are needed to compute the steady-state probabilities, while $\exp(-x\mathbf{B})\boldsymbol{\varepsilon}'$ can be calculated recursively by the algorithm given in Section 3.1.4.

The mean time remaining for the one in service is

$$\overline{x_r}(n; N) = \frac{\Psi[\mathbf{U}^n\mathbf{V}]}{\Psi[\mathbf{U}^n]} \qquad \text{for } 0 \le n < N \quad (4.3.3a)$$

and

$$\overline{x_r}(N; N) = \frac{\Psi[\mathbf{U}^{N-1}\mathbf{V}^2]}{\Psi[\mathbf{U}^{N-1}\mathbf{V}]}. \quad (4.3.3b)$$

We have thus let $\overline{x_r}(0; N) = \Psi[\mathbf{V}] = \bar{x}$.

The mean residual time [from (3.3.12b)] is of interest because it can differ enormously from the mean service time. It is not hard to find examples where the queue-length dependent residual times differ as much from $\overline{x_r}$ and each other as $\overline{x_r}$ differs from $\bar{x}$.

Example 4.3.1: As can be seen in Figure 4.3.1, for even the simplest nonexponential distribution (the Erlangian-2), $\overline{x_r}(n; N)$ can vary greatly. The value at $n = 0$ corresponds to the mean service time, which we have set equal to 1. The average value of the queue-length times, weighted over the $r(n; N)$'s, we will show below, turns out to be equal to the mean residual time, $\overline{x_r}$. Therefore, the weighted average is independent of ρ and N. The big drop in all curves between 19 and 20 is real. Note that *all* of these numbers would be equal to each other and to $\bar{x}$ if this were an M/M/1 queue. ●

The procedure we have applied to $\overline{x_r}(n; N)$ and $\overline{b_r}(x; n; N)$ can also be applied to the Laplace transform. Thus,

$$B_r^*(s; n; N) := \boldsymbol{\pi}_\mathbf{r}(n; N)[\mathbf{I} + s\mathbf{V}]^{-1}\boldsymbol{\varepsilon}'. \quad (4.3.3c)$$

This function actually has an interesting physical meaning. Since $B^*(\lambda)$ is the probability that a customer who has just started service will finish before the next customer comes, [see discussion after Theorem 3.1.1, and (2.1.1b)], $B_r^*(\lambda; n; N)$ must be the probability that the customer in service at S_1 when a random observer starts looking (and sees n customers there), will finish service before the next customer arrives. Let's state this more precisely. Let $X_r(n;N)$

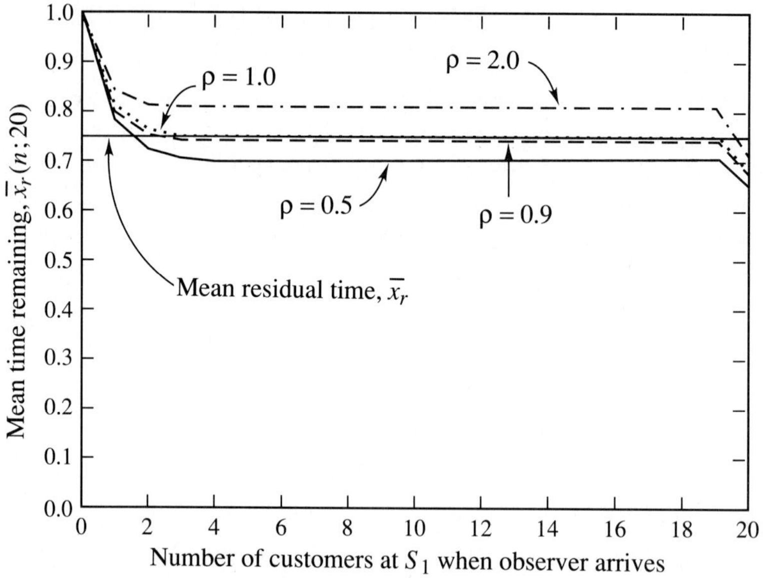

Figure 4.3.1: Mean time remaining for the customer in service, as seen by a random observer, as a function of the number of customers at S_1 when she starts observing. The service distribution is an E_2 function, with a mean service time of $\bar{x} = 1.0$. Thus we have an $M/E_2/1//20$ loop. The curves for four different values of ρ are presented. In all cases, the expected time until completion for a customer who just began service is 1.0. If the random observer takes no notice of the length of the queue, the mean time until completion is $\overline{x_r} = 0.75$. If she *does* note the length of the queue, the mean time to completion is given by $\overline{x_r}(n;N)$.

be the random variable representing the time remaining when our observer appears, and let X_2 be the time until the next customer arrives. Then

$$C_r(n;N) := \Pr[X_r(n;N) < X_2] = B_r^*(\lambda; n; N).$$

Exercise 4.3.1: Let S_1 be an Erlangian-2 server with mean service time equal to 1. Calculate $b_r(0;n;N)$ $(0 < n \le N)$, for the four systems with ρ = 0.5, 0.9, 1.0, and 2.0. Plot all your answers on the same graph [$b_r(0;n;N)$ versus n]. Also draw as a horizontal line $b_r(0) = \mathbf{pVB\epsilon'}/\Psi[\mathbf{V}] = 1/\bar{x}$, which in this case is equal to 1. Note that $b(0) = 0$.

> **Exercise 4.3.2:** Let S_1 be the H_2 server of Exercise 4.1.2. Do the same as in Exercise 4.2.3. In this case, the mean residual time is 1090/361 = 3.019391 (over three times greater than the mean service time of S_1).

As can be seen from (4.3.1a), (4.3.2a), and (4.3.3a), or should be obvious from the last two exercises, the queue-length-dependent behavior does not depend on the total number of customers in the system, as long as they are not all at S_1. We next show that the "average" residual time, given that S_1 was busy when observation began, is none other than the mean residual time, again independent of N. In fact, the average density function is the same as the mean residual distribution given in (3.3.13). We do this by examining the "average" of any matrix operator. First define for any square matrix, **F**,

$$\overline{F_r}(n;N) := \boldsymbol{\pi}_{\mathbf{r}}(n;N)\mathbf{F}\boldsymbol{\varepsilon}' = \frac{\Psi[\mathbf{U}^n\mathbf{F}]}{\Psi[\mathbf{U}^n]} \quad \text{for} \quad 0 \le n < N$$

and

$$\overline{F_r}(N;N) := \boldsymbol{\pi}_{\mathbf{r}}(N;N)\mathbf{F}\boldsymbol{\varepsilon}' = \frac{\Psi[\mathbf{U}^{N-1}\mathbf{V}\mathbf{F}]}{\Psi[\mathbf{U}^{N-1}\mathbf{V}]}.$$

Then, for instance, from (4.3.3), $\overline{x_r}(n;N) = \overline{V_r}(n;N)$. Next, by virtue of (4.1.6d), (4.2.2a), (4.2.2b), and some rearrangement, we have

$$\mathbf{K}(N) = \lambda\mathbf{A}\mathbf{V} + \frac{\lambda^2}{1-\rho}\mathbf{A}\mathbf{V}\mathbf{Q}\mathbf{V} - \frac{\lambda^2}{1-\rho}\mathbf{U}^N\mathbf{A}\mathbf{V}\mathbf{Q}\mathbf{V}. \tag{4.3.4a}$$

Then from Lemma 4.1.1, it easily follows that

$$\mathbf{p}[\mathbf{K}(N) - \mathbf{I}] = \frac{\lambda}{1-\rho}\left(1 - \Psi\left[\mathbf{U}^{N-1}\lambda\mathbf{V}\right]\right)\mathbf{p}\mathbf{V}. \tag{4.3.4b}$$

(In case you were not able to answer it, the equation in Exercise 4.1.3 follows directly from this.) Note that the expression in the large parentheses is a scalar, and that **pV** is proportional to $\boldsymbol{\pi}_{\mathbf{r}}$ as defined by (3.3.10b), so (4.3.4b) can be rewritten as

$$\mathbf{p}[\mathbf{K}(N) - \mathbf{I}] = \Psi[\mathbf{K}(N) - \mathbf{I}]\boldsymbol{\pi}_{\mathbf{r}}. \tag{4.3.4c}$$

We can now state the following theorem.

Theorem 4.3.1: Let **F** be any matrix operator for any residual properties of a steady-state M/ME/1//N queue [e.g., **V** or exp($-x$**B**)], the weighted average of **F**, given that S_1 is busy, is independent of N and is equal to $\boldsymbol{\pi}_{\mathbf{r}}\mathbf{F}\boldsymbol{\varepsilon}'$, where $\boldsymbol{\pi}_{\mathbf{r}}$ is given by (3.3.10b). That is, the *expected value of* **F**, E_r[**F**], is

$$\mathrm{E}_r[\mathbf{F}] := \frac{\sum_{n=1}^{N} \overline{F_r}(n;N)\, r(n;N)}{1-r(0;N)} = \frac{\mathbf{pVF\varepsilon'}}{\mathbf{pV\varepsilon''}} = \frac{\Psi[\mathbf{VF}]}{\Psi[\mathbf{V}]} = \boldsymbol{\pi}_{\mathbf{r}}\mathbf{F\varepsilon'}. \tag{4.3.5a}$$

Since the result is independent of N, it is also true for open systems (i.e., when $N \to \infty$). ■

Proof: First note that $1-r(0;N) = \Psi[\mathbf{K}(N)-\mathbf{I}]/\Psi[\mathbf{K}(N)]$, and for $n < N$ that

$$r(n;N)\overline{F_r}(n;N) = \frac{\Psi[\mathbf{U}^n\mathbf{F}]}{\Psi[\mathbf{K}(N)]},$$

with a similar expression for $n = N$. Then

$$\sum_{n=1}^{N} \overline{F_r}(n;N) r(n;N) = \frac{\Psi[(\mathbf{K}(N)-\mathbf{I})\mathbf{F}]}{\Psi[\mathbf{K}(N)-\mathbf{I}]}.$$

The theorem follows directly from (4.3.4c). QED

We next state as a corollary that the "average" residual time and density are the same as the mean residual time and density discussed in Chapter 3.

Corollary 4.3.1: The mean time (appropriately averaged over the steady-state queue-length probabilities) a randomly arriving observer of an M/G/1 queue (either open or closed) will have to wait for the customer who is presently in service at S_1 to complete service is given by $\mathrm{E}_r[\mathbf{V}]$ and is equal to $\overline{x_r}$, the mean residual time of S_1. Furthermore, the time remaining is distributed according to (3.3.13). Finally, the mean residual vector [Equation (3.3.10b)] is the same for all N and satisfies

$$\boldsymbol{\pi}_{\mathbf{r}} = \frac{\sum_{n=1}^{N} r(n;N)\boldsymbol{\pi}_{\mathbf{r}}(n;N)}{1-r(0;N)}. \tag{4.3.5b}$$

■

Proof: Let $\mathbf{F} = \mathbf{V}$ in (4.3.5a) to get the mean time, and let $\mathbf{F} = \mathbf{B}\exp(-x\mathbf{B})$ for the distribution, then compare with (3.3.10b) and (3.3.13). QED

This applies to the Laplace transform as well. If our observer knows nothing about S_1, except that someone is in service, then the probability, C_r, that service will complete before another customer arrives is given by

$$C_r = \boldsymbol{\pi}_{\mathbf{r}}[\mathbf{I} + \lambda\mathbf{V}]^{-1}\boldsymbol{\varepsilon}' = \frac{1}{\bar{x}}\Psi[\mathbf{V}(\mathbf{I}+\lambda\mathbf{V})^{-1}].$$

But $\lambda\mathbf{V}(\mathbf{I}+\lambda\mathbf{V})^{-1} = \mathbf{I} - (\mathbf{I}+\lambda\mathbf{V})^{-1}$, so

$$C_r = \frac{1-C}{\rho}, \tag{4.3.5c}$$

where C is $B^*(\lambda) = \Psi[(\mathbf{I} + \lambda\mathbf{V})^{-1}]$. Of course, C is also the probability that S_1 will finish before S_2, given that they started at the same time.

We have seen that even if a random observer does not know when a customer started service, she can get some inkling of the internal state of S_1 by observing the number of customers in its queue. Let us suppose that our random observer wishes to pass through S_1 without spending any time there, and also without preempting the customer presently in service. Then she must wait a mean time of $\overline{x_r}(n; N)$. If, in addition she must wait at the end of the queue, she will have to wait an additional time of $(n-1)\overline{x}$. She also knows that the customer in service will finish before the next customer arrives with probability, $B_r^*(\lambda; n; N)$. If this readily available information (i.e., n, the queue length at time of first observation) is ignored, she is left with the "residual results" of renewal theory, $C_r(0)$ and the P-K formula.

4.3.2. Waiting Time as Seen by an Arriving Customer

In the preceding section we viewed the M/G/1//N queue as an observer who did not affect the behavior of the system. It is perhaps more interesting to view the system from the customers' vantage point. The two viewpoints are different in principle, since a customer both observes those in front and affects those behind. (See e.g., [MELA90] and [WOLF82] for other thoughts on the subject.)

The items we need to examine this question were set up in Section 4.1.3. We already know $a(n; N)$ and $\mathbf{a}(n; N)$ (Theorem 4.1.4), the scalar and vector probabilities that, upon arriving at S_1, a customer will find n other customers already there. Since $\mathbf{a}(n; N)\boldsymbol{\varepsilon}' = a(n; N)$, the ith component of the unit vector, $[\mathbf{a}(n; N)]/a(n; N)$, is the conditional probability that an arriving customer will find S_1 in state i, given that n customers are there already. But this normalized vector is identical to (4.3.1a), so aside from the fact that an arriving customer cannot find N customers ahead of him at S_1, we have proven

Theorem 4.3.2: In an M/G/1 queue (both open or closed) and $n < N$ given, a newly arrived customer and a random observer will find S_1 in the same state. Thus (4.3.1a), (4.3.2a), and (4.3.3a) are valid for the arriving customer as well as for the random observer. ■

Be reminded that although a customer has the same probability of finding n customers in S_1 when he arrives as when he leaves, the internal state will be different. The internal state seen by the arriving customer is proportional to $\pi(n; N)$, while that for the departing customer is always $\mathbf{p}$ (the next customer in the queue enters S_1).

Now that we know that our random observer sees the same thing as the customers in the system, it pays to elaborate some on the equations of the preceding section. On the one hand, the outsider cannot make any use of the system's facilities without changing the steady-state solution. On the other hand, the customers cannot refuse to make use of the facilities without destroying the steady state. Therefore, even though both may have "inside information" as to expected waiting times, they cannot act upon it without changing the system's subsequent behavior. In any case it is good to know what one is in for. So, when a customer arrives at S_1 with n customers already there, he must wait for the one in service to complete, and for $n-1$ additional customers to start and finish. The distribution of time that he must wait is identical to that for the nth renewal epoch, Y_n, of a generalized renewal process, as discussed in Section 3.3. All n customers have pdf's generated by the same matrix, **B**, but the first one has a starting vector given by (4.3.1a) as opposed to **p** for the other customers. Thus the mean waiting time for the new customer conditioned on the number already in the queue [call it $\overline{x_w}(n)$] is [see (4.3.3a)]

$$\overline{x_w}(n) = \overline{x_r}(n) + (n-1)\overline{x}, \quad 0 < n < N. \tag{4.3.6a}$$

The total time he will spend in the system averages to $\overline{x_w}(n) + \overline{x}$. Continuing in this way, we see that the variance of his waiting time can be written as

$$\sigma_w^2(n) = \frac{\Psi[\mathbf{U}^n(2\mathbf{V}^2 - \mathbf{VQV})]}{\Psi[\mathbf{U}^n]} + (n-1)\Psi[(2\mathbf{V}^2 - \mathbf{VQV})], \tag{4.3.6b}$$

where the term multiplying $(n-1)$ is the variance of S_1's service time, σ^2. To get the variance of his total system time (again, conditioned on n), simply add one more σ^2 to (4.3.6b). These equations are easy enough to compute, especially if one is calculating the steady-state queue length probabilities anyway. The higher moments are also accessible, but more difficult.

Exercise 4.3.3: Continuing Exercise 4.2.3, calculate $x_w(n)$, $\sigma_w^2(n)$, and the coefficient of variation, $C_w^2(n) := \sigma_w^2(n)/[x_w(n)]^2$. Plot your answers for $C_w^2(n)$ versus n $(0 < n < N)$. Also plot the equivalent points for the M/M/1//N queue for comparison $[C_w^2(n) = 1/x]$.

Exercise 4.3.4: Do the same as in Exercise 4.3.3, except let S_1 be the hyperexponential server of Exercise 4.1.2.

4.3.3. System Time of an Arriving Customer

We saw in Section 4.2.3 that since a departing customer left behind the same number of customers as a random observer finds, we were able to derive the system time distribution. We should also be able to derive the same expression from the arriving customer's point of view. At present we cannot quite make it, but we do come up with some interesting results.

First we shall derive the mean system time, $\overline{x_s}$, which we already know from the P-K formula to be $\overline{T}$ (4.2.5d). If a customer arrives with no one at S_1 [which he does with probability $(1-\rho)$], he can expect to spend an average of $\overline{x}$ $(=\Psi[\mathbf{V}])$ units of time before leaving. However, if there are n customers there already, $[r(n)]$, he must first wait for the one in service to finish $[\overline{x_r}(n)]$, and then n more customers (counting himself) must finish $[n\,\overline{x}]$. That is,

$$\overline{x_s} = \overline{T} = (1-\rho)\overline{x} + \sum_{n=1}^{\infty} r(n)\overline{x_r}(n) + \overline{x}\sum_{n=1}^{\infty} n\,r(n).$$

But from Corollary 4.3.1, for the open system ($N=\infty$), the middle term must be equal to the probability that S_1 is already busy [ρ], times the mean residual time (i.e., $\rho\,\overline{x_r}$), while the last sum is $\overline{q}$ [Equation (4.2.5a)]. So

$$\overline{x_s} = (1-\rho)\overline{x} + \rho\,\overline{x_r} + \overline{x}\,\overline{q} = (1-\rho)\overline{x} + \rho\,\overline{x_r} + \rho\,\overline{T},$$

where we have used the fact that $\overline{q} = \lambda\overline{T}$ and $\rho = \lambda\overline{x}$. Next, replacing $\overline{x_s}$ with its equivalent and solving for $\overline{T}$, we get

$$\overline{T} = \overline{x} + \frac{\rho}{1-\rho}\overline{x_r},$$

which does, indeed, yield the same answer as (4.2.5c).

It is not easy to get the distribution of the system time this way because, as we saw in Chapter 3, it is not easy to work with the convolutions of functions. We can, however, get an expression for the Laplace transform (LT) of $b_s(x)$. Recall that the LT of a convolution of two pdfs is equal to the product of their LTs, and that the LT of the sum of two functions is equal to the sum of their LTs. First, let $B_r^*(s;n)$ be the LT of $b_r(x;n)$, then by an argument analogous to the one used in this section to get $\overline{T}$,

$$B_s^*(s) = (1-\rho)B^*(s) + \sum_{n=1}^{\infty} r(n)\pi_{\mathbf{r}}(n)(\mathbf{I}+s\mathbf{V})^{-1}\boldsymbol{\varepsilon}'[B^*(s)]^n.$$

This can be simplified using (4.2.4a) and (4.3.1a) to

$$B_s^*(s) = (1-\rho)B^*(s) + (1-\rho)\Psi\left[\sum_{n=1}^{\infty}[B^*(s)]^n\mathbf{U}^n(\mathbf{I}+s\mathbf{V})^{-1}\right].$$

On the one hand, for any $\mathbf{F}$, $\sum_{n=1}^{\infty}\mathbf{F}^n = \mathbf{F}\sum_{n=0}^{\infty}\mathbf{F}^n = \mathbf{F}(\mathbf{I}-\mathbf{F})^{-1}$, so this can be summed as

$$B_s^*(s) = (1-\rho)B^*(s) + (1-\rho)\Psi\left[B^*(s)\mathbf{U}(\mathbf{I}-B^*(s)\mathbf{U})^{-1}(\mathbf{I}+s\mathbf{V})^{-1}\right]$$

$$= (1-\rho)B^*(s)\left(1+\Psi\left[\mathbf{U}(\mathbf{I}-B^*(s)\mathbf{U})^{-1}(\mathbf{I}+s\mathbf{V})^{-1}\right]\right).$$

Now the system time for our customer is equal to the convolution of his waiting time, with the time for him to receive service, so the LT of his waiting (or queueing) time is

$$B_w^*(s) = (1-\rho)\left(1 + \Psi\left[\mathbf{U}(\mathbf{I} - B^*(s)\mathbf{U})^{-1}(\mathbf{I} + s\mathbf{V})^{-1}\right]\right).$$

On the other hand, for any $\mathbf{F}$, $\sum_{n=1}^{\infty} \mathbf{F}^n = \sum_{n=0}^{\infty} \mathbf{F}^n - \mathbf{I}$, so

$$B_s^*(s) = (1-\rho)\Psi\left[[\mathbf{I} - B^*(s)\mathbf{U}]^{-1}(\mathbf{I} + s\mathbf{V})^{-1}\right].$$

This equation looks fairly simple, but it is not nearly as easy to use as (4.2.7b) or (4.2.9c), since s appears in the expression in an implicit way through $B^*(s)$. As mentioned earlier, we cannot at present show the two expressions to be equal by purely algebraic means.

Exercise 4.3.5: Use this expression to show that $B^*(0) = 1$, and that its derivative evaluated at $s = 0$ does indeed yield $\overline{T}$.

4.4. RELATION TO STANDARD SOLUTION

From this book's point of view, Theorems 4.1.2 and 4.2.2 [Equations (4.1.7) and (4.2.4)] are quite sufficient for studying the steady-state M/ME/1 queue. However, it is always informative to connect to the formulas used by other methods. We will do that here, after establishing some matrix relations. Some of these relations are important, some are interesting, while some are just true. Since queues have never been analyzed in the way presented here, it is not clear just which formulas will prove to be useful ultimately. Therefore, we are including as many as we run across as we go.

4.4.1. Connection to Laguerre Polynomials

Look at Equations (4.2.6b) and (4.2.6c). On the one hand, since $\mathbf{D}(s) = (\mathbf{I} + s\mathbf{V})^{-1}$,

$$\frac{d}{ds}\mathbf{D}(s) = -\mathbf{V}(\mathbf{I} + s\mathbf{V})^{-2} = -\mathbf{V}\mathbf{D}(s)^2$$

and

$$\frac{d}{ds}d(s) = -\Psi[\,\mathbf{V}\mathbf{D}(s)^2\,],$$

or, in general,

$$\left(\frac{d}{ds}\right)^k d(s) = (-1)^k k!\,\Psi[\mathbf{V}^k \mathbf{D}(s)^{k+1}].$$

On the other hand,

$$\left(\frac{d}{ds}\right)^k d(s) = \left(\frac{d}{ds}\right)^k \int_0^\infty e^{-sx} b(x)\,dx = (-1)^k \int_0^\infty x^k\, e^{-sx}\,dx.$$

Define the *exponential moments* (we are using the notation given in [KLEI75]),

$$\alpha_k(s) := \int_0^\infty \frac{(sx)^k}{k!} e^{-sx}\, b(x)\,dx \tag{4.4.1a}$$

and

$$d_k(s) := \Psi[\mathbf{D}(s)^k]. \tag{4.4.1b}$$

[Note that from (4.2.6c), $d(s) = d_1(s) = \alpha_0(s)$, and that $d_0(s) = 1$.] When no confusion is likely to arise, the dependence of **D** on the parameter, s, will be suppressed [i.e., $\mathbf{D}(s)$ and **D** are the same thing]. Then

$$\alpha_k(s) := \Psi[(s\mathbf{V}\mathbf{D})^k \mathbf{D}]. \tag{4.4.1c}$$

But a little manipulation lets us see that

$$s\mathbf{V}\mathbf{D}(s) = \mathbf{I} - \mathbf{D}(s),$$

so (using the binomial expansion) we get a relationship between the α's and the d's,

$$\alpha_k(s) = \Psi[(\mathbf{I}-\mathbf{D})^k\mathbf{D}] = \Psi\left[\sum_{j=0}^{k}\binom{k}{j}(-1)^j \mathbf{D}^{j+1}\right] = \sum_{j=0}^{k}\binom{k}{j}(-1)^j d_{j+1}(s). \tag{4.4.2a}$$

The inverse relation leads to a more interesting result. Since $\mathbf{D} = \mathbf{I} - s\mathbf{V}\mathbf{D}$, we can write for $d_k(s)$,

$$d_{k+1}(s) = \Psi[(\mathbf{I} - s\mathbf{V}\mathbf{D})^k\mathbf{D}] = \sum_{j=0}^{k}\binom{k}{j}\Psi[(-s\mathbf{V}\mathbf{D})^j\mathbf{D}]$$

$$= \sum_{j=0}^{k}\binom{k}{j}(-1)^j \alpha_j(s). \tag{4.4.2b}$$

Next substitute the original definition for α_j from (4.4.1a) to get

$$d_{k+1}(s) = \int_0^\infty \left[\sum_{j=0}^{k}\binom{k}{j}\frac{(-sx)^j}{j!}\right] e^{-sx} b(x)\,dx. \tag{4.4.2c}$$

It is somewhat surprising to find that the expression in brackets is a *Laguerre polynomial* of order j $[L_j(sx)]$, which satisfies the orthogonality condition (see a book such as [ABRA64] for full information)

$$\int_0^\infty L_j(x)L_k(x)e^{-2x}\,dx = \delta_{jk}. \tag{4.4.3}$$

The Laguerre polynomials form a complete set, in that any "appropriately well-behaved" function of x can be expanded by them in much the same way that periodic functions can be expanded in a Fourier series of sines and cosines. That is, we can say the following. Equation (4.4.2c) can be rewritten as

$$d_{k+1}(s) = \int_0^\infty L_k(sx)e^{-sx}b(x)\,dx, \tag{4.4.4a}$$

which by the completeness property of orthogonal polynomials, lets us formally write

$$b(x) = s\sum_{k=0}^{\infty} d_{k+1}(s)L_k(sx)e^{-sx}. \tag{4.4.4b}$$

This leads to the sum rule,

$$\int_0^\infty b^2(x)\,dx = s\sum_{k=0}^{\infty}\Big(d_k(s)\Big)^2. \tag{4.4.4c}$$

These equations are true for any $s > 0$. This allows us to make the statement that every theorem proved by the method of Laguerre functions is automatically true here, too.

Exercise 4.4.1: Prove that (4.4.4b) is identically true in the formal sense. That is, replace d_k by (4.4.1b), substitute for $L_k(sx)$, and manipulate to get (3.1.7d). Similarly, use (4.4.3) and (4.4.4b) to prove (4.4.4c).

There is one last set of functions of s to be defined. Since $s\,\mathbf{VD}(s) = \mathbf{I} - \mathbf{D}(s)$, we can write

$$\gamma_n(s) := \Psi\big[(s\,\mathbf{VD})^n\big] = \Psi\big[(\mathbf{I}-\mathbf{D})^n\big]. \tag{4.4.5a}$$

Just as with (4.4.2a), we use the binomial theorem to get a relation between the d's and the γ's,

$$\gamma_k(s) = \Psi\left[\sum_{j=0}^{k}\binom{k}{j}(-1)^j\mathbf{D}^j\right] = \sum_{j=0}^{k}\binom{k}{j}(-1)^j d_j(s). \tag{4.4.5b}$$

The equivalent to (4.4.2b) is

$$d_k(s) = \Psi\left[\sum_{j=0}^{k}\binom{k}{j}(-s\,\mathbf{VD})^j\right] = \sum_{j=0}^{k}\binom{k}{j}(-1)^j\gamma_j(s). \tag{4.4.5c}$$

The relation between the α's and the γ's is found by replacing the last **D** in (4.4.1c) with $\mathbf{I} - s\,\mathbf{VD}$ to get

$$\alpha_k(s) = \gamma_k(s) - \gamma_{k+1}(s). \tag{4.4.6a}$$

It should be clear from (4.4.1a) that $\sum_{k=0}^{\infty}\alpha_k = 1$. It is not hard to prove by induction that

$$\gamma_k(s) = 1 - \sum_{j=0}^{k-1}\alpha_j(s) = \sum_{j=k}^{\infty}\alpha_j(s). \tag{4.4.6b}$$

It is left for Exercise 4.4.2 to prove that

$$\gamma_{k+1}(s) = s\int_0^{\infty}\frac{(sx)^k}{k!}e^{-sx}R(x)\,dx. \tag{4.4.6c}$$

Exercise 4.4.2: Prove (4.4.6c) by first substituting (4.4.1a) into (4.4.6b), identifying the sum as the *incomplete gamma function*, $\Gamma(k, x)$, which can then be replaced by its integral representation,

$$\Gamma(k, x) := \int_x^{\infty} e^{-t}t^{k-1}\,dt.$$

Finally, change the order of integration and end up with (4.4.6c). By the way, $\Gamma(k, x)$, by its definition, must be proportional to the reliability function for the Erlangian-k distribution. That is, $\Gamma(k, x) = k!\,E_k(x)$.

Exercise 4.4.3: Use their matrix definitions [Equations (4.4.1c), (4.4.1b), and (4.4.5a)] to prove the following sum rules for α, d, and γ:

$$\sum_{k=0}^{\infty}\alpha_k(s) = 1, \tag{4.4.7a}$$

$$\sum_{k=0}^{\infty}d_k(s) = 1 + \frac{1}{s}b(0), \tag{4.4.7b}$$

and

$$\sum_{k=0}^{\infty}\gamma_k(s) = 1 + s\,\bar{x}. \tag{4.4.7c}$$

These equations can also be proven using their integral definitions, (4.4.1a), (4.4.4a), and (4.4.6c) if one is careful about $k = 0$.

Exercise 4.4.4: Using any of the formulas just established, find explicit expressions for $d_k(s)$, $\alpha_k(s)$, and $\gamma_k(s)$ for (1) an exponential server with service rate β, (2) a server with hyperexponential-2 distribution, and (3) an Erlangian-2 server. Verify explicitly that in all three cases the sum rules of Equations (4.4.7) are valid.

4.4.2. Connection to Standard Solution

The standard solution of the steady-state M/G/1 queue is actually an algorithm involving the α_n's. To reproduce those results, we must rearrange (4.1.7) and (4.2.4a), which means that we must do something with $\Psi[\mathbf{U}^n]$. First recall that $\mathbf{V} = \mathbf{B}^{-1}$, and from (4.1.4) and (4.2.6b) that $\mathbf{A} = \mathbf{I} + \mathbf{B}/\lambda - \mathbf{Q}$, $\mathbf{U} = \mathbf{A}^{-1}$, and $\mathbf{D}(\lambda) = (\mathbf{I} + \lambda\mathbf{V})^{-1}$. Then

$$\mathbf{A} = \frac{1}{\lambda}\mathbf{B}[\mathbf{I} + \lambda\mathbf{V} - \lambda\mathbf{VQ}] = \frac{1}{\lambda}\mathbf{B}[\mathbf{I} + \lambda\mathbf{V}]\,[\mathbf{I} - \lambda\mathbf{DVQ}].$$

With the aid of Lemma 4.2.1, we can write

$$\mathbf{U} = [\mathbf{I} - \lambda\mathbf{DVQ}]^{-1}\lambda\mathbf{DV} = \left(\mathbf{I} + \frac{1}{1-\gamma_1(\lambda)}\lambda\mathbf{DVQ}\right)\lambda\mathbf{DV}.$$

Let

$$\mathbf{C} := \lambda\mathbf{DV} = \mathbf{I} - \mathbf{D}, \tag{4.4.8a}$$

which means [from (4.4.5a)] that $\gamma_n(\lambda) = \Psi[\mathbf{C}^n]$; then

$$\mathbf{U} = \mathbf{C} + \frac{1}{1-\gamma_1}\mathbf{CQC} \tag{4.4.8b}$$

and ($\mathbf{pCQ} = \mathbf{pC}\boldsymbol{\varepsilon}'\mathbf{p} = \Psi[\mathbf{C}]\mathbf{p} = \gamma_1\mathbf{p}$)

$$\mathbf{pU} = \mathbf{pC} + \frac{\gamma_1}{1-\gamma_1}\mathbf{pC} = \frac{1}{1-\gamma_1}\mathbf{pC}. \tag{4.4.8c}$$

It simply follows that $\Psi[\mathbf{U}] = \gamma_1/(1-\gamma_1)$. Next, look at

$$\mathbf{pU}^2 = \mathbf{pUU} = \frac{1}{1-\gamma_1}\mathbf{pC}\left[\mathbf{C} + \frac{1}{1-\gamma_1}\mathbf{CQC}\right] = \frac{1}{1-\gamma_1}\mathbf{pC}^2 + \frac{\gamma_2}{(1-\gamma_1)^2}\mathbf{pC},$$

which yields $\Psi[\mathbf{U}^2] = \gamma_2/(1-\gamma_1)^2$. From this it can be seen that successive applications of (4.4.7a) to $\mathbf{pU}^n$ will lead to a series expansion in terms of $\mathbf{pC}^k$. Therefore, let

$$\mathbf{pU}^n = \sum_{k=1}^{n} a_k^{(n)} \mathbf{pC}^k, \tag{4.4.9a}$$

which leads to

$$\Psi[\mathbf{U}^n] = \sum_{k=1}^{n} a_k^{(n)} \gamma_k. \tag{4.4.9b}$$

On the one hand,

$$\mathbf{pU}^{n+1} = \sum_{k=1}^{n+1} a_k^{(n+1)} \mathbf{pC}^k,$$

while on the other hand,

$$\mathbf{pU}^{n+1} = \mathbf{pU}^n \mathbf{U} = \sum_{k=1}^{n} a_k^{(n)} \mathbf{pC}^k \left[\mathbf{C} + \frac{1}{1-\gamma_1} \mathbf{CQC} \right]$$

$$= \sum_{k=1}^{n} a_k^{(n)} \mathbf{pC}^{k+1} + \sum_{k=1}^{n} a_k^{(n)} \gamma_{k+1} \mathbf{pC}.$$

For these two expressions to be identically equal, we must have

$$a_{k+1}^{(n+1)} = a_k^{(n)} \quad \text{for} \quad 1 \le k \le n \tag{4.4.9c}$$

and

$$a_1^{(n+1)} = \frac{1}{1-\gamma_1} \sum_{k=1}^{n} a_k^{(n)} \gamma_{k+1}. \tag{4.4.9d}$$

The penultimate equation implies that $a_k^{(n)} = a_j^{(m)}$ as long as $n-k = m-j$. In particular,

$$a_k^{(n)} = a_1^{(n-k+1)} \tag{4.4.9e}$$

We come up with an interesting relation by putting (4.4.9c) into (4.4.9d),

$$a_1^{(n+1)} = \frac{1}{1-\gamma_1} \sum_{k=1}^{n} a_{k+1}^{(n+1)} \gamma_{k+1} = \frac{1}{1-\gamma_1} \left(\sum_{k=2}^{n+1} a_k^{(n+1)} \gamma_k + a_1^{(n+1)} \gamma_1 - a_1^{(n+1)} \gamma_1 \right)$$

$$= \frac{1}{1-\gamma_1} \left(\sum_{k=1}^{n+1} a_k^{(n+1)} \gamma_k - a_1^{(n+1)} \gamma_1 \right).$$

Next bring the extra term on the right- to the left-hand side of the equation, clear fractions, and cancel like terms to get, for $n > 1$,

$$a_1^{(n+1)} = \sum_{k=1}^{n+1} a_k^{(n+1)} \gamma_k = \Psi[\mathbf{U}^{n+1}]. \tag{4.4.10a}$$

The relationship with $\Psi[\mathbf{U}^{n+1}]$ comes from comparing with (4.4.9b). For convenience, let $u_1 := 1/(1-\gamma_1)$, and for $n > 1$,

$$u_n := \Psi[\mathbf{U}^n],$$

then with the aid of (4.4.9b) and (4.4.9e), (4.4.10a) can be rewritten as

$$u_n = \sum_{k=1}^{n} \gamma_k \, u_{n-k+1} \quad \text{for} \quad n > 1, \tag{4.4.10b}$$

while, as previously noted, $\Psi[\mathbf{U}] = \gamma_1/(1-\gamma_1) = \gamma_1 u_1$. The standard solution as given in standard texts such as [ALLE90] is expressed in terms of the $\alpha_n(\lambda)$'s. Therefore, manipulate the above (n is replaced by $n+1$), and get

$$u_{n+1} = \sum_{k=1}^{n+1} \gamma_k \, u_{n-k+2} = \gamma_1 \, u_{n+1} + \sum_{k=2}^{n+1} \gamma_k \, u_{n-k+2} = \gamma_1 \, u_{n+1} + \sum_{k=1}^{n} \gamma_{k+1} \, u_{n-k+1}.$$

Next bring the loose term over to the left, recall that $\alpha_0 = 1 - \gamma_1$, and substitute (4.4.6a) to get

$$\alpha_0 u_{n+1} = \sum_{k=1}^{n} u_{n-k+1}(\gamma_k - \alpha_k) = u_n - \sum_{k=1}^{n} u_{n-k+1} \, \alpha_k.$$

This is next rearranged to give

$$u_n = \alpha_0 u_{n+1} + \sum_{k=1}^{n} u_{n-k+1}\alpha_k = \sum_{k=0}^{n} u_{n-k+1}\alpha_k.$$

We are almost there now. Note that for an open system, $r(n) = r(0) u_n$, for $n > 1$, and $r(1) = r(0)\gamma_1 u_1$. So, upon multiplying by $r(0)$, we get

$$r(n) = r(0)\alpha_n u_1 + \sum_{k=0}^{n-1} r(n-k+1)\alpha_k = r(0)\alpha_n u_1 + \sum_{k=0}^{n} r(n-k+1)\alpha_k - r(1)\alpha_n.$$

But $r(0)\alpha_n u_1 - r(1)\alpha_n = r(0)\alpha_n [1/\alpha_0 - \gamma_1/\alpha_0] = r(0)\alpha_n$. Therefore,

$$r(n) = r(0)\alpha_n + \sum_{k=0}^{n} r(n-k+1)\alpha_k = r(0)\alpha_n + \sum_{k=0}^{n+1} r(k)\alpha_{n-k+1}. \tag{4.4.10c}$$

This, together with the fact that $r(0) = 1 - \rho$, is the recursive formula given in most books.

The closed system is somewhat more difficult. Since it is true that for all $n < N$, $r(n; N)$ is proportional to $r(n)$, we need only evaluate $\Psi[\mathbf{U}^{N-1}\mathbf{V}]$ by a separate means, and renormalize. That is, since Equations (4.1.7) are true, $r(N; N) = \lambda r(0; N)\Psi[\mathbf{U}^{N-1}\mathbf{V}]$, and $r(n; N) = r(0; N) u_n$ for $n < N$, so

$$\frac{1}{r(0; N)} = \Psi[\mathbf{U}^{N-1}\mathbf{V}] + 1 + \sum_{n=1}^{N-1} u_n.$$

Finally, note that $\lambda\mathbf{AV} = \mathbf{I} + \lambda\mathbf{V} - \lambda\mathbf{QV}$, so

$$b_n := \Psi[\mathbf{U}^{n-1}\lambda\mathbf{V}] = \Psi[\mathbf{U}^n \lambda\mathbf{AV}] = \Psi[\mathbf{U}^n(\mathbf{I} + \lambda\mathbf{V} - \lambda\mathbf{QV})]$$

$$= u_n + b_{n+1} - \rho u_n,$$

which yields the simple recursive relation, starting with $b_1 = \rho$,

$$b_{n+1} = (1-\rho)u_n - b_n. \tag{4.4.11}$$

This last formula is not usually included in standard texts but is valid for all M/G/1//N queues.

As a final thought on this subject, note that the b_N's not only give the steady-state probabilities that all N customers are at S_1, $[r(N;N)]$, they also yield the residual waiting times for $n < N$ in the queue. That is, from (4.3.3a),

$$\overline{x_r}(n < N; N) = \frac{b_{n+1}}{\lambda u_n}. \tag{4.4.12}$$

Equation (4.3.3b) must still be calculated by a different algorithm.

4.4.3. M/M/X//*N* Approximations to M/ME/1//*N* Loops

It is not our purpose here to search for approximations to the equations we have worked so hard to derive. Rather, we wish to explore the extent of *robustness* of Jackson networks (see, e.g., [BASK75], [LIPS77], [LAZO84] and the entire issue containing [DENN78]). The loop with two load-dependent servers, which we described in Section 2.1.4 and assigned the symbol M/M/X//N, can be viewed as a Jackson network with two service centers. In fact, Equations (2.1.11) were deliberately written in a form that reflects the product-form solution one sees in more general networks. One of the great attributes of Jackson networks is their ability to describe the steady-state behavior of a whole network, based on the properties of the individual service centers (the set of load-dependent service rates). Most important, the properties ascribed to each service center do *not* depend on the properties of other servers or the system as a whole. Of course, we are accomplishing the same thing in this book, but we have found it necessary to give each subsystem properties that must be expressed by a non-trivial matrix rather than a simple set of scalars. We point out that the M/M/X//N loop, unlike our M/ME/1//N formulation, simply does not have the structure to distinguish *residual* processes, or other transient properties, from those of the steady state. Therefore, we shall only compare the steady-state behaviors here.

Suppose that we observe a system which is exactly described by an M/G/1//N loop over a very long period of time. How would one measure the *load dependence* of a server? A natural and self-consistent definition, or defining measurement procedure, would be as follows (thanks to Victor Wallace for the underlying idea). Let t be the total time that the system has been under observation, and as we did in Section 1.1.1, let $N_i(t)$ be the number of customers who have left S_i in that time. If t is indeed very large, we would expect the ratio of $N_1(t)$ to $N_2(t)$ to be very close to 1, close enough so that we can assume that they are equal to each other, and drop the subscript. Then the system throughput is measured as

$$\Lambda(N) \approx \frac{N(t)}{t}. \tag{4.4.13a}$$

Next, we define the following measurable quantities.

Definition 4.4.1

$N_i(n;t)$:= *number of departures from* S_i *in the time interval, t, which occurred while there were n customers there (counting the one who left).* Every time a customer leaves S_i, the observer, noting how many customers were there just before the departure, increments that counter by 1. ◊

Then we can say that

$$\sum_{n=1}^{N} N_i(n;t) = N(t). \tag{4.4.13b}$$

Definition 4.4.2

$T_i(n;t)$:= *total time that there were n customers at* S_i. Every time a customer enters or leaves a subsystem, the observer notes how many customers were there just before that event and adds the amount of time since the previous arrival or departure to the appropriate counter. Of course, an arrival to one subsystem occurs at the same time as the departure from the other subsystem, so two counters are modified simultaneously. ◊

We then have

$$\sum_{n=0}^{N} T_i(n;t) = t \qquad \text{for } i = 1, 2. \tag{4.4.13c}$$

The best we can say about load-dependent service rates is to describe them as the *rate at which customers leave a subsystem for a given load.* So we use the following, consistent with the definitions we gave in Section 2.1.4 (for arbitrary load dependence):

$$\mu(n) \approx \frac{N_1(n;t)}{T_1(n;t)} \tag{4.4.14a}$$

and

$$\lambda(k) \approx \frac{N_2(k;t)}{T_2(k;t)}, \tag{4.4.14b}$$

where $n + k = N$. These parameters are similar to those considered by J. P. Buzen's *operations analysis*, (e.g., [DENN78]). We assume that t is so large that the steady-state probabilities we have previously derived for various events are very close to the measured relative frequencies of those events. From our rules and definitions, $T_1(n;t)$ must be approximately equal to the probability that there are n customers at S_1, multiplied by the total time that the system was observed. That is,

$$T_1(n;t) \approx r(n;N)t. \tag{4.4.15a}$$

Also, after some thought, the reader should be able to accept the following formula:

$$N_1(n;t) = d(n-1;N)N(t). \tag{4.4.15b}$$

The d's and π's were defined in Section 4.1. Remember that the d's were defined as *leaving behind*, while N_i was defined as *including the departing customer*, hence the $n-1$. The equivalent formulas for S_2 are (k is the number at S_2)

$$T_2(k;t) \approx r(N-k;N)t \tag{4.4.16a}$$

and

$$N_2(k;t) = d_2(k-1;N)N(t).$$

The symbol $d_2(k-1;N)$ is borrowed from Section 5.1.2 and is the probability that a customer when departing S_2 will leave behind $k-1$ customers. Clearly, that same customer will arrive at S_1, finding $N-k$ customers already there. (Let's see: there are $k-1$ at S_2, $N-k$ at S_1, and 1 traveling, giving a total of $k-1+N-k+1=N$, right.) Therefore, $d_2(k-1;N) = a(N-k,N)$, and we can write

$$N_2(k;t) = a(N-k;N)N(t). \tag{4.4.16b}$$

We are now ready to put things together. Using Equations (4.4.13a) and (4.4.15) in (4.4.14a) yields

$$\mu(n) = \frac{d(n-1;N)N(t)}{r(n;N)t} = \frac{d(n-1;N)}{r(n;N)}\Lambda(N).$$

But from (4.1.13c) and (4.1.8b), and noting that $1/\bar{x}_2 = \lambda$, we have

$$\mu(n) = \frac{r(n-1;N)}{r(n;N)}\,\frac{\Lambda(N)}{1-r(N;N)} = \lambda\frac{r(n-1;N)}{r(n;N)}. \tag{4.4.17a}$$

Similarly, we can work on S_2.

$$\lambda(k) = \frac{a(N-k;N)\,N(t)}{r(N-k;N)\,t} = \lambda. \tag{4.4.17b}$$

We have again made use of (4.1.13c). Thus we see that S_2 is indeed a load-independent server, just as we theorized, but S_1 is more complicated, as we would have expected.

Before we look at (4.4.17a) more closely, let us see what these service rates give us for probabilities when we use them in Equations (2.1.9). For the moment, let us call these probabilities $r_a(n;N)$ (a is for *approximation*). Then

$$\lambda(N-n)r_a(n;N) = \mu(n+1)r_a(n+1;N).$$

Put in Equations (4.4.17), to get

$$\lambda r_a(n;N) = \lambda \frac{r(n;N)}{r(n+1;N)} r_a(n+1;N)$$

or

$$\frac{r(n;N)}{r_a(n;N)} = \frac{r(n+1;N)}{r_a(n+1;N)} \quad \text{for } 0 \le n < N. \tag{4.4.18}$$

Since the set of r_a's is proportional to the set of r's, and both sets sum to 1, they must be equal, term by term. Thus we have proven that for any M/G/1//N loop, we can find appropriate μ's that yield identical steady-state probabilities for an M/M/X//N loop. Is this truly miraculous? **NO**. All we did was find N numbers, $[\mu(n)]$, which would let us generate $N-1$ other numbers, $[r(n;N)]$, constrained to sum to 1. In other words, the *product-form solution* has so many unspecified parameters that it can fit anything. The real test of the model comes when we see if the same parameters can be used to model a system that has been changed slightly.

Let us look again at (4.4.17a). From (4.1.7a) and (4.1.7b) we can write the following:

$$\mu(n) = \lambda \frac{\Psi[\mathbf{U}^{n-1}]}{\Psi[\mathbf{U}^{n}]} \quad \text{for } 1 \le n < N, \tag{4.4.19a}$$

but

$$\mu(N) = \frac{\Psi[\mathbf{U}^{N-1}]}{\Psi[\mathbf{U}^{N-1}\mathbf{V}]}. \tag{4.4.19b}$$

These equations tell us that if we tested a system with, say, four customers, and then asked how the system would behave if we added one more customer, we could use some of the same parameters, but we would have to change μ(4), as well as find a value for μ(5). In other words, from an M/M/C//N point of view, the properties ascribed to S_1 depend on the system's population to a certain extent.

We can ask what happens if we change the behavior of another server somewhat. In our M/ME/1//N loop the only thing we can do to S_2 is change λ. Suppose that we again tested a system with four customers, and then asked how the system behaved if λ were changed somewhat. Everything seems to be constant, *but* the matrix **U** depends on λ in a nontrivial way.

Example 4.4.1: Let us look at μ(1) from (4.4.19a), for an M/E_2/1//N queue. In that case (see Exercise 3.1.1)

$$<\mathbf{p}, \mathbf{B}> = <[1\ 0],\ \mu\begin{bmatrix} 1 & -1 \\ 0 & 1 \end{bmatrix}>,$$

It is not known whether the form for $\mathbf{H_u}(n)$ that contains $\mathbf{A}$ will ultimately be more convenient than that which contains $\mathbf{B}$, so we will use whichever seems more useful.

The formula, when only one customer is at S_1, is slightly different. If the customer should leave, S_1 remains idle until another customer arrives and enters, $[\mathbf{p}]$. After that, the system eventually gets to $n = 2$. The equation for this is

$$\mathbf{H}(1) = \lambda(\lambda\mathbf{I} + \mathbf{M})^{-1} + (\lambda\mathbf{I} + \mathbf{M})^{-1}\mathbf{MPH}(1) + (\lambda\mathbf{I} + \mathbf{M})^{-1}\mathbf{Mq'pH}(1).$$

[Compare this with (4.5.2a) before going on.] We easily solve for $\mathbf{H_u}(1)$, getting

$$\mathbf{H_u}(1) = \lambda[\lambda\mathbf{I} + \mathbf{B} - \mathbf{BQ}]^{-1} = [\mathbf{A} + \mathbf{Q} - \mathbf{AQ}]^{-1}. \tag{4.5.2d}$$

As with all recursive relations, we must start with a nonrecursive equation, which we easily get by noting that $\mathbf{H_u}(0) = \mathbf{p}$. But we can make believe that state $\{\cdot\,;0\}$ has internal states [we already did this in (4.1.1)]; then we can use

$$\mathbf{H_u}(0) = \mathbf{Q}, \tag{4.5.2e}$$

and thus since $\mathbf{Q}^2 = \mathbf{Q}$, (4.5.2d) becomes a special case of (4.5.2c).

We next show that $\mathbf{H_u}(1)\boldsymbol{\varepsilon}' = \boldsymbol{\varepsilon}'$, and by induction, using (4.5.2c), prove that (4.5.1) is true for all n. First note that for any matrix $\mathbf{D}$, if $\mathbf{D}\boldsymbol{\varepsilon}' = \boldsymbol{\varepsilon}'$, then its inverse satisfies $\mathbf{D}^{-1}\boldsymbol{\varepsilon}' = \boldsymbol{\varepsilon}'$. We say, then, that if $\mathbf{D}$ is isometric, so is $\mathbf{D}^{-1}$. Therefore, if $[\mathbf{A} + \mathbf{Q} - \mathbf{AQ}]\boldsymbol{\varepsilon}' = \boldsymbol{\varepsilon}'$, its inverse, $\mathbf{H}(1)$, satisfies $\mathbf{H}(1)\boldsymbol{\varepsilon}' = \boldsymbol{\varepsilon}'$. But the *if* condition is obviously true, since $\mathbf{Q}\boldsymbol{\varepsilon}' = (\boldsymbol{\varepsilon}'\mathbf{p})\boldsymbol{\varepsilon}' = \boldsymbol{\varepsilon}'(\mathbf{p}\boldsymbol{\varepsilon}') = \boldsymbol{\varepsilon}'$. Next assume that $\mathbf{H}(k)\boldsymbol{\varepsilon}' = \boldsymbol{\varepsilon}'$ for all $k = 1, 2, \cdots, n-1$, and rewrite (4.5.2c) as

$$\mathbf{H}(n)[\mathbf{A} + \mathbf{Q} - \mathbf{AQH}(n-1)] = \mathbf{I}.$$

Then multiply both sides on the right with $\boldsymbol{\varepsilon}'$. The left-hand side of the resulting equation gives

$$\begin{aligned}\mathbf{H}(n)[\mathbf{A}\boldsymbol{\varepsilon}' + \boldsymbol{\varepsilon}' - \mathbf{AQH}(n-1)\boldsymbol{\varepsilon}'] &= \mathbf{H}(n)[\mathbf{A}\boldsymbol{\varepsilon}' + \boldsymbol{\varepsilon}' - \mathbf{AQ}\boldsymbol{\varepsilon}'] \\ &= \mathbf{H}(n)[\mathbf{A}\boldsymbol{\varepsilon}' + \boldsymbol{\varepsilon}' - \mathbf{A}\boldsymbol{\varepsilon}'] = \mathbf{H}(n)\boldsymbol{\varepsilon}',\end{aligned}$$

while the right-hand side yields $\mathbf{I}\boldsymbol{\varepsilon}' = \boldsymbol{\varepsilon}'$. Therefore, we have proven our assertion that $\mathbf{H}(n)$ is isometric for all n. In effect, we have proven the following (perhaps obvious) theorem:

Theorem 4.5.1: For any M/G/1//N queue, and for any ρ, if at any time there are $n < N$ customers in the queue of S_1, then given enough time, the queue will eventually have $n+1$ customers in it (i.e., $\mathbf{H_u}(n)\boldsymbol{\varepsilon}' = \boldsymbol{\varepsilon}'$). ∎

This might be called the *pessimist's theorem*, since it implies that no matter how bad things are now (long queue), if our random observer waits long enough, she will certainly see it get worse some day (longer queue). There are at least two weaknesses to this argument, however. First of all, the theorem assumes that conditions will remain the same for time immemorial, the *homogeneous* assumption. Second, the pessimist is assuming that the random observer will live long enough to see things get worse. This is an important reason for studying non-steady-state behavior. For if *some day* is longer than say, the age of the universe, who cares? In Chapter 2 we calculated what this time would be for an exponential queue [Equations (2.3.2)], and saw that this could be long indeed if $\rho < 1$. We will show how to calculate mean first-passage times for general queues in the next section, after saying some final remarks about the first-passage matrices.

Equation (4.5.2d) seems simple enough, so we might be encouraged to substitute it into (4.5.2b) or (4.5.2c) to get an explicit formula for $\mathbf{H_u}(2)$, but the resulting expression does not simplify greatly. For higher n it is even messier. It is better to think of these formulas as a recursive definition of the $\mathbf{H_u}$'s, and to use (4.5.2b) or (4.5.2c) to numerically compute them recursively when explicit examples are needed. Note that in general, the $\mathbf{H_u}$'s are all different, although they do approach a limit for large n.

From these matrices one can also find the probability matrices of first passage from n to $n+l$, for any n and l.

Definition 4.5.2

$\mathbf{H_u}(n \to n+l)$:= *probability matrix of first passage from* n *to* $n+l$, $l \geq 1$. That is, $[\mathbf{H_u}(n \to n+l)]_{ij}$ is the probability that S_1 will be in state j when its queue goes from n to $n+l$ for the first time, given that it started in state i with n customers. In particular, $\mathbf{H_u}(n \to n+1) = \mathbf{H_u}(n)$. ◊

The first-passage matrix of going from n to $n+2$ is simply

$$\mathbf{H_u}(n \to n+2) = \mathbf{H_u}(n)\mathbf{H_u}(n+1),$$

and in general,

$$\mathbf{H_u}(n \to n+l+1) = \mathbf{H_u}(n \to n+l)\mathbf{H_u}(n+l)$$

$$= \mathbf{H_u}(n)\,\mathbf{H_u}(n+1) \cdots \mathbf{H_u}(n+l). \tag{4.5.3}$$

Would the author be presumptuous in declaring it obvious that $\mathbf{H_u}(n \to n+l)$ is isometric?

A particularly interesting matrix (it is actually a vector) is the probability of first passage from $0 \to n$. It is given by

$$\mathbf{p_u}(n) := \mathbf{p}\mathbf{H_u}(1)\mathbf{H_u}(2) \cdots \mathbf{H_u}(n-1). \tag{4.5.4}$$

Here too, it is clear that $\mathbf{p_u}(n)\boldsymbol{\varepsilon}' = 1$ for all n, so Theorem 4.5.1 extends to the statement: *Given enough time, every possible queue length will be experienced at least once.* But what is *enough time*? We discuss this vector further when we actually define it in Definition 4.5.4.

The first-passage matrices may not appear to be very interesting in their own right, but they are needed for calculating first-passage times, as shown in the next section.

Mean First-Passage Time for Queue to Reach Given Length

This section is a direct generalization of the material in Section 2.3.1. By arguments similar to those required to derive (4.5.2), one can derive the mean time for the queue to grow from n to $n+1$ for the first time. First define the vector $\boldsymbol{\tau}'_{\mathbf{u}}(n)$.

Definition 4.5.3

$\boldsymbol{\tau}'_{\mathbf{u}}(n) :=$ *mean first-passage time vector from n to $n+1$.* The ith component is the mean time it takes for the queue at S_1 to have $n+1$ customers for the first time, having started in state $\{i; n\}$. ◊

Look once more at Figure 4.5.1. Suppose that there are n customers in the queue at S_1, and the active customer is at phase i. From that figure, one of three things will happen next. The mean time until the next event is given by $1/(\lambda+\mu_i) = [(\lambda\mathbf{I}+\mathbf{M})^{-1}\boldsymbol{\varepsilon}']_i$. If the event that occurs is an arrival from S_2, [path I], the process is over. If, however, the event is internal to S_1, [path II], the system will go to state $\{j; n\}$ with probability given by $[(\lambda\mathbf{I}+\mathbf{M})^{-1}\mathbf{MP}]_{ij}$, and then will take another $[\boldsymbol{\tau}'_{\mathbf{u}}(n)]_j$ to accomplish the task. Worse yet, if the event results in a departure from S_1, as shown in path III, the system, finding itself in some state $\{j; n-1\}$ with probability $[(\lambda\mathbf{I}+\mathbf{M})^{-1}\mathbf{Mq}'\mathbf{p}]_{ij}$, must first get back to length n in time $[\boldsymbol{\tau}'_{\mathbf{u}}(n-1)]_j$ and then on to $n+1$. But this long excursion of going down and back up puts the system into state k with probability $[(\lambda\mathbf{I}+\mathbf{M})^{-1}\mathbf{Mq}'\mathbf{p}\mathbf{H}_{\mathbf{u}}(n-1)]_{ik}$. (At last we see the need for a first-passage matrix.) The three processes together lead to the following vector equation:

$$\boldsymbol{\tau}'_{\mathbf{u}}(n) = (\lambda\mathbf{I}+\mathbf{M})^{-1}\boldsymbol{\varepsilon}' + (\lambda\mathbf{I}+\mathbf{M})^{-1}\mathbf{MP}\boldsymbol{\tau}'_{\mathbf{u}}(n)$$

$$+ (\lambda\mathbf{I}+\mathbf{M})^{-1}\mathbf{Mq}'\mathbf{p}[\boldsymbol{\tau}'_{\mathbf{u}}(n-1) + \mathbf{H}_{\mathbf{u}}(n-1)\boldsymbol{\tau}'_{\mathbf{u}}(n)].$$

Next, premultiply both sides of the equation by $(\lambda\mathbf{I}+\mathbf{M})$, bring all terms proportional to $\boldsymbol{\tau}'_{\mathbf{u}}(n)$ to the left-hand side, and get

$$[\lambda\mathbf{I}+\mathbf{M}-\mathbf{MP}-\mathbf{Mq}'\mathbf{p}\mathbf{H}_{\mathbf{u}}(n-1)]\boldsymbol{\tau}'_{\mathbf{u}}(n) = \boldsymbol{\varepsilon}' + \mathbf{Mq}'\mathbf{p}\boldsymbol{\tau}'_{\mathbf{u}}(n-1).$$

This formula has several familiar components. Recall that $\mathbf{M}-\mathbf{MP}=\mathbf{B}$, $\mathbf{Mq}'\mathbf{p}=\mathbf{BQ}$, and thus from (4.5.2b) the term in brackets is λ times the inverse of $\mathbf{H}_{\mathbf{u}}(n)$. This, then, gives us the important recursive equation for the $\boldsymbol{\tau}'_{\mathbf{u}}(n)$'s.

$$\boldsymbol{\tau}'_{\mathbf{u}}(n) = \frac{1}{\lambda}\boldsymbol{\varepsilon}' + \frac{1}{\lambda}\mathbf{H}_{\mathbf{u}}(n)\mathbf{BQ}\boldsymbol{\tau}'_{\mathbf{u}}(n-1), \qquad \text{with} \quad \boldsymbol{\tau}'_{\mathbf{u}}(0) := \frac{1}{\lambda}\boldsymbol{\varepsilon}'. \tag{4.5.5}$$

To get these formulas we had to divide by λ, premultiply both sides by $\mathbf{H_u}(n)$, and make use of the isometric property of $\mathbf{H_u}(n)$.

Before going on, let us use the following theorem to summarize what we have done so far.

Theorem 4.5.2: For any M/ME/1//N queue, and for any ρ, the first-passage matrices, $\mathbf{H_u}(n)$, and mean first-passage time vectors, $\boldsymbol{\tau}'_{\mathbf{u}}(n)$, are recursively given by (4.5.2) and (4.5.5), and can be calculated efficiently in the following way:

$$\mathbf{H_u}(0) = \mathbf{Q}, \quad \boldsymbol{\tau}'_{\mathbf{u}}(0) = \frac{1}{\lambda}\boldsymbol{\varepsilon}'.$$

For $n = 1, 2, \cdots$,

$$\mathbf{H_u}(n) = \lambda[\lambda\mathbf{I} + \mathbf{B} - \mathbf{BQH_u}(n-1)]^{-1}$$

and

$$\boldsymbol{\tau}'_{\mathbf{u}}(n) = \frac{1}{\lambda}\boldsymbol{\varepsilon}' + \frac{1}{\lambda}\mathbf{H_u}(n)\mathbf{BQ}\boldsymbol{\tau}'_{\mathbf{u}}(n-1).$$

These objects are the same for all N as long as $n < N$, and thus the theorem is true for the open system ($N \to \infty$) as well. The first-passage matrices are isometric, (i.e., $\mathbf{H_u}(n)\boldsymbol{\varepsilon}' = \boldsymbol{\varepsilon}'$). ■

Example 4.5.1: We have computed $\boldsymbol{\tau}'_{\mathbf{u}}(n)$ of the M/E_2/1 queue for various values of ρ and have plotted the results in Figure 4.5.2. Note that $\boldsymbol{\tau}'_{\mathbf{u}}(n)$ has two components, so there are two curves for each ρ. The most obvious feature is that for a given value of n, smaller ρ leads to longer times for the queue at S_1 to grow by 1. Next, for a given ρ, the mean time it takes to grow by 1 increases with n, and if $\rho < 1$, the increase is exponential. This is caused by the fact that for large n, the queue can drop much further before it finally goes up (remember, the first-passage times include possible excursions down to 0). For $\rho > 1$, the curve approaches a constant, since the queue is not likely to drop very far before going up. The curve for $\rho = 1$ appears to be linear, just as it is in the M/M/1 case.

The third feature we see is the separation of the two components of $\boldsymbol{\tau}'_{\mathbf{u}}(n)$. The mean first-passage time to grow by 1 is a weighted average of the two components, depending on the state the system was in when the process began. If the process begins at the moment a customer enters S_1, then $\mathbf{p} = [1\ ,\ 0]$, and the growth time follows the curve labeled $[\boldsymbol{\tau}'_{\mathbf{u}}(n)]_1$. Note that the curves for the two components actually diverge as n gets bigger. If a customer starts at phase 2, he has a higher probability of leaving before another customer arrives, thereby leaving the queue with only $n-1$ customers. Therefore, it will take longer to recover if n is larger. ●

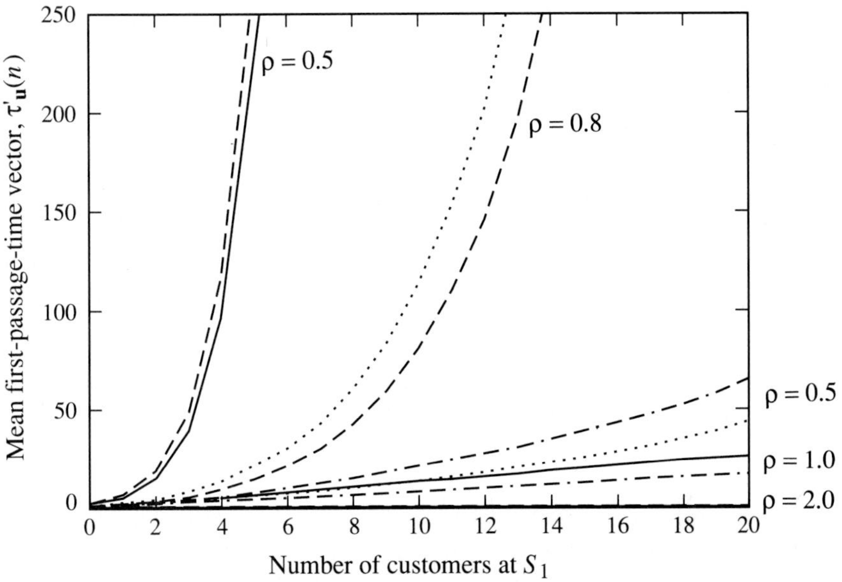

Figure 4.5.2: The two components of mean first-passage time vector, $\tau'_{\mathbf{u}}(n)$, as a function of the number of customers at S_1, for the $M/E_2/1$ queue. There are five sets of curves, corresponding to $\rho = 0.5$, 0.8, 0.95, 1.0, and 2.0. If the process starts when a customer first enters, he goes to phase 1, and when finished there, goes to phase 2, after which he leaves. If a new customer arrives before the active customer leaves, the process ends (the queue has grown by 1). Otherwise, the process continues, with possibly many events occurring. For all ρ and n, $[\boldsymbol{\tau}']_1$ lies below $[\boldsymbol{\tau}']_2$. The curves corresponding to $\rho = 2$ are not negligible. For $n \geq 20$, their values are constant at 0.809 and 1.118, respectively.

It would be most interesting to study first-passage times for other distributions, since very little is known about this type of behavior.

The first-passage time vectors do not by themselves give us the times we are looking for, since we must first decide what state we are in when the process begins. In Chapter 2 this was no problem, since we only had to know the number in the queue. Now unfortunately, we must make some statement as to the initial internal state of S_1. Once we do this (whether the system is open or closed, irrespective of whether ρ is less than, equal to, or greater than 1), we can then calculate such things as:

1. The mean first-passage time of going from n to $n+1$, given that the customer in service has just begun.

2. The mean first-passage time to $n+1$, as seen by a random observer who sees n customers there initially.

3. The mean first-passage time to $n+1$, given that there are n customers in the queue, and the nth customer just arrived.

4. The mean first-passage time from n to $n+1$, given that the queue was originally empty.

5. The mean time for a queue to grow to n for the first time, given that a customer has just arrived at an empty queue.

(Note that items 1 to 4 yield identical results for the M/M/1 queue.) For instance, the internal state of S_1 immediately after a customer departs and a new customer enters is the entrance vector, **p**. Therefore, the first item of the list above can be calculated as follows.

Corollary 4.5.2a: The mean time for an M/ME/1 queue to grow to $n+1$ for the first time, given that n customers are there at the beginning, and the customer in service at S_1 has just begun, is given by

$$\mathbf{p}\boldsymbol{\tau}'_{\mathbf{u}}(n).$$

This is the same as the *mean first-passage time to* $n+1$, *given that a customer has just left behind n customers.* It is also the same as the *mean time for* S_1 *to return to queue length* $n+1$ *for the first time, given that it just dropped to* n. There are, no doubt, other equivalent statements. The state the system will be in when this occurs is given by

$$\mathbf{pH_u}(n).$$

Thus the expression

$$\mathbf{p}\boldsymbol{\tau}'_{\mathbf{u}}(n) + \mathbf{pH_u}(n)\boldsymbol{\tau}'_{\mathbf{u}}(n+1)$$

is the mean first-passage time to $n+2$, given that service has just begun with $n \le N-2$ customers. ■

Another interesting passage time is given by item 2 above. The condition as stated there is insufficient to derive an expression. After all, what was the history of the queue before the random observer arrived? We could assume that the system has been in operation for a long time, long enough to be near its steady state. This was discussed in Section 4.3.1 in analyzing residual times. We follow that section here. Thus the random observer will find the system in the composite state described by the vector [see (4.3.1a)]

$$\frac{1}{a(n;N)}\mathbf{a}(n;N) = \boldsymbol{\pi}_{\mathbf{r}}(n) = \frac{1}{\Psi[\mathbf{U}^n]}\mathbf{pU}^n.$$

The residual vector appears again. Note that this vector does not depend on N, except that n must be less than N. If $n = N$, the queue can never rise above N anyway. We can state the time for this process by the following.

Corollary 4.5.2b: The mean time for a steady-state M/ME/1 queue to grow to $n+1$ for the first time, as seen by a random observer who finds n customers there already, is given by

$$\pi_{\mathbf{r}}(n)\tau'_{\mathbf{u}}(n) \quad \text{for} \quad 0 \le n < N$$

(yes, it is true for $n = 0$). After the system finally gets to $n+1$, it will be in state

$$\pi_{\mathbf{r}}(n)\mathbf{H}_{\mathbf{u}}(n).$$

The expression

$$\pi_{\mathbf{r}}(n)\tau'_{\mathbf{u}}(n) + \pi_{\mathbf{r}}(n)\mathbf{H}_{\mathbf{u}}(n)\tau'_{\mathbf{u}}(n+1)$$

is the mean first-passage time from n to $n+2$, as seen by a random observer. ■

In Section 4.3.2 we showed that an arriving customer will see much the same thing as a random observer if he remembers not to count himself as a member of the queue. Thus a newly arriving customer will find S_1 in state $\pi_{\mathbf{r}}(n)$, given that there are *already* n customers there. In other words, he becomes the $(n+1)$st customer. Thus he will *not* see the same mean passage times as did the random observer, because he is part of the action. In fact, he may not even be around long enough to see the queue grow longer than it was when he first arrived. For instance, suppose that a customer arrives at an empty queue. Then he himself enters S_1 and puts it in state $\pi_{\mathbf{r}}(0) = \mathbf{p}$. If he finishes service before the next customer arrives, he will not be around to see the queue grow to 2, even though it will eventually happen [in mean time $\mathbf{p}\tau'_{\mathbf{u}}(1)$]. We state this result as yet another corollary. The reader should compare this with the previous one to be sure that the differences are clear.

Corollary 4.5.2c: The mean time for a steady-state M/ME/1 queue to grow to $n+1$ for the first time, given that the nth customer has just arrived, is given by

$$\pi_{\mathbf{r}}(n-1)\tau'_{\mathbf{u}}(n) \quad \text{for} \quad 0 < n < N$$

(no, it is not true for $n = 0$). After the queue length finally reaches $n+1$, it will be in state

$$\pi_{\mathbf{r}}(n-1)\mathbf{H}_{\mathbf{u}}(n).$$

The expression

$$\pi_{\mathbf{r}}(n-1)\tau'_{\mathbf{u}}(n) + \pi_{\mathbf{r}}(n-1)\mathbf{H}_{\mathbf{u}}(n)\tau'_{\mathbf{u}}(n+1)$$

is the mean first-passage time for the queue to grow from n to $n+2$, given that a customer has just arrived. ■

The most important variation on the theme of this section is the first-passage time starting with an empty subsystem, or starting with the arrival of a customer to an empty subsystem. We will assume the former, but the two differ only by the mean time until a customer arrives, which is $1/\lambda$. This is the process that corresponds to the queue growth discussed in Section 2.3.1 for the M/M/1 queue. To do this, we need two new types of objects.

Definition 4.5.4

$\mathbf{p_u}(n)$:= *probability vector for first passage from 0 to n.* Its ith component $[\mathbf{p_u}(n)]_i$, is the probability that a customer will be in state i when the queue at S_1 reaches length n for the first time, given that the queue was initially empty. ◊

This vector was actually introduced in (4.5.4), but we were not ready to use it then.

Definition 4.5.5

$t_u(n)$:= *mean first-passage time for the queue at* S_1 *to grow from n to* $n+1$*, given that the queue was originally empty, and a customer has just arrived.* The process begins at the moment the queue reaches length n. ◊

This more or less corresponds to Definition 2.3.1 for $\tau_u(n)$, but since an M/M/1 queue has no internal states (or rather, only one internal state), it did not make any difference when service began.

We can describe this process through the eyes of the random observer. At some time in the past she observed that the queue at S_1 was empty (no one was being served). She then watched the queue, and when it finally reached the length n, she turned on her timer. At that moment, the system was in state $\mathbf{p_u}(n)$. When the queue finally reaches length $n+1$, she turns off the timer. The mean time that her timer will show is $t_u(n)$.

Let us suppose that there is no one at S_1 initially, then in mean time, $t_u(0) = 1/\lambda$, the first customer will arrive, putting the system into internal state $\mathbf{p_u}(1) = \mathbf{p}$. Eventually, the queue will grow to 2 for the first time, in mean time, $t_u(1) = \mathbf{p}\boldsymbol{\tau}'_\mathbf{u}(1)$, at which time the system will be in internal state, $\mathbf{p_u}(2) = \mathbf{p}\mathbf{H_u}(1)$. At some time in the future the queue will get to 3 for the first time, taking on average $t_u(2) = \mathbf{p_u}(2)\boldsymbol{\tau}'_\mathbf{u}(2)$ units of time. At that moment the system will find itself in internal state $\mathbf{p_u}(3) = \mathbf{p}\mathbf{H_u}(1)\mathbf{H_u}(2) = \mathbf{p_u}(2)\mathbf{H_u}(2)$. The sequence continues until the number of customers at S_1 reaches N. The total time it takes to go from 0 to n is the sum of the t_u's.

Example 4.5.2: In Figure 4.5.3, we have plotted the components of $\mathbf{p_u}(n)$ as a function of n for the M/E_2/1 queue. It is not easy to understand what is going on here, since the process is so complicated. The residual vector (Definition 4.3.1), $\boldsymbol{\pi}_\mathbf{r}(n)$, is quite different from this. In the residual process, a random observer (and for the M/G/1 queue, an arriving customer) will find a steady-state system in vector state, $\boldsymbol{\pi}_\mathbf{r}(n)$, as given by (4.3.1), assuming that there were n customers there already. The vector $\mathbf{p_u}(n)$ refers only to the customer whose arrival brings the queue to length n for the first time, given that the queue started at 0. Thus this special customer found $n-1$ customers there already when he arrived. ●

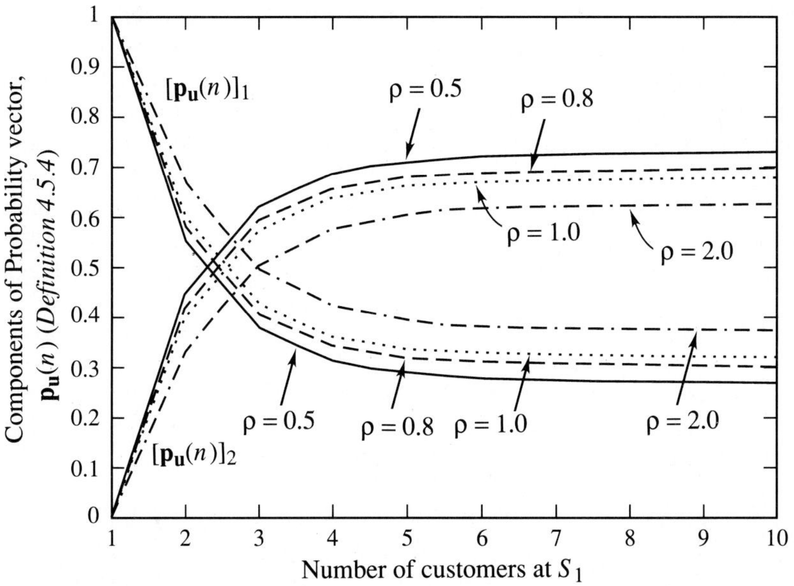

Figure 4.5.3: Components of $\mathbf{p_u}(n)$, the probability vector of first passage from 0 to n, versus n, for the $M/E_2/1$ queue. If a customer arrives at an empty queue, he will certainly go to phase 1. Thus all curves start with [1 , 0]. The two sets of curves are mirror images of each other about the line ½, since for any n, the sum of the two components is 1. For all ρ, the vectors reach their asymptotic values before $n = 10$.

Definition 4.5.6

$t(0 \to n)$:= *mean first-passage time from 0 to n.* This is the mean time it will take the queue at S_1 to grow to length n, given that S_1 was initially empty. This is the same as Definition 2.3.2. ◊

This process is summarized by the final corollary of this section.

Corollary 4.5.2d: The mean time for an M/ME/1 queue (open or closed) to grow from n to $n+1$ for the first time, given that S_1 was initially empty, starting with $t_u(0) = 1/\lambda$, is

$$t_u(n) := \mathbf{p_u}(n)\tau'_{\mathbf{u}}(n), \quad n = 1, 2, \cdots . \tag{4.5.6a}$$

Starting with $\mathbf{p_u}(1) := \mathbf{p}$, the internal state of the system at the moment of first passage to n, $\mathbf{p_u}(n)$, is given recursively by

$$\mathbf{p_u}(n) = \mathbf{p_u}(n-1)\mathbf{H_u}(n-1), \quad n = 1, 2, \cdots \tag{4.5.6b}$$

The mean first-passage time from 0 to n is

$$t(0 \to n) = \sum_{l=0}^{n-1} t_u(l). \tag{4.5.7}$$

Compare this with (2.3.3a). These formulas are independent of N and are true as long as $n \le N$. ■

Example 4.5.3: The overall behavior of $t(0 \rightarrow n)$ for the M/E_2/1 queue is similar to that for the M/M/1 queue, given in Figure 2.3.2. In Figure 4.5.4 we have plotted the two types on the same graph so they can be compared. Although similar, they have distinct differences. As n gets larger, the two curves separate from each other, and the closer ρ gets to 1, the greater the separation, with the E_2 curve lying below the M/M/1 curve. When ρ is much greater than 1, the growth is dominated by the difference of the arrival and service rates, but there *is* a difference, depending on the pdf of S_1. For the M/H_2/1 queue, the differences are more pronounced, and the corresponding curves end up *above* those for the M/M/1 queue. ●

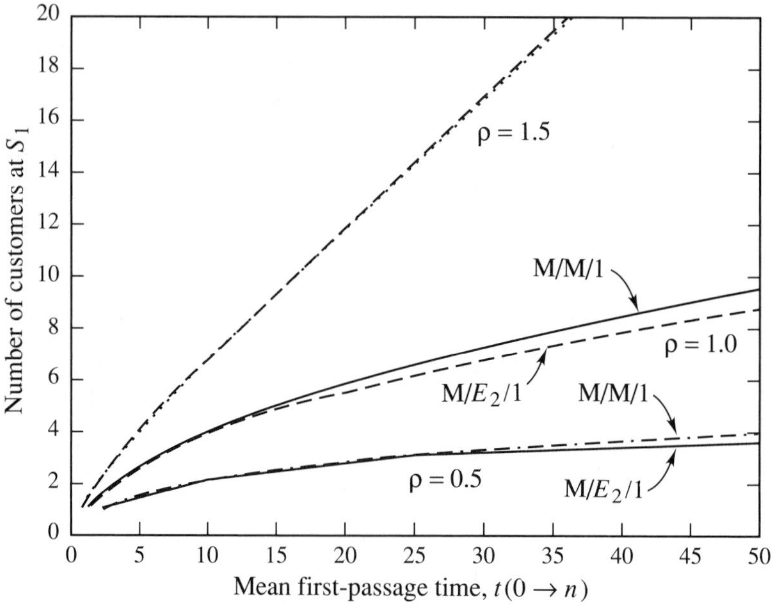

Figure 4.5.4: Comparison of the mean first-passage times from 0 to n, $t(0 \rightarrow n)$, between the M/M/1 queue and the M/E_2/1 queue, for $\rho = 0.5$, 1.0, and 1.5. In all cases, the curve corresponding to E_2 ends up lower, but the two do cross.

Is this property dependent on C^2? The reader should explore the behavior of this process for other distributions, since so little is known about this subject.

Note that the first-passage processes we have been discussing allow the queue at S_1 to empty any number of times before finally reaching its goal. In the third subsection of Section 4.5.3 we will study the first excursion to n during a *busy period*. In that case we will find the *probability* that the queue will reach length n before it empties, since it is not certain to do so. We have to develop some other expressions first.

4.5.2. Formal Procedure for Finding System Parameters

After reading the preceding section, the reader must have become familiar with how to set up the expressions needed to calculate various system parameters. We will now outline a formal procedure for doing this. First, based on the given conditions, or initial assumptions, one finds the *initial vector*, $\mathbf{p_i}$, that describes the internal state of the system initially (e.g., the $\mathbf{p_u}$'s). Then, depending on the process of interest, *propagation matrices*, $\mathbf{S_p}$, are found (e.g., the first-passage matrices). Finally, the *final vector*, $\mathbf{v'_f}$, that contains the kind of information desired (e.g., $\boldsymbol{\tau}'$ for mean times and $\boldsymbol{\varepsilon}'$ for probabilities), is found. The desired scalar property (call it g) is then given by

$$g = \mathbf{p_i} \mathbf{S_p} \mathbf{v'_f}. \tag{4.5.8a}$$

The initial and final vectors are commonly made of propagation matrices post- or premultiplying other initial or final vectors, while propagation matrices are usually products of other propagation matrices. In this way one can build up an unlimited sequence of conditions and results without difficulty. Also, the boundary between **i** and **p**, or **p** and **f**, is not necessarily unique, nor is it important to try to find a definition that makes them unique.

The reader should peruse through the material already covered, to see if this scheme holds true everywhere. In doing so you will notice that almost always, an initial vector can be written as the entrance vector of S_1, [**p**], postmultiplied by a propagation matrix (call it $\mathbf{S_i}$). Also, almost always, the final vector can be written as some other propagation matrix (call it $\mathbf{S_f}$), premultiplying $\boldsymbol{\varepsilon}'$. This leads to

$$g = \mathbf{p} \mathbf{S_i} \mathbf{S_p} \mathbf{S_f} \boldsymbol{\varepsilon}' = \Psi[\mathbf{S_i} \mathbf{S_p} \mathbf{S_f}]. \tag{4.5.8b}$$

We now see why the $\Psi[\cdot]$ operator appears in so many places and why it is such a useful object.

4.5.3. Properties of the *k*-Busy Period

Everything that goes up must come down – well, almost everything. Whenever S_1 is emptied of customers ($n = 0$), we say that *a busy period has ended.* If observation began with some initial conditions (call them collectively $\{\cdots\}$), then we have a $\{\cdots\}$*-busy period.* As described in Section 2.3.2, if a customer has just arrived to an empty subsystem, we simply have the beginning of a *busy period*. Clearly, studying busy periods requires studying queue-length reduction.

We will proceed in analogy with Section 4.5.1. However, not all objects will work out in the same way. On the one hand, the $\mathbf{H_d}$ matrices are much simpler than the $\mathbf{H_u}$'s. On the other hand, the $\boldsymbol{\tau}'_\mathbf{d}$ vectors depend explicitly on N, as well as n; thus length reduction processes are somewhat more complicated to express. The reason for this, as we shall presently see, is that in its attempt to go up, the queue can never drop below 0, and thus is bounded by its own length. In trying to shrink, the queue can falter and grow to any length before finally coming down. A ceiling of N is imposed by the system's finite population, and the higher the ceiling, the longer it takes to get down to 0. Actually, first-passage processes to go from n to $n-1$ depend on $N-n$, so unless S_2 is load dependent, the problem is not that bad.

We have one last point before going on. In dealing with a growing queue, we had to start at 0. Similarly, in studying the decreasing queue, the recursive equations must start at N. But where does one start in an open system?

Conditional State Probabilities When Queue Decreases by One

The first-passage matrices from n to $n-1$ can be written down with some thought and no algebra. A decrease in queue length can only occur after a departure. Immediately after that a new customer (provided that n was greater than 1 originally) enters S_1, putting the system in internal state $\mathbf{p}$. Since this is independent of the state the system was in initially, our desired matrix must be $\mathbf{Q}$. However, we will go through a full algebraic derivation, since we will have do it anyway when we derive the corresponding first-passage times. Define the following matrix for an M/ME/1//N loop:

Definition 4.5.7

$\mathbf{H_d}(n;N)$:= *probability matrix of first passage from n to n* -1. Its ijth component, $[\mathbf{H_d}(n;N)]_{ij}$, is the probability of finding the system in state $\{j;n-1;N\}$, given that the queue at S_1 has reached length $n-1$ for the first time, after starting in state $\{i;n;N\}$. ◊

Next look at Figure 4.5.5. This diagram is similar to Figure 4.5.1, but here the wavy lines go from higher to lower n. It also includes the possibilities when all customers are already at S_1 ($n=N$). Clearly, if all customers are at S_1, the next event must be there, and either the customer in service stays in S_1, [$\mathbf{P}$], and then eventually leaves, [$\mathbf{H_d}(N;N)$], or leaves directly, and is replaced by the next customer, [$\mathbf{q'p}$]. Thus

$$\mathbf{H_d}(N;N) = \mathbf{P}\mathbf{H_d}(N;N) + \mathbf{q'p}.$$

Now solve for $\mathbf{H_d}(N;N)$ to get, using (3.1.1b):

$$\mathbf{H_d}(N;N) = (\mathbf{I}-\mathbf{P})^{-1}\mathbf{q'p} = \boldsymbol{\varepsilon}'\mathbf{p} = \mathbf{Q}, \tag{4.5.9a}$$

as expected. For any other $n>0$ (we drop the $\mathbf{d}$), Figure 4.5.5 implies that

$$\mathbf{H}(n;N) = (\lambda\mathbf{I}+\mathbf{M})^{-1}\mathbf{Mq'p} + (\lambda\mathbf{I}+\mathbf{M})^{-1}\mathbf{MPH}(n;N)$$
$$+\,\lambda(\lambda\mathbf{I}+\mathbf{M})^{-1}\mathbf{H}(n+1;N)\mathbf{H}(n;N).$$

Probability That Queue Will Reach at Least Length n

We closely parallel the discussion in the second part of Section 2.3.2, with the added complication that we must keep track of internal states. The consequent matrices, $\mathbf{W_u}(n)$, are similar to the $\mathbf{H_u}$'s. In fact, they satisfy the same recursive equations, *but* they are not isometric. Define for $k < n < N$:

Definition 4.5.11

$\mathbf{W_u}(n;k) :=$ *probability matrix that the queue will go from n to n+1 without dropping to k.* $\mathbf{W_u}(n;k)$ assumes the following initial conditions: Given an M/ME/1//N loop, there are $n < N$ customers at S_1, and the active customer is at phase i. The process ends when either the queue grows to $n+1$, or shrinks to $k < n$. If it is the former, then $[\mathbf{W_u}(n;k)]_{ij}$ is the probability that the system will be in state $\{j; n+1; N\}$. ◊

In other words, $[\mathbf{W_u}(n;k)]_{ij}$ is the probability that the system will go from $\{i; n; N\}$ to $\{j; n+1; N\}$ for some (any) j without going to $\{\cdot, k; N\}$. The ith component of vector, $\mathbf{W_u}(n;k)\boldsymbol{\varepsilon}'$, is the probability of going from $\{i; n; N\}$ to any internal state of $n+1$ customers, without dropping to k. Since this process is not certain to happen, it cannot be equal to 1, thus $\mathbf{W_u}(n;k)$ is not isometric.

This process, and in fact *all* first-passage processes, fall into the class of *taboo processes*. For such processes, the entire state space of the system is partitioned into two (disjoint) subsets. The process begins with the system in a single state in one of the subsets, and ends when the system finds itself in any one of the states of the other subset. In our case, subset 1 consists of *all* internal states corresponding to queue lengths of $k+1, k+2, \cdots, n-1$, and n, while subset 2 is everything else. The initial state is $\{i; n; N\}$, which is an element of subset 1. This concept is much broader than we need. In fact, it obscures the underlying view in LAQT, that all internal states belonging to one queue length should always be treated as a whole.

There is a natural scalar that goes with matrix $\mathbf{W_u}(n;k)$, which we now define.

Definition 4.5.12

$W_u(n;k) :=$ *probability that the queue at S_1 will rise from n to n+1 without first dropping to k, given that the active customer has just begun service.* We will call this the scalar probability associated with $\mathbf{W_u}(n;k)$. From its definition it is clear that

$$W_u(n;k) := \mathbf{p}\mathbf{W_u}(n;k)\boldsymbol{\varepsilon}' = \Psi[\mathbf{W_u}(n;k)]. \tag{4.5.17}$$

Each of the $\mathbf{W}$ matrices we will be presently introducing has an analogous W scalar counterpart. ◊

Fortunately, Figure 4.5.1 is applicable to deriving the relationships among the $\mathbf{W_u}(n;\,k)$'s. Let us first look at $\mathbf{W_u}(k{+}1;\,k)$. [$\mathbf{W_u}(k;\,k)$ must be $\mathbf{O}$, because the system is already at its lower bound.] There are two successful paths available in this case. Either an arrival occurs, putting the system in state $\{i;\,n+1\}$ immediately, or there is an internal transition, after which the queue eventually rises to $n+1$ without ever having a departure. Thus

$$\mathbf{W_u}(k{+}1;\,k) = \lambda(\lambda\mathbf{I}+\mathbf{M})^{-1} + (\lambda\mathbf{I}+\mathbf{M})^{-1}\mathbf{M}\mathbf{P}\mathbf{W_u}(k{+}1;\,k).$$

Multiply by $(\lambda\mathbf{I}+\mathbf{M})$, collect terms, and get

$$\mathbf{W_u}(k{+}1;\,k) = \left(\mathbf{I}+\frac{1}{\lambda}\mathbf{B}\right)^{-1} = \lambda(\lambda\mathbf{I}+\mathbf{B})^{-1}. \tag{4.5.18a}$$

We are now ready to treat the general case. Here all three types of events can occur, since the queue at S_1 can drop by 1 and still go back up. We write, for $n > k+1$ (while momentarily dropping the $\mathbf{u}$'s),

$$\mathbf{W}(n;\,k) = \lambda(\lambda\mathbf{I}+\mathbf{M})^{-1} + (\lambda\mathbf{I}+\mathbf{M})^{-1}\mathbf{M}\mathbf{P}\mathbf{W}(n;\,k)$$

$$+ (\lambda\mathbf{I}+\mathbf{M})^{-1}\mathbf{M}\mathbf{q}'\mathbf{p}\mathbf{W}(n{-}1;\,k)\mathbf{W}(n;\,k).$$

The usual manipulations lead to the following recursive formulas. For fixed k, and $n > k+1$,

$$\mathbf{W_u}(n;\,k) = \lambda[\lambda\mathbf{I}+\mathbf{B}-\mathbf{B}\mathbf{Q}\mathbf{W_u}(n{-}1;\,k)]^{-1}. \tag{4.5.18b}$$

So, for instance,

$$\mathbf{W_u}(k{+}2;\,k) = \lambda\left[\lambda\mathbf{I}+\mathbf{B}-\mathbf{B}\mathbf{Q}(\lambda\mathbf{I}+\mathbf{B})^{-1}\right]^{-1}.$$

A comparison of (4.5.18b) with (4.5.2b) shows that $\mathbf{W_u}(n;\,k)$ and $\mathbf{H_u}(n)$ satisfy the same recursive formula, yet they are not equal. In particular, the $\mathbf{W_u}(n;\,k)$'s are not isometric, even though the $\mathbf{H_u}(n)$'s are. This apparent dilemma is easily resolved when we recognize that the two sets of matrices have different first matrices in their recursive construction, Equations (4.5.2e) and (4.5.18a). Recall that we proved the $\mathbf{H_u}(n)$'s to be isometric by induction. First we showed that $\mathbf{H_u}(0)\boldsymbol{\varepsilon}' = \boldsymbol{\varepsilon}'$, and second, showed that if it was true for $\mathbf{H_u}(n)$, then it must be true for $\mathbf{H_u}(n{+}1)$. We could show the second part of the proof for $\mathbf{W_u}(n;\,k)$, *but* we cannot satisfy the first condition. For $\mathbf{W_u}(k{+}1;\,k)$ to be isometric, we must have $\mathbf{B}\boldsymbol{\varepsilon}' = \mathbf{o}'$, which in turn implies that $b(x)$, the pdf for S_1, is identically 0 everywhere, an impossibility.

Now we are prepared to find the object described in the title of this section. Let us define the following multistep matrix:

Definition 4.5.13

$\mathbf{W_u}(n \to n{+}l;\,k) :=$ *probability matrix that the number of customers at* S_1 *will* *grow from n to $n+l$ without dropping to k.* The process starts with the system in state $\{i; n\}$, and stops when the system is either in some state with $n+l$ customers, or in some state with $k < n$ customers. $[\mathbf{W_u}(n \to n+l; k)]_{ij}$ is the probability that the system will be in state $\{j; n+l\}$ when the process ends. $W_u(n \to n+l; k) := \Psi[\mathbf{W_u}(n \to n+l; k)]$ is the associated scalar probability. ◊

The ith component of $[\mathbf{W_u}(n \to n+l; k)\boldsymbol{\varepsilon}']$ is the probability that the process will end with $n+l$ customers at S_1, and the ith component of $[\mathbf{I} - \mathbf{W_u}(n \to n+l; k)]\boldsymbol{\varepsilon}'$ is the probability that the process will end with only k customers in the queue.

By their very definitions, we know that

$$\mathbf{W_u}(n \to n+1; k) = \mathbf{W_u}(n; k). \tag{4.5.19a}$$

In analogy with the discussion surrounding (2.3.10a), we see that in order to go up two steps without dropping to k, we must first go up one step without dropping to k, and then go the second step. Therefore,

$$\mathbf{W_u}(n \to n+2; k) = \mathbf{W_u}(n; k)\, \mathbf{W_u}(n+1, k) = \mathbf{W_u}(n \to n+1; k)\, \mathbf{W_u}(n+1; k),$$

or in general,

$$\mathbf{W_u}(n \to n+l+1; k) = \mathbf{W_u}(n \to n+l; k)\, \mathbf{W_u}(n+l; k). \tag{4.5.19b}$$

This recursive expression is all that is needed to calculate everything, but it can also be written in the alternative form

$$\mathbf{W_u}(n \to n+l+1; k) = \mathbf{W_u}(n; k)\mathbf{W_u}(n+1; k) \cdots \mathbf{W_u}(n+l; k). \tag{4.5.19c}$$

Keep in mind that these matrices do not commute with each other, so the order of multiplication is important.

It is time to summarize the results of this section in a theorem.

Theorem 4.5.3: For any k and n such that $0 \le k < n < N$, the matrices, $\mathbf{W_u}(n; k)$ and $\mathbf{W_u}(n \to n+l; k)$, are recursively given by the following procedure. From (4.5.18a)

$$\mathbf{W_u}(k+1; k) = \lambda(\lambda\mathbf{I} + \mathbf{B})^{-1}.$$

Next, from (4.5.18b), for $l = 1, 2, \cdots, N-k-1$,

$$\mathbf{W_u}(k+l+1; k) = \lambda[\lambda\mathbf{I} + \mathbf{B} - \mathbf{BQW_u}(k+l; k)]^{-1}.$$

For any $n > k$, [Equation (4.5.19a)], set

$$\mathbf{W_u}(n \to n+1; k) = \mathbf{W_u}(n; k),$$

and for $l = 1, 2, \cdots, N-n-1$,

$$\mathbf{W_u}(n \to n+l+1; k) = \mathbf{W_u}(n \to n+l; k)\mathbf{W_u}(n+l; k)$$

[from (4.5.19b)]. Given any initial vector $\mathbf{p_I}$, the conditional scalar probabilities of queue growth as defined in this section are given by

$$\mathbf{p_I W_u}(n;\, k)\boldsymbol{\varepsilon}' \qquad \text{and} \qquad \mathbf{p_I W_u}(n \rightarrow n{+}l;\, k)\boldsymbol{\varepsilon}'.$$

All of these equations are valid for any ρ, and for both open and closed M/ME/1 systems (as long as $n+l \leq N$). ■

There are numerous variations that one can pursue, but by far the most important and most interesting is the *probability that the queue at S_1 will grow to AT LEAST n during a (k = 1) busy period.* Here we want to see the queue reach n without going to 0. Since the busy period is so special, we will provide special treatment.

Definition 4.5.14

$\mathbf{W_u}(n) :=$ *probability matrix that the queue at S_1 will rise from n to $n+1$ customers during a busy period.* This is the same as the probability that the queue will get to some state with $n + 1$ customers before it empties, given that the system started with n customers, i.e., $\mathbf{W_u}(n) = \mathbf{W_u}(n;\, 0)$. ◊

Definition 4.5.15

$\mathbf{W_u}(1 \rightarrow n) :=$ *probability matrix that the queue length at S_1 will reach at least n during a busy period.* The component definitions are the same as those given in Definitions 4.5.11 and 4.5.13, with $k = 0$. $W_u(1 \rightarrow n) = \Psi[\mathbf{W_u}(1 \rightarrow n)]$ is the associated scalar probability. ◊

That is, these matrices satisfy the following equations:

$$\mathbf{W_u}(n) := \mathbf{W_u}(n;\, 0), \tag{4.5.20a}$$

$$\mathbf{W_u}(1 \rightarrow n) := \mathbf{W_u}(1 \rightarrow n;\, 0), \tag{4.5.20b}$$

and $W_u(1 \rightarrow n)$ is the same as that defined in the second part of Section 2.3.2. That is, it is the (scalar) probability that the queue at S_1 will grow at least to n during a busy period. Then, by definition, $W_u(1;\, 1) = 1$. Recall that a busy period begins with the arrival of a customer to an empty server, thus the system is initially put into internal state $\mathbf{p}$, with queue length $n = 1$. Then we can state the busy-period corollary to Theorem 4.5.3.

Corollary 4.5.3: For any n such that $1 < n < N$, the probability that the queue at S_1 will reach at least n during a busy period can be calculated in the following way. Let

$$\mathbf{W_u}(1) = \lambda(\lambda\mathbf{I} + \mathbf{B})^{-1}; \quad \mathbf{W_u}(1 \rightarrow 2) = \mathbf{W_u}(1); \quad W_u(1 \rightarrow 2) = \Psi[\mathbf{W_u}(1 \rightarrow 2)].$$

Then for $n = 2, 3, \cdots, N-1$,

$$\mathbf{W_u}(n) = \lambda\Big[\lambda\mathbf{I} + \mathbf{B} - \mathbf{BQW_u}(n-1)\Big]^{-1}, \tag{4.5.21a}$$

For this process to occur, the queue must grow from 0 to k without ever returning to 0, [$\mathbf{pW_u}(1 \to k)$], and then it must reduce to 0 without ever exceeding k, [$\mathbf{w'_d}(k \to 0)$]. Therefore, we have

$$W_m(k; N) = \mathbf{pW_u}(1 \to k)\, \mathbf{w'_d}(k \to 0), \tag{4.5.27a}$$

which upon using (4.5.26) yields

$$W_m(k; N) := \left(\prod_{l=1}^{k-1} Z(l) \right) \Psi\left[\mathbf{W_u}(1 \to k)\, (\mathbf{I} + \lambda \mathbf{V})^{-1} \right], \tag{4.5.27b}$$

and from (4.5.24b) and (4.5.21c),

$$W_m(N; N) := \Psi\left[\mathbf{W_u}(1 \to N) \right] = W_u(1 \to N). \tag{4.5.27c}$$

With some practice, one can figure out what process is going on just by looking at the terms in the equation. For instance, in (4.5.27b), the term inside the Ψ brackets is the probability that the queue will get to k, [$\mathbf{W_u}(1 \to k)$], and then drop back to $k-1$, [$(\mathbf{I} + \lambda \mathbf{V})^{-1}$]†. Then the queue works its way back to 0 without ever exceeding k [$Z(1), Z(2), \cdots, Z(k-1)$].

We are indeed fortunate to find Z's that are so simple. For more complicated systems (either load-dependent, M/ME/C, or ME/ME/1), we will not find such simple equations for these processes, although the general formalism is the same. We are still left with the question of how Equation (2.3.16b) connects *up* and *down* operators. Recall that

$$W_m(k; N) = W_u(1 \to k) - W_u(1 \to k+1),$$

thus, from (4.5.21b), (4.5.27a), and (4.5.26b), we can write

$$W_m(k; N) = \mathbf{pW_u}(1 \to k)\mathbf{w'_d}(k \to 0) = \mathbf{pW_u}(1 \to k)[\mathbf{I} - \mathbf{W_u}(k)]\boldsymbol{\varepsilon}'. \tag{4.5.28a}$$

A sufficient condition for this to be true is for the following equation to be true:

$$\begin{aligned}[\mathbf{I} - \mathbf{W_u}(k)]\boldsymbol{\varepsilon}' &= \mathbf{w'_d}(k \to 0) \\ &= \left(Z(1)\, Z(2) \cdots Z(k-1) \right) (\mathbf{I} + \lambda \mathbf{V})^{-1} \boldsymbol{\varepsilon}'. \end{aligned} \tag{4.5.28b}$$

It would be nice if we could prove this equality algebraically, for it might give us some further insights into *up* and *down* processes.

We will now show how this all fits together in the following summary algorithm/theorem.

Theorem 4.5.4: For any M/ME/1//*N* loop, and for any ρ, the busy-period queue-length probabilities of Definitions 4.5.15 and 4.5.18 can be calculated using Equations (4.5.21), (4.5.23), and (4.5.27) [or (4.5.28a)], in the following way.

† Remember, this is the operator which finds the probability that S_1 will finish before S_2.

BEGIN PROCEDURE

* Set

$$\mathbf{Z}(0) = (\mathbf{I} + \lambda\mathbf{V})^{-1},$$
$$\mathbf{W_u}(0) = \mathbf{O},$$
$$\mathbf{w_u}(1 \to 1) = \mathbf{p}$$
$$z(1) = 1,$$
$$W_m(1) = \Psi\left[(\mathbf{I} + \lambda\mathbf{V})^{-1}\right],$$
$$W_u(1 \to 1) = 1.$$

FOR $n = 1$ TO $NMAX$, DO

$$\mathbf{Z}(n) := [\mathbf{I} + \lambda\mathbf{V} - \lambda\mathbf{V}\mathbf{Z}(n-1)\mathbf{Q}]^{-1},$$
$$\mathbf{W_u}(n) = \lambda[\lambda\mathbf{I} + \mathbf{B} - \mathbf{B}\mathbf{Q}\mathbf{W_u}(n-1)]^{-1},$$
$$\mathbf{w_u}(1 \to n+1) = \mathbf{w_u}(1 \to n)\, \mathbf{W_u}(n),$$
$$Z(n) = \Psi\left[\mathbf{Z}(n)\right]$$
$$z(n+1) = z(n)\, Z(n)$$
$$W_u(1 \to n+1) = \mathbf{w_u}(1 \to n+1)\boldsymbol{\varepsilon}'$$
$$W_m(n+1) = z(n+1)\, \mathbf{w_u}(1 \to n+1)(\mathbf{I} + \lambda\mathbf{V})^{-1}\boldsymbol{\varepsilon}'.$$

END FOR

END PROCEDURE

Therefore, for any M/ME/1//N loop, with $N \geq 2$, the probability that the queue at S_1 will contain at least n customers at one time during a busy period is given by the sequence

$$1,\ W_u(1 \to 2),\ W_u(1 \to 3),\ \cdots,\ W_u(1 \to N-1),\ W_u(1 \to N).$$

The probability that the largest queue length during a busy period will be n is given by the sequence

$$W_m(1),\ W_m(2),\ \cdots,\ W_m(N-1),\ W_u(1 \to N).$$

(Note: The last term in the sequence is correct.) ■

In both sequences, the dependence on N is determined by how one ends the sequence. For the open queue, the sequences never end, but when $\rho \leq 1$ they tend to 0. If $\rho > 1$, then $W_u(1 \to N)$ will *not* approach 0, but instead will approach

$$\lim_{N \to \infty} W_u(1 \to N) = \text{probability that a busy period will never end.}$$

Example 4.5.6: We have calculated the W_m's, once again for the M/E_2/1/20 queue, and compared them with the corresponding values for the M/M/1/20 queue for three different values of ρ. The results are presented in Figure 4.5.9. For any given ρ, when n is small, the curve for the exponential distribution is higher than that for the E_2. But since the sum over all integer points must be 1, the two curves cross somewhere, and for large n, the curve for E_2 is higher. We give the same warning here as we did for the other curves: Although the two distributions yield similar results, one should look for other distributions that could give radically different results. ●

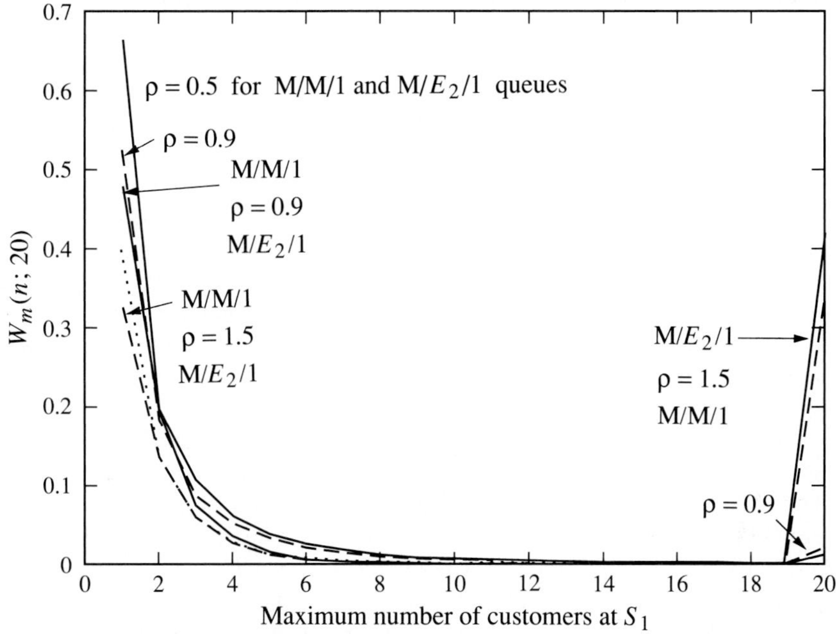

Figure 4.5.9: Probability, $W_m(n; 20)$ as a function of n, for both the M/M/1 and M/E_2/1 queues. Three sets of curves are shown, for $\rho = 0.5$, 0.9, and 1.5. The curves for the M/M/1 queue are the same as those in Figure 2.3.4 as long as $n < 20$. The value at $n = 20$ corresponds to the probability that the queue will exceed 19 during a busy period for any loop where $N \geq 20$, so for $\rho > 1$ it is quite significant. Even for $\rho = 0.9$, this probability is not negligible. The sum over all integer points must be 1.

As a final comment, note that $N = 1$ is a trivial case, for then the queue will always grow to 1, and never grow further, before the busy period ends. For $N = 2$, we have

$$W_m(1) = \Psi[(\mathbf{I} + \lambda\mathbf{V})^{-1}]$$

and

$$W_m(2) = \lambda\Psi[(\lambda\mathbf{I} + \mathbf{B})^{-1}].$$

The first equation, as we have noted before, is the probability that S_1 will finish before S_2, and of course, the second term is the probability that S_2 will finish before S_1. Their sum must be 1, which is easily shown by the following:

$$(\mathbf{I}+\lambda\mathbf{V})^{-1}+\lambda(\lambda\mathbf{I}+\mathbf{B})^{-1}=(\mathbf{I}+\lambda\mathbf{V})^{-1}+\lambda\mathbf{V}(\mathbf{I}+\lambda\mathbf{V})^{-1}$$
$$=[\mathbf{I}+\lambda\mathbf{V}](\mathbf{I}+\lambda\mathbf{V})^{-1}=\mathbf{I},$$

followed by $\Psi[\mathbf{I}]=1$.

4.5.4. Mean Time to Failure with Backup and Repair

Our emphasis so far has been on viewing customers as individuals who go around in circles demanding service, one at a time, from two different subsystems. An increasingly important application, with a completely different emphasis than we usually see in queueing theory texts, occurs in reliability theory, where one asks such questions as: *How long it will take before a subsystem has fewer functional components than is acceptable?* We are ready to set up the procedure by which such questions can be analyzed, using the material already discussed in this chapter. We are even prepared to solve many of the simpler problems, although the question of how one deals with multiple components functioning simultaneously must wait until Chapter 6. It is most important to note that the procedures we discuss now generalize directly once we have set up the structure for parallel processing.

Consider the following. Suppose that we have several identical appearing devices (terminals, computers, automobiles, VLSI chips, etc.). Once one of them is turned on, it continues to run until it fails (breaks down, or something). Assume that the lifetime of one of these devices is described by the function $R_1(t)$. As you already know, this is the *reliability function* for S_1, which is where the name came from. That is, $R_1(t)$ is the probability that the device will be functioning t units of time after it was first turned on, and $b_1(t)=-R'_1(t)$ is the pdf of the failure time. If only one device is available, the mean time until failure is the expected life of the device, namely,

$$T_1 := \int_0^\infty t\, b_1(t)\, dt.$$

Let there now be several devices available, and as soon as the first one fails, a second one is started up. The second one is referred to as a (cold) backup (cold – because it does not start up until the first one fails). If the first one is discarded, the pdf of the time until both have failed is the convolution of $b_1(t)$ with itself, with the mean time to failure of $2T_1$. Suppose, instead, that the broken one is immediately sent to the repair shop (with only one repairman), where the time it takes to fix it is distributed according to the pdf $b_2(x)$. As soon as it is repaired, it is returned to the pool of available devices, as good as new [its reliability function, $R_1(t)$ is the same as it was the first time through]. The question to be answered is: *How long will it take until all the devices are in the repair shop?* Thus we have described the title of this section.

Let us call the process above, scheme 1. There are numerous variations that one can play on this scheme, some of which are: (2) failure occurs when only one (or in general, k) device(s) is(are) still functioning; (3) a backup must always be running, whether it is being used or not, even while the primary device is still functioning (hot backup, or parallel redundancy); and (4) the system has been running for some unknown time before questions are asked (residual times). Schemes 2 and 4 can be treated with material that we already have prepared in this chapter, but scheme 3 must wait until Chapter 6 and the M/G/C//N queue.

By now it should be clear to the reader that if we let S_1 represent $b_1(t)$, and $b_2(x)$ is exponentially distributed with mean, $1/\lambda$, then we are looking at an M/ME/1//N loop, where N is the total number of devices, (i.e., $N-1$ backups). Following scheme 1, suppose that initially all devices are functional, and then one of them is started. Then the initial vector is $\mathbf{p}$ itself. The mean time to failure in this circumstance is the same as the mean time for the N-busy period, $t(N \to 0; N)$, as given in Definition 4.5.10. The utilization parameter ρ is less meaningful in this context. It is still λT_1, which is now the ratio of the mean lifetime to the mean repair time of a single device. We are not particularly interested in systems where ρ is close to 1, nor do *open systems* have much relevance (an infinite number of backups? – well, in *inventory problems* where new parts are being manufactured continuously). Instead, we might expect ρ to be much greater than 1, since it usually takes much less time to repair a device than it did for it to break in the first place (retail commercial products such as children's toys, excepted).

Let us first examine our equations for $N = 1$. Here repair time is of no significance (once you start falling, if you do not have a spare parachute, it is no use telling you that your failed parachute can be mended in *no time at all*, after you land), so as we said before, $MTTF(1) = T_1$.

The case where $N = 2$ is most enlightening. As before, the mean time for the first one to fail is T_1, but now the race is on to see if the first device can be repaired before the second one fails. According to (4.5.15),

$$MTTF(2) = t_d(1;2) + t_d(2;2).$$

But from (4.5.11a),

$$t_d(2;2) = \mathbf{p}\boldsymbol{\tau}'_{\mathbf{d}}(2;2) = \mathbf{pV}\boldsymbol{\varepsilon}' = T_1$$

(of course), while from (4.5.12),

$$t_d(1;2) = \mathbf{p}\boldsymbol{\tau}'_{\mathbf{d}}(1;2) = \frac{1}{\lambda}\Psi[\mathbf{U}(\mathbf{I}+\lambda\mathbf{V})].$$

We played with expressions like this in Section 4.4.2. Look at (4.4.8b), where $\mathbf{C} = \lambda\mathbf{VD} = \mathbf{I} - \mathbf{D}$, $\mathbf{D} = (\mathbf{I}+\lambda\mathbf{V})^{-1}$, $\gamma_1 = \Psi[\mathbf{C}] = 1 - \Psi[\mathbf{D}]$, and

$$\mathbf{pU}(\mathbf{I}+\lambda\mathbf{V}) = \frac{1}{1-\gamma_1}\mathbf{pCD}^{-1} = \frac{\lambda}{\Psi[\mathbf{D}]}\mathbf{pV}.$$

Therefore, $t_d(1;\,2) = T_1 / \Psi[\mathbf{D}]$, so

$$MTTF(2) = T_1 \left(1 + \frac{1}{\Psi[\mathbf{D}]} \right). \qquad (4.5.29a)$$

As expected, the MTTF is proportional to the mean uptime of one device, but it also depends on the term, $1/\Psi[\mathbf{D}]$, which can be interpreted as the expected number of times the broken device will be repaired before the good one fails, given that both processes began simultaneously. First, we show that $\Psi[\mathbf{D}]$ is truly the probability that repair will occur before backup failure.

Given that two processes (call them S_1 and S_2), begin simultaneously,

$$X := \Pr(S_1 \text{ will finish before } S_2) = \int_0^\infty b_1(t) R_2(t)\, dt.$$

But in our case, S_2 is exponentially distributed, so $R_2(t) = \exp(-\lambda t)$, and from (3.1.10),

$$X = \int_0^\infty e^{-\lambda t} b_1(t)\, dt = B^*(\lambda) = \Psi[(\mathbf{I} + \lambda \mathbf{V})^{-1}] = \Psi[\mathbf{D}].$$

Thus we have shown that the Laplace transform and the definition of X are the same. Which interpretation is more basic to our understanding of this process? Well, (4.5.29a) is also the expression for the MTTF of a **G**/G/1//2 queue, but in that case, while the expression for X will still hold, there will be no Laplace transform to interpret, since in that more general case R_2 is *not* exponential.

Let us look at (4.5.29a) one more time before going on to $N > 2$. Note that if there is no repair ($\lambda = 0$), then $X = 1$, and $MTTF(2) = 2T_1$, as already predicted. On the other hand, if repair is instantaneous (*and* breakdown can never occur instantaneously), then $X = 0$ and $MTTF(2) = \infty$, also as expected. A third possibility, implied by the parenthetical statement, is the probability that breakdown *can* occur instantaneously. This would happen, for instance, if the backup part was faulty to begin with. We have almost completely ignored this possibility in our discussions, but it is easily handled. It corresponds to $R_1(0) = 1 - \alpha < 1$, and to a service time matrix, $\mathbf{V}$, which has a 0 eigenvalue. Interestingly enough, such distributions are referred to as *defective*. Such behavior can be handled by a pdf of the form

$$b_1(x) = \alpha\, \delta(x - 0_+) + (1 - \alpha) f_1(x),$$

where α is the probability that a part is faulty to begin with, and f_1 is the pdf for parts that are not faulty. δ is the Dirac delta function, which will be described in detail in (5.1.12a) and following. If we put this into the equation for X, and note that at least $R_2(0) = 1$, we get

$$X = \alpha + (1 - \alpha) \int_0^\infty f_1(x) R_2(x)\, dx. \qquad (4.5.29b)$$

From this we see that even if repair is almost instantaneous (assume that instantaneous breakdown occurs before instantaneous repair), X must be greater than α, and

$$MTTF(2) \le T_1 \left(1 + \frac{1}{\alpha} \right) < \infty.$$

This implies that the behavior of $b_1(x)$ for very small x (even if there is no instantaneous breakdown) could be critical for estimating the mean time to failure of a system.

We were able to find a convenient expression for $MTTF(2)$, but for $N > 2$ it becomes more tedious. Since (4.5.12) is fairly simple, we now will seek a general expression that is not recursive. From (4.5.12) and (4.5.13) we know that $t_d(N-l; N) = \Psi[\mathbf{UK}(l)]/\lambda$. Therefore from (4.5.15),

$$MTTF(N) = t(N \to 0; N) = \frac{1}{\lambda}\sum_{k=1}^{N} t_d(k; N) = \frac{1}{\lambda}\sum_{k=1}^{N} \Psi[\mathbf{UK}(N-k)].$$

Now let $l = N - k$; then

$$MTTF(N) = \frac{1}{\lambda}\sum_{l=0}^{N-1} \Psi[\mathbf{UK}(l)]. \tag{4.5.30a}$$

We actually have worked with something like this already, in Section 4.3.1. There, in (4.3.4b) we showed that (we have replaced N with $l+1$)

$$\mathbf{p}[\mathbf{K}(l+1) - \mathbf{I}] = \frac{1}{1-\rho}\Big(1 - \Psi[\mathbf{U}^l \lambda \mathbf{V}]\Big)\lambda \mathbf{pV}.$$

But $\mathbf{K}(l+1) - \mathbf{I} = \mathbf{UK}(l)$, so if we postmultiply with $\boldsymbol{\varepsilon}'$, we get

$$\Psi[\mathbf{UK}(l)] = \frac{\rho}{1-\rho}\Big(1 - \lambda\Psi[\mathbf{U}^l \mathbf{V}]\Big).$$

When this is placed in (4.5.30a), and we use the fact that [see (4.1.6f)]

$$\sum_{l=0}^{N-1} \mathbf{U}^l = \mathbf{K}(\mathbf{I} - \mathbf{U}^N),$$

with $\mathbf{K} = (\mathbf{I} - \mathbf{U})^{-1}$ [from (4.2.3a)], we get

$$MTTF(N) = \frac{1}{\lambda}\frac{\rho}{1-\rho}\left(N - \lambda\Psi\left[\sum_{l=0}^{N-1} \mathbf{U}^l \mathbf{V} \right] \right) = \frac{N T_1}{1-\rho} - \frac{\rho}{1-\rho}\Psi[\mathbf{K}(\mathbf{I} - \mathbf{U}^N)\mathbf{V}].$$

Now, from its definition, we know that $\mathbf{KU} = \mathbf{K} - \mathbf{I}$, and by postmultiplying (4.2.3d) with U, we know that $\mathbf{pKU} = [\lambda/(1-\rho)\mathbf{pV}]$, so with some awkward manipulation, we get the following expression:

$$MTTF(N) = \frac{N-\rho}{1-\rho}T_1 - \frac{\lambda}{1-\rho}\Psi[\mathbf{V}^2] + \frac{\rho}{1-\rho}\Psi[\mathbf{KU}^N\mathbf{V}]. \tag{4.5.30b}$$

This expression is deceptive in that it seems to be telling us that the MTTF depends on $\Psi[\mathbf{V}^2]$, when it really does not, at least not for small N. When N is small, the last term can be manipulated so that it cancels the middle term, as well as the dependence on $1/(1-\rho)$, as can clearly be seen from the expressions we already derived for $MTTF(1)$ and $MTTF(2)$. However, it does tell us this much for $\rho < 1$. For then, $\mathbf{U}^N$ gets to be negligibly small for large N, and thus $MTTF(N)$ grows as $N T_1/(1-\rho)$. Anyway, either (4.5.30a) or (4.5.30b) can be used to calculate $MTTF(N)$ in general.

From what we have seen in this section, there are unlimited variations one can pursue based on what has been done. We have already suggested a few. We elaborate further here. For instance, suppose that a system has N devices, and one has just failed, leaving behind k good ones. What is the *MTTF* then? The answer is $t(k \to 0; N)$, from (4.5.15). But what if you, as the new manager have just arrived, and do not know when the last breakdown occurred, what is the *MTTF* then? You are the random observer, and the system was in state, $\pi_{\mathbf{r}}(k)$ (see Corollary 4.5.2b) when you arrived. Thus the mean time to failure is the mean time to drop from k to $k-1$, and thence to 0:

$$MTTF = \pi_{\mathbf{r}}(k)\tau'_{\mathbf{d}}(k; N) + t(k-1 \to 0; N).$$

Suppose, instead, that you must change your plans once you are down to your last device; what is the MTTF then? Just subtract $t_d(1; N)$ from the above.

Now take a different viewpoint. What is the *probability* that the system will fail (down to your last device) before it ever gets back to full strength? Maybe you should quit now. This probability is given in Definition 4.5.17 and is

$$\pi_{\mathbf{r}}(k)\mathbf{W}_{\mathbf{d}}(k \to 1; N-1)\boldsymbol{\varepsilon}'.$$

Note the $N-1$. By definition, $\mathbf{W_d}$ deals with, *not exceeding*, while we are seeking the probability of *not reaching*. So we did find an additional use for the $\mathbf{W_d}$'s.

The open system also has some application in this context. Suppose that instead of repairing devices, you go out and buy new ones when old ones break. There is an unlimited supply of these devices on the market, but it takes time to do this. If you work for a public university, the longest part of this task is getting the purchase order approved. Because of the uncertain delay, you try to have k devices on hand. The mean time until you run out of devices is given by the k-busy period, which for the M/G/1 system is [from (4.5.16b)]

$$\lim_{N \to \infty} t(k \to 0; N) = \frac{k T_1}{1-\rho},$$

and so on and on and on.

We have seen an inkling of the power of LAQT in being able to separate the initial conditions from the transition period from the final result. Now what remains is for us to extend the procedure to include other, and more general systems, which we do in the following chapters.

CHAPTER 5
G/M/1 QUEUE

"Thou com'st in such a questionable shape."
– Hamlet, Act I, Scene IV

In Chapter 4 we talked about a closed loop made up of two subsystems, S_1 and S_2, where each subsystem was equivalent to a matrix representation of some general distribution, $b_i(x)$. The notation for such a loop is $G_2/G_1/1//N$, where the N stands for the number of customers in the loop. However, we only treated the case where $G_2 = M$ [i.e., $b_2(x)$ is an exponential function]. In that case we found that an arriving customer would find n customers already in S_1 with the same probability as he would leave n behind. Furthermore, we showed that except for the fact that $d(N;N) = a(N;N) = 0$, these probabilities are proportional to the random observer's probability of finding n customers there. We also argued that the "finite waiting room," M/G/1/N queue (i.e., where S_1 could hold no more than N customers, thereby forcing all extra arrivals to disappear), yielded the same results as M/G/1//N, by virtue of the memoryless property of S_2. The behavior of the open M/G/1 system came easily (provided that the utilization parameter, ρ, was less than 1) by letting N become unboundedly large. In that case, since S_2 was assumed to be the slower server, the probability that it would ever be idle went to zero. Then it became a "constant" source of customers to S_1, with independent, exponentially distributed interarrival times, that is, a Poisson process. Finally, we showed that the three queue length probabilities, $a(n)$, $d(n)$, and $r(n)$, are all equal.

In this chapter we turn things upside down by letting ρ be greater than 1. Now, S_1 is the slower server, and in the limit as N goes to infinity, becomes a non-Poisson source of customers to S_2, with interarrival times distributed according to $b_1(x)$. We will find that the limit, which yields the GI/M/1 open queue (at S_2), does not come so easily, that the finite waiting room GI/M/1/N does *not* give the same results as the closed G/M/1//N loop, nor do the arriving or departing customers see the same thing as our random observer. The formulas are sufficiently simple that we can hope to gain physical insight into the behavior of steady-state queues generally.

5.1. STEADY-STATE OPEN ME/M/1 QUEUE

Since we will be making considerable use of the equations of Chapter 4, we retain the definition of ρ as $\lambda\bar{x}$, which properly should be written as $\bar{x}_1/\bar{x}_2$, since

$1/\lambda = \bar{x}_2$. Clearly, the utilization of S_2 in an open GI/M/1 queue is given by $\bar{x}_2/\bar{x}_1$, which is $1/\rho$. One must remember to replace ρ by $1/\rho$ when comparing with formulas given in the general literature. To emphasize this difference, and because it sort of looks like an upside-down ρ, we will often use the greek letter, *zeta*, as the utilization parameter. Therefore,

$$\zeta := \frac{1}{\rho} = \frac{\bar{x}_2}{\bar{x}_1}.$$

We will have to make some other notational changes; the first, referring to $\{\cdot;\cdot;\cdot\}$, we give now.

Definition 5.1.1

$\{i; k; N\}$ *describes the state of the system, where N is the total number of customers in the system, k is the number of customers at* S_2 *(therefore there are* $N-k$ *customers at* S_1*), and i is the phase in* S_1 *that is busy.* We might say that $i \in \Xi$ is an *internal state* of the system, and that $\{k; N\}$ (or $\{k, N-n\}$) is an *external state* of the system. ◊

The only change in notation from Section 4.1.1 is that now the second argument stands for the number of customers at S_2, rather than S_1. This will be the notation throughout this chapter.

Rather than introduce a collection of new notations, we will modify previous symbols. For instance, the $\mathbf{d}(n; N)$ of Chapter 4 (the vector probability that a customer will leave n customers behind when departing S_1) will now be written as $\mathbf{d}_1(n; N)$, while we make three new definitions.

Definition 5.1.2

$\mathbf{d}_2(k; N)$:= *steady-state vector probability that a customer will leave k customers behind when departing* S_2. Given that a customer has just left S_2 in an ME/M/1//*N* loop with nothing else known (another viewpoint of the steady state), $[\mathbf{d}_2(k; N)]_i$ is the probability that k customers are still at S_2, while the active customer at S_1 is at phase i. $d_2(k; N) := \mathbf{d}_2(k; N)\boldsymbol{\varepsilon}'$ is the steady-state scalar probability that a customer will leave k customers behind when departing S_2. ◊

We also say that this is the probability that the system will be in state $\{i; k; N\}$ immediately after a departure from S_2.

Definition 5.1.3

$\mathbf{a}_2(k; N)$:= *steady-state vector probability that a customer, upon arriving at* S_2, *will find k customers already there.* $[\mathbf{a}_2(k; N)]_i$ is the probability that there are now $k+1$ customers at S_2, while the active customer in S_1 is at phase i. $a_2(k;N) := \mathbf{a_2}(k;N)\boldsymbol{\varepsilon}'$ is the associated scalar probability. ◊

We also say that this is the probability that the system will be in state $\{\, i\,; k+1\,; N \,\}$ immediately after an arrival to S_2.

Definition 5.1.4

$\boldsymbol{\pi_2}(k;N)$:= *steady-state vector probability that there are k customers at* S_2 *in an ME/M/1//N loop.* A random observer will find a long-running system in state $\{\, i\,; k\,; N \,\}$ with probability $[\boldsymbol{\pi_2}(k;N)]_i$. $r_2(k;N) := \boldsymbol{\pi_2}(k;N)\boldsymbol{\varepsilon}'$ is the associated scalar probability. ◊

As long as we have a closed system (i.e., N is finite), an arrival to one queue corresponds exactly to a departure from the other queue, so for $n + k = N$,

$$\mathbf{d_1}(n;N) = \mathbf{a_2}(k-1;N), \tag{5.1.1a}$$

$$\mathbf{a_1}(n;N) = \mathbf{d_2}(k-1;N), \tag{5.1.1b}$$

and

$$r_1(n;N) = r_2(k;N). \tag{5.1.1c}$$

The sum of the first argument on the left-hand side of (5.1.1a) and (5.1.1b), $[n]$, and the first argument on the right-hand side of those equations, $[k-1]$, is $N-1$. The "1" missing is our customer-observer. There may be some confusion when dealing with vector probabilities. As before, $[\boldsymbol{\pi_1}(n;N)]_i$ is the steady-state probability of there being n customers at S_1 with phase i (in S_1) busy. But the corresponding probability that there are k customers at S_2, and phase i in S_1(!) is busy, given by the ith component of the equation

$$\boldsymbol{\pi_2}(k;N) := \boldsymbol{\pi_1}(N-k;N), \tag{5.1.1d}$$

still refers to the internal status of S_1. After all, S_2 is represented by only one state, so it has no internal status. This becomes a bit sticky when $\rho > 1$ (or when $\zeta < 1$) and we go to the open system ($N \to \infty$), for then we would like to think that S_1 has somehow disappeared, while the arriving customers are of their own volition selecting their interarrival times from some nonexponential distribution. Conceptually, it might be more useful to view S_1 as being *upstream* from S_2, with an inexhaustible supply of customers trying to get through its gates, one at a time, of course.

It would seem that this notational change is unnecessary, and indeed it is, but only as long as we are dealing with a closed system. In Chapter 4, with $\rho < 1$, we let N go to infinity, holding n constant. In this chapter, with $1/\rho = \zeta < 1$, we want to let N go to infinity, holding k constant. This subtle difference is best handled by our change of notation. Note that under these conditions, with n

fixed, $\mathbf{d_1}(n;N)$, $\mathbf{a_1}(n;N)$, $r_1(n;N)$, and $\pi_1(n;N)$ all go to 0 as N increases to infinity, just as $\mathbf{d_1}(N-n;N)$, and so on, did when ρ was less than 1.

5.1.1. Steady-State Probabilities

We can see from (4.2.2a) and (4.2.2b) that $\mathbf{K}$ exists whether ρ is greater than or less than 1 (it only lacks definition when $\rho = 1$, in which case neither the M/G/1 nor G/M/1 queues have a steady-state solution). The problem is that when $\rho > 1$, the limit of $\mathbf{K}(N)$ [Equation (4.2.1a)] does *not* exist! We must be more careful in taking the limit. Let $\{ s_i \}$, $\{ \mathbf{u_i} \}$, and $\{ \mathbf{v_i}' \}$ be the sets of eigenvalues, left and right eigenvectors of $\mathbf{A}$, respectively. Define s to be the eigenvalue of smallest magnitude, with corresponding eigenvectors $\mathbf{u}$ and $\mathbf{v}'$. That is,

$$|s| = \min_{i=1}^{m} |s_i|.$$

For simplicity, assume that the eigenvalues are distinct (although what follows only needs the fact – known from other sources – that s is unique, positive, and less than 1). Then from the spectral decomposition theorem [Equation (1.3.8a)],

$$\mathbf{A} = \sum_{i=1}^{m} s_i \mathbf{v_i}' \mathbf{u_i},$$

so (recall that $\mathbf{U}$ is the inverse of $\mathbf{A}$)

$$\mathbf{U}^N = \sum_{i=1}^{m} \left(\frac{1}{s_i} \right)^N \mathbf{v_i}' \mathbf{u_i}. \tag{5.1.2a}$$

Then it follows that

$$s^N \mathbf{U}^N = \mathbf{v}'\mathbf{u} + \sum_{i=1}^{m}{}' \left(\frac{s}{s_i} \right)^N \mathbf{v_i}' \mathbf{u_i}, \tag{5.1.2b}$$

where $\sum'$ stands for the sum of all terms excluding the term that corresponds to s. The limit is now straightforward, since $|s/s_i|$ is less than 1 for all i,

$$\lim_{N \to \infty} s^N \mathbf{U}^N = \mathbf{v}'\mathbf{u}. \tag{5.1.2c}$$

We are almost ready to move ahead, but first look at [from (4.1.6d)]

$$s^N \mathbf{K}(N) = s^N [\mathbf{I} + \mathbf{U} + \mathbf{U}^2 + \cdots + \mathbf{U}^{N-1} + \lambda \mathbf{U}^{N-1} \mathbf{V}].$$

Note that for all N greater than 0,

$$[\mathbf{I} + \mathbf{U} + \mathbf{U}^2 + \cdots + \mathbf{U}^{N-1}](\mathbf{I} - \mathbf{U}) = \mathbf{I} - \mathbf{U}^N,$$

and since $\mathbf{K} = (\mathbf{I} - \mathbf{U})^{-1}$ exists, we have

$$s^N \mathbf{K}(N) = s^N [(\mathbf{I} - \mathbf{U}^N)\mathbf{K} + \mathbf{U}^N(\lambda \mathbf{A} \mathbf{V})].$$

At last we are ready to let N go to infinity. Note that since (we are assuming that) $0 < s < 1$, the term $s^N \mathbf{K}$ goes to 0, leaving us with

$$\mathbf{F} := \lim_{N\to\infty} s^N \mathbf{K}(N) = \mathbf{v}'\mathbf{u}[-\mathbf{K} + \lambda\mathbf{AV}]$$

$$= -\frac{\lambda^2}{1-\rho}\mathbf{v}'\mathbf{uAVQV} = -\frac{\lambda^2}{1-\rho}\mathbf{v}'\mathbf{uAV}\boldsymbol{\varepsilon}'\mathbf{pV}, \tag{5.1.3a}$$

where we have made use of (4.2.3b). Equation (4.2.3c) finally comes in handy, for it allows us to replace $\lambda\mathbf{AV}\boldsymbol{\varepsilon}'$ with $(1-\rho)\mathbf{K}\boldsymbol{\varepsilon}'$ to get

$$\mathbf{F} = -\lambda\mathbf{v}'\mathbf{uK}\boldsymbol{\varepsilon}'\mathbf{pV}.$$

Ah, but $\mathbf{K}$ is a function of $\mathbf{U}$, so it has $\mathbf{u}$ as a left eigenvector, and

$$\mathbf{uK} = \left(1 - \frac{1}{s}\right)^{-1}\mathbf{u} = -\frac{s}{1-s}\mathbf{u}.$$

That leaves us with the simple expression (it really is simple)

$$\mathbf{F} = \left(\frac{\lambda s}{1-s}(\mathbf{u}\boldsymbol{\varepsilon}')\right)\mathbf{v}'\mathbf{pV}, \tag{5.1.3b}$$

where the expression in large parentheses is a scalar. The last preliminary step is to find $\Psi[\mathbf{F}]$, which is (remember that $1/\rho = \zeta < 1$)

$$\Psi[\mathbf{F}] = \left(\frac{\lambda s}{1-s}(\mathbf{u}\boldsymbol{\varepsilon}')\right)\mathbf{pv}'\mathbf{pV}\boldsymbol{\varepsilon}' = \frac{s\,(\mathbf{u}\boldsymbol{\varepsilon}')(\mathbf{pv}')}{(1-s)\zeta}. \tag{5.1.3c}$$

From Theorem 4.1.2 and (5.1.1d), for $k > 0$,

$$\pi_2(k;N) = \frac{1}{\Psi[\mathbf{K}(N)]}\mathbf{pU}^{N-k} = \frac{1}{\Psi[\mathbf{K}(N)]}\mathbf{pU}^N\mathbf{A}^k.$$

Now multiply and divide by s^N and take the limit on N, while holding k fixed,

$$\pi_2(k) := \frac{\lim_{N\to\infty} s^N \mathbf{pU}^N\mathbf{A}^k}{\lim_{N\to\infty} s^N \Psi[\mathbf{K}(N)]} = \frac{1}{\Psi[\mathbf{F}]}\mathbf{pv}'\mathbf{uA}^k.$$

Remember now that $\mathbf{u}\mathbf{A} = s\mathbf{u}$, and use (5.1.3c)

$$\pi_2(k) = (1-s)\zeta\, s^{k-1}\frac{\mathbf{u}}{\mathbf{u}\boldsymbol{\varepsilon}'}, \quad k > 0. \tag{5.1.4a}$$

How interesting – all the vector probabilities for $k > 0$ are proportional to the same isometric vector. Since it will appear often, this vector is given the special symbol

$$\hat{\mathbf{u}} := \frac{\mathbf{u}}{\mathbf{u}\boldsymbol{\varepsilon}'}, \tag{5.1.4b}$$

where $\hat{\mathbf{u}}\boldsymbol{\varepsilon}' = 1$.

The same game can be played with $k = 0$, except that now we have

$$\pi_2(0) = \frac{\lambda(1-s)}{\rho s}\hat{\mathbf{u}}\mathbf{A}\mathbf{V} = \frac{1}{\bar{x}_1}(1-s)\,\hat{\mathbf{u}}\mathbf{V}. \tag{5.1.4c}$$

We must think about this for a moment before going on. Here we have a simple exponential server (S_2), with no customers present, yet it has some memory of when the last customer came. That is, the vector, $\pi_2(0)$, has nontrivial components and is proportional to $\hat{\mathbf{u}}\mathbf{V}$, not $\hat{\mathbf{u}}$. (There is an analogy to this in quantum electrodynamics, where empty space – the "vacuum" – has nontrivial properties, as well as in the pre-Einsteinian view of the *ether*.) Now we can appreciate the view that has S_1 upstream, busily generating customers for S_2, even though $k = 0$. We must never lose sight of the fact that it is the system as a whole that is in one state or another, not the subsystems by themselves. This is particularly true in the steady state, where the two subsystems have been exchanging customers for a long time. In our treatment of transient behavior, we see (perhaps) that the two subsystems gradually become interdependent as they exchange more and more customers.

The scalar probabilities, $r_2(k)$, can now be found from Equations (5.1.4), since $r_2(k) = \pi_2(k)\boldsymbol{\varepsilon}'$. The formulas are summarized in

Theorem 5.1.1: The steady-state probabilities of (a random observer) finding k customers in an open ME/M/1 queue as given by (5.1.4a) to (5.1.4c) can be written in the form ($\zeta = 1/\rho$):

$$\pi_2(0) = (1-\zeta)\frac{\hat{\mathbf{u}}\mathbf{V}}{\hat{\mathbf{u}}\mathbf{V}\boldsymbol{\varepsilon}'} \tag{5.1.5a}$$

and

$$\pi_2(k) = (1-s)\zeta s^{k-1}\hat{\mathbf{u}} \quad \text{for} \quad k > 0. \tag{5.1.5b}$$

The associated scalar probabilities are given by the following equations:

$$r_2(0) = 1-\zeta \tag{5.1.5c}$$

and

$$r_2(k) = (1-s)\zeta s^{k-1} \quad \text{for} \quad k > 0. \tag{5.1.5d}$$

The parameter s (with its associated left eigenvector, $\hat{\mathbf{u}}$) is the smallest positive eigenvalue satisfying $\hat{\mathbf{u}}\mathbf{A} = s\hat{\mathbf{u}}$, and $\hat{\mathbf{u}}\boldsymbol{\varepsilon}' = 1$. ■

Proof: Note from (4.2.3c) that $\lambda\mathbf{A}\mathbf{V}\boldsymbol{\varepsilon}' = (1-\rho)\mathbf{K}\boldsymbol{\varepsilon}'$, so on premultiplying by $\hat{\mathbf{u}}$ and rearranging,

$$\hat{\mathbf{u}}\mathbf{V}\boldsymbol{\varepsilon}' = \frac{1-\zeta}{(1-s)\zeta\lambda}. \tag{5.1.5e}$$

The rest follows directly. QED

So the probabilities are geometrically distributed, just as in the M/M/1 queue, but with s instead of ζ. Also, $r_2(0)$ does not satisfy the general expression but is what it should be, namely 1 minus the utilization, $[\zeta]$, of S_2. This well-known result is simple in form but is deceptively complicated in that the dependence of s on ζ is not easy to get in general. Only when $b_1(x)$ is exponentially distributed does $s = \zeta$. It is known from other sources that s satisfies the following implicit relation. We will state it as a corollary to Theorem 5.1.1 and prove it by purely algebraic means.

Corollary 5.1.1: The eigenvalue s is the smallest positive root of the following implicit equation:

$$B^*[\lambda(1-s)] = s, \tag{5.1.6a}$$

where $B^*(\sigma)$ is the Laplace transform of $b_1(x)$, the pdf of S_1. The associated eigenvector satisfies the equation

$$\hat{\mathbf{u}} = \lambda \mathbf{p} \mathbf{V}\left(\mathbf{I} + \lambda(1-s)\mathbf{V}\right)^{-1}. \tag{5.1.6b}$$

Proof: First we prove (5.1.6b). From its definition,

$$\hat{\mathbf{u}}\mathbf{A} = s\hat{\mathbf{u}} = \hat{\mathbf{u}}\left(\mathbf{I} + \frac{1}{\lambda}\mathbf{B} - \mathbf{Q}\right) = \hat{\mathbf{u}} + \frac{1}{\lambda}\hat{\mathbf{u}}\mathbf{B} - \mathbf{p}.$$

We have used the fact that $\hat{\mathbf{u}}\mathbf{Q} = \hat{\mathbf{u}}\boldsymbol{\varepsilon}'\mathbf{p} = \mathbf{p}$. Next separate all terms that contain $\hat{\mathbf{u}}$ from those that do not.

$$\hat{\mathbf{u}}\left((1-s)\mathbf{I} + \frac{1}{\lambda}\mathbf{B}\right) = \mathbf{p}.$$

Now multiply both sides of the equation by $\lambda\mathbf{V}$:

$$\hat{\mathbf{u}}\left(\mathbf{I} + \lambda(1-s)\mathbf{V}\right) = \lambda\mathbf{p}\mathbf{V}. \tag{5.1.6c}$$

Multiplying both sides by the inverse of the matrix expression in large parentheses yields (5.1.6b). Equation (5.1.6a) follows directly by multiplying (5.1.6b) on the right with the vector $\boldsymbol{\varepsilon}'$, noting that

$$\lambda\mathbf{V}\left(\mathbf{I} + \lambda(1-s)\mathbf{V}\right)^{-1} = \frac{1}{1-s}\left(\mathbf{I} - \left(\mathbf{I} + \lambda(1-s)\mathbf{V}\right)^{-1}\right),$$

and recalling (3.1.10), the matrix definition of the Laplace transform. ■

To find s, one must either solve an eigenvalue problem, or find the smallest positive root of (5.1.6a). In either case, numerical techniques are usually required. Once s is known, (5.1.6b) gives us $\hat{\mathbf{u}}$. As with many other objects we encounter in this book, $\hat{\mathbf{u}}$ has more information in it than that for which it was derived. In particular, it contains information regarding the arrival of the next customer. We discuss this further in the next section, after deriving the departure probabilities.

It remains to verify that the probabilities sum to 1 and to find the mean queue length and system time.

Exercise 5.1.1: Show that $\sum_{k=0}^{\infty} r_2(k) = 1$.

Since the $r_2(k)$'s are of geometric form, it is just as easy to get the z-transform of $\{ r_2(k) \mid k \geq 0 \}$ as it is to get $\bar{q}_2$ directly. By definition,

$$Q_2(z) = \sum_{k=0}^{\infty} z^k r_2(k) = 1 - \zeta + \frac{(1-s)\zeta}{s} \sum_{k=1}^{\infty} (zs)^k = 1 - \zeta + \frac{(1-s)\zeta z}{1-zs}.$$

We rewrite this in the form

$$Q_2(z) = 1 + \frac{(z-1)\zeta}{1-zs}. \tag{5.1.7a}$$

Obviously, $Q_2(1) = 1$, while the derivative evaluated at $z = 1$ yields the mean queue length,

$$\bar{q}_2 = \frac{\zeta}{1-s}. \tag{5.1.7b}$$

As in Chapter 4 [Equation (4.2.5c)], we use Little's theorem to get the mean system time (in this case, the arrival rate to S_2 is $1/\bar{x}_1$, and $\bar{x}_2$ is $1/\lambda$),

$$\bar{T}_2 = \bar{x}_1 \bar{q}_2 = \frac{\bar{x}_2}{1-s}. \tag{5.1.7c}$$

Again, this formula looks very similar to (2.1.6b) for the M/M/1 queue, except that s appears instead of ζ. $\bar{T}_2$ becomes unbounded when s approaches 1. The graph of (5.1.7c) for the D/M/1 queue was given in Figure 1.1.2, together with various M/G/1 queues.

It should be comforting to know that s/ζ goes to 1 as ζ approaches 1 from below. We explore the relation between ζ and s further in Section 5.1.3.

Exercise 5.1.2: Verify that (5.1.7a) and (5.1.7b) are indeed true.

5.1.2. Arrival and Departure Probabilities

The hard work has already been done in preparing to take the limit as N goes to infinity of the arrival and departure probabilities. We have the following string of equalities:

$$\mathbf{a_2}(k;N) = \mathbf{d_1}(N-k-1;N) = \frac{r_1(N-k-1;N)}{1-r_1(N;N)}\mathbf{p} = \frac{r_2(k+1;N)}{1-r_2(0;N)}\mathbf{p}.$$

We already found the limits of both numerator and denominator for the last expression, and they are each finite [Equations (5.1.5c) and (5.1.5d)], so

$$\mathbf{a_2}(k) := \lim_{N\to\infty} \mathbf{a_2}(k;N) = \frac{1}{1-(1-\zeta)}[\zeta(1-s)]s^k\mathbf{p} = (1-s)s^k\mathbf{p}. \quad (5.1.8a)$$

The scalar probabilities obviously satisfy

$$a_2(k) = (1-s)s^k. \quad (5.1.8b)$$

We point out that (5.1.8a) and (5.1.8b) are valid for all k, even $k=0$, which is not the case for $r_2(k)$ and $\pi_2(k)$. Also, note that (not merely at $k=0$) $a_2(k)$ does not equal $r_2(k)$, and $\mathbf{a_2}(k)$ is not even parallel to $\pi_2(k)$! Well, it is not all that bad, since the $a_2(k)$'s ($k \neq 0$) are proportional to the $r_2(k)$'s [i.e., $a_2(k) = s\,r_2(k)/\zeta$, for all k greater than 0].

The $\mathbf{d_2}(k)$'s can be found in a manner identical to that for $\mathbf{a_2}(k)$. From (5.1.1b) and (4.1.13a),

$$\mathbf{d_2}(k;N) = \frac{1}{1-r_2(0;N)}\pi_2(k+1;N).$$

The limit follows directly. The different formulas are collected in the following theorem.

Theorem 5.1.2: The steady-state probabilities of queue lengths as seen by customers arriving to, and departing from an open ME/M/1 queue are given for all $k \geq 0$ by [repeating Equation (5.1.8a)]

$$\mathbf{a_2}(k) = (1-s)s^k\mathbf{p},$$
$$\mathbf{d_2}(k) = (1-s)s^k\hat{\mathbf{u}}, \quad (5.1.9a)$$

and

$$d_2(k) = a_2(k) = (1-s)s^k. \quad (5.1.9b)$$

Thus we have shown for this simple system that except for the M/M/1 queue, $\mathbf{a_2}(k)$, $\mathbf{d_2}(k)$, and $\pi_2(k)$ are distinctly different. They are similar, but nonetheless different. ■

The form of (5.1.9b) is so familiar by now that one can truly say "it is obvious that" the sum of the $a_2(k)$'s is 1, and the mean queue length seen by both a departing and an arriving customer is $s/(1-s)$. Although the difference seems minor, it is important to recognize that this quantity is not equal to the mean queue length as seen by our random observer, $\bar{q}_2$ [Equation (5.1.7b)]. As with the a_2's and r_2's, they differ by the factor s/ζ. It is (5.1.7b) which one uses in Little's theorem to get the mean system time, as we did in (5.1.7c). We will now use $\{a_2(k)\}$ to find $\overline{T}_2$ from its definition. Since S_2 is an exponential server, there is no distinction between its mean time and its residual time, so the care we had to take in Section 4.3.3 is not necessary here. If there are k customers at S_2 (including none) when a customer arrives, he will have to wait an average of $(k+1)\,\bar{x}_2$ units of time before leaving. The mean time averaged over all queue lengths is

$$\overline{T}_2 = \sum_{k=0}^{\infty} a_2(k)(k-1)\bar{x}_2 = (1-s)\bar{x}_2 \sum_{k=0}^{\infty} (k+1)s^k = \frac{(1-s)\bar{x}_2}{s} \sum_{k=1}^{\infty} ks^k = \frac{\bar{x}_2}{1-s},$$

the same as (5.1.7c). Our purpose here was to prepare the reader to derive the system time distribution.

The time for a customer to go through an exponential server $k+1$ times, or equivalently, of $k+1$ customers going through one at a time, is distributed according to the Erlangian-(k+1) distribution, $[E_{k+1}(x;\,\lambda)]$, whose pdf is given in (3.2.1a), and is $\lambda(\lambda x)^k e^{-\lambda x} / (k!)$. The weighted average over all k is, then,

$$b_{2s} := \sum_{k=0}^{\infty} a_2(k) E_{k+1}(x;\lambda) = (1-s)\lambda \sum_{k=0}^{\infty} s^k \frac{(\lambda x)^k}{k!} e^{-\lambda x}$$

$$= (1-s)\lambda \left(\sum_{k=0}^{\infty} \frac{(\lambda s x)^k}{k!} \right) e^{-\lambda x},$$

or finally,

$$b_{2s} = (1-s)\lambda e^{-(1-s)\lambda x}. \tag{5.1.10}$$

So the system time is exponentially distributed, with mean time equal to $1/[(1-s)\lambda]$ (but we already knew $\overline{T}_2$). Is it surprising that it should be exponential? Not really, for we showed in Lemma 3.3.1 that the average of all convolutions of an exponential function with itself, weighted over geometric probabilities, is exponential.

5.1.3. Properties of Geometric Parameter *s*

Theorem 5.1.1 showed us that the behavior of the GI/M/1 queue is dominated by the geometric parameter s. Even $\hat{\mathbf{u}}$ can be evaluated from (5.1.6c) if we know s. The value of s can be found from any one of the three equations: (1) (5.1.6a); (2) (3.1.10); or (3) $\mathbf{A}\hat{\mathbf{u}} = s\hat{\mathbf{u}}$, by a root-finding or other numerical technique. That is, for a given arrival process, with interarrival times generated by $\langle \mathbf{p}, \mathbf{B} \rangle$, and given λ, s is uniquely determined by any of these equations. The properties of s are best understood by thinking of it as a function of λ, or ρ, or better, ζ. How one should calculate numerical values for s is a matter of taste and numerical analysis and is by and large outside the interests of this book, but we will bring out some points so that the reader may avoid possible pitfalls.

For convenience of description, in the rest of this section, we make the following symbol changes. Recall that $\rho = \lambda T$, where

$$T := \bar{x}_1 = \Psi[\mathbf{V}].$$

We have already defined ζ, which can now be written as

$$\zeta := \frac{1}{\rho} = \frac{1}{\lambda T}.$$

Therefore, we will replace λ whenever it is to our convenience with $1/T\zeta$.

Let us start slowly and see what the three formulas tell when S_1 is an exponential server. In that case, **B** becomes $\mu = 1/T$, **V** becomes T, **Q** becomes 1, and $\mathbf{A} = 1 + (1/\lambda)\mu - 1 = \zeta$. Thus the eigenvalue equation, (3), $\hat{\mathbf{u}}\mathbf{A} = s\hat{\mathbf{u}}$, reduces to $\zeta = s$. This obvious result tells us that the G/M/1 queue reduces to the M/M/1 queue when G is (M) (that *is* obvious). Equation (1), (5.1.6a), is not quite so simple. From that equation, $s = B^*[(1-s)\lambda] = B^*[(1-s)/(\zeta T)]$, so

$$s = \int_0^\infty \exp[-x(1-s)/\zeta T]\frac{e^{-x/T}}{T}dx = \frac{1/T}{1/T + (1-s)/(\zeta T)} = \frac{\zeta}{\zeta + (1-s)}.$$

After we clear fractions, we get $s\zeta + s(1-s) = \zeta$, or

$$(1-s)\zeta = (1-s)s.$$

Notice that although we get the root we are looking for, $s = \zeta$, we also get the meaningless, extraneous root, $s = 1$, for all ζ. This extraneous root always appears for any distribution, since it reflects the fact that the integral of $b(x)$ is 1. It can get in the way when ζ is close to 1, and can be a real drag when one is looking for heavy traffic performance, as we shall see presently. The third equation has the same difficulty, but we can get around that. First, (3.1.10) gives us the following when S_1 is exponential.

$$s = \Psi[(\mathbf{I} + (1-s)\lambda\mathbf{V})^{-1}] = \frac{1}{1 + (1-s)T/(\zeta T)},$$

which, indeed, leads to the same awkward equation we had before. Now let us play a little trick for any ME distribution, by noting that $s = \Psi[s\mathbf{I}]$ in (3.1.10), then (keep λ for the moment)

$$0 = \Psi\Big[(\mathbf{I} + (1-s)\lambda\mathbf{V})^{-1} - s\mathbf{I}\Big] = \Psi\Big[[\mathbf{I} + (1-s)\lambda\mathbf{V}]^{-1}[\mathbf{I} - s\mathbf{I} - (1-s)s\lambda\mathbf{V}]\Big],$$

or

$$(1-s)\Psi\Big[[\mathbf{I} + (1-s)\lambda\mathbf{V}]^{-1}[\mathbf{I} - \lambda s\mathbf{V}]\Big] = 0.$$

Therefore, we can throw away the term $(1-s)$ before we begin. Finally, replace λ, clear fractions, and after some other trickery, get the following alternative equation:

$$\Psi\Big[\mathbf{V}[\zeta T\mathbf{I} + (1-s)\mathbf{V}]^{-1}\Big] = 1. \tag{5.1.11}$$

This form is about as good as we can get for the purposes we have in mind. In particular, we can see that when $\zeta = 0$, we must have $s = 0$, while if $s = 1$, ζ must be 1 also. This is true for every G/M/1 queue, *except* for those which are *defective*, or equivalently, have an *initial impulse*, which we now discuss.

Distributions with an initial impulse are those for which $R(0) < 1$, or equivalently, $B(0) > 0$. When a customer finally gets to be served, he decides with probability $p = B(0)$ that he does not need any service. Such distributions are not uncommon. For instance, in reliability theory, this is the probability that a device will be faulty even though it is brand new, an important problem to worry about. Any distribution that has this property has a pdf of the form

$$b(x) = p\,\delta(x) + (1-p)\,b_a(x), \qquad (5.1.12a)$$

where $b_a(x)$ is the pdf of those devices that were *not* faulty initially [i.e., $\int b_a(x)\,dx = 1$], and $\delta(x)$ is the *Dirac delta function*, which has these properties:

$$\int_{-\varepsilon}^{\varepsilon} \delta(x)\,dx = 1 \quad \text{for any } \varepsilon > 0,$$

or

$$f(0) = \int_{-\infty}^{\infty} f(x)\,\delta(x)\,dx$$

for all $f(x)$ which are continuous at $x = 0$, or

$$f(t) = \int_{-\infty}^{\infty} f(x)\,\delta(x-t)\,dx.$$

It can also be viewed as the derivative of the *unit step function*, which satisfies

$$\Delta(t) = \int_{-\infty}^{t} \delta(x)\,dx$$

$$\Delta(t) = \begin{cases} 0 & \text{if } t < 0 \\ ½ & \text{if } 0 \\ 1 & \text{if } t > 0 \end{cases}$$

Pictorially think of $\delta(x)$ as a spike of infinite height with unit area and 0 width, or the limit of a family of very high but very narrow functions. One such example was given in Section 3.2.4 as the limit of the set of Erlangian-k distributions with the same mean. There are other ways to look at it.

Anyway, we can also write

$$B(x) = p + (1-p)\,B_a(x) \qquad (5.1.12b)$$

and

$$R(x) = (1-p)\,R_a(x). \qquad (5.1.12c)$$

This distribution has the following Laplace transform:

$$B^*(s) = \int_{0-}^{\infty} e^{-sx}\,[p\,\delta(x) + (1-p)b_a(x)]\,dx = p + (1-p)B_a^*(s).$$

The LAQT treatment is as follows. Let $b_a(x)$ be generated by $<\mathbf{p_a}\,,\,\mathbf{V_a}>$; then

$$\mathbf{V} = \begin{bmatrix} 0 & \mathbf{o} \\ \mathbf{o}' & \mathbf{V_a} \end{bmatrix} \quad \text{and} \quad \mathbf{p} = [p\;,\,(1-p)\,\mathbf{p_a}], \qquad (5.1.13)$$

where $\mathbf{p_a}\boldsymbol{\varepsilon}' = 1$. Be careful, $\mathbf{V}$ does not have an inverse, but luckily, (5.1.11), which we worked so hard to get, does not have $\mathbf{B}$ in it. The mean time, T, for the process represented by $<\mathbf{p}\,,\mathbf{V}>$ is related to the mean time, T_a, for the process $<\mathbf{p_a}\,,\mathbf{V_a}>$ by the following:

$$T = \Psi[\mathbf{V}] = \mathbf{pV}\boldsymbol{\varepsilon}' = (1-p)\,\mathbf{p_a V_a}\boldsymbol{\varepsilon}' = (1-p)\Psi_a[\mathbf{V_a}] = (1-p)T_a.$$

If some customers take no time at all, the rest must take more time than the overall average, so $T_a > T$ if $p > 0$.

When (5.1.13) is substituted into (5.1.11), the following expression results.

$$(1-p)\Psi_a\left[\mathbf{V_a}[\zeta T\mathbf{I} + (1-s)\mathbf{V_a}]^{-1}\right] = 1. \qquad (5.1.14a)$$

For $s = 1$, this equation yields

$$\frac{(1-p)}{\zeta T}\Psi_a[\mathbf{V_a}] = \frac{(1-p)T_a}{\zeta T} = \frac{T}{\zeta T} = 1$$

(i.e., $\zeta = 1$, as before). But if $\zeta = 0$, then (5.1.14a) yields

$$(1-p)\Psi_a\left[\mathbf{V_a}[(1-s)\mathbf{V_a}]^{-1}\right] = \frac{1-p}{1-s} = 1,$$

or $s = p$. What this means is the following. Even though arrivals are infrequent (after all, ζ is 0), when a customer *does* arrive, there is a finite probability, (p), that he will be followed immediately by a second customer, and this second customer will have to wait for the first one to finish. It is even possible for a third (p^2) or fourth (p^3) or more customers to arrive together. This implies that even if the arrival rate is negligible, the waiting time (the time a customer must wait from the moment he arrives at S_2 until he begins to be served) will be greater than 0. From (5.1.7c), and (2.1.7c), the mean waiting time for an G/M/1 queue is given by

$$\overline{T}_{2w} = \overline{T}_2 - \bar{x}_2 = \frac{s}{1-s}\bar{x}_2,$$

where the reader should remember that we are now looking at S_2, whereas in Chapter 2 we were looking at S_1. In the limit as the arrival rate goes to 0, we see that

$$\lim_{\zeta\to 0}\overline{T}_{2w} = \frac{p}{1-s}\bar{x}_2 > 0 \quad \text{as long as } p > 0. \qquad (5.1.14b)$$

Notice that this is equivalent to a *bulk arrival process* which is geometrically distributed; that is, given that an arrival has occurred, $(1-p)p^{j-1}$ is the probability that $j \geq 1$ customers have arrived together (in bulk).

What we have discovered so far can best be summarized by Figure 5.1.1, where s is plotted as a function of ζ. Since we will be examining several distributions, we will use the notation $s(\zeta; X)$, to indicate the dependence of s on ζ for the distribution symbolized by X. For the M/M/1 queue [$X = M$],

$s(\zeta; M) = \zeta$, corresponding to the straight line from (0, 0) to (1, 1). If there is an initial impulse $[X = M_p]$, then $s(0) = p$. The M_p/M/1 queue with initial impulse p corresponds to the straight line from $(0, p)$ to (1, 1), or $s(\zeta; M_p) = p + (1-p)\zeta$. For general interarrival distributions $[D = G]$, $s(\zeta; G)$ also increases monotonically until it reaches (1, 1). We know that the larger s is, the longer will be the system time, from (5.1.7c). We also know that the system time can be reduced by regulating arrivals, and the most regular arrival pattern is the one where the time between arrivals is constant. In other words, the *deterministic distribution* $[X = D]$, given by

$$b_D(x) = \delta(x-T) \quad \text{or} \quad B_D(x) = \Delta(x-T),$$

should yield the smallest s for a given ζ. Said yet another way, the D/M/1 queue has the shortest mean system time among all G/M/1 queues with the same ζ. Unfortunately, there is no finite-dimensional representation of $B_D(x)$; therefore we shall have to resort to (3.1.10) to find the dependence of s on ζ. From that equation (remembering that $\lambda T = 1/\zeta$),

$$s = \int_0^\infty e^{-\lambda(1-s)x}\, \delta(x-T)\, dx = e^{-(1-s)/\zeta}.$$

Notice that $s = 1$ is a solution to this equation for all ζ, but as we stated previously, this root has no physical significance except when $\zeta = 1$ also. It turns out that one can draw the graph of the relation between s and ζ by solving for ζ (since one cannot solve explicitly for s). The function $\zeta = (s-1)/\log s$ yields the graph labeled D on Figure 5.1.1. This curve is the greatest lower bound for all possible distributions, for every ζ between 0 and 1. That is, let $s(\zeta; G)$ be the geometric parameter corresponding to some general PDF, $B_G(x)$. Then

$$s(\zeta; G) \geq s(\zeta; D) \qquad \text{for all } \ 0 \leq \zeta \leq 1.$$

Example 5.1.1: For comparison, we have plotted the geometric parameter for the uniform distribution $[X = U]$, the Erlangian-k, for $k = 2, 4, 8$ $[E_k]$, the Erlangian-2 with initial impulse $p = 0.1$ $[E_{2p}]$, and a hyperexponential distribution with coefficient of variation [recall that $C^2 = \sigma^2/(\bar{x})^2$], equal to 10 $[H]$. Again, the uniform distribution required the use of (3.1.10), but the others, being proper ME distributions, were best suited for (5.1.14a).

Note that, indeed, all the curves satisfy the bound theorem just stated, and in fact, $s(\zeta; E_k)$ approaches $s(\zeta; D)$ from above for all ζ, as $k \to \infty$:

$$\lim_{k\to\infty} [s(\zeta; E_k) - s(\zeta; D)] = 0_+ \quad \text{for all } \ 0 \leq \zeta \leq 1.$$

This equation indicates how the deterministic distribution is approximated arbitrarily closely by a family of finite-dimensional ME distributions. We can say that

$$\delta(x-T) = \lim_{n\to\infty} E_n(x; n/T),$$

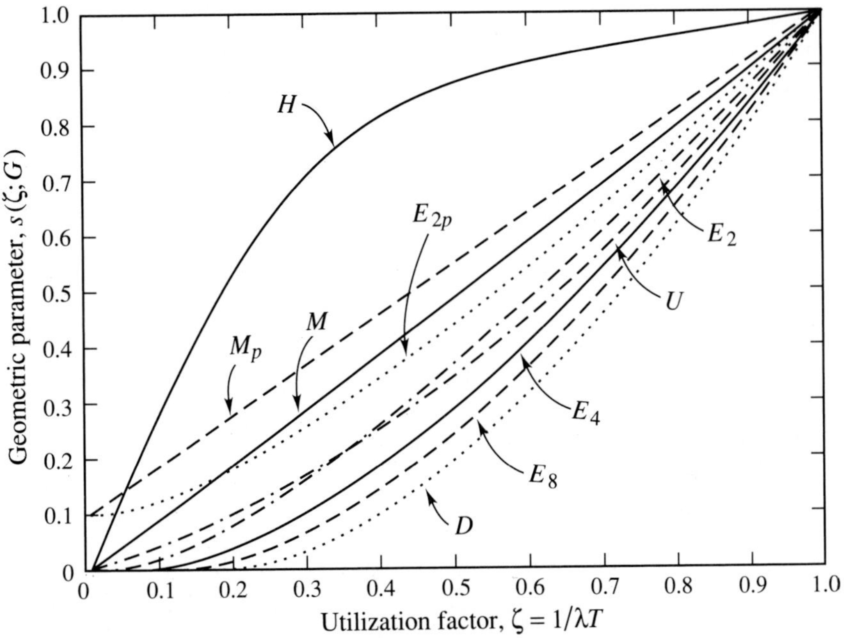

Figure 5.1.1: Dependence of s, the geometric parameter of the steadystate G/M/1 queue, on the utilization, $\zeta = 1/\lambda T = \bar{x}_2/\bar{x}_1$, for various interarrival distributions. The distributions and their labels are (M), exponential; (M_p), exponential with initial impulse, $p = 0.1$; (E_k), Erlangian-k, for $k = 2, 4, 8$; (E_{2p}), Erlangian-2 with initial impulse; (D), Deterministic; (U), uniform; and (H), hyperexponential, with $C^2 = 10$. All possible geometric parameters must be monotonically nondecreasing functions of ζ, bounded from below by $s(\zeta; D)$, bounded from above by $s = 1$. For all distributions $s(1; G) = 1$.

where $E_n(x; \alpha)$ is defined in (3.2.1a). Therefore, in some sense, the deterministic distribution is an ME distribution, since it is a member of the closure set. Further discussion in this vein requires more advanced mathematics than we ask for understanding this book, so we leave it to the experts.

It is commonly accepted that if the coefficient of variation for a given distribution is greater than 1, its geometric parameter will be greater than that for the M/M/1 queue [i.e., $s(\zeta; G) > \zeta$], and if $C_G^2 < 1$, we would expect $s(\zeta; G) < \zeta$. While this is true for ζ sufficiently close to 1, it need not be true for all ζ. The function $s(\zeta; H)$, with $C_H^2 = 10$, clearly satisfies this rule. So does $s(\zeta; U)$, with $C_U^2 = 1/3$. However, $s(\zeta; 2p)$, for which $C_{2p}^2 = 2/3$, clearly does *not*, as seen by the crossing of the two curves corresponding to $2p$ and M. ●

We cannot in general show how s varies explicitly with ζ, except by direct numerical computation. However, we can see their relation more clearly near $\zeta = 0$ and 1, by expanding s in a Taylor series about each of those points. To do this, we need to know the derivatives of s with respect to ζ there. Since we do not have an explicit relation between the two variables, we must perform the differentiation implicitly. Consider the function, taken from (5.1.14a),

$$g(s\,;\zeta) := (1-p)\Psi_a\Big[\mathbf{V_a}[\zeta T\mathbf{I} + (1-s)\mathbf{V_a}]^{-1}\Big] - 1. \qquad (5.1.15a)$$

For a given ζ, the geometric parameter for the G/M/1 queue satisfies the equation: $g(s\,;\zeta) = 0$. In particular, we know that

$$g(p;0) = 0 \quad \text{and} \quad g(1;1) = 0. \qquad (5.1.15b)$$

Next we can write

$$\frac{dg}{d\zeta} = \frac{\partial g}{\partial s}\frac{ds}{d\zeta} + \frac{\partial g}{\partial \zeta} = 0.$$

Therefore, we have

$$\frac{ds}{d\zeta} = -\frac{\dfrac{\partial g}{\partial \zeta}}{\dfrac{\partial g}{\partial s}}. \qquad (5.1.16)$$

The higher derivatives can be computed by differentiating this expression over and over again. We will spare the reader this tedium, and only quote the results. However, we will go part of the way, so as to prove that all the derivatives of s evaluated at $\zeta = 1$ depend only on the moments of the distribution function. We will also prove that all the derivatives of s evaluated at $\zeta = 0$ depend only on the value of $B(x)$ [or $R(x)$] and its derivatives at $x = 0_+$.

First, from Equations (5.1.12), we see that the kth derivatives of $R(x)$ and $b(x)$ are related to the derivatives of $R_a(x)$ and $b_a(x)$ by the following:

$$B^{(k)}(x) := \frac{d^k B(x)}{dx^k} = (1-p)B_a^{(k)}(x) = -(1-p)R_a^{(k)}(x), \quad k \ge 1, \qquad (5.1.17a)$$

and

$$b^{(k)}(x) = (1-p)b_a^{(k)}(x), \quad k \ge 1 \;\text{ and }\; x > 0. \qquad (5.1.17b)$$

Next note from its definition that $\Psi[\mathbf{V}^n] = (1-p)\Psi_a[\mathbf{V_a}^n]$, so, using (3.1.8b) and (3.1.9), we can write for $k, n \ge 0$

$$\begin{aligned} b^{(k)}(0_+) &= -R^{(k+1)}(0_+) = (1-p)b_a^{(k)}(0) \\ &= (-1)^k(1-p)\Psi_a[\mathbf{B_a}^{k+1}] \end{aligned} \qquad (5.1.17c)$$

and (pick any pair)

$$\mathrm{E}(x^n) = n!\,\Psi[\mathbf{V}^n] = n!(1-p)\Psi_a[\mathbf{V_a}^n] = (1-p)\mathrm{E}_a(x^n). \qquad (5.1.17d)$$

Now we are going to show the utility of (5.1.15a) (and LAQT) for finding the derivatives of functions. As we do so often, we define an auxiliary function that seems to be more general than we need, and then come up with simpler expressions than we otherwise would. Let

$$G(k,l\,;\zeta) := (1-p)\Psi_a\Big[\mathbf{V_a}^k\,[\zeta T\mathbf{I} + (1-s)\mathbf{V_a}]^{-l}\Big], \quad k,l \geq 1, \quad (5.1.18a)$$

where we have suppressed the dependence on s, since it, in turn, also depends on ζ. Then from (5.1.15a),

$$g(s\,;\zeta) = G(1,1\,;\zeta) - 1. \quad (5.1.19)$$

Now we are ready to take partial derivatives.

$$G_s(k,l\,;\zeta) := \frac{\partial}{\partial s}G(k,l\,;\zeta) = l\,(1-p)\Psi_a\Big[\mathbf{V_a}^{k+1}\,[\zeta T\mathbf{I} + (1-s)\mathbf{V_a}]^{-(l+1)}\Big].$$

Thus

$$G_s(k,l\,;\zeta) = l\,G(k{+}1,l{+}1\,;\zeta). \quad (5.1.20a)$$

Similarly, we can show that

$$G_\zeta(k,l\,;\zeta) := \frac{\partial}{\partial\zeta}G(k,l\,;\zeta) = -T\,G(k,l{+}1\,;\zeta). \quad (5.1.20b)$$

We can use these equations to differentiate over and over again. For instance, applying (5.1.20b) twice, we get

$$G_{\zeta\zeta}(k,l\,;\zeta) := T^2 G(k,l{+}2\,;\zeta),$$

and applying (5.1.20a) and (5.1.20b) once each, we get

$$G_{\zeta s}(k,l\,;\zeta) := G_{s\zeta}(k,l\,;\zeta) := -l\,T\,G(k{+}1,l{+}2\,;\zeta).$$

Notice that if we start with $l \geq k$, then no matter how many partial derivatives we take of both kinds, we will always end up with an expression where the second argument of G is greater than, or equal to, the first argument. Since our object is to differentiate $g(s\,;\zeta)$ (where $k = l = 1$), this will always be the case.

Actually, we are only interested in the G's when $\zeta = 0$ ($s = p$) and $\zeta = 1$ ($s = 1$). Thus (use $\mathbf{V_a} = \mathbf{B_a}^{-1}$)

$$\begin{aligned} G(k,l;0) &= (1-p)\Psi_a\Big[\mathbf{V_a}^k\,[(1-p)\mathbf{V_a}]^{-l}\Big] \\ &= (1-p)^{-(l-1)}\Psi_a\Big[\mathbf{B_a}^{l-k}\Big] \end{aligned} \quad (5.1.21a)$$

(we have assumed that $l \geq k$) and

$$G(k,l;1) = (1-p)T^{-l}\,\Psi_a[\,\mathbf{V_a}^k\,] = T^{-l}\,\Psi[\,\mathbf{V}^k\,]. \quad (5.1.21b)$$

Notice that all the G's, evaluated at $\zeta = 0$, depend only on the scalars, $\Psi_a[\,\mathbf{B_a}^j\,]$ (for $j \geq 0$), which from (5.1.17c) tells us that they depend on $R(x)$ and its derivatives at $x = 0$ only. Also notice from (5.1.21b) that the values of the G's at $\zeta = 1$ do *not* explicitly depend on the initial impulse, as represented by p and a, and in fact, depend only on the moments of the interarrival distribution. Although (5.1.21a) and (5.1.21b) are only valid for ME distributions, (5.1.17c) and

(5.1.17d) allow us to extend the equations to *any* distribution for which the appropriate objects exist:

$$G(k,l;1) = \frac{k!\,\mathrm{E}(x^k)}{T^l}, \tag{5.1.21c}$$

and

$$G(k,l;0) = \frac{(-1)^{l-k}}{(1-p)^l} R^{(l-k)}(0_+). \tag{5.1.21d}$$

Such relations could have been derived without the aid of the ME formulas, but the mathematical difficulties would have been enormous. For instance, *l'Hôpital's* rule must be applied $k+1$ times just to get the kth derivative. The reader should try it and see.

Okay, let us see what all this has done for us. Return to (5.1.16), and use (5.1.19) and (5.1.20), to get

$$s'(\zeta) := \frac{ds}{d\zeta} = -\frac{G_\zeta(1,1;\zeta)}{G_s(1,1;\zeta)} = \frac{T\,G(1,2;\zeta)}{G(2,2;\zeta)}. \tag{5.1.22a}$$

For $\zeta = 1$, using (5.1.21b), we have (remember, $\Psi[\mathbf{V}] = T$)

$$s'(1) = T\,\frac{\Psi[\mathbf{V}]}{T^2} \times \frac{T^2}{\Psi[\mathbf{V}^2]} = \frac{T^2}{\Psi[\mathbf{V}^2]}. \tag{5.1.22b}$$

But $\Psi[\mathbf{V}^2] = \mathrm{E}(x^2)/2$, $\mathrm{E}(x^2) = T^2 + \sigma^2$, and $C^2 = \sigma^2/T^2$, so

$$s'(1) = \frac{2}{1+C^2}. \tag{5.1.22c}$$

In a similar fashion we can show that (pick one)

$$s'(0) = T\Psi[\mathbf{B_a}] = T\,b_a(0) = \frac{T\,b(0_+)}{1-p} = (1-p)T_a\,b_a(0). \tag{5.1.22d}$$

Since $s'(1;M) = 1$, (5.1.22c) tells us that any interarrival distribution that has a coefficient of variation less than (greater than) 1 will have a slope greater than (less than) 1, and thus its geometric parameter must be below (above) that for the M/M/1 queue as they both approach (1, 1). The largest slope attainable occurs for the deterministic distribution for which $C^2 = 0$ and $s'(1;D) = 2$; thus all other curves must lie above it (at least for ζ near 1).

A Taylor series expansion near $\zeta = 0$ gives us

$$s(\zeta) \approx p + s'(0)\zeta + \frac{1}{2}s''(0)\zeta^2 + \cdots \tag{5.1.23a}$$

and for ζ near 1,

$$s(\zeta) \approx 1 - s'(1)(1-\zeta) + \frac{1}{2}s''(1)(1-\zeta)^2 + \cdots. \tag{5.1.23b}$$

Equation (5.1.23a) tells us that if $p > 0$, then $s(\zeta; G) > s(\zeta; M)$ near $\zeta = 0$, and if $p = 0$, (5.1.22d) tells us that $s(\zeta; G)$ is greater (less) than $s(\zeta; M)$ if $T\,b(0) > 1$ [$T\,b(0) < 1$]. It can be shown that

$$s(\zeta; D) \approx e^{1/\zeta} \quad \text{for small } \zeta;$$

therefore, all its derivatives are 0 at $\zeta = 0_+$. This does not actually violate Taylor's theorem, since $s(0_-; D)$ does not exist, so (5.1.23a) does not hold, but it does tell us that $s(\zeta; D)$ is *very* flat near 0, and thus bounds all other s's from below (at least for ζ near 0).

Note that the conditions required for s to be smaller than ζ near 0, and the requirements that s be larger than ζ near 1, are completely unrelated; thus it is possible to construct distribution functions whose geometric parameters cross the line $s = \zeta$, at least once, even with $p = 0$.

We have done everything we can to state and prove the following theorem, which summarizes this section.

Theorem 5.1.3: For any steady-state G/M/1 queue, with utilization factor, $\zeta = 1/\rho = 1/\lambda T$, and geometric parameter $s(\zeta; G)$, given by Theorem 5.1.1, the following statements are true:

(a) $s(\zeta; G)$ depends only on $B(x)$ [or $R(x)$] and its derivatives near $\zeta = 0$.

(b) $s(\zeta; G)$ depends only on the moments of $b(x)$ near $\zeta = 1$.

(c) $s(\zeta; G)$ is bounded from below by $s(\zeta; D)$ for $0 \le \zeta \le 1$.

We have not actually proven (c). ■

This theorem has an important implication for dealing with approximations to density functions when applied to heavy traffic queues and reliability theory. Heavy traffic queues occur when ρ, or in this chapter, ζ, is close to 1, and the common belief that our approximations should fit the first few moments is vindicated here. However, as we have already pointed out in Section 4.5.4 and will do again in Section 6.4.3, MTTF is more interested in small ζ, for one expects the time to repair a device to much less than the time it takes for it to break. Therefore, the behavior of the pdf near $x = 0$ plays a more important role than the moments! Furthermore, in real-life situations, decisions are usually made before problems become serious (well, they should be), so the intermediate region should be the most important. Conclusion? Both moments *and* derivatives are important.

For the record, we give explicit formulas for the second derivatives without forcing the reader to go through the tedious derivations (just looking at the formulas is bad enough):

$$s''(1) = \frac{2T^2}{\left(\Psi[\mathbf{V}^2]\right)^3}\left(\left(\Psi[\mathbf{V}^2]\right)^2 - T\,\Psi[\mathbf{V}^3]\right)$$

$$= \frac{4T^2}{E(x^2)} - \frac{8}{3}\left(\frac{T}{E(x^2)}\right)^3 E(x^3), \qquad (5.1.24a)$$

(depends only on the moments), and

$$s''(0) = \frac{T^2}{1-p}\left[\Psi_a[\mathbf{B_a}]^2 - \Psi_a[\mathbf{B_a}^2]\right] = \frac{T^2}{1-p}\left[[b_a(0)]^2 + (1-p)b'_a(0)\right]$$

$$= \frac{T^2}{(1-p)^3}\left[[b(0_+)]^2 + (1-p)b'(0_+)\right] \qquad (5.1.24b)$$

[depends only on $R(0)$ and its derivatives].

Finally, we note that one can automate the numerical calculation of all the derivatives of $s(\zeta; G)$ at 0 and 1, with the aid of (5.1.22a), (5.1.20), and (5.1.21), but you probably have already been exposed to more information about the geometric parameter than you care to know.

5.2. REPRESENTATIONS OF DEPARTURE PROCESSES

We now turn our attention to the behavior of customers *leaving* a service center. We already looked at this to some extent in Sections 2.1.5 and 4.2.4, in looking at the M/M/1 and M/G/1 queues. We will do the same here for the G/M/1 queue, but first we will look at arrivals to S_2 conditioned by departures from S_2. From the closed-loop point of view, arrivals to S_2 are the same as departures from S_1. There was no point in examining the equivalent question for the M/G/1 queue, because arrivals to the "G" queue (S_1) were governed according to the Poisson process, and thus no conditions could change that.

5.2.1. Arrival Time Distribution Conditioned by a Departure

We saw in Theorem 5.1.2 that all the steady-state vector departure probabilities $[\mathbf{d}_2(n)]$ are proportional to the same vector, $\hat{\mathbf{u}}$. Thus at the moment a customer leaves S_2, S_1 will be found in that same state. We conclude, then, that the time until the next arrival to S_2 is generated by the vector-matrix pair $<\hat{\mathbf{u}}, \mathbf{B}>$. We must say a few words to distinguish this process from the interarrival process to S_2. The interarrival process refers to the distribution of times between arrivals to S_2, or the time until the next arrival, given that a customer has just arrived. It is the same as the time between departures from S_1, which is generated by

<**p** , **B**>. In this section we are interested in the time to the next arrival, given that a customer has just *departed* S_2 thus the change from **p** to $\hat{\mathbf{u}}$ as the initial, or startup, vector. For lack of a better symbol, we will denote all properties of this process with the subscript ω. Thus $E_\omega(X)$ denotes the mean time until the next arrival, given that a customer has just departed S_2, and $b_\omega(x)$ is its density function.

We will be consistent with Theorems 3.1.1 and 4.2.4 and describe this latest process by the following theorem.

Theorem 5.2.1: The arrival times for an open ME/M/1 queue, given that a customer has just left, is generated by the vector-matrix pair <$\hat{\mathbf{u}}$, **B**> (or <$\hat{\mathbf{u}}$, **V**>), where $\hat{\mathbf{u}}$ is given by Corollary 5.1.1, and **B** is the service rate matrix for S_1. It then follows that (where $\Psi_\omega[\mathbf{D}] := \hat{\mathbf{u}}\,\mathbf{D}\,\boldsymbol{\varepsilon}'$)

$$E_\omega(X^n) = \overline{x^n_\omega} = n!\,\Psi_\omega[\mathbf{V}^n], \tag{5.2.1a}$$

$$b_\omega(x) = \Psi_\omega[\mathbf{B}\exp(-x\mathbf{B})], \tag{5.2.1b}$$

and

$$B^*_\omega(s) = \Psi_\omega[(\mathbf{I} + s\mathbf{V})^{-1}]. \tag{5.2.1c}$$

The proof follows from the definition of $\hat{\mathbf{u}}$ and Theorem 3.1.1. ■

Let us now examine this distribution further by calculating its mean and variance, and then see what we can do with its pdf. We can find $E_\omega(X)$ by multiplying (5.1.6c) on the right with $\boldsymbol{\varepsilon}'$. This process yields

$$1 + (1-s)\lambda\hat{\mathbf{u}}\mathbf{V}\boldsymbol{\varepsilon}' = \lambda\mathbf{p}\mathbf{V}\boldsymbol{\varepsilon}' = 1 + (1-s)\lambda E_\omega(X) = \lambda E_1(X) = \rho.$$

Upon solving for $E_\omega(X)$ and letting $1/\rho = \zeta$, we get

$$E_\omega(X) = \frac{\rho - 1}{(1-s)\lambda} = \frac{1-\zeta}{(1-s)\zeta\lambda} = \frac{1-\zeta}{1-s}E_1(X). \tag{5.2.2a}$$

In general, $E_\omega(X)$ is not equal to $E_1(X)$. Of course, for the M/M/1 queue $s = \zeta$, so the two are equal. They are also equal in the limit as $\zeta \to 0$ (i.e., in the no-load limit), if there is no initial impulse. For then s also becomes 0. The heavy load limit is not so easy to find, since now both s and ζ go to 1, and we are left with the indeterminate, $0/0$. We can take the limit by going back to (5.1.6b). Now, remember that since $\rho = \lambda E_1(X)$, when ζ goes to 1, λ and $1/E_1(X)$ become equal. Also, recall the definition of the mean residual vector, $\boldsymbol{\pi}_\mathbf{r}$, from (3.3.10b). What we get is

$$\lim_{s\to 1} \hat{\mathbf{u}} = \mathbf{p}(\mathbf{I} + 0\,\mathbf{V})^{-1}\frac{1}{E_1(X)}\mathbf{V} = \frac{1}{\Psi[\mathbf{V}]}\mathbf{p}\mathbf{V} = \boldsymbol{\pi}_\mathbf{r}. \tag{5.2.2b}$$

This is what a random observer would see upon visiting S_1 without noting its previous behavior. We would expect this, since a customer departing S_2 sort of randomly arrives at S_1. It is surprising, then that this does not happen for all ρ (or all s).

Recall from Section 4.3.2 that a customer arriving at an M/G/1 queue will observe that S_1 is in the probability state $\pi_\mathbf{r}(n)$ given by (4.3.1a), which happens to be the same thing a random observer would see. Now, this is dependent on the queue he sees upon his arrival. If he takes no note of how many customers are in the queue (or equivalently, if a weighted average is taken over all queue lengths), then for all ρ, as we proved in (4.3.5b), he will see that S_1 is in state i with probability $[\pi_\mathbf{r}]_i$. Now, a customer departing S_2 is the same as a customer arriving at S_1, so why the apparent contradiction between Chapter 4 and here? The resolution of this dilemma follows. As long as ρ is less than 1, as was the case in Chapter 4, a customer has a chance of finding a small number of customers at S_1. But when ρ is greater than 1, as is the case in this chapter, we have already argued several times that we can imagine there are an infinite number of customers at S_1. Therefore, our customer departing S_2 will find S_1 in the state

$$\lim_{n\to\infty} \pi_\mathbf{r}(n) = \lim_{n\to\infty} \frac{\mathbf{p}\mathbf{U}^n}{\Psi[\,\mathbf{U}^n\,]} = \hat{\mathbf{u}}.$$

The last step follows from (5.1.2c). So consistency of view is reestablished.

Returning to the calculation of $\mathrm{E}_\omega(X)$ near $s = 1$, we see from the discussion of residual times (5.5.2b) that

$$\lim_{s\to 1} \mathrm{E}_\omega(X) = \lim_{s\to 1} \hat{\mathbf{u}}\mathbf{V}_1\boldsymbol{\varepsilon}' = \pi_\mathbf{r}\mathbf{V}_1\boldsymbol{\varepsilon}' = \frac{\mathrm{E}_1(X^2)}{2\mathrm{E}_1(X)}.$$

So, as s (and ζ) approach 1, the mean time until the next arrival after a departure from S_2 is equal to the mean residual time of S_1.

We now find an expression for the variance and coefficient of variation for the process. First we need $\Psi_\omega[\,\mathbf{V}^2\,]$. We get this in the same way we found $\mathrm{E}_\omega(X)$. Multiply (5.1.6c) from the right by $\mathbf{V}\boldsymbol{\varepsilon}'$, and then solve for the desired term.

$$\lambda^2\Psi_\omega[\,\mathbf{V}^2\,] = \frac{1}{1-s}\left(\lambda^2\Psi[\,\mathbf{V}^2\,] - \lambda\mathrm{E}_\omega(X)\right). \tag{5.2.3a}$$

We know that $\mathrm{E}_\omega(X^2) = 2\Psi_\omega[\,\mathbf{V}^2\,]$, and that $\sigma_\omega^2 = \mathrm{E}_\omega(X^2) - [\mathrm{E}_\omega(X)]^2$. Therefore,

$$\lambda^2\sigma_\omega^2 = \frac{1}{1-s}\left(\lambda^2\mathrm{E}(X^2) - 2\lambda\mathrm{E}_\omega(X)\right) - \lambda^2\mathrm{E}_\omega(X)^2$$

$$= \frac{1}{1-s}\left(\lambda^2\sigma_1^2 + \rho^2 - 2\lambda\,\mathrm{E}_\omega(X) - (1-s)\lambda^2[\mathrm{E}_\omega(X)]^2\right).$$

Next let $\lambda\mathrm{E}_\omega(X) = (\rho-1)/(1-s)$, from (5.2.2a), carry out some minimal algebra and end up with

$$\lambda^2\sigma_\omega^2 = \frac{\lambda^2\sigma_1^2}{1-s} + \frac{1-s\rho^2}{(1-s)^2}. \tag{5.2.3b}$$

Recall that the coefficient of variation is the dimensionless ratio of variance to mean squared of any distribution. Therefore,

$$C_\omega^2 = \frac{\lambda^2\sigma_\omega^2}{\lambda^2[\mathrm{E}_\omega(X)]^2} = \left(\frac{\lambda^2\sigma_1^2}{1-s} + \frac{1-s\rho^2}{(1-s)^2}\right)\left(\frac{1-s}{\rho-1}\right)^2.$$

After replacing ρ by $1/\zeta$, and performing a little cleanup, the following expression emerges:

$$C_\omega^2 = \frac{(1-s)C_1^2 + \zeta^2 - s}{(1-\zeta)^2}, \tag{5.2.3c}$$

not very memorable, but the best we can do at the moment. We can tell this much, though. If S_1 is exponential, then $C_1^2 = 1$, $s = \zeta$, and $C_\omega^2 = 1$. But you knew that already. We also know from previous discussion that for low load $(s \to 0)$, $C_\omega^2 \to C_1^2$. Under heavy load, the more detailed relation between s and ζ which we did in Section 5.1.3 is needed before we can get a reasonable expression. Alternatively, by (5.2.2b) we can find what $b_\omega(x, 1)$ itself is, and calculate C_ω^2 from it. That is what we will do at the end of this section.

Our last task in this subsection is to find an expression for $b_\omega(x; s)$ itself. There are two approaches we will take, neither of which yields analytically useful results, although both can be used computationally. First multiply (5.1.6b) on the right with $\mathbf{B}\exp(-x\mathbf{B})\boldsymbol{\varepsilon}'$; then we get an explicit equation:

$$b_\omega(x; s) = \Psi_\omega\big[\mathbf{B}\exp(-x\mathbf{B})\big] = \lambda\Psi\big[[\mathbf{I} + \lambda(1-s)\mathbf{V}]^{-1}\exp(-x\mathbf{B})\big]. \tag{5.2.4a}$$

Our other approach is to use (5.1.6c). Again we multiply on the right with $\mathbf{B}\exp(-x\mathbf{B})$ to get the following:

$$b_\omega(x; s) + (1-s)\lambda R_\omega(x; s) = \lambda R_1(x), \tag{5.2.4b}$$

where we have made use of the definition of the reliability function given in (3.1.7b). This can be viewed as a differential equation in R_ω, since b_ω is its negative derivative, namely,

$$\frac{d}{dx}R_\omega(x; s) = (1-s)\lambda R_\omega(x; s) - \lambda R_1(x). \tag{5.2.4c}$$

The inhomogeneous term $R_1(x)$ is a known function of x, and $R_\omega(0; s) = 1$. For those who know something about solving differential equations, this formula has an interesting, if disconcerting property. Note that the coefficient of the homogeneous term is positive, namely, $(1-s)\lambda$. Thus the homogeneous solution $[R_H(x; s)]$ is a positive exponential, which increases unboundedly for large x. But $R_\omega(x; s)$ must go to zero for large x; therefore, the inhomogeneous solution $[R_I(x; s)]$ must have the value 1 at $x = 0$, which then makes the homogeneous term drop out. Elaborating further, the general solution of (5.2.4c) must be of the form

$$R_\omega(x; s) = AR_H(x; s) + R_I(x; s),$$

where R_H is the general solution of (5.2.4c) with R_1 removed, A is an arbitrary constant, and R_I is any solution of the entire equation. The constant A is fixed by making $R_\omega(0; s) = 1$. Such a solution does not exist for arbitrary s, but only for that unique s less than 1 which satisfies (5.1.6a).

Exercise 5.2.1: Solve the differential equation (5.2.4c) for the case that S_1 is exponential and $R_\omega(0; s) = 1$, for any s [i.e., let $R_1(x) = e^{-x\mu}$]. Show that only if $s = \mu/\lambda$ does there exist a solution for which R_ω goes to 0 as x goes to infinity.

The following expression can best summarize what we have discovered in this subsection. As s increases,

$$b_1(x) = b_\omega(x; 0) \rightarrow b_\omega(x; s) \rightarrow b_\omega(x; 1) = \frac{R_1(x)}{E_\omega(X)}.$$

That is, as ζ (and thus s) increases from 0 to 1, the distribution of arrival times for customers to an open G/M/1 queue which has just experienced a departure changes gradually from the interarrival process to the residual distribution.

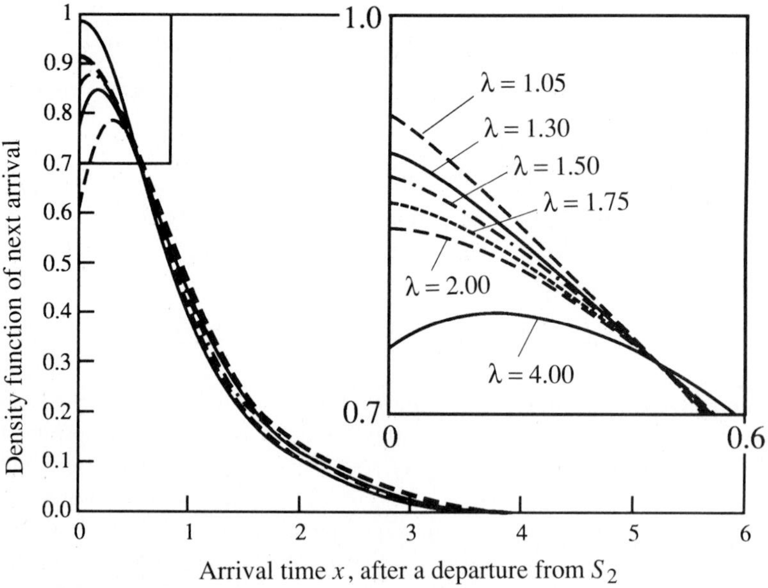

Figure 5.2.1: Distribution of arrival times, conditioned by departures from an E_2/M/1 queue, for various values of λ, the mean service rate of the lone exponential server in S_2. The mean interarrival time for all cases is held fixed at $\bar{x}_1 = 1$. Thus in all cases, $\lambda = \rho = 1/\zeta$. Note the multiple crossing, which is shown in detail in the inset.

Example 5.2.1: It is no problem to calculate the pdf for the ω process, since we know its generator; therefore, we have done so for the Erlangian-2 distribution, and show the results in Figure 5.2.1 for various values of $\zeta < 1$. We have held the interarrival times constant at $\bar{x}_1 = 1$ and have varied λ, which equals $1/\zeta$. The smaller ζ (and therefore s) is, the smaller b_ω at $x = 0$. This agrees with the relation we described above, since $b_1(0) = 0$ for all Erlangians. The curve labeled, $\lambda = 4$, is already close to $b_1(x) = 4x\exp(-2x)$. When ζ is close to 1, $b_\omega(x)$ is close to $R_1(x) = (1 + 2x)\exp(-2x)$. There is one obvious unusual feature. All the curves seem to cross each other at the same point. We have expanded the box surrounding the crossing, and show it in the inset. The curves do indeed cross, and *exactly* at $x = 0.5$, with $b_\omega(0.5; s) = 2/e$. This happens to be exactly where $b_1(x)$ and $R_1(x)$ cross. That is, for all ζ, $b_1(0.5) = R_1(0.5) = 2/e$. Any explanations? ●

We can also calculate $C_\omega^2(s{=}1)$ since we already know all the moments of $R_1(x)/\bar{x}_1$ from (3.3.12c) and (3.3.13). The result is not particularly interesting, but can be expressed in terms of the moments $\mathrm{E}_1(X^k)$, of the pdf for S_1, and comes to

$$\left(C_\omega^2\right)_{s=1} = \frac{2}{3}\frac{\mathrm{E}_1(X^3)}{\mathrm{E}_1(X)^3}\left(\frac{2}{1+C_1^2}\right) - 1.$$

Note the appearance of the third moment of S_1 in the expression. If S_1 is an exponential server, then $C_1^2 = 1$, and $\mathrm{E}_1(X^3)/\mathrm{E}_1(X)^3 = 6$, so $C_\omega^2 = 1$ as well, (of course).

5.2.2. Distribution of Interdeparture Times

We spent considerable space in the preceding section discussing a process that does not seem to be of enormous interest to queueing practitioners. However, several of the formulas derived there will be useful here, as we explore the behavior of customers departing the GI/M/1 queue. We have already discussed this process twice before, in Sections 2.1.5 and 4.2.4, in conjunction with the M/M/1 and M/G/1 queues. The method presented here is similar to that already used in those sections; however, the results are considerably different and thus warrant a fresh analysis.

Let us follow the argument we used in Section 4.2.4 in examining Figure 5.2.2. We will use the subscripts **2d** and *2d* to denote the **d**_eparture_ process from S_2. For instance, $b_{2d}(x; s)$ is the density function for the process. Our ubiquitous observer is now sitting just downstream from S_2, watching customers go by. Assuming that C_0 has just left, what can we tell her about customer C_1? Well, either he is at S_2, with probability $1 - d_2(0)$, which from (5.1.9b) equals s, or S_2 is empty, and C_1 is at S_1, already in the process of being served. The vector $\hat{\mathbf{u}}$ gives her the probability of where in S_1 he is at the moment C_0 left S_2. (Remember, *she* is the observer, and *he* is C_1.) Thus the startup vector for the interdeparture process is

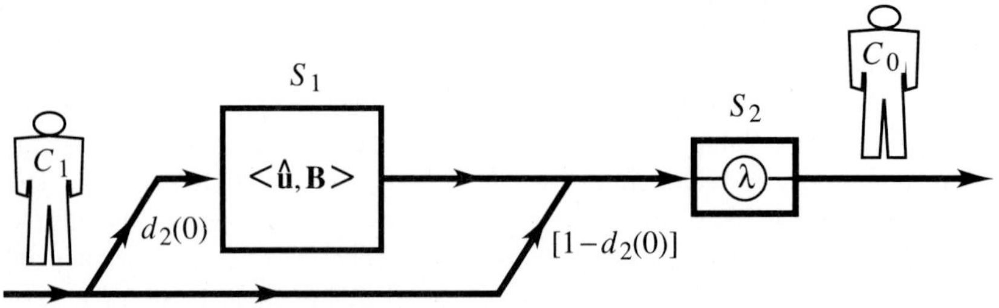

Figure 5.2.2: Pictorial representation of the departure process from S_2, in a G/M/1 open queue. Dependence on the number of customers is implicitly given through the steady-state probabilities at departure times. Given that customer C_0 has just left, C_1 must enter S_2, and be served before leaving [λ], or if S_2 is empty, C_1 must finish being served by S_1 [<$\hat{\mathbf{u}}$, $\mathbf{B}$>], and then go to S_2 to be served. The probability that no one is at S_2 at the moment of a departure, $d_2(0)$, is given by (5.1.9c).

$$\mathbf{p_{2d}} := [(1-s)\hat{\mathbf{u}}, s] . \tag{5.2.5a}$$

In words, the process starts with C_1 either being at phase i in S_1 with probability $(1-s)[\hat{\mathbf{u}}]_i$, or at S_2 with probability s. Clearly, since $\hat{\mathbf{u}}\boldsymbol{\varepsilon}' = 1$, it follows that $\mathbf{p_{2d}}\boldsymbol{\varepsilon}' = 1$ also. Note that we have changed the ordering of our states from that in Chapter 4, by placing S_1 first. Now the numbering of the states goes from 1 to $m+1$, where the state corresponding to being in S_2 is $m+1$ rather than 0. Figure 5.2.1 is descriptive enough for us to write down the completion rate and transition matrices for the process.

$$\mathbf{M_{2d}} = \begin{bmatrix} \mathbf{M} & \mathbf{o}' \\ \mathbf{o} & \lambda \end{bmatrix} \quad \text{and} \quad \mathbf{P_{2d}} = \begin{bmatrix} \mathbf{P} & \mathbf{q}' \\ \mathbf{o} & 0 \end{bmatrix},$$

where $\mathbf{o}$ ($\mathbf{o}'$) is a row (column) vector with the same dimension as $\mathbf{M}$ and $\mathbf{P}$, namely, m. Then in direct analogy with (4.2.11b), we can write down the process rate matrix.

$$\mathbf{B_{2d}} = \mathbf{M_{2d}}(\mathbf{I_{2d}} - \mathbf{P_{2d}}) = \begin{bmatrix} \mathbf{B} & -\mathbf{B}\boldsymbol{\varepsilon}' \\ \mathbf{o} & \lambda \end{bmatrix}. \tag{5.2.5b}$$

The process time matrix also follows easily:

$$\mathbf{V_{2d}} = \mathbf{B_{2d}}^{-1} = \begin{bmatrix} \mathbf{V} & \frac{1}{\lambda}\boldsymbol{\varepsilon}' \\ \mathbf{o} & \frac{1}{\lambda} \end{bmatrix}. \tag{5.2.5c}$$

We now know enough to state the following theorem concerning interdeparture times.

Theorem 5.2.2: The distribution of times between departures from a steady-state open G/M/1 queue is generated by the vector-matrix pair, $\langle \mathbf{p_{2d}}, \mathbf{B_{2d}} \rangle$ (or $\langle \mathbf{p_{2d}}, \mathbf{V_{2d}} \rangle$), as given by Equations (5.2.5). The following equations must be true (where $\Psi_{2d}[\mathbf{D}] := \mathbf{p_{2d}} \mathbf{D} \varepsilon'_{\mathbf{2d}}$):

$$E_{2d}(X^n) = \overline{x_{2d}^n} = n!\,\Psi_{2d}[(\mathbf{V_{2d}})^n], \qquad (5.2.6a)$$

$$b_{2d}(x) = \Psi_{2d}[\mathbf{B_{2d}} \exp(-x\mathbf{B_{2d}})], \qquad (5.2.6b)$$

and

$$B^*_{2d}(s) = \Psi_{2d}[(\mathbf{I} + s\mathbf{V_{2d}})^{-1}]. \qquad (5.2.6c)$$

The proof follows from Theorem 3.1.1. ∎

Before calculating the mean interdeparture time, we use (5.2.5a) and (5.2.5c) to find the following row vector:

$$\mathbf{p_{2d}} \mathbf{V_{2d}} = \left[(1-s)\hat{\mathbf{u}}\mathbf{V} \, , \, \frac{1}{\lambda} \right].$$

Then, since $E_{2d}(X) = \mathbf{p_{2d}} \mathbf{V_{2d}} \varepsilon'_{\mathbf{2d}}$, the mean is

$$E_{2d}(X) = (1-s)\Psi_{\omega}[\mathbf{V}] + \frac{1}{\lambda} = (1-s)\frac{\rho-1}{(1-s)\lambda} + \frac{1}{\lambda} = \bar{x}_1, \qquad (5.2.7a)$$

certainly not a surprising result.

En route to finding the variance, we need $\Psi_{2d}[(\mathbf{V_{2d}})^2]$, which can be written as $(\mathbf{p_{2d}} \mathbf{V_{2d}})(\mathbf{V_{2d}} \varepsilon'_{\mathbf{2d}})$, so first calculate the column vector:

$$\mathbf{V_{2d}} \varepsilon'_{\mathbf{2d}} = \begin{bmatrix} \left(\mathbf{V} + \frac{1}{\lambda}\mathbf{I}\right)\boldsymbol{\varepsilon}' \\ \frac{1}{\lambda} \end{bmatrix}.$$

We can put $\mathbf{p_{2d}} \mathbf{V_{2d}}$ and $\mathbf{V_{2d}} \varepsilon'_{\mathbf{2d}}$ together to get the second moment of $b_{2d}(x; s)$, making use of (5.2.2a) and (5.2.3a):

$$E_{2d}(X^2) = 2\Psi_{2d}[(\mathbf{V_{2d}})^2] = \frac{2\rho}{\lambda^2} + E_1(X^2) - \frac{2}{\lambda^2}\frac{\rho-1}{1-s}.$$

We know from (5.2.7a) that $b_{2d}(x)$ and $b_1(x)$ have the same mean, so with some algebraic steps left out,

$$\sigma_{2d}^2 = E_{2d}(X^2) - E_1(X)^2 = \sigma_1^2 + \frac{2(1-\rho s)}{\lambda^2(1-s)}. \qquad (5.2.7b)$$

We simply divide both sides of the equation by $[E_1(X)]^2$ to find the coefficient of variation.

$$C_{2d}^2 = C_1^2 + \frac{2(1-\rho s)}{\rho^2(1-s)}. \qquad (5.2.7c)$$

It will be helpful for the discussion that follows, to replace ρ with $1/\zeta$ in (5.2.7c). Then

$$C_{2d}^2 = C_{\mathrm{I}}^2 + \frac{2\zeta(\zeta-s)}{1-s}. \tag{5.2.7d}$$

We know from Section 5.1.3 that we can view s as a function of ζ, and as such, when $\zeta = 0$ or 1, $s = 0$ or 1, also. Therefore, when $\zeta = 0$, we get

$$\left(C_{2d}\right)^2_{\zeta=0} = C_{\mathrm{I}}^2.$$

Its value at $\zeta = 1$ is trickier and requires the functional dependence of s with respect to ζ near 1. From (5.1.22c) and (5.1.23b), we are able to say that $s = 1 - \alpha(1-\zeta) + \cdots$, with $\alpha = 2/(1+C_{\mathrm{I}}^2)$. We put this into (5.2.7d), move some things around, and come up with an expected result.

$$\left(C_{2d}\right)^2_{\zeta=1} = 1.$$

Now in general, we see that C_{2d}^2 is greater than (less than) C_{I}^2 whenever ζ is greater than (less than) s. We also know that for Erlangian distributions, C_{I}^2 is less than 1, and s is less than ζ in the entire range 0 to 1. Furthermore, for hyperexponential distributions, C_{I}^2 is greater than 1 and s is greater than ζ. We might thus conclude that C_{2d}^2 always lies between C_{I}^2 and 1, just as it did for the M/G/1 case in the discussion following (4.2.13d) in Section 4.2.4. But our conclusion would be *wrong*. In fact, we can find distributions in which s and ζ switch around several times between 0 and 1. All we can say is that this is true for ζ sufficiently close to 1.

We continue to follow our procedure in Section 4.2.4 to get the interdeparture distribution itself in terms of $\mathbf{B}$, $\mathbf{p}$ and $\hat{\mathbf{u}}$. Keep in mind, though, that $<\mathbf{p_{2d}}, \mathbf{B_{2d}}>$ can be used directly, in calculating the distribution. We can see that $\mathbf{B_{2d}}$, (5.2.5b), and $\mathbf{B_d}$, (4.2.11b), are quite similar, so we can immediately guess what $(\mathbf{B_{2d}})^n$ is for all n. Its proof, as always, can be shown by induction.

$$(\mathbf{B_{2d}})^n = \begin{bmatrix} \mathbf{B}^n & \mathbf{g}'(n) \\ \mathbf{o} & \lambda^n \end{bmatrix}, \tag{5.2.8a}$$

where

$$\mathbf{g}'(n) := -\lambda^{n-1}\mathbf{B}\left[\mathbf{I} + \frac{1}{\lambda}\mathbf{B} + \left(\frac{1}{\lambda}\mathbf{B}\right)^2 + \cdots + \left(\frac{1}{\lambda}\mathbf{B}\right)^{n-1}\right]\boldsymbol{\varepsilon}',$$

satisfying the recurrence relation

$$\mathbf{g}'(n+1) = -\lambda^n\mathbf{B}\boldsymbol{\varepsilon}' + \mathbf{B}\,\mathbf{g}'(n).^\dagger$$

For convenience we let

$$\mathbf{X} = (\mathbf{I} - \lambda\mathbf{V})^{-1}.$$

† The ' reminds us that $\mathbf{g}'$ is a column vector of dimension m.

We use the now-familiar summation formula for the finite geometric series, and carry out some further algebra to get the following:

$$\mathbf{g}'(n) = \mathbf{X}(\lambda^n \mathbf{I} - \mathbf{B}^n)\boldsymbol{\varepsilon}'. \qquad (5.2.8b)$$

We can almost exactly follow the steps leading up to (4.2.15b) to find the reliability matrix for this departure process, giving us

$$\mathbf{R_{2d}}(x) := \exp(-x\mathbf{B_{2d}}) = \begin{bmatrix} \exp(-x\mathbf{B}) & \mathbf{X}[e^{-x\lambda}\mathbf{I} - \exp(-x\mathbf{B})]\boldsymbol{\varepsilon}' \\ \mathbf{o} & e^{-x\lambda} \end{bmatrix}$$

$$= \begin{bmatrix} \mathbf{R_1}(x) & \mathbf{X}[e^{-x\lambda}\mathbf{I} - \mathbf{R_1}(x)]\boldsymbol{\varepsilon}' \\ \mathbf{o} & e^{-x\lambda} \end{bmatrix}. \qquad (5.2.8c)$$

To get $b_{2d}(x;s)$, we first must find $\mathbf{B_{2d}}\mathbf{R_{2d}}(x)$. This turns out to be

$$\mathbf{B_{2d}}\mathbf{R_{2d}}(x) = \begin{bmatrix} \mathbf{B}\mathbf{R_1}(x) & \mathbf{X}[\lambda e^{-x\lambda}\mathbf{I} - \mathbf{B}\,\mathbf{R}_1(x)]\boldsymbol{\varepsilon}' \\ \mathbf{o} & e^{-x\lambda} \end{bmatrix}. \qquad (5.2.9a)$$

Our next step is to evaluate $\mathbf{B_{2d}}\mathbf{R_{2d}}(x)\boldsymbol{\varepsilon}'_{\mathbf{2d}}$. Because $\mathbf{X}$, $\mathbf{B}$, and $\mathbf{R_1}(x)$ all commute with each other, and $\mathbf{I} - \mathbf{X} = \lambda\mathbf{X}\mathbf{V}$, this turns out to be the following column vector:

$$\mathbf{B_{2d}}\mathbf{R_{2d}}(x)\boldsymbol{\varepsilon}'_{\mathbf{2d}} = \begin{bmatrix} \lambda\mathbf{X}[e^{-x\lambda}\mathbf{I} - \mathbf{B}\mathbf{R_1}(x)]\boldsymbol{\varepsilon}' \\ \lambda e^{-x\lambda} \end{bmatrix}. \qquad (5.2.9b)$$

Finally, since $b_{2d}(x;s) = \Psi_{2d}[\mathbf{B_{2d}}\mathbf{R_{2d}}(x)] = \mathbf{p_{2d}}\mathbf{B_{2d}}\mathbf{R_{2d}}(x)\boldsymbol{\varepsilon}'_{\mathbf{2d}}$, and $\mathbf{p_{2d}}$ is given by (5.2.5a), we have the density function for the steady-state departure process:

$$b_{2d}(x;s) = \lambda e^{-x\lambda}\hat{\mathbf{u}}\mathbf{X}(\mathbf{I} - s\lambda\mathbf{V})\boldsymbol{\varepsilon}' - (1-s)\lambda\hat{\mathbf{u}}\mathbf{X}\mathbf{R_1}(x)\boldsymbol{\varepsilon}'. \qquad (5.2.9c)$$

Although this expression looks rather complicated, it is expressed in terms of m-dimensional matrices, whereas the original representation is $(m+1)$-dimensional. It can be used as a practical way to get the pdf for any specific examples, particularly if they are of small dimension. Also, note the striking similarity with its M/G/1 counterpart, $b_{1d}(x)$ [called $b_d(x)$ in (4.2.16)]. These formulas have not been known until very recently, so not many researchers have worked with them. Therefore, we have no way of knowing if they can be manipulated into simpler or more interesting forms.

Whether $b_{2d}(x)$ can be manipulated into a convenient form or not, we know its generator, $<\mathbf{p_{2d}}, \mathbf{B_{2d}}>$, given by (5.2.5a) and (5.2.5b). Therefore, there is little effort to computing the function once the once the interarrival distribution is given.

Example 5.2.2: We have calculated $b_{2d}(x; s)$ for an E_2/M/1 queue, and plotted it in Figure 5.2.3, for several values of ζ, all less than 1. We already know that when $\zeta \geq 1$ the interdeparture times must look like the service time distribution. Even when ζ is close to, but less than 1, they look very much like the exponential function. Of course, when ζ is small, b_{2d} looks like $E_2(x)$, the interarrival distribution. Notice the rapid change from one to the other when ζ goes from 0.25 to 0.50. The reader might compare this figure with its M/E_2/1 counterpart in Figure 4.2.3. ●

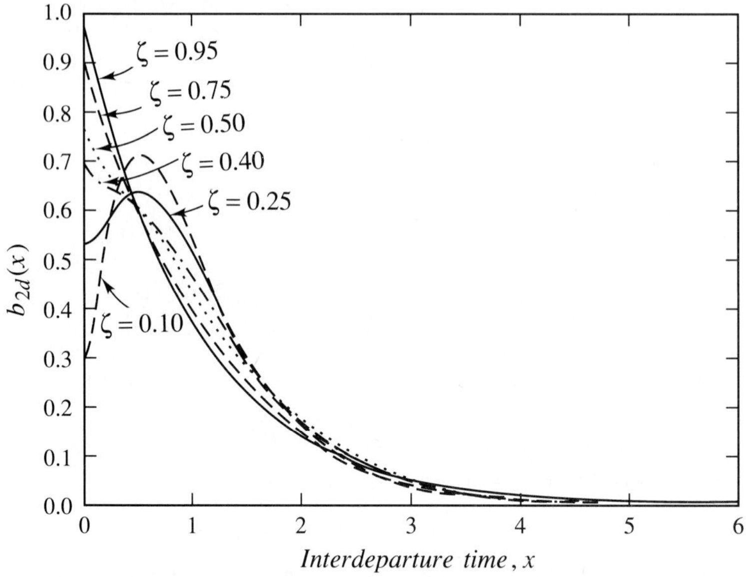

Figure 5.2.3: Distribution of interdeparture times, $b_{2d}(x)$, of an E_2/M/1 queue, for $\zeta = 0.10, 0.25, 0.40, 0.50, 0.75$, and 0.95. When ζ is small, the interdeparture distribution looks like the interarrival distribution, $E_2(x)$, and when ζ is near (or greater than) 1, it looks like the service distribution, $\lambda \exp(-\lambda x)$.

5.3. ME/M/1/*N* AND ME/M/1//*N* QUEUES COMPARED

Now is the time for you to ask what the difference is between one and two slashes. In Chapter 4 we brushed the question aside, explaining that in an M/ME/1 queue, they yielded identical results. But for ME/M/1 queues, they do not. For definiteness, let us adhere to the conventions of this chapter. The first position (ME) refers to the service time distribution at S_1, the second (M) refers to S_2, and the third position refers to the maximum number of customers (1) at S_2 who can be active at the same time. The fourth position in this notation refers to the amount of space available at S_2, including the customers in service (finite waiting room, or buffer). If that position has J there, then when J customers are at S_2, new arrivals are turned away (discounted or killed) until someone leaves, at which time there are then only $J-1$ customers there. If that position is blank, it is assumed to be infinite. The fifth position refers to the total number of customers in the system, k of whom are at S_2 and the remaining n are at S_1. If that space is blank, or nonexistent, then we have an open system, (or N is infinite).

Consider the following string: $G_1/G_2/C/J/N$. All systems with $J \geq N$ are equivalent, since it does not pay to have more space than there are customers. Similarly, all systems with $C \geq \min[J, N]$ are the same. Somewhat less obvious is the equivalence of all systems with $N > J$. We only need one more customer than there is buffer space, for if a customer is turned away because of a full waiting room, there is no difference between his returning immediately to S_1 or being replaced by another customer. We can say that if the inequality string, $C < J < N$, is not satisfied, the violating integer can be replace by ∞ (or any integer greater than or equal to the next symbol in the sequence).

Although the $G_1/G_2/1/N$ queue is usually classified as an open system, the equivalent (but closed) loop, $G_1/G_2/1/N/(N+1)$, may be easier to visualize. In this case, the last customer loops on S_1 until room is made available at S_2. However, even after space becomes available, he must still complete service at S_1 before finally being admitted to S_2. In the $G_1/G_2/1//N$ (or $G_1/G_2/1/N/N$) case, when all N customers are at S_2, S_1 is idle until the customer in service finishes. Only then does S_1 begin processing a customer to generate the next arrival. That is, in the former case S_1 is already processing the next arrival to S_2 when room becomes available, while in the latter case, processing at S_1 begins at the moment a completion occurs at S_2. Only when the residual distribution time for S_1 is the same as the overall distribution time (i.e., when S_1 is exponential) will the customer return to S_2 at the right time to make both systems identical. The algebraic analysis in the next section will make this clear.

5.3.1. Steady-State Solution of the ME/M/1/*N* Queue

First let us define (the subscript **f** stands for "finite buffer") the steady-state probability vector:

Definition 5.3.1

$\boldsymbol{\pi}_{\mathbf{f}}(k; N) :=$ *steady-state probability vector that there are k customers at* S_2 *in an* ME/M/1/*N* *system*. There are at least $N+1$ customers in this system. If a customer arrives at S_2 and finds N other customers already there, he immediately returns to S_1. A random observer, with probability $[\boldsymbol{\pi}_{\mathbf{f}}(k; N)]_i$, will find the system in state $\{ i; k; N \}$. As usual, $r_f(k; N) := \boldsymbol{\pi}_{\mathbf{f}}(k; N)\boldsymbol{\varepsilon}'$ is the associated scalar probability. ◊

The more traditional view is that there are an infinite number of customers waiting to be served by S_1. A customer who completes service there and finds S_2 to be full, immediately self-destructs. The two views are mathematically equivalent, but if nothing else, our view is more humane.

Except when $k = N$, the balance equations for $\pi_f(k; N)$ are identical to those for $\pi_2(k; N)$. Remember to replace the vector $\pi(n; N)$ with $\pi_f(N-k; N)$ [see (5.1.1d)] when you examine the equations in Section 4.1.1. The equation for $k = N$ differs from (4.1.3b). Now, even though there are N customers at S_2, S_1 is *not* idle. Therefore, the vector probability of leaving the state $\{\cdot; N; N\}$ is proportional to $(\lambda\mathbf{I} + \mathbf{M})$; that is, something can happen in either S_1 or S_2. There are three ways to enter state $\{\cdot; N; N\}$. One is to be in some state $\{\cdot; N-1; N\}$, $[\pi_f(N-1; N)]$, and have a completion, $[\mathbf{M}]$, which results in a departure from S_1, $[\mathbf{Mq}']$, while simultaneously the next customer enters S_1, $[\mathbf{p}]$. The second way is for there to be N customers at S_2, $[\pi_f(N; N)]$, with an event again occurring in S_1, $[\mathbf{M}]$, with that customer going to another phase in S_1, $[\mathbf{P}]$. The third way is similar to the second, except that now the customer in S_1 [the lonesome $(N+1)$st customer] leaves, $[\mathbf{q}']$, but since the buffer at S_2 is full, he immediately returns to S_1 and starts up again, $[\mathbf{p}]$. In total, we have

$$\pi_f(N; N)(\lambda\mathbf{I} + \mathbf{M}) = \pi_f(N-1; N)\mathbf{Mq}'\mathbf{p} + \pi_f(N; N)\mathbf{M}(\mathbf{P} + \mathbf{q}'\mathbf{p}).$$

Upon regrouping terms, and recognizing yet once again that $\mathbf{Mq}'\mathbf{p} = \mathbf{BQ}$, we get

$$\pi_f(N; N)(\lambda\mathbf{I} + \mathbf{B} - \mathbf{BQ}) = \pi_f(N-1; N)\mathbf{BQ}. \tag{5.3.1a}$$

The equation equivalent to (4.1.3c) gives

$$\pi_f(0; N)\mathbf{B} = \pi_f(1; N)\lambda, \tag{5.3.1b}$$

which in a manner identical to Section 4.1.2 recursively leads to results equivalent to (4.1.5b),

$$\pi_f(k; N)\mathbf{U} = \pi_f(k-1; N), \quad 2 \le k \le N. \tag{5.3.1c}$$

But (5.3.1a) is yet to be satisfied. Equation (5.3.1c) with $k = N$ must be made consistent with (5.3.1a). Upon combining the two, we get

$$\pi_f(N; N)(\lambda\mathbf{I} + \mathbf{B} - \mathbf{BQ}) = \pi_f(N; N)\mathbf{UBQ}.$$

But from Lemma 4.1.1, $\mathbf{UBQ} = \lambda\mathbf{Q}$, so we bring everything to the left side of the equation to get

$$\pi_f(N; N)(\lambda\mathbf{I} + \mathbf{B} - \lambda\mathbf{Q} - \mathbf{BQ}) = [\pi_f(N; N)(\lambda\mathbf{I} + \mathbf{B})]\,(\mathbf{I} - \mathbf{Q}) = \mathbf{o}$$

($\mathbf{o}$ is the null row vector). This is an eigenvector equation which says that the vector in brackets is a left eigenvector of $(\mathbf{I} - \mathbf{Q})$ with eigenvalue 0. Can this be satisfied? It had better be. Note that $\mathbf{C} := \mathbf{I} - \mathbf{Q}$ is idempotent, just like $\mathbf{Q}$. That is, $\mathbf{C}^2 = \mathbf{C}$. (See Lemma 3.3.1.) Therefore, all of $\mathbf{C}$'s eigenvalues are either 0 or 1. Now, $\mathbf{Q}$ is of rank 1, so it has only one eigenvalue with value 1. Therefore, $\mathbf{C}$ is of rank $m-1$ and has only one zero eigenvalue. The corresponding left and right eigenvector pair are our old companions $\mathbf{p}$ and ε'. The vector in brackets must, then, be proportional to $\mathbf{p}$. Write

$$\pi_f(N; N)(\lambda\mathbf{I} + \mathbf{B}) = c\,\mathbf{p},$$

where c is an undetermined constant. Recall from the definition of $\mathbf{A}$ [Equation (4.1.4a)], that $\lambda\mathbf{I} + \mathbf{B} = \lambda(\mathbf{A} + \mathbf{Q})$. Also, multiply both sides of the equation by

U to get

$$\lambda \boldsymbol{\pi}_{\mathbf{f}}(N;N)(\mathbf{I}+\mathbf{QU}) = c\,\mathbf{pU},$$

but $\boldsymbol{\pi}_{\mathbf{f}}(N;N)\mathbf{QU} = \boldsymbol{\pi}_{\mathbf{f}}(N;N)\boldsymbol{\varepsilon}'\mathbf{pU} = c'\mathbf{pU}$, where c' is another constant. We regroup, divide by λ, and get

$$\boldsymbol{\pi}_{\mathbf{f}}(N;N) = g(N)\mathbf{pU}, \tag{5.3.1d}$$

where $g(N)$ is yet another constant, which we *will* evaluate. This time we have noted its dependence on N.

We can now combine (5.3.1b), (5.3.1c), and (5.3.1d) to get the explicit matrix geometric solution to the ME/M/1/N queue:

$$\boldsymbol{\pi}_{\mathbf{f}}(k;N) = g(N)\mathbf{pU}^{N+1-k} \quad \text{for} \quad 1 \le k \le N,$$

and

$$\boldsymbol{\pi}_{\mathbf{f}}(0;N) = g(N)\lambda\mathbf{pU}^{N}\mathbf{V}.$$

The scalar probabilities are, by now, easy to write down. For $k > 0$,

$$r_f(k;N) := \boldsymbol{\pi}_{\mathbf{f}}(k;N)\boldsymbol{\varepsilon}' = g(N)\Psi[\,\mathbf{U}^{N+1-k}\,]$$

and the probability that S_2 is idle is given by

$$r_f(0;N) = g(N)\lambda\Psi[\,\mathbf{U}^{N}\mathbf{V}\,].$$

These formulas look very familiar [look at (4.1.6a) and (4.1.6b)], and we will relate them to the ME/M/1//(N+1) queue after we have found $g(N)$. We calculate this constant by requiring that the sum of the $r_f(k;N)$'s be 1. Then $g(N)$ satisfies the relation

$$\frac{1}{g(N)} = \Psi[\,\lambda\mathbf{U}^{N}\mathbf{V}\,] + \sum_{k=1}^{N}\Psi[\,\mathbf{U}^{N+1-k}\,] = \Psi[\,\mathbf{U}+\mathbf{U}^2+\cdots+\mathbf{U}^N+\lambda\mathbf{U}^N\mathbf{V}\,].$$

We need only compare this with the definition of $\mathbf{K}(N+1)$ in (4.1.6c) to see that

$$\frac{1}{g(N)} = \Psi[\,\mathbf{UK}(N)\,] = \Psi[\,\mathbf{K}(N+1)\,] - 1. \tag{5.3.2a}$$

We will next summarize these equations in the following theorem so that they can all be found in one place.

Theorem 5.3.1: The steady-state vector probabilities of (a random observer) finding k customers in an ME/M/1/N queue are given below.

$$\boldsymbol{\pi}_{\mathbf{f}}(0;N) = \lambda g(N)\mathbf{pU}^{N}\mathbf{V} \tag{5.3.2b}$$

and

$$\boldsymbol{\pi}_{\mathbf{f}}(k;N) = g(N)\mathbf{pU}^{N+1-k} \quad \text{for} \quad 0 < k \le N. \tag{5.3.2c}$$

The associated scalar probabilities are given by the next two formulas.

$$r_f(0;N) = g(N)\Psi[\,\lambda\mathbf{U}^{N}\mathbf{V}\,] \tag{5.3.2d}$$

and

$$r_f(k;N) = g(N)\Psi[\mathbf{U}^{N+1-k}] \quad \text{for} \quad 0 < k \le N, \tag{5.3.2e}$$

where $1/g(N) = \Psi[\mathbf{UK}(N)] = \Psi[\mathbf{K}(N+1)] - 1$. ■

This theorem is very similar to Theorem 4.1.2 with N replaced by $N+1$, so we can see by inspection, upon invoking (5.1.1c) and (5.1.1d), that the quantities, $\pi_{\mathbf{f}}(k;N)$ and $\pi_{\mathbf{2}}(k;N+1)$ and their scalar counterparts satisfy the following corollary.

Corollary 5.3.1: The steady-state probabilities of (a random observer) finding k customers in an ME/M/1/N queue are related to the steady-state probabilities of finding k customers in an ME/M/1//(N+1) loop by the following formulas:

$$\pi_{\mathbf{f}}(k;N) = c(N)\pi_{\mathbf{2}}(k;N+1) \quad \text{for} \quad 0 \le k \le N \tag{5.3.3a}$$

and

$$r_f(k;N) = c(N)r_2(k;N+1) \quad \text{for} \quad 0 \le k \le N, \tag{5.3.3b}$$

where

$$c(N) := \frac{\Psi[\mathbf{K}(N+1)]}{\Psi[\mathbf{K}(N+1) - \mathbf{I}]} = \frac{1}{1 - r_2(N+1;N+1)}. \tag{5.3.3c}$$

The probabilities for the two queues differ only by the fact that $r_2(N+1;N+1)$ exists but $r_f(N+1;N)$ does not. Therefore, they must be multiplied by different constants so that they each sum to 1. If $\zeta < 1$, then as N becomes unboundedly large, the two systems yield identical results. ■

Our view of the ME/M/1/N queue as a closed loop with $N+1$ customers, where a lone customer at S_1 circles until room is made for him at S_2, would seem to be better than we would have expected. Thinking of queues with finite waiting rooms as open systems is certainly not nearly as helpful.

5.4. STEADY-STATE G/M/C-TYPE QUEUES

We are now prepared to give more properties to S_2. It will still have a one-dimensional internal representation, but we will allow its service rate to vary with its queue length, k. This has the obvious application to systems in which several (C) exponential servers are fed by a single queue. Another, potentially important application is in the study of complex networks. In this case, one server is singled out to be S_1, the nonexponential server, and the rest of the network is approximated by S_2, with suitably chosen flow rates, $\lambda(k)$, to represent customer flow. Thus one can combine the power of the product-form solutions in constructing (maybe) reasonable λ's with the correct representation of one nonexponential server. This technique has been tried but not enough is known as yet to decide under what conditions it will give realistic results.

In Section 2.1.4 we discussed load-dependent exponential servers. We viewed a subsystem in either of two ways. Either there were multiple servers available to handle more than one customer at a time, or a single server worked faster when more customers were present. Since exponential subsystems have only one internal state, the two views are mathematically equivalent. For instance, if there is one customer present, let the probability rate of completion be λ, and if two are present, let the probability rate be 2λ. There is no way to tell if two servers are each processing a customer at the rate of λ, or one server is working twice as fast.

Actually, there is a way to tell the difference, by *marking* the customers. In the first case, if a customer is in service when a second arrives and begins service, there is a distinct possibility that the second will finish before the first (in fact, the probability is 0.5 for exponential servers). In the second case, the FCFS ordering is always maintained. If the customers are marked according to their order of arrival, an observer can tell the difference, since the two-server option will allow customers to leave in a different order from which they arrived. We have been and will continue to take the view that all customers are alike, and unmarkable. To do otherwise would greatly increase the amount of information required, even of exponential subsystems.

In many applications, the customers present share the single server on equal terms. For instance, a customer may be given a small amount of service and whether or not he is finished, the next customer is given an equal amount. After all customers present have been given a share, the first one is given another increment of service, and so on in *round-robin* fashion. If the time accorded each in turn is very small compared to the mean service time then we have *processor sharing*. There is a related queueing discipline known as *time slicing* in which each *potential* customer is given an increment of time, whether or not he uses it (e.g., a rotary switch on a multiplexed cable). Only the processor sharing discipline fits easily into our scheme of things. Conceptually we have multiple servers which are load-dependent. If there is one customer present, then he gets the whole server. If two customers are present, then each one gets his own server, but the servers go at half speed. Once again, if the server is exponential then there is no easy way to tell the difference between this and the simple FCFS queue.

If a subsystem has multiple internal states (i.e., is nonexponential), the three views described in the preceding paragraphs are distinctly different. Modifying the service (actually, completion) rates corresponds to changing $\mathbf{M}$ as a function of queue length but leaving $(\mathbf{I}-\mathbf{P})$ alone. Serving two customers at a time requires keeping track of both customers, for even when one of them leaves, the other is still in some phase of service. The latter view (for *processor sharing*, as well as *multiple servers*) will be reserved for Chapter 6, since it requires an increase in complexity of our formalism.

5.4.1. Steady-State ME/M/X//N Loops

Since S_2 has only one phase, the two views are still equivalent. Also, recall that solution of the M/G/1 and G/M/1 queues depends almost completely on the matrix

$$\mathbf{A} = \mathbf{I} + \frac{1}{\lambda}\mathbf{B} - \mathbf{Q}.$$

We see that λ and $\mathbf{B} = \mathbf{M}(\mathbf{I} - \mathbf{P})$ always appear together. Therefore, changing λ (modifying S_2), or modifying $\mathbf{M}$ by a constant factor, yields the same result. We will assume that $\mathbf{M}$ is fixed. The difference amounts to deciding whether the load dependence is a function of the number of customers at S_1, $[n]$, or the number of customers at S_2, $[k]$. In a closed loop it does not make any mathematical difference, since $n + k = N$, but if we look at the same system for many values of N, there is an algorithmic difference. There is also a difference from a modeling viewpoint. For instance, if we are interested in the behavior at S_1, and the load factor depends on n, we can think of this as an arrival rate that varies according to the number of customers already at S_1. In the literature this is known as a queue with *discouraged arrivals* (although arrivals could also be *encouraged*). We will not be interested in this view here.

Let us take the view that S_2 has a service rate which depends on its queue length, and as in Chapter 2, call it $\lambda(k)$. For now, we will make no further assumptions concerning the values of the $\lambda(k)$'s. Therefore, following the notational comments at the end of Section 2.1.4, we will be looking at ME/M/**X**//N loops. The steady-state balance equations can be taken directly from Equations (4.1.3) by replacing λ with $\lambda(k)$, where k corresponds to the queue number in the matching π_2. The reader can check this by comparing the steady-state transition diagram in Figure 5.4.1 with Figure 4.1.2. Using the notation of this chapter, we have

$$\lambda(N)\pi_2(N; N) = \pi_2(N-1; N)\mathbf{BQ}; \tag{5.4.1a}$$

$$\pi_2(0; N)\mathbf{M} = \pi_2(1; N)\lambda(1) + \pi_2(0; N)\mathbf{MP}; \tag{5.4.1b}$$

and for $0 < k < N$,

$$\pi_2(k; N)[\mathbf{B} + \lambda(k)\mathbf{I}] = \pi_2(k-1; N)\mathbf{BQ} + \pi_2(k+1; N)\lambda(k+1). \tag{5.4.1c}$$

(Remember that we still maintain the notation $\mathbf{B} = \mathbf{V}^{-1} = \mathbf{M}[\mathbf{I} - \mathbf{P}]$.) The π_2's and associated r_2's retain Definition 5.1.4, including the standard notational assumption that

$$\pi_2(N; N) := r_2(N; N)\mathbf{p}. \tag{5.4.1d}$$

Following the procedure we used in Chapter 4, we would like to solve for the π_2's in terms of $r_2(N; N)$, but (5.4.1a) does not allow us to do that directly, since $\mathbf{BQ}$ does not have an inverse. So we must start at the other end. Equation (5.4.1c) can be rewritten as

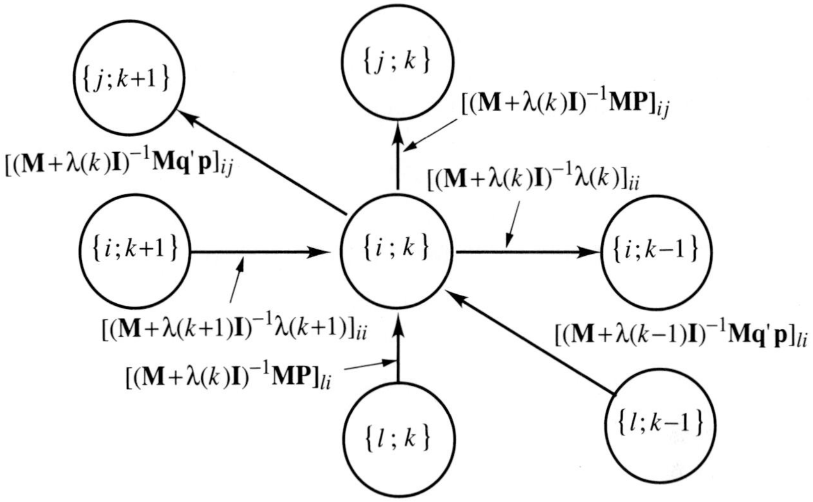

Figure 5.4.1: Steady-state transition diagram for state $\{i; k; N\}$ of an ME/M/$X//N$ closed loop. An arrow pointing to the left (either horizontally or diagonally) represents a customer finishing at phase i in S_1, $\{[(\mathbf{M}+\lambda(k)\mathbf{I})^{-1}\mathbf{M}]_{ii}\}$, and leaving to go to S_2, $[q'_i]$, followed by another customer entering and going to j, $[p_j]$. A vertical arrow corresponds to a customer finishing at phase i, in S_1 $[(\mathbf{M}+\lambda(k)\mathbf{I})^{-1}\mathbf{M}]$, and going to phase j, $[P_{ij}]$. An arrow to the right (no diagonal arrows allowed) corresponds to a customer finishing at S_2, $[(\lambda(k)\mathbf{I}+\mathbf{M})^{-1}\lambda(k)]$, and immediately going to S_1, without changing the internal state. In all cases, the argument of $\lambda(\cdot)$ matches the value of the queue length of S_2 at the tail of the arrow. Compare with Figure 4.1.2.

$$\pi_2(0; N)\mathbf{M}[\mathbf{I} - \mathbf{P}] = \pi_2(1; N)\lambda(1),$$

or

$$\pi_2(0; N) = \pi_2(1; N)\mathbf{U}(0), \tag{5.4.2a}$$

where

$$\mathbf{U}(0) := \lambda(1)\mathbf{V}. \tag{5.4.2b}$$

Next we look at (5.4.1c) for $k = 1$, while making use of (5.4.2a),

$$\begin{aligned}\pi_2(1; N)[\mathbf{B} + \lambda(1)\mathbf{I}] &= \pi_2(0; N)\mathbf{BQ} + \pi_2(2; N)\lambda(2)\\ &= \lambda(1)\pi_2(1; N)\mathbf{Q} + \lambda(2)\pi_2(2; N),\end{aligned}$$

or

$$\pi_2(1; N) = \pi_2(2; N)\mathbf{U}(1), \tag{5.4.2c}$$

where

$$\mathbf{A}(1) := \frac{\lambda(1)}{\lambda(2)}\left[\mathbf{I} + \frac{1}{\lambda(1)}\mathbf{B} - \mathbf{Q}\right] = [\mathbf{U}(1)]^{-1}. \tag{5.4.2d}$$

In preparation for the general solution by induction, first define

$$\mathbf{A}(k) := \frac{\lambda(k)}{\lambda(k+1)}\left[\mathbf{I} + \frac{1}{\lambda(k)}\mathbf{B} - \mathbf{Q}\right] = [\mathbf{U}(k)]^{-1}, \tag{5.4.3}$$

of which (5.4.2d) is a special case. Next observe the following lemma.

Lemma 5.4.1: For matrices $\mathbf{A}(k)$ and $\mathbf{U}(k)$, defined by (5.4.3), the following are matrix identities for all $k \geq 0$:

$$\mathbf{B}\boldsymbol{\varepsilon}' = \lambda(k+1)\mathbf{A}(k)\boldsymbol{\varepsilon}' \quad \text{and} \quad \mathbf{U}(k)\mathbf{B}\boldsymbol{\varepsilon}' = \lambda(k+1)\boldsymbol{\varepsilon}', \tag{5.4.4a}$$

and

$$\mathbf{B}\mathbf{Q} = \lambda(k+1)\mathbf{A}(k)\mathbf{Q} \quad \text{and} \quad \mathbf{U}(k)\mathbf{B}\mathbf{Q} = \lambda(k+1)\mathbf{Q}, \tag{5.4.4b}$$

exactly analogous to Lemma 4.1.1, for which this is a generalization. ■

Proof: Since $\mathbf{I}\boldsymbol{\varepsilon}' = \mathbf{Q}\boldsymbol{\varepsilon}' = \boldsymbol{\varepsilon}'$, all follows directly from the definition of $\mathbf{A}(k)$ and $\mathbf{U}(k)$ in (5.4.3). QED

Now assume that (it is certainly true for $j = 0$ and 1)

$$\pi_2(j; N) = \pi_2(j+1; N)\mathbf{U}(j) \quad \text{for} \quad j = 0, 1, \cdots, k-1, \tag{5.4.5}$$

and use it in (5.4.1c) to get

$$\begin{aligned}\pi_2(k; N)[\mathbf{B} + \lambda(k)\mathbf{I}] &= \pi_2(k; N)\mathbf{U}(k-1)\mathbf{B}\mathbf{Q} + \pi_2(k+1; N)\lambda(k+1) \\ &= \pi_2(k; N)\lambda(k)\mathbf{Q} + \pi_2(k+1; N)\lambda(k+1),\end{aligned}$$

where we have used (5.4.4b). Solving for $\pi_2(k+1; N)$ yields (5.4.5) for $j = k$, from the definition of $\mathbf{U}(k)$ given in (5.4.3). Thus we have proven our assertion by induction, for all j up to $j = N-1$. From Lemma 5.4.1, (5.4.1a) is also satisfied.

We are now prepared to put all the above together in the following theorem.

Theorem 5.4.2: The steady-state solution to the ME/M/X//N loop is given by the following formulas, taken from (5.4.2b), (5.4.3), and (5.4.5). For arbitrary $\lambda(k) > 0$, let

$$\mathbf{U}(k) = \frac{\lambda(k+1)}{\lambda(k)}\left[\mathbf{I} + \frac{1}{\lambda(k)}\mathbf{B} - \mathbf{Q}\right]^{-1}.$$

Then

$$\begin{aligned}\pi_2(N; N) &= r_2(N; N)\mathbf{p}, \\ \pi_2(N-1; N) &= r_2(N; N)\mathbf{p}\mathbf{U}(N-1), \\ \pi_2(N-2; N) &= r_2(N; N)\mathbf{p}\mathbf{U}(N-1)\mathbf{U}(N-2), \\ \pi_2(k; N) &= r_2(N; N)\mathbf{p}\mathbf{U}(N-1)\mathbf{U}(N-2) \cdots \mathbf{U}(k), \\ \pi_2(0; N) &= r_2(N; N)\mathbf{p}\mathbf{U}(N-1)\mathbf{U}(N-2) \cdots \mathbf{U}(k) \cdots \mathbf{U}(0),\end{aligned}$$

and

$$r_2(k; N) = \pi(k; N)\boldsymbol{\varepsilon}' \quad \text{for all } k.$$

The equivalent recursive formula is given by (5.4.5). $r_2(N; N)$ is evaluated by normalization, or let

$$\mathbf{K}(N) = \mathbf{I} + \mathbf{U}(N-1) + \mathbf{U}(N-1)\mathbf{U}(N-2) + \cdots + \mathbf{U}(N-1)\mathbf{U}(N-2) \cdots \mathbf{U}(N-k)$$
$$+ \mathbf{U}(N-1)\mathbf{U}(N-2) \cdots \mathbf{U}(N-k) \cdots \mathbf{U}(0). \tag{5.4.6a}$$

Then, with $\mathbf{K}(1) = \mathbf{I} + \mathbf{U}(0)$, it follows recursively that

$$\mathbf{K}(N) = \mathbf{I} + \mathbf{U}(N-1)\mathbf{K}(N-1) \tag{5.4.6b}$$

and

$$[r_2(N; N)]^{-1} = \Psi\big[\,\mathbf{K}(N)\,\big]. \tag{5.4.6c}$$

Compare this with the load-independent case, Theorem 4.1.2, and (4.1.6d), (4.1.6e) and (4.1.6g). ■

Proof: The equations are obvious once we note that (5.4.6b) comes from (5.4.6a) by grouping all terms that are left-multiplied by $\mathbf{U}(N-1)$. QED

As far as we know, there is no way to take advantage of a simplification such as $\lambda(k) = k\lambda$. As long as $\lambda(k) \neq \lambda(k+1)$, we are stuck with this complexity. The same order of complexity occurred for the M/M/X//N loop. Also, the formulas given in Theorem 5.4.2 require families of matrices that require recursive multiplication from both the left and right. This limits our ability to find a recursive procedure that is efficient in both space and time, in studying ME/M/X//N loops for a sequence of values of N. The algorithm we are about to present is not necessarily the most efficient, but it shows how the matrices fit together.

Define the auxiliary matrices for any $n \geq 0$,

$$\mathbf{X}(n, n) := \mathbf{I}, \tag{5.4.7a}$$

and for $k < n$

$$\mathbf{X}(k, n) := \mathbf{U}(n-1)\mathbf{U}(n-2) \cdots \mathbf{U}(k) = \mathbf{U}(n-1)\,\mathbf{X}(k, n-1). \tag{5.4.7b}$$

This can be helpful in dealing with various objects. For instance, (5.4.6a) can be rewritten as

$$\mathbf{K}(N) = \sum_{k=0}^{N} \mathbf{X}(k, N). \tag{5.4.7c}$$

Then the vector and scalar probabilities, $\pi_2(k; N)$ and $r_2(k; N)$, can be computed in the following way.

Corollary 5.4.2: (Algorithm) To compute the vector and scalar queue-length probabilities of an ME/M/X//N loop for all $N = 1, 2, \cdots, N_{\max}$, do the following:

$$\mathbf{X}(0,0) = \mathbf{I}, \quad \mathbf{K}(0) = \mathbf{I}.$$

FOR $N = 1$ TO $N_{\max}$;

$$\mathbf{X}(N,N) = \mathbf{I},$$
$$\mathbf{K}(N) = \mathbf{I} + \mathbf{U}(N-1)\mathbf{K}(N-1),$$
$$r_2(N;N) = 1/\Psi[\mathbf{K}(N)],$$

FOR $k = 0$ TO $N-1$;

$$\mathbf{X}(k,N) = \mathbf{U}(N-1)\mathbf{X}(k,N-1),$$
$$\pi_2(k;N) = r_2(N;N)\mathbf{p}\mathbf{X}(k,N),$$
$$r_2(k;N) = \pi_2(k;N)\boldsymbol{\varepsilon}' = r(N;N)\Psi[\mathbf{X}(k,N)],$$

END FOR(k).

END FOR(N).

The mean queue length and other performance characteristics can be found by computing them directly. ■

There are no further insights we can gain without becoming more specific about the properties of the λ's. Letting $\lambda(k) = k\lambda$ will not tell us much unless we do the calculations. If we let N become infinite, we can say very little unless

$$\lambda_\infty := \lim_{N\to\infty} \lambda(N) < \infty,$$

and $\rho = \bar{x}_1 \lambda_\infty < 1$. In that case we would revert back to the steady-state M/ME/1 open queue of Section 4.2, with arrival rate λ_∞. If the inequality is the other way around, (i.e., if $\rho > 1$), we have a problem. We do not even know how to start without more information. However, we *can*, and in the next section, will solve those systems for which the load-dependent service rates are constant above a certain queue length.

5.4.2. Steady-State ME/M/C Queue

Let us assume that $N > C > 1$, and

$$\lambda(k) = \lambda(C) \quad \text{for} \quad k \geq C. \tag{5.4.8}$$

What the values of $\lambda(1), \lambda(2), \cdots$, and $\lambda(C)$ actually are does not seem to be helpful for finding simpler solutions, so we will leave them unspecified. Then by our own definition at the end of Section 2.1.4, this is an ME/M/C//N-type loop, but we will not emphasize that here. However, in Chapter 6, when we examine the M/ME/C//N loop, the generalization *will* be significant, and it will be emphasized.

Given our assumption, we see that (5.4.3) becomes

$$\mathbf{A}(k) = \frac{\lambda(C)}{\lambda(C)}\left[\mathbf{I} + \frac{1}{\lambda(C)}\mathbf{B} - \mathbf{Q}\right] = \mathbf{A}(C) \quad \text{for all } k \geq C. \tag{5.4.9a}$$

Then, of course,

$$\mathbf{U}(k) = \mathbf{U}(C) \quad \text{for all } k \geq C. \tag{5.4.9b}$$

Since this matrix will play a dominant role in this section, we call it by the more concise symbol

$$\mathbf{U_c} := \mathbf{U}(C) \quad \text{and} \quad \mathbf{A_c} := \mathbf{U_c}^{-1}. \tag{5.4.9c}$$

Every formula we derived in Section 5.4.1 is still valid here, but now we can say something more about the various matrices. From assumption (5.4.8), Equations (5.4.7) become

$$\mathbf{X}(k, N) = [\mathbf{U_c}]^{N-k} \quad \text{for } k \geq C \tag{5.4.10a}$$

and for $k \leq C$,

$$\mathbf{X}(k, N) = [\mathbf{U_c}]^{N-C}\, \mathbf{U}(C-1) \cdots \mathbf{U}(k) = [\mathbf{U_c}]^{N-C}\, \mathbf{X}(k, C). \tag{5.4.10b}$$

Also, from (5.4.7c) (remember, $N > C$),

$$\mathbf{K}(N) = \sum_{k=0}^{C} \mathbf{X}(k, N) + \sum_{k=C+1}^{N} \mathbf{X}(k, N) \tag{5.4.11a}$$

$$= \mathbf{U_c}^{N-C}\, \mathbf{K}(C) + \sum_{k=C+1}^{N} \mathbf{U_c}^{N-k},$$

where $\mathbf{K}(C)$ is the same as it was before, namely

$$\mathbf{K}(C) = \mathbf{I} + \mathbf{U}(C-1) + \cdots + \mathbf{U}(C-1) \cdots \mathbf{U}(0). \tag{5.4.11b}$$

From our knowledge of the partial geometric series, we can rewrite this equation as

$$\mathbf{K}(N) = \left(\mathbf{I} - \mathbf{U_c}^{N-C}\right)[\mathbf{I} - \mathbf{U_c}]^{-1} + \mathbf{U_c}^{N-C}\, \mathbf{K}(C). \tag{5.4.11c}$$

These simplifications may be of some help in solving for systems with finite populations, but otherwise they are not particularly enlightening. Their real use comes in solving for the open system, to which we devote the rest of this section. Let $\lambda_c := \lambda(C)$ and $\rho_c := \lambda_c \bar{x}_1$. Then if $\rho_c < 1$, the limit of $\mathbf{K}(N)$, as N goes to infinity exists, and

$$\lim_{N\to\infty} \mathbf{U_c}^N = \mathbf{O} \quad (\text{for } \rho_c < 1).$$

Then from either (5.4.11b) or (5.4.6b), we have

$$\lim_{N\to\infty} \mathbf{K}(N) = [\mathbf{I} - \mathbf{U_c}]^{-1} \quad (\text{for } \rho_c < 1),$$

which is identical to the results in Section 4.2 for the M/ME/1 queue, with λ_c, ρ_c, and $\mathbf{U_c}$ replacing λ, ρ, and $\mathbf{U}$, respectively.

The interesting case occurs when the limit does not exist, presumably when $\rho_C > 1$, so we will assume that for the rest of this section. Proceeding in a manner similar to Section 5.1, for which we will get similar but not identical results, define s_C, $\mathbf{u_c}$, and $\mathbf{v'_c}$ to be the smallest eigenvalue in magnitude of $\mathbf{A_c}$, with corresponding eigenvectors, satisfying $\mathbf{u_c v'_c} = 1$. That is,

$$\mathbf{u_c A_c} = s_C \mathbf{u_c} \quad \text{and} \quad \mathbf{A_c v'_c} = s_C \mathbf{v'_c}, \tag{5.4.12}$$

and $|1/s_C|$ is the largest among all eigenvalues of $\mathbf{U_c}$. Using (5.1.2c), with $\mathbf{U_c}$ replacing $\mathbf{U}$, Equations (5.4.10) become for very large N and $\rho_C > 1$,

$$\lim_{N\to\infty} s_C{}^N \mathbf{X}(k,N) = s_C{}^k \mathbf{v'_c u_c} \quad \text{for} \quad k \ge C \tag{5.4.13a}$$

and

$$\lim_{N\to\infty} s_C{}^N \mathbf{X}(k,N) = s_C{}^C \mathbf{v'_c u_c X}(k,C) \quad \text{for} \quad k < C. \tag{5.4.13b}$$

Also, from (5.4.11b) (using $s_C{}^N \mathbf{I} \to \mathbf{O}$), in a manner very similar to that used in deriving (5.1.3a),

$$\begin{aligned}\lim_{N\to\infty} s_C{}^N \mathbf{K}(N) &= [\mathbf{I} - \mathbf{U_c}]^{-1}[\mathbf{O} - s_C{}^C \mathbf{v'_c u_c}] + s_C{}^C \mathbf{v'_c u_c K}(C) \\ &= \mathbf{v'_c u_c}\, s_C{}^C \left[\mathbf{K}(C) + \frac{s_C}{1-s_C}\right].\end{aligned} \tag{5.4.13c}$$

Next, take Ψ of the above to get, in analogy with (5.1.3b),

$$\lim_{N\to\infty} \Psi\left[s_C{}^N \mathbf{K}(N)\right] = s_C{}^C (\mathbf{pv'_c})(\mathbf{u_c \varepsilon'})\left[\frac{s_C}{1-s_C} + \hat{\mathbf{u}}_\mathbf{c}\mathbf{K}(C)\boldsymbol{\varepsilon}'\right], \tag{5.4.13d}$$

where we have made the definition analogous to (5.1.4b):

$$\hat{\mathbf{u}}_\mathbf{c} := \frac{\mathbf{u_c}}{\mathbf{u_c \varepsilon'}}. \tag{5.4.14}$$

Then we have $\hat{\mathbf{u}}_\mathbf{c}\boldsymbol{\varepsilon}' = 1$. Also, note that $\hat{\mathbf{u}}_\mathbf{c}\mathbf{K}(C)\boldsymbol{\varepsilon}'$ is a scalar.

Before going on, we must make a slight addition to our notation, so that the symbols we use will explicitly reflect their dependence on C. Remember that in Section 5.4.1 there was no C, so we used the same notation as we did in the preceding sections. But now, if one wishes to examine systems that are identical except for differing values of C, the symbols must show it. So, for the rest of this chapter, we use the following.

Definition 5.4.1

$\pi_2(k; N \mid C)$ = *steady-state vector probability of finding k customers at* S_2, *in an* ME/M/C//N*-type queue. The associated scalar probability is denoted by* $r_2(k; N \mid C) = \pi_2(k; N \mid C)\boldsymbol{\varepsilon}'$. This change of notation carries over to the open ME/M/C queue as follows:

$$\mathbf{d}_2(k;N) = C(N)G(N)\lambda(k+1)\boldsymbol{\pi}_2(k+1;N) \qquad \text{for } 0 \le k \le N-1.$$

The sum, $\sum_{k=0}^{N-1}\mathbf{d}_2(k;N)\boldsymbol{\varepsilon}' = 1$, yields the same result as we got for the arrival; thus we have the following theorem.

Theorem 5.4.5: The steady-state arrival and departure vector probabilities for the ME/M/X//N loop are given by the following equations. For $0 \le k \le N-1$,

$$\mathbf{a}_2(k;N) = \frac{\lambda(k+1)}{\Lambda(N)} r_2(k+1;N)\mathbf{p}, \tag{5.4.22a}$$

$$\mathbf{d}_2(k;N) = \frac{\lambda(k+1)}{\Lambda(N)} \boldsymbol{\pi}_2(k+1;N). \tag{5.4.22b}$$

The corresponding scalar probabilities are

$$a_2(k;N) = d_2(k;N) = \frac{\lambda(k+1)}{\Lambda(N)} r_2(k+1;N). \tag{5.4.22c}$$

The $\boldsymbol{\pi}_2$'s are given by Theorem 5.4.2 or Corollary 5.4.2, and $\Lambda(N)$ is given by normalization through (5.4.21). Clearly, $\mathbf{a}_2$ and $\mathbf{d}_2$ are related by the formula $\mathbf{a}_2(k;N) = \mathbf{d}_2(k;N)\mathbf{Q}$. ■

These equations are as close to the M/ME/1//N results of Theorem 4.1.4, as one could hope. All the arrival vectors (departure vectors in Chapter 4) are proportional to $\mathbf{p}$, while the scalar arrival and departure probabilities are equal to each other. Even the vector departure probabilities are the same as the load-independent arrival probabilities, but remember that $\mathbf{d}_2(k;N) = \mathbf{a}_1(n-1;N)$ in general. Why, you may ask, did we bother studying the simpler case in the first place? The author would ask in return if you would have had much greater difficulty understanding this section had you not gone through it first in Chapter 4.

Once again we cannot say much more about these equations without providing more information about the λ's. We can say this much though: If

$$\lim_{N\to\infty} \lambda(N) > \frac{1}{\bar{x}_1},$$

then S_1 is surely the bottleneck, and the system throughput approaches

$$\lim_{N\to\infty} \Lambda(N) = \frac{1}{\bar{x}_1}. \tag{5.4.23}$$

We turn our attention once again to the ME/M/C queue, and assume that (5.4.8) holds.

Definition 5.4.3

$\mathbf{a}_2(k;N \mid C)$ and $\mathbf{d}_2(k;N \mid C)$ *are the steady-state vector arrival and departure probabilities for the* ME/M/C//N*-type queue. The associated scalar probabilities are denoted by* $a_2(k;N \mid C)$ and $d_2(k;N \mid C)$. This change of notation carries over to the open ME/M/C queue. For instance,

$$\mathbf{a_2}(k \mid C) := \lim_{N\to\infty} \mathbf{a_2}(k; N \mid C),$$

with similar expressions for $\mathbf{d_2}(k \mid C)$, $a_2(k \mid C)$, and $d_2(k \mid C)$. ◊

The limiting expressions for these entities are just a little tricky. Let us define

$$\zeta_c := \frac{1}{\rho_c},$$

and assume that $\zeta_c < 1$. First look at departures. Using (5.4.22a), (5.4.23), and (5.4.16a), we have

$$\mathbf{d_2}(k \mid C) = \lim_{N\to\infty} \frac{\lambda_c}{\Lambda(N)} \pi_2(k+1; N \mid C) = \lambda_c \bar{x}_1\, \pi_2(k+1 \mid C)$$

$$= \frac{1}{\zeta_c} g(C)\, s_c{}^{k+1-C}\, \hat{\mathbf{u}}_\mathbf{c} \quad \text{for } k \geq C-1,$$

where $g(C)$ is defined by (5.4.15). But for $k < C-1$, from (5.4.16b),

$$\mathbf{d_2}(k \mid C) = \frac{1}{\zeta_c} g(C)\, \hat{\mathbf{u}}_\mathbf{c}\mathbf{X}(k+1, C) \quad \text{for } k < C-1.$$

Theorem 5.4.2 tells us that $\mathbf{a_2}(k; N) = \mathbf{d_2}(k; N)\mathbf{Q}$, so the arrival vectors require no further effort. We thus summarize.

Theorem 5.4.6: The steady-state departure and arrival vector probabilities for the ME/M/C queue are given by the following equations. For departures,

$$\mathbf{d_2}(k \mid C) = \frac{g(C)}{\zeta_c} s_c{}^{k+1-C}\, \hat{\mathbf{u}}_\mathbf{c} \quad \text{for } k \geq C-1 \tag{5.4.24a}$$

and

$$\mathbf{d_2}(k \mid C) = \frac{g(C)}{\zeta_c} \hat{\mathbf{u}}_\mathbf{c}\mathbf{X}(k+1, C) \quad \text{for } k < C-1. \tag{5.4.24b}$$

For arrivals,

$$\mathbf{a_2}(k \mid C) = \frac{g(C)}{\zeta_c} s_c{}^{k+1-C}\, \mathbf{p} \quad \text{for } k \geq C-1 \tag{5.4.24c}$$

and

$$\mathbf{a_2}(k \mid C) = \frac{g(C)}{\zeta_c} [\hat{\mathbf{u}}_\mathbf{c}\mathbf{X}(k+1, C)\boldsymbol{\varepsilon}']\,\mathbf{p} \quad \text{for } k < C-1. \tag{5.4.24d}$$

The associated scalar probabilities are equal to each other, and satisfy

$$a_2(k \mid C) = d_2(k \mid C) = \frac{1}{\zeta_c} r_2(k+1 \mid C) \quad \text{for all } k \geq 0, \tag{5.4.24e}$$

where $r_2(k \mid C)$ is given by either (5.4.17a) or (5.4.17b), depending on the value of k. ■

Once again we have a family of geometrically distributed distributions. Comparing with Equations (5.4.17), we see that the $\mathbf{a}_2$'s and the $\mathbf{d}_2$'s for a given k are related to the π_2's (and r_2's) for $k+1$. Therefore, the geometric form starts at $k = C$ for the steady-state probabilities, but starts at $k = C - 1$ for the arrivals and departures. This compares with the ME/M/1 queue, where $\mathbf{a}_2(0)$ and $\mathbf{d}_2(0)$ are part of a general geometric sequence, but $\pi_2(0)$ is not. Remember that by the notation of this section, $\pi_2(k \mid 1)$ is the same as $\pi_2(k)$ of Section 5.1.

It is only fair to ask why one would want to know the vector details of these systems. A perusal of Section 4.5 will give some idea of how they could be useful. For instance, if we wanted to know the mean time for the arrival of the next customer, given that one has just departed from S_2, leaving behind $C - 1$ or more customers, one would use $\hat{\mathbf{u}}_c\mathbf{V}\boldsymbol{\varepsilon}'$. If, however, there were fewer than $C - 1$ customers left behind, the mean time until the next arrival would be

$$\frac{\hat{\mathbf{u}}_c\mathbf{K}(k+1,\, C)\mathbf{V}\boldsymbol{\varepsilon}'}{\hat{\mathbf{u}}_c\mathbf{K}(k+1,\, C)\boldsymbol{\varepsilon}'}.$$

We could do the same for random observers, or just arrived customers, or more extended combinations. We will discuss a little more transient behavior for ME/M/1 queues in the next and final section of this chapter.

5.5. TRANSIENT PROPERTIES OF ME/M/1 QUEUES

Much of what we do in this section will be a copy of what was done in Section 4.5, but from an upside-down point of view. What was there d is $2u$ here, and so on. We should be able to move much more quickly, but there will be some new difficulties as well as some new insights. First we examine how the ME/M/1 queue grows with time. Afterward we will study how long it takes for a queue to drain, or equivalently, the k-busy period. We need not recommend that you reread Section 4.5, since you will have to do it to retrieve formulas we need here. We note, finally, that all procedures are directly generalizable to the ME/M/X queue.

5.5.1. First-Passage Times for Queue Growth

We have already seen several times that before we can discuss how a queue grows by k we must examine how it grows by 1. And we must see how the state of the system has changed after a unit growth. So we define the equivalent of Definition 4.5.1, which is really the inverse of Definition 4.5.7.

Definition 5.5.1

$\mathbf{H}_{2\mathbf{u}}(k)$:= *probability matrix of first passage from k to $k+1$.* $[\mathbf{H}_{2\mathbf{u}}(k)]_{ij}$ is the probability that S_1 will be in state j when the queue at S_2 goes from k *to* $k+1$ for the first time, given that the system started in state $\{i;\, k\}$. The subscript, **2u** stands for "S_2 goes up." ◊

Note that this matrix only has one argument (k), since going up always implies that there is a *bottom*, namely an empty queue at S_2. We have already derived these in finding $\mathbf{H_d}(n, N)$ in Equations (4.5.10). It was clear, then (in the notation of this chapter), that

$$\mathbf{H_{2u}}(k) = \mathbf{Q} \quad \text{for all } k \geq 0. \tag{5.5.1}$$

Why do we bother defining this matrix yet again, when it is of such simple form? We answer that in the next chapter it will not be so simple, even though the concept will be the same.

First we must define the upside-down version of Definition 4.5.8.

Definition 5.5.2

$\boldsymbol{\tau}'_{\mathbf{2u}}(k)$:= *mean first-passage time vector from* k *to* $k+1$. $[\boldsymbol{\tau}'_{\mathbf{2u}}(k)]_i$ is the mean time for the queue at S_2 to reach $k+1$ for the first time, given that the system started in state $\{i; k\}$. ◊

We can write down the equations for the $\boldsymbol{\tau}'_{\mathbf{2u}}$'s directly from their equivalents in Chapter 4. Thus (4.5.11a) and (4.5.11b) convert to

$$\boldsymbol{\tau}'_{\mathbf{2u}}(0) = \mathbf{V}\boldsymbol{\varepsilon}' \tag{5.5.2a}$$

and

$$\boldsymbol{\tau}'_{\mathbf{2u}}(k) = \frac{1}{\lambda}\mathbf{U}\boldsymbol{\varepsilon}' + \mathbf{U}\boldsymbol{\tau}'_{\mathbf{2u}}(k-1), \tag{5.5.2b}$$

where $\mathbf{U}$ is defined by (4.1.4). The solution of these equations comes directly from (4.5.12):

$$\boldsymbol{\tau}'_{\mathbf{2u}}(k) = \frac{1}{\lambda}\mathbf{U}\mathbf{K}(k)\boldsymbol{\varepsilon}' = \frac{1}{\lambda}[\mathbf{K}(k+1) - \mathbf{I}]\boldsymbol{\varepsilon}', \tag{5.5.2c}$$

where $\mathbf{K}(k)$ is the normalization matrix for the M/ME/1//k queue, defined in (4.1.6).

Now we are ready to set up the formulas for time of queue growth, or are we? To do this, we must know the state the system is in originally. That is, we must know the initial vector $\mathbf{p_i}$. If the queue (at S_2) is empty, what does that tell us about the system (i.e., what state is S_1 in)? The initial vector certainly *cannot* be $\mathbf{p}$, since that would imply that a customer just left S_1, which in turn means that the same customer has just arrived at S_2, contradicting our assumption that S_2 is empty. If ζ (which you recall is $1/\lambda\bar{x}_1$) is less than 1, there are two possibilities of immediate interest. We could assume that a customer has just departed S_2 in a system that has been running for a long time. This corresponds to

$$\mathbf{p_i} = \frac{\mathbf{d_2}(0)}{d_2(0)} = \hat{\mathbf{u}}.$$

Then the mean time until the first arrival to the empty queue conditioned by a departure, is

$$\mathbf{p_i}\tau'_{2\mathbf{u}}(0) = \hat{\mathbf{u}}\mathbf{V}\boldsymbol{\varepsilon}' = \frac{1-\zeta}{1-s}\bar{x}_1 \qquad (\zeta, s < 1),$$

where we got the last part from (5.1.5e) and remembered that $\rho = 1/\zeta$. As an aside, this formula gives us some idea of what the difference between s and ζ means. If we are looking at an M/M/1 queue, then $\zeta = s$, and the mean time until the next arrival is $\bar{x}_1$, as expected. For other systems, we might look at Figure 5.1.1 for insight. If s is greater than (less than) ζ, we would expect to wait longer than (less than) $\bar{x}_1$ for the first customer to arrive.

The other possibility is to ask what a random observer would see. This corresponds to

$$\mathbf{p_i} = \frac{\pi_2(0)}{r_2(0)} = \frac{\hat{\mathbf{u}}\mathbf{V}}{\hat{\mathbf{u}}\mathbf{V}\boldsymbol{\varepsilon}'}.$$

Thus the mean time for the first customer to arrive, as seen by a random observer, is [call it $t_r(0)$]

$$t_r(0) := \mathbf{p_i}\tau'_{2\mathbf{u}}(\mathbf{0}) = \frac{\hat{\mathbf{u}}\mathbf{V}^2\boldsymbol{\varepsilon}'}{\hat{\mathbf{u}}\mathbf{V}\boldsymbol{\varepsilon}'}.$$

We can actually get an interesting expression for this by noting that on the one hand, $\lambda\hat{\mathbf{u}}\mathbf{A}\mathbf{V} = \lambda s\,\hat{\mathbf{u}}\mathbf{V}$, and on the other hand,

$$\lambda\hat{\mathbf{u}}\mathbf{A}\mathbf{V} = \hat{\mathbf{u}}[\lambda\mathbf{V} + \mathbf{I} - \lambda\mathbf{Q}\mathbf{V}].$$

By equating both expressions, noting that $\hat{\mathbf{u}}\mathbf{Q} = \mathbf{p}$, and solving for $\hat{\mathbf{u}}\mathbf{V}$, we get

$$\hat{\mathbf{u}}\mathbf{V} = \frac{1}{\lambda(1-s)}[\lambda\mathbf{p}\mathbf{V} - \hat{\mathbf{u}}] \qquad (\zeta < 1), \tag{5.5.3a}$$

and

$$\frac{\hat{\mathbf{u}}\mathbf{V}}{\hat{\mathbf{u}}\mathbf{V}\boldsymbol{\varepsilon}'} = \frac{\zeta}{1-\zeta}[\lambda\mathbf{p}\mathbf{V} - \hat{\mathbf{u}}]. \tag{5.5.3b}$$

We put this all together to get

$$t_r(0) = \frac{\Psi[\mathbf{V}^2]}{(1-\zeta)\bar{x}_1} - \frac{1}{(1-s)\lambda}.$$

It is difficult to see what is going on here, since near $\zeta = 1$ both terms are unboundedly large, so their difference can be anything. Actually, their difference is finite. We leave it as an exercise to show [using (5.1.22b), (5.1.22c) and (5.1.23b)] that (we are 99.44% sure)

$$\lim_{\zeta\to 1_-} t_r(0) = \bar{x}_1\frac{1+C^2}{2}\left(1 - \frac{s''(1)}{2}\frac{1+C^2}{2}\right).$$

Even at the distinct value of $\zeta = 1_-$, this expression is not simple, particularly when one notes, from (5.1.24a), that $s''(1)$ is quite complicated. Only when we consider the M/M/1 queue does this simplify, for then, $s''(1) = 0$ and $C^2 = 1$, so $t_r(0) = \bar{x}_1$. One should compare these results with the *residual time* as given in (3.3.12d).

There are, of course, any number of other possibilities that one could consider in setting up $\mathbf{p_i}$, all of which would require more information about the history of the system. Now, if $\zeta > 1$, even though (5.5.2a) is still valid, we have nothing to go on for *preparing* the initial vector. In this case there is no such thing as the *steady state*, and after a *long period of time*, the probability that no one will be at S_2 is 0. Therefore, any initial condition must be based on some transient events. We must be given some special information ("S_1 just woke up" – or something).

Once there is someone at S_2 ($k > 0$), we can say something even if $\zeta > 1$. We can talk about the state of the system immediately after a customer arrives, and in fact this is the most important situation. After all, every increase in queue length is the result of an arrival (at S_2), so after the first increase, all subsequent increases begin their epics with the initial vector, $\mathbf{p}$ (at S_1). The other two cases we described for $k = 0$ are still applicable. The initial vector for the time to rise from k to $k+1$, conditioned on a departure, is

$$\mathbf{p_i} = \frac{\mathbf{d_2}(k)}{d_2(k)} = \hat{\mathbf{u}},$$

the same as for the empty queue. Similarly, the random observer, from (5.1.5b) and (5.1.5d), will see the *same* initial vector as a departing customer! Only when the queue is empty will she see something different. An arriving customer, however, will always see something different. (Speaking from a purely physical point of view, the arriving customer will see nothing special, since the initial vector refers to S_1, the subsystem he left behind.) Thus we see that there are cases when departing customers, arriving customers, and random observers will all see different behavior.

We could continue piling variations upon variations, but let us merely consider the mean time for the queue at S_2 to grow to k, given that the first customer has just arrived. Let us call this $t_2(1 \to k)$. Then [using the obvious convention that $t_2(1 \to 1) = 0$]

$$t_2(1 \to k) = \sum_{l=1}^{k-1} \mathbf{p}\boldsymbol{\tau}'_{\mathbf{2u}}(l) = \frac{1}{\lambda} \sum_{l=1}^{k-1} \Psi[\mathbf{UK}(l)].$$

This formula is easy enough to compute, based on what we know from Chapter 4; however, we will indicate how the queue grows for large k. We know that for $\zeta > 1$, $\mathbf{K}(l)$ approaches a limit for large l. Specifically [from (4.2.3)]

$$T := \lim_{l \to \infty} \mathbf{p}\boldsymbol{\tau}'_{\mathbf{2u}}(l) = \frac{1}{\lambda}\Psi[\mathbf{UK}] = \frac{1}{\lambda}\Psi[(\mathbf{I} - \mathbf{U})^{-1} - \mathbf{I}]$$

$$= \frac{1}{\lambda}\left[\frac{1}{1 - 1/\zeta} - 1\right] = \frac{1/\lambda}{\zeta - 1} \quad \text{for } \zeta > 1.$$

The expression is independent of everything except ζ and λ, and tells us that once the queue grows large enough, it will continue to grow linearly with time. Each incremental increase will take the same amount of time (on average, of course). Since T is the mean time for the queue to grow by 1, its reciprocal can be considered to be the *rate* at which the queue grows. This leads to

$$\frac{1}{T} = \lambda\zeta - \lambda = \frac{1}{\bar{x}_1} - \lambda \qquad \text{for } \zeta > 1.$$

So we have the perfectly reasonable result that the rate of queue growth is equal to the rate of arrivals minus the rate of departures from S_2. In other words, the queue-length growth curves for all ME/M/1 queues approach straight lines, and have the same slope. You should compare this result with that for the M/M/1 queue [Equation (2.3.2a)] to see that, in fact, the two are asymptotically equal. Keep in mind, however, that this is true only for $\zeta > 1$! When $\zeta < 1$, asymptotic behavior is completely different. We must be very careful when conceptually replacing *probability flow rates* with *physical flow rates*. This is meaningful only in very heavy traffic.

For $\zeta < 1$, the normalization matrix, $\mathbf{K}(l)$, does not approach a limit, but rather, grows geometrically as $(1/s)^l$. We saw this in deriving the steady-state solution for the ME/M/1 queue. For large l, and from (5.1.3),

$$\lambda \mathbf{p}\boldsymbol{\tau}'_{2\mathbf{u}}(l) \approx \Psi[\mathbf{UF}]\left(\frac{1}{s}\right)^l = \frac{(\mathbf{u}\boldsymbol{\varepsilon}')(\mathbf{p}\mathbf{v}')}{\zeta(1-s)}\left(\frac{1}{s}\right)^l \qquad (\text{for } \zeta < 1).$$

Now, let us define

$$\mathbf{D}(\sigma) := [\mathbf{I} + \sigma\mathbf{V}]^{-1}.$$

[We have actually used this useful matrix before, in (4.2.6b). It is the generator of the Laplace transform of $b(x)$.] Then it can be shown, when $\sigma = \lambda(1-s)$, that

$$\hat{\mathbf{u}} = \lambda \mathbf{p}\mathbf{V}\mathbf{D}(\sigma), \tag{5.5.4a}$$

$$\mathbf{v}' = c\lambda\mathbf{V}\mathbf{D}\boldsymbol{\varepsilon}', \tag{5.5.4b}$$

and

$$c = (\mathbf{u}\boldsymbol{\varepsilon}')(\mathbf{p}\mathbf{v}') = \frac{1}{\Psi[(\lambda\mathbf{V}\mathbf{D})^2]} = \frac{(1-s)^2}{1-2s+\Psi[\mathbf{D}^2]}. \tag{5.5.4c}$$

From Equations (4.4.1), c can also be written as

$$c = \frac{1-s}{1-\lambda\int_0^\infty x\, e^{-\sigma x}\, b(x)\, dx}. \tag{5.5.4d}$$

We put this all together, coming up with a form that is valid for G/M/1 queues:

$$\mathbf{p}\boldsymbol{\tau}'_{2\mathbf{u}}(l) \approx \frac{\bar{x}_1}{1-\lambda\int_0^\infty x\, e^{-\lambda(1-s)x}\, b(x)\, dx}\left(\frac{1}{s}\right)^l, \qquad \zeta < 1.$$

Compare this equation with (2.3.2a) for the M/M/1 queue. Since both formulas are of geometric form, we know from the arguments given in deriving (2.3.4a) that the queue length grows logarithmically with time. However, the *rates* of growth vary enormously, depending on what distribution function is generating the arrival process.

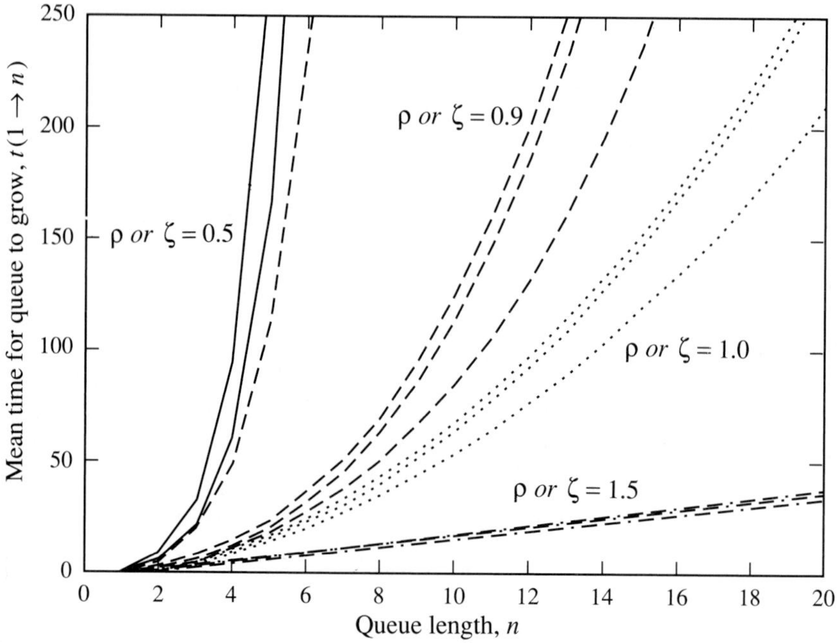

Figure 5.5.1: Mean time for queue growth, $t_2(1 \to n)$ for the E_2/M/1 queue, and the equivalent function, $[t(1 \to n)]$, for the M/M/1 and M/E_2/1 queues, as a function of queue length, n. Four values of ρ were used, (0.5, 0.9, 1.0, and 1.5) in calculating the M/ME/1 queues, and the same values for ζ were used for the ME/M/1 queue. In all cases E_2/M/1 lies above M/E_2/1, which lies above M/M/1.

Example 5.5.1: We have combined the numbers required to generate Figure 4.5.4, for the M/M/1 and M/E_2/1 queues with $t_2(1 \to n)$ for the E_2/M/1 queue, and plotted them in Figure 5.5.1. Even without the extra curves, the two figures are different. First of all, the x- and y-axes are interchanged. Second, the Chapter 4 curves give the mean time to grow from 0 to n, while here all curves give the time to go from 1 to n. Looking at this figure, when ρ or ζ is much greater than 1, all three queues give virtually the same growth curve. This agrees with our previous argument that $1/T$, the asymptotic growth rate, depends only on the difference between the arrival and the service rates. However, when ρ or ζ is less than 1, the time for growth is exponential (or the growth rate is logarithmic), and all three queues differ from each other. It is somewhat surprising that near saturation, the M/E_2/1 and E_2/M/1 queues are much closer to each other than they are to the M/M/1 queue. ●

It would be interesting to see how queues with other distributions behave.

5.5.2. The *k*-Busy Period

We are now ready to study the draining of an ME/M/1 queue, the final topic of this chapter. As in the preceding section, we can take definitions and formulas directly from their equivalents in Section 4.5 and redress them in the notation of this chapter.

Definition 5.5.3

$\mathbf{H_{2d}}(k;N)$:= *probability matrix of first passage from k to k* -1 *in an* ME/M/1//N *loop.* $[\mathbf{H_{2d}}(k;N)]_{ij}$ is the probability that S_1 will be in state j when the queue at S_2 goes from k *to* $k-1$ for the first time, given that the system started in state $\{i;k;N\}$. The subscript, **2d** stands for "S_2 goes down." ◊

This is almost identical to Definition 4.5.7, but is equal to (4.5.2) in value. We can say directly that

$$\mathbf{H_{2d}}(N;N) = \mathbf{Q} \tag{5.5.5a}$$

and

$$\mathbf{H_{2d}}(k;N) = \lambda[\lambda\mathbf{I} + \mathbf{B} - \mathbf{BQH_{2d}}(k+1;N)]^{-1} \quad \text{for } k < N. \tag{5.5.5b}$$

Clearly, $\mathbf{H_{2d}}(k;N)$ is isometric and invertible. For the open system we need,

$$\mathbf{H_{2d}} := \lim_{N\to\infty} \mathbf{H_{2d}}(k;N) = \lambda[\lambda\mathbf{I} + \mathbf{B} - \mathbf{BQH_{2d}}]^{-1}. \tag{5.5.6a}$$

Pre-multiply both sides with the matrix in square brackets, and get

$$[\lambda\mathbf{I} + \mathbf{B} - \mathbf{BQH_{2d}}]\,\mathbf{H_{2d}} = \lambda\mathbf{I}$$

which can also be expressed as

$$\mathbf{BQ}[\mathbf{H_{2d}}]^2 - (\lambda\mathbf{I} + \mathbf{B})\,\mathbf{H_{2d}} + \lambda\mathbf{I} = \mathbf{O}. \tag{5.5.6b}$$

Thus $\mathbf{H_{2d}}$ is independent of k and satisfies a matrix quadratic equation. It would be nice if we had an explicit form for this, but in general, only iterative methods are available to find $\mathbf{H_{2d}}$ or its inverse. For instance, one can start with $\mathbf{H_{2d}} = \mathbf{Q}$, and by brute force, substitute iteratively into (5.5.6a) until convergence is reached. There are other methods available that can find the solution much more efficiently. Interestingly enough, for all ζ, greater than, as well as less than 1, a physical solution to this equation always exists, since the arguments that created it are physical, and thus each iterate must exist. (Remember, each iterate was really $\mathbf{H_{2d}}(k-1;\infty)$, coming down the recursive ladder, and each one must be isometric.)

Definition 5.5.4

$\boldsymbol{\tau}'_{2\mathbf{d}}(k;N) :=$ *mean first-passage time vector from k to k − 1.* $[\boldsymbol{\tau}'_{2\mathbf{d}}(k;N)]_i$ is the mean time for the queue at S_2 to reach $k-1$ for the first time, given that the system started in state $\{i;\, k;\, N\}$. ◊

Using (4.5.5) directly, we can write for $1 \le k < N$

$$\boldsymbol{\tau}'_{2\mathbf{d}}(k;N) = \frac{1}{\lambda}\boldsymbol{\varepsilon}' + \frac{1}{\lambda}\mathbf{H}_{2\mathbf{d}}(k;N)\mathbf{B}\mathbf{Q}\boldsymbol{\tau}'_{2\mathbf{d}}(k{+}1;N) \tag{5.5.7a}$$

and for $k = N$

$$\boldsymbol{\tau}'_{2\mathbf{d}}(N;N) := \frac{1}{\lambda}\boldsymbol{\varepsilon}'. \tag{5.5.7b}$$

We find the mean first-passage vector for the open queue by letting N go to infinity, so

$$\boldsymbol{\tau}'_{2\mathbf{d}} := \lim_{N\to\infty} \boldsymbol{\tau}'_{2\mathbf{d}}(k;N) = \frac{1}{\lambda}\boldsymbol{\varepsilon}' + \frac{1}{\lambda}\mathbf{H}_{2\mathbf{d}}\mathbf{B}\mathbf{Q}\boldsymbol{\tau}'_{2\mathbf{d}}.$$

This vector is also independent of k, but unlike its first-passage matrix, we can solve for it directly, to get,

$$\boldsymbol{\tau}'_{2\mathbf{d}} := [\lambda\mathbf{I} - \mathbf{H}_{2\mathbf{d}}\mathbf{B}\mathbf{Q}]^{-1}\boldsymbol{\varepsilon}'. \tag{5.5.7c}$$

This can be rewritten in another form with the use of Lemma 4.2.1:

$$\boldsymbol{\tau}'_{2\mathbf{d}} := \frac{1}{\lambda}\left[\mathbf{I} + \frac{1}{\lambda - \Psi[\mathbf{H}_{2\mathbf{d}}\mathbf{B}]}\mathbf{H}_{2\mathbf{d}}\mathbf{B}\mathbf{Q}\right]\boldsymbol{\varepsilon}', \tag{5.5.7d}$$

so if we can find $\mathbf{H}_{2\mathbf{d}}$, then we get $\boldsymbol{\tau}'_{2\mathbf{d}}$. The mean time for the queue to drop by one, given that a customer has just arrived is, using (5.5.7d),

$$\mathbf{p}\boldsymbol{\tau}'_{2\mathbf{d}} = \frac{1}{\lambda}\left[1 + \frac{\Psi[\mathbf{H}_{2\mathbf{d}}\mathbf{B}]}{\lambda - \Psi[\mathbf{H}_{2\mathbf{d}}\mathbf{B}]}\right] = \frac{1}{\lambda - \Psi[\mathbf{H}_{2\mathbf{d}}\mathbf{B}]}. \tag{5.5.7e}$$

This is also the mean time for the busy period. But this is valid only when $\zeta < 1$, for otherwise, the busy period may never end. Even though $\mathbf{H}_{2\mathbf{d}}$ is meaningful for all ζ, when $\zeta = 1$, the term $\Psi[\mathbf{H}_{2\mathbf{d}}\mathbf{B}] = \lambda$ also, so (5.5.7e) is infinite. When $\zeta > 1$ the term, $\Psi[\mathbf{H}_{2\mathbf{d}}\mathbf{B}] > \lambda$, and (5.5.7e) yields a negative, or nonphysical, value, as it should. In any case, observe that this formula is very different from the busy period time of an M/G/1 queue as given in (4.5.16b).

Example 5.5.2: To take a closer look at the differences between M/ME/1 and ME/M/1 queues, we have calculated the mean time for the busy periods for several different distributions and plotted them in Figure 5.5.2. Since all M/ME/1 queues have the same busy period time, they are all represented by the same curve, labeled M/M/1. Yet the different ME/M/1 queues vary all over the place. There is some order here, since for a given ζ the bigger the coefficient of variation, the larger the mean time. The functions we chose are not necessarily truly representative of what can be expected in real-world applications, so we cannot be sure that this observation is universally valid, but the figure certainly shows that the arrival pattern is critically important, at least in studying busy periods. ●

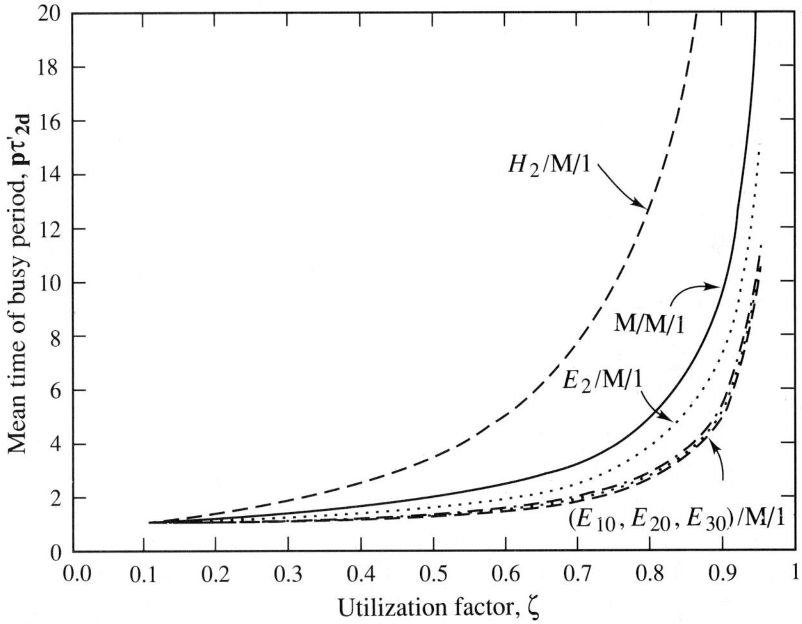

Figure 5.5.2: Mean time of a busy period for various ME/M/1 queues, as a function of ζ. All M/G/1 queues have the same busy period time, so all are equal to the curve labeled M/M/1. For a given ζ, the busy periods for the functions chosen seem to rank themselves according to C^2. The hyperexponential function has $C^2 = 5.0388$, while the Erlangians have $C^2 = 1/n$, for $n = 2$, 10, 20, and 30.

As our last subject in this chapter, we discuss the k-busy period. If we knew the state the system was in at the beginning, the first-passage time for the queue at S_2 to drop by 1 is simply $\mathbf{p_i}\tau'_{2d}$. At that moment, the system is in state $\mathbf{p_i H_{2d}}$, so the mean first-passage time to drop by one more is simply $\mathbf{p_i H_{2d}}\tau'_{2d}$, and the time to drop by two is the sum of the two terms. In general, then, the time it takes for an ME/M/1 queue to drop by k is given by the expression

$$\sum_{l=0}^{k-1} \mathbf{p_i}(\mathbf{H_{2d}})^l \tau'_{2d}.$$

As with the M/G/1 queue, if there were k customers in the queue in the first place, this is the time for the k-busy period, conditioned by $\mathbf{p_i}$. If the queue was longer than k at the start, this would still be the time for the queue to decrease by k for the first time.

It is hoped that the reader is sufficiently skilled by now to be able to set up the equations for the probabilities of queue growth, the $\mathbf{W_x}$, where $\mathbf{x}$ is one of **2u**, **2m**, or **2d**. Therefore, we leave those items as exercises.

CHAPTER 6
M / G / C – TYPE SYSTEMS

"Having two bathrooms ruined the capacity to cooperate." – Margaret Mead

In previous chapters when dealing with nonexponential distributions, we always assumed that only one customer was active at a time at S_1. We did look at multiple servers, but only if S_i was exponential, introducing the idea of a load-dependent server (Sections 2.1.4 and 5.4). In doing this, it was not necessary to distinguish between:

1. A subsystem containing a single server that works twice as fast on one customer when a second one is present.
2. A subsystem that has two active servers, one for each customer.

In fact, the only way the two cases can be distinguished is by *marking* the customers so as to tell if they left in the same order in which they arrived. This has become of interest in recent years, and is called the *resequencing problem.* LAQT has been used successfully in analyzing the departure process of an M/M/C queue where customers *must* leave in the same order in which they arrived [DING91]. Since we have made our customers indistinguishable, we have not bothered to consider this at all, nor can we consider it here without expanding our state space.

We cannot get away so easily when dealing with nonexponential servers. Case 1 is not very realistic but can be used in studying queues with *discouraged arrivals*, and is modeled by multiplying the completion rate matrix, **M** by a constant factor when a second customer arrives, leaving the dimension and internal state description of S_1 otherwise unchanged. This turns out to be formally identical to the description of ME/M/C//N loops in Section 4.4, except that the load dependence factor depends on the number of customers at S_1 instead of the number at S_2.

The second case is much more complicated. When a second customer arrives, as always, he begins service by going to phase i with probability p_i, but the first customer is already in service and is at some other phase. Furthermore, when one of the customers leaves, the other customer is still in service, in some phase determined by the system's past. Put differently, a departing customer does *not* leave behind the empty state. We must therefore set up a formalism that keeps track of where *both* customers are.

Since we will have to introduce new symbols and concepts, it is best to start with the simplest extension possible. Therefore, in the following section we will set up the formalism, and find the steady-state solution of a system where S_1 has exactly two identical ME servers (i.e., the M/ME/2//*N* loop). In doing this, we will use a three-phase ME server as an example. In Section 6.2 we will extend this to C servers, for by then it will be easier for the reader to follow the notation.

After that, we will show that the formulas are actually applicable to a more general class of systems, which we call *generalized* M/G/*C*//*N systems.* With little more than a change of notation and a slight generalization of some parameters, we will see that we are suddenly dealing with a network of queues. When $N \le C$, our generalized network turns out to be equivalent to the single-class Jackson network, and we will spend some time discussing the connection. We will then extend the model further to allow S_2 to be a load-dependent server, as we did in Section 5.4. This is potentially an important extension, since it is the correct treatment of *timesharing systems with population-size constraints* (i.e., when $N > C$).

In doing all this, we will find that the equations are still algebraically manipulable but too complex to reduce to simple formulas. Thus we will describe in detail algorithms that allow the user to get computational results for particular systems. Our formalism reduces the dimensions of all relevant matrices to their bare minima. (At present, at least, it is impossible to do it with smaller matrices.) Even so, problems can quickly become intractable if C and/or m become too large, so we will discuss the computational complexity of the algorithms.

In Section 6.3 we will study the open M/G/*C* system in the usual way by letting *N* become unboundedly large. The reverse game, which considers systems where S_1, even at full capacity, is slower than S_2 (i.e., the generalization of $\rho > 1$), leads to a *semi-Markov arrival process* to S_2. This is treated in Section 6.4.2. In the rest of Section 6.4 we will look at some transient phenomena, including those related to the busy periods. Some of these are potentially important in studying the reliability of systems and *rush-hour traffic.*

Because of time and space limitations, we have foregone the pleasure of developing the formulas for departure and arrival times, even though it should prove quite interesting, with several new insights. Its treatment should be a straightforward extension of what we have already done and what we will do in this chapter – maybe in the next edition. In the meantime, the reader is welcome to take this adventurous trip unguided. We mention that much of this chapter is an outgrowth of material in Aby Tehranipour's Ph.D. Thesis [TEHR83], [LIPS85a] and [TEHR90].

6.1. STEADY-STATE M/ME/2//*N* LOOP

Consider the queueing system in Figure 6.1.1. It is identical to Figure 4.1.1 except that now S_1 contains *two* identical ME servers. Previously, each subsystem contained one ME server, so there was no real distinction between S_1 and a server. Thus the statement, "The subsystem is in state i" meant the same as the statement, "The active customer is at phase i." Now that S_1 contains two servers, and *each* server is made up of m phases, we must describe where *both* customers are if there are two or more customers at S_1.

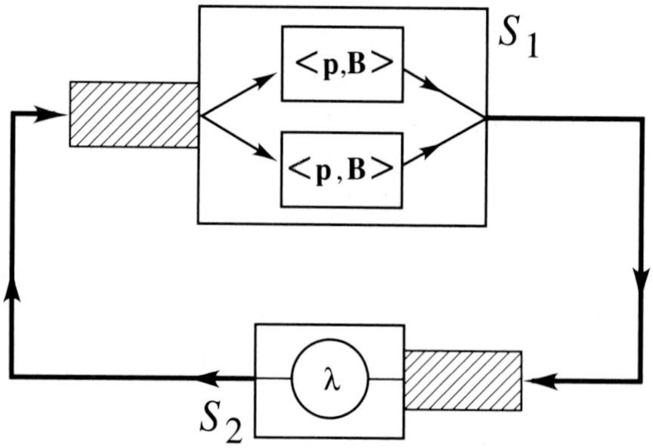

Figure 6.1.1: Two subsystem loop where S_2 is an exponential server, but S_1 is made up of two identical nonexponential ME servers, each made up of m phases, and represented by the vector-matrix pair, $<\mathbf{p}\,,\mathbf{B}>$.

6.1.1. Definitions

The process is as follows. No more than two customers can be active in subsystem S_1 at any one time, one being in each of the servers. When a customer completes service at either server, he leaves S_1 and joins the queue at S_2 (still exponential), while another customer (if one is available) takes its place at the momentarily idle server. Any other customer who was active in S_1 at the time the first one finished continues unperturbed. Each of the servers is described by the same objects introduced in Chapter 3: **p**, **q**', **ε**', **P**, **M**, **B**, and **V**. As before, N is the number of customers in the system, and n is the number of customers at S_1, counting the ones who are being served. Then we have the following definition.

Definition 6.1.1

$\Xi_1 = \{\, i \mid i = 1, 2, \cdots, m \,\}$ *is the set of internal states of* S_1 *when* $n = 1$. $D(1) = m$ *is the dimension, or number of elements, of* Ξ_1. The external state of the system is still $\{\, 1; N \,\}$. Thus we would say that "the system is in internal state $i \in \Xi_1$, and external state, $\{\, 1\,; N \,\}$." Alternatively, we could simply say that "the system is in state $\{\, i;\, 1;\, N \,\}$," with it being understood that $i \in \Xi_1$. ◊

Definition 6.1.1 is the same as in previous chapters, but with two or more customers at S_1 we must have an extended definition. It would seem that since each of two customers can be in any one of m places, we need m^2 states to describe this. This is actually more than is necessary if we do not care to distinguish one customer from the other or one server from the other. Thus the statement, "Customer 1 is at phase i_1 of server 1, and customer 2 is at phase i_2 in server 2" is more than we want to know. All we need to know is that phase i_1 in one of the servers and i_2 in the other server are busy. This leads to the specification of an internal state by the unordered pair $\{\, i_1, i_2 \,\}$. Then $\{\, i_1, i_2 \,\}$ and $\{\, i_2, i_1 \,\}$ are the *same* state. Therefore, think of $\{\cdot,\cdot\}$ as a set containing two integers. For the sake of definiteness, we will assume that $i_1 \le i_2$. Then we have

Definition 6.1.2

$\Xi_2 = \{ i = \{i_1, i_2\} \mid i_1 \le i_2 = 1, 2, \cdots, m\}$ *is the set of internal states of* S_1 when $n \ge 2$. $D(2) = m(m+1)/2$ is the number of elements of Ξ_2. The external state of the system is still $\{n\,; N\}$. Thus we would say that "the system is in internal state $i \in \Xi_2$, and external state, $\{n\,; N\}$." Alternatively, we could simply say that "the system is in state $\{i; n; N\}$," with the understanding that $i \in \Xi_2$ if $n \ge 2$ (i.e., $i = \{i_1, i_2\}$). ◊

There is an alternative state space definition that is equivalent to this. Instead of listing the phases that are active, we can enumerate the number of customers at each phase. This requires an m-vector of nonnegative integers whose sum is 2. Although this convention seems more verbose, it will prove to be more useful when we consider systems in which $C > 2$ or when we go to the generalized system. For now, we will use the definition as given.

Next we define the probability vectors, and their associated scalars, needed for the work of this chapter. We will use the obvious notation, $\boldsymbol{\varepsilon}'_{\mathbf{k}}$ for the $D(k)$-dimension column vector of all 1's. We will also use $\mathbf{I_k}$ for the identity matrix of dimension $D(k)$.

Definition 6.1.3

$[\boldsymbol{\pi}_{\mathbf{1}}(1; N)]_i :=$ *probability that only one customer is at* S_1 *(with* $N-1$ *at* S_2*), and the system is in internal state* $i \in \Xi_1$ *(i.e., the customer is at phase* i *in either server in* S_1*).* $\boldsymbol{\pi}_{\mathbf{1}}(1; N)$ is a vector with $D(1)$ components. $r(1; N) = \boldsymbol{\pi}_{\mathbf{1}}(1; N)\boldsymbol{\varepsilon}'_{\mathbf{1}}$ is the associated scalar probability. ◊

Definition 6.1.4

$[\boldsymbol{\pi}_{\mathbf{2}}(n; N)]_i :=$ *probability that there are* $n \ge 2$ *customers at* S_1 *(with* $N-n$ *at* S_2*), and the system is in internal state* $i = \{i_1, i_2\} \in \Xi_2$ *(i.e., the two active customers are at phases* i_1 *and* i_2 *one in each of the two servers in* S_1*).* $\boldsymbol{\pi}_{\mathbf{2}}(n; N)$ is a $D(2)$-vector, and $r(n; N) = \boldsymbol{\pi}_{\mathbf{2}}(n; N)\boldsymbol{\varepsilon}'_{\mathbf{2}}$ is the associated scalar probability. ◊

Note that the subscript k on the $\boldsymbol{\pi}_{\mathbf{k}}$'s stands for the number of customers active in S_1. Thus our convention differs from the one we used in Chapter 5. Strictly speaking, the subscripts are unnecessary because they correlate with the first integer in the argument ($k = n$ for $n \le 2$, and $k = 2$ for $n \ge 2$); however, since they denote objects of different dimensions, we include them both for emphasis. According to this convention, in this chapter we use the subscript k to denote any object that applies only to space Ξ_k, or, if it is an object connecting two spaces (nonsquare matrices), k will correspond to the higher-numbered space.

Although we wish to avoid sounding too abstract, we must say something about the connection between our *state space*, Ξ_k, and the vector spaces on which our matrices operate. We have defined Ξ_k to be a set with $D(k)$ elements. We could think of each element as being a *unit vector* in a $D(k)$-dimensional vector space. Consider the set of row vectors with $D(k)$ components. Then each state, $i \in \Xi_k$ corresponds to one such vector with a 1 in one position, and a 0 in all the other positions. We can say, then that these *states* form a complete basis for the $D(k)$ dimensional vector space, since every vector in that space can be written as a linear combination of the *basis states*, or *basis vectors*. This is such a natural correspondence, that we can usually get away without having to make a distinction between the set, and the vector space its members generate. Remember, though, that i stands for several things itself, when $k > 1$.

Consistent with the above, we proceed to rename several of our previously known operators. First, the completion-rate matrix for states in Ξ_1, $\mathbf{M_1} := \mathbf{M}$, is a diagonal matrix whose iith element, $[\mathbf{M_1}]_{ii}$, is the probability rate of leaving state $i \in \Xi_1$ by way of a completion inside S_1. The transition matrix for Ξ_1 is $\mathbf{P_1} := \mathbf{P}$, and $\mathbf{V_1} = \mathbf{B_1^{-1}} := \mathbf{V}$. We proved in Section 3.3.4, in the discussion surrounding Lemma 3.3.2, that an exponential server with feedback is equivalent to an exponential server without feedback, but with service rate reduced to $(1-\theta)\mu$, where θ is the feedback probability. This implies that we can assume without loss of generality that $[\mathbf{P}_1]_{ii} = 0^{\dagger}$. We will do that here, since it simplifies our examples. However, in Section 6.2.2 we will allow P_{ii} to be nonzero. There is no need to relabel $\mathbf{Q}$, $\mathbf{p}$ and $\mathbf{q}'$, but we must define a new set of operators that operate on vectors in Ξ_2. First we have the completion rate matrix.

Definition 6.1.5

$\mathbf{M_2}$ *is a diagonal matrix whose* iith *element is the probability rate of leaving the state* $i \in \Xi_2$ *by way of a completion in* S_1. By our definition of this system, $[\mathbf{M_2}]_{ii} = \mu_{i_1} + \mu_{i_2}$. ◊

The following is another Ξ_2-space object, the transition matrix.

Definition 6.1.6

$[\mathbf{P_2}]_{ij}$ *:= probability of going to state* $j \in \Xi_2$ *upon a completion in* S_1*, given that the system is in internal state* $i \in \Xi_2$. $\mathbf{P_2}$ is a (non-isometric) transition matrix for states in Ξ_2, (i.e., $\mathbf{P_2}\boldsymbol{\varepsilon}'_2 \neq \boldsymbol{\varepsilon}'_2$). ◊

† Suppose that $\mathbf{P}$ is a transition matrix with $\mathbf{P}_{ii} \neq 0$ for some i. Then construct the new matrix $\mathbf{P}$, with *new* $\mathbf{P}_{ii} = 0$ and *new* $\mathbf{P}_{ij} = \textit{old}\ \mathbf{P}_{ij} / (1 - \textit{old}\ \mathbf{P}_{ii})$, and *new* $\mathbf{P}_{kj} = \textit{old}\ \mathbf{P}_{kj}$ for $j, k \neq i$. Also, replace μ_i with $(1 - \textit{old}\ \mathbf{P}_{ii})\mu_i$ in $\mathbf{M}$. All other components remain the same. See Lemma 3.3.5.

Assume that $[\mathbf{P}_1]_{ii} = 0$, and recall that only one customer can change his phase at a time. Then the elements of $\mathbf{P_1}$ are related to $\mathbf{P_2}$ in the following way:

$$[\mathbf{P_2}]_{ij} = 0 \text{ unless } \{i_1, i_2\} \cap \{j_1, j_2\} \text{ contains exactly one element}^{\dagger}.$$

Let the common element be called h. Call the other member of the i-pair, γ, and the other member of the j-pair, ν. Then, if there *is* exactly one element in common,

$$[\mathbf{P_2}]_{ij} = [\mathbf{P_1}]_{\gamma\nu} \frac{\mu_\gamma}{[\mathbf{M_2}]_{ii}}, \qquad i_1 \neq i_2,$$

and

$$[\mathbf{P_2}]_{ij} = [\mathbf{P_1}]_{\gamma\nu}, \qquad i_1 = i_2.$$

The reason for this is as follows. If $i_1 \neq i_2$, since only one active customer can move at a time, the probability that it is the customer at phase γ is $\mu_\gamma / (\mu_\gamma + \mu_h)$, and the probability that he will go to ν, given that he *is* the one who will move is $P_{\gamma\nu}$. If $i_1 = i_2$, it makes no difference which one moves. As an example, let $m = 3$, then

$$\Xi_1 = \{1, 2, 3\},$$

and

$$\Xi_2 = \left\{ \{1,1\}, \{1,2\}, \{1,3\}, \{2,2\}, \{2,3\}, \{3,3\} \right\}.$$

The transition matrix is given by the following, where the ordering is the same as that given in the list above:

$$\mathbf{P_2} = \begin{bmatrix} 0 & P_{12} & P_{13} & 0 & 0 & 0 \\ \dfrac{\mu_2 P_{21}}{\mu_1+\mu_2} & 0 & \dfrac{\mu_2 P_{23}}{\mu_1+\mu_2} & \dfrac{\mu_1 P_{12}}{\mu_1+\mu_2} & \dfrac{\mu_1 P_{13}}{\mu_1+\mu_2} & 0 \\ \dfrac{\mu_3 P_{31}}{\mu_1+\mu_3} & \dfrac{\mu_3 P_{32}}{\mu_1+\mu_3} & 0 & 0 & \dfrac{\mu_1 P_{12}}{\mu_1+\mu_3} & \dfrac{\mu_1 P_{13}}{\mu_1+\mu_3} \\ 0 & P_{21} & 0 & 0 & P_{23} & 0 \\ 0 & \dfrac{\mu_3 P_{31}}{\mu_2+\mu_3} & \dfrac{\mu_2 P_{21}}{\mu_2+\mu_3} & \dfrac{\mu_3 P_{32}}{\mu_2+\mu_3} & 0 & \dfrac{\mu_2 P_{23}}{\mu_2+\mu_3} \\ 0 & 0 & P_{31} & 0 & P_{32} & 0 \end{bmatrix}.$$

The Ξ_2-space equivalent of $\mathbf{p}$ is $\mathbf{R_2}$, which we now define.

† Note: If we had not assumed that $[\mathbf{P}_1]_{ij} := 0$, we would have had to consider the additional possibility that $i = j$.

Definition 6.1.7

$[\mathbf{R}_2]_{ij}$:= *probability that a customer, upon entering* S_1 *will put it in internal state* $j \in \Xi_2$, *given that it was in internal state* $i \in \Xi_1$. This is a $D(1) \times D(2)$-dimensional matrix. ◊

A matrix element of $\mathbf{R}_2$ is 0 unless either j_1 or $j_2 = i$. As before, let ν be the member of the pair $\{j_1, j_2\}$ that is not i; then

$$[\mathbf{R}_2]_{ij} = p_\nu \quad \text{if} \quad j_1 \text{ or } j_2 = i,$$

and

$$[\mathbf{R}_2]_{ij} = 0 \quad \text{otherwise.}$$

Since the system must end up in some state after a customer enters S_1, we have

$$\mathbf{R}_2 \varepsilon'_2 = \varepsilon'_{\mathbf{1}}. \tag{6.1.1}$$

For our example with $m = 3$,

$$\mathbf{R}_2 = \begin{bmatrix} p_1 & p_2 & p_3 & 0 & 0 & 0 \\ 0 & p_1 & 0 & p_2 & p_3 & 0 \\ 0 & 0 & p_1 & 0 & p_2 & p_3 \end{bmatrix}.$$

This clearly satisfies (6.1.1).

We define the last matrix of this set, $\mathbf{Q}_2$, the Ξ_2 equivalent of $\mathbf{q}'$.

Definition 6.1.8

$[\mathbf{Q}_2]_{ij}$:= *probability that upon a completion in* S_1 *from internal state* $i \in \Xi_2$, *a customer leaves, putting the system in internal state* $j \in \Xi_1$. This is a $D(2) \times D(1)$-dimensional matrix. This matrix has no direct relation to either $\mathbf{Q} = \varepsilon'_{\mathbf{1}}\mathbf{p}$, defined in Chapter 3, or Q defined by (1.3.2c), but is the 2-space equivalent of $\mathbf{q}'$. ◊

Again, unless $j = i_1$ or i_2, $[\mathbf{Q}_2]_{ij} = 0$. For those matrix elements that are not 0, define ν to be the member of the pair that does not equal j, then

$$[\mathbf{Q}_2]_{ij} = \frac{\mu_\nu q_\nu}{[\mathbf{M}_2]_{ii}} \quad \text{for} \quad i_1 \neq i_2$$

and

$$[\mathbf{Q}_2]_{ij} = q_\nu \quad \text{for} \quad i_1 = i_2.$$

Note that $[\mathbf{Q}_2\mathbf{R}_2]_{ij}$ is the probability that upon a completion from internal state $i \in \Xi_2$, one customer leaves and another enters, putting the system in internal state $j \in \Xi_2$. This process can only happen if there are initially $n > 2$ customers at S_1, in which case the system is left in external state $\{n - 1; N\}$. Thus we have the equivalent to (3.1.1a), or rather (3.3.11a),

$$(\mathbf{P}_2 + \mathbf{Q}_2\mathbf{R}_2)\varepsilon'_2 = \varepsilon'_2. \tag{6.1.2}$$

So $(\mathbf{P}_2 + \mathbf{Q}_2\mathbf{R}_2)$ is isometric. Since $\mathbf{R}_2\varepsilon'_2 = \varepsilon'_1$ from (6.1.1), we have a relation that compares with $\mathbf{q}' = (\mathbf{I} - \mathbf{P})\varepsilon'$ from (3.1.1b), namely,

$$\mathbf{Q}_2\varepsilon'_1 = (\mathbf{I}_2 - \mathbf{P}_2)\varepsilon'_2. \tag{6.1.3}$$

We are just beginning to see the problems we will be having. Whereas (3.1.1b) permitted us to express $\mathbf{q}'$ uniquely in terms of $\mathbf{P}$, (6.1.3) only yields a partial relation between $\mathbf{Q}_2$ and $\mathbf{P}_2$.

For our specific example we have

$$\mathbf{Q}_2 = \begin{bmatrix} q_1 & 0 & 0 \\ \dfrac{\mu_2 q_2}{\mu_1+\mu_2} & \dfrac{\mu_1 q_1}{\mu_1+\mu_2} & 0 \\ \dfrac{\mu_3 q_3}{\mu_1+\mu_3} & 0 & \dfrac{\mu_1 q_1}{\mu_1+\mu_3} \\ 0 & q_2 & 0 \\ 0 & \dfrac{\mu_3 q_3}{\mu_2+\mu_3} & \dfrac{\mu_2 q_2}{\mu_2+\mu_3} \\ 0 & 0 & q_3 \end{bmatrix}.$$

> **Exercise 6.1.1:** Verify that the example matrices, $\mathbf{P}_2$, $\mathbf{Q}_2$, and $\mathbf{R}_2$ for $m = 3$ do indeed satisfy (6.1.1) to (6.1.3).

6.1.2. Balance Equations

At last we can write the balance equations. The terms on the left-hand side of each equation represent the rate of leaving a given state, while the right-hand side is the rate of entering that state. First let $n = 0$; then there is only one internal state (empty), so we will play the same notational game as before, namely,

$$\pi_1(0; N) := r(0; N)\mathbf{p},$$

where $r(0; N)$ is the steady-state probability that there is no one at S_1. The balance equation is identical to (4.1.3a), since there is no way to make use of the second server in S_1. We rewrite it in the notation of this chapter.

$$\lambda\pi_1(0; N) = \pi_1(1; N)\mathbf{B}_1\mathbf{Q}. \tag{6.1.4a}$$

Its interpretation should be clear.

The vector equation corresponding to entering and leaving external state $\{1; N\}$ provides us with something new.

$$\pi_1(1; N)(\mathbf{M}_1 + \lambda\mathbf{I}_1)$$
$$= \lambda r(0; N)\mathbf{p} + \pi_1(1; N)\mathbf{M}_1\mathbf{P}_1 + \pi_2(2; N)\mathbf{M}_2\mathbf{Q}_2. \tag{6.1.4b}$$

The interpretation is as follows. The system can leave state $\{i; 1; N\}$ by being in that state, $[\pi_1(1; N)]$, and either have the active customer at S_2 finish, $[\lambda\mathbf{I_1}]$, or have the lone customer in S_1 finish at phase i, $[\mathbf{M_1}]$. The system can enter state $\{i; 1; N\}$ by any of three ways. One path starts with no one initially at S_1, $[r(0; N)]$, a completion occurs at S_2, $[\lambda]$, and that customer immediately goes to S_1 and enters, $[\mathbf{p}]$. Another path starts with one customer already at S_1, $[\pi_1(1; N)]$, a completion occurs in S_1, $[\mathbf{M_1}]$, and that customer moves to another phase, $[\mathbf{P_1}]$. The third path starts with two customers in S_1, $[\pi_2(2; N)]$, one of those two customers has a completion, $[\mathbf{M_2}]$, and leaves, $[\mathbf{Q_2}]$; thus S_1 remains with one customer, since there is no one in its queue to replace the customer who just left.

The balance equation for $n = 2$ requires the matrix $\mathbf{R_2}$ for the first time.

$$\pi_2(2; N)(\mathbf{M_2} + \lambda\mathbf{I_2}) = \pi_2(3; N)\mathbf{M_2Q_2R_2} + \pi_2(2; N)\mathbf{M_2P_2} + \lambda\pi_1(1; N)\mathbf{R_2}. \qquad (6.1.4c)$$

Two of the terms might require some clarification. When there are at least three customers at S_1, and a departure occurs, $[\mathbf{M_2Q_2}]$, a new customer *is* available in the queue to enter, $[\mathbf{R_2}]$, and put the system back into an internal state of Ξ_2. Similarly, if originally there is only one customer at S_1 and a customer arrives, $[\lambda\pi_1(1; N)]$, he immediately enters, taking the system from some state $i \in \Xi_1$ to some state $j \in \Xi_2$, $[\mathbf{R_2}]$.

The general vector balance equation for $2 < n < N$ follows.

$$\pi_2(n; N)(\mathbf{M_2} + \lambda\mathbf{I_2}) = \pi_2(n{+}1; N)\mathbf{M_2Q_2R_2} + \pi_2(n; N)\mathbf{M_2P_2} + \lambda\pi_2(n{-}1; N). \qquad (6.1.4d)$$

The only difference between this equation and the one for $n = 2$ is the missing matrix, $\mathbf{R_2}$, in the last term. For $n > 2$, when a customer arrives at S_1, there already are two customers active, so his arrival does not change the internal state of the system.

The final balance equation, for external state $\{N; N\}$, is

$$\pi_2(N; N)\mathbf{M_2} = \pi_2(N; N)\mathbf{M_2P_2} + \lambda\pi_2(N{-}1; N). \qquad (6.1.4e)$$

We next define a new matrix and its inverse

$$\mathbf{B_2} = \mathbf{V_2}^{-1} := \mathbf{M_2}(\mathbf{I_2} - \mathbf{P_2}), \qquad (6.1.5)$$

and combine like terms in the balance equations to get

$$\pi_1(1; N)(\mathbf{B_1} + \lambda\mathbf{I_1}) = \pi_2(2; N)\mathbf{M_2Q_2} + \lambda\pi_1(0; N), \qquad (6.1.6a)$$

$$\pi_2(2; N)(\mathbf{B_2} + \lambda\mathbf{I_2}) = \pi_2(3; N)\mathbf{M_2Q_2R_2} + \lambda\pi_1(1; N)\mathbf{R_2}, \qquad (6.1.6b)$$

$$\pi_2(n; N)(\mathbf{B_2} + \lambda\mathbf{I_2}) = \pi_2(n{+}1; N)\mathbf{M_2Q_2R_2} + \lambda\pi_2(n{-}1; N), \qquad (6.1.6c)$$

and

$$\pi_2(N; N)\mathbf{B_2} = \lambda\pi_2(N{-}1; N). \qquad (6.1.6d)$$

Equation (6.1.5) looks quite familiar, and it should. The matrix, $\mathbf{B_2}$, does indeed generate the distribution of the time until one customer leaves S_1, given that it started with two customers in some vector state of Ξ_2. We will postpone looking into this until we discuss transient behavior.

6.1.3. Solution of Probability Vectors

The balance equations are quite similar to those for the ME/M/C//N system of Section 5.4, and the solution is also similar. The additional complication we have here is that we are dealing with several different sizes of matrices. As always, we would like to express everything in terms of $r(0; N)\mathbf{p}$, since that depends on only one number. Unfortunately, we cannot start with Equation (6.1.4a) since $\mathbf{B}_1\mathbf{Q}$ does not have an inverse. We have already faced this problem in solving for the M/ME/1//N queue in (4.1.2), so we copy that method, and start at the top, at (6.1.6d). Note that if we had attempted to solve the open system directly, we would have no *top* to start at. Anyway, (6.1.6d) can be rewritten as

$$\pi_2(N; N) = \lambda\pi_2(N-1; N)\mathbf{V}_2. \tag{6.1.7a}$$

Next we use (6.1.6c) for $n = N-1$, and substitute for $\pi_2(N; N)$ to get

$$\pi_2(N-1; N)(\mathbf{B}_2 + \lambda\mathbf{I}_2) = \lambda[\pi_2(N-1; N)\mathbf{V}_2]\mathbf{M}_2\mathbf{Q}_2\mathbf{R}_2 + \lambda\pi_2(N-2; N),$$

or, collecting terms and solving for $\pi_2(N-1; N)$,

$$\pi_2(N-1; N) = \pi_2(N-2; N)\left[\mathbf{I}_2 + \frac{1}{\lambda}\mathbf{B}_2 - \mathbf{V}_2\mathbf{M}_2\mathbf{Q}_2\mathbf{R}_2\right]^{-1}. \tag{6.1.7b}$$

This equation is of the form

$$\pi_2(N-1; N) = \pi_2(N-2; N)\mathbf{U}_2(1),$$

where

$$\mathbf{U}_2(1) := \left[\mathbf{I}_2 + \frac{1}{\lambda}\mathbf{B}_2 - \mathbf{V}_2\mathbf{M}_2\mathbf{Q}_2\mathbf{R}_2\right]^{-1}.$$

This matrix, and the family of $\mathbf{U}_2$'s we are about to introduce are the direct generalizations of the $\mathbf{U}$ of (4.1.4) and the $\mathbf{U}(k)$'s of Section 5.4.

Now suppose that all the π_2's can be written in the form

$$\pi_2(N-j; N) = \pi_2(N-j-1; N)\mathbf{U}_2(j), \tag{6.1.8}$$

and that $\mathbf{U}_2(j)$, $j = 1, 2, \cdots, n$ are already known. Then put this into (6.1.6c) to get

$$\begin{aligned}\pi_2(N-n-1; N)(\lambda\mathbf{I}_2 + \mathbf{B}_2) &= \pi_2(N-n; N)\mathbf{M}_2\mathbf{Q}_2\mathbf{R}_2 + \lambda\pi_2(N-n-2; N)\\ &= \pi_2(N-n-1; N)\mathbf{U}_2(n)\mathbf{M}_2\mathbf{Q}_2\mathbf{R}_2 + \lambda\pi_2(N-n-2; N),\end{aligned}$$

or

$$\pi_2(N-n-1; N)\left[\mathbf{B}_2 + \lambda\mathbf{I}_2 - \mathbf{U}_2(n)\mathbf{M}_2\mathbf{Q}_2\mathbf{R}_2\right] = \lambda\pi_2(N-n-2; N).$$

This implies, starting with $\mathbf{U}_2(0) := \lambda\mathbf{V}_2$, that all the $\mathbf{U}_2$'s are recursively defined and can be computed by the following equation:

$$\mathbf{U}_2(n) = \left[\frac{1}{\lambda}\mathbf{B}_2 + \mathbf{I}_2 - \mathbf{U}_2(n-1)\mathbf{M}_2\mathbf{Q}_2\mathbf{R}_2\right]^{-1} \quad \text{for} \quad n = 1, 2, \cdots. \tag{6.1.9}$$

Note that the $\mathbf{U}_2$'s do not depend on N, and therefore the same set can be used for all N. This point will be explored further when we discuss algorithms for evaluating everything. In the meantime, we still have not satisfied (6.1.6a) and (6.1.6b). But with (6.1.8), (6.1.6b) becomes†

$$\pi_2(2; N)\left[\frac{1}{\lambda}\mathbf{B}_2 + \mathbf{I}_2 - \frac{1}{\lambda}\mathbf{U}_2(N-3)\mathbf{M}_2\mathbf{Q}_2\mathbf{R}_2\right] = \pi_1(1; N)\mathbf{R}_2$$

or

$$\pi_2(2; N) = \pi_1(1; N)\mathbf{R}_2\mathbf{U}_2(N-2). \tag{6.1.10a}$$

Similarly, (6.1.6a) becomes†

$$\pi_1(1; N)\left[\frac{1}{\lambda}\mathbf{B}_1 + \mathbf{I}_1 - \frac{1}{\lambda}\mathbf{R}_2\mathbf{U}_2(N-2)\mathbf{M}_2\mathbf{Q}_2\right] = \pi_1(0; N). \tag{6.1.10b}$$

We must show that (6.1.10b) is consistent with (6.1.4a), since they both connect states $\{i; 1; N\}$ with $\{\cdot; 0; N\}$. To do this we must first prove the following, which is analogous to Lemma 4.1.1.

Lemma 6.1.1: Let $\mathbf{B}_2$ be defined by (6.1.5), then multiplying (6.1.3) by $\mathbf{M}_2$ yields

$$\mathbf{M}_2\mathbf{Q}_2\varepsilon'_1 = \mathbf{B}_2\varepsilon'_2, \tag{6.1.11a}$$

and since $\mathbf{R}_2\varepsilon'_2 = \varepsilon'_1$, we also have

$$\mathbf{M}_2\mathbf{Q}_2\mathbf{R}_2\varepsilon'_2 = \mathbf{B}_2\varepsilon'_2. \tag{6.1.11b}$$

Furthermore, let $\mathbf{U}_2(0) = \lambda\mathbf{V}_2$, and let $\mathbf{U}_2(n)$ satisfy (6.1.9) for $n \geq 1$. Then

$$\mathbf{U}_2(n)\mathbf{B}_2\varepsilon'_2 = \lambda\varepsilon'_2 \quad \text{for all} \quad n \geq 0. \tag{6.1.11c}$$

In other words, $[(1/\lambda)\mathbf{U}_2(n)\mathbf{B}_2]$ is isometric. ■

Proof: We shall actually prove that its inverse is isometric. First, observe [$\mathbf{U}_2(0)$ was defined to be $\lambda\mathbf{V}_2$] that

$$\mathbf{U}_2(0)\mathbf{B}_2\varepsilon'_2 = \lambda\mathbf{V}_2\mathbf{B}_2\varepsilon'_2 = \lambda\mathbf{I}_2\varepsilon'_2 = \lambda\varepsilon'_2.$$

Now assume that (6.1.11c) is true for all $k = 0, 1, \cdots, n$, and let

$$\mathbf{A}_2(k) := [\mathbf{U}_2(k)]^{-1};$$

† Note that $\mathbf{Q}_2(n)\mathbf{X}_1\mathbf{R}_2(n)$ is a $D(2) \times D(2)$ matrix, where $\mathbf{X}_1$ is any matrix of dimension $D(1) \times D(1)$. Also, $\mathbf{R}_2(n)\mathbf{X}_2\mathbf{Q}_2(n)$ is a $D(1) \times D(1)$ matrix, where $\mathbf{X}_2$ is any matrix of dimension $D(2) \times D(2)$.

then

$$\left[\frac{1}{\lambda}\mathbf{U}_2(n+1)\mathbf{B}_2\right]^{-1} = \lambda\mathbf{V}_2\mathbf{A}_2(n+1) = \lambda\mathbf{V}_2\left[\mathbf{I}_2 + \frac{1}{\lambda}\mathbf{B}_2 - \frac{1}{\lambda}\mathbf{U}_2(n)\mathbf{M}_2\mathbf{Q}_2\mathbf{R}_2\right]$$

$$= \lambda\mathbf{V}_2 + \mathbf{I}_2 - \mathbf{V}_2\mathbf{U}_2(n)\mathbf{M}_2\mathbf{Q}_2\mathbf{R}_2.$$

Next postmultiply both sides of the equation by ε'_2 and get

$$\lambda\mathbf{V}_2\mathbf{A}_2(n+1)\varepsilon'_2 = \lambda\mathbf{V}_2\varepsilon'_2 + \varepsilon'_2 - \mathbf{V}_2\mathbf{U}_2\left[\mathbf{M}_2\mathbf{Q}_2\mathbf{R}_2\varepsilon'_2\right]$$

$$= \lambda\mathbf{V}_2\varepsilon'_2 + \varepsilon'_2 - \mathbf{V}_2\left[\mathbf{U}_2(n)\mathbf{B}_2\varepsilon'_2\right].$$

But the expression in the second set of brackets is equal to $\lambda\varepsilon'_2$ by assumption; thus by induction we have

$$\lambda\mathbf{V}_2\mathbf{A}_2(n)\varepsilon'_2 = \varepsilon'_2 \quad \text{for all } n.$$

Premultiplying both sides of this equation by $\mathbf{U}_2(n)\mathbf{B}_2$ yields our lemma. QED

We now return to (6.1.10b). When we postmultiply both sides by ε'_1, the right-hand side becomes $r(0; N)$, while the left-hand side becomes

$$\frac{1}{\lambda}\pi_1(1; N)\mathbf{B}_1\varepsilon'_1 + r(1; N) - \pi_1(1; N)\mathbf{R}_2\left[\frac{1}{\lambda}\mathbf{U}_2(N-2)\mathbf{M}_2\mathbf{Q}_2\varepsilon'_1\right]$$

$$= \frac{1}{\lambda}\pi_1(1; N)\mathbf{B}_1\varepsilon'_1 + r(1; N) - \pi_1(1; N)\mathbf{R}_2\varepsilon'_2 = \frac{1}{\lambda}\pi_1(1; N)\mathbf{B}_1\varepsilon'_1.$$

The two parts together reproduce (6.1.4a), so we have proven our case.

We now define the matrix implied by (6.1.10b),

$$\overline{\mathbf{U}}_1(N) := \left[\frac{1}{\lambda}\mathbf{B}_1 + \mathbf{I}_1 - \mathbf{R}_2\mathbf{U}_2(N-2)\mathbf{M}_2\mathbf{Q}_2\right]^{-1}.$$

Note that this a Ξ_1-space matrix, and it satisfies the equation

$$\overline{\mathbf{U}}_1(N)\mathbf{B}_1\varepsilon'_1 = \lambda\varepsilon'_1.$$

We now list the solution vectors, which should help the reader make sense of what we have derived so far in this chapter:

$$\pi_1(0; N) = r(0; N)\mathbf{p},$$

$$\pi_1(1; N) = r(0; N)\mathbf{p}\overline{\mathbf{U}}_1(N),$$

$$\pi_2(2; N) = r(0; N)\mathbf{p}\overline{\mathbf{U}}_1(N)\mathbf{R}_2\mathbf{U}_2(N-2),$$

$$\cdots \qquad \cdots \qquad \cdots$$

$$\pi_2(n; N) = r(0; N)\mathbf{p}\overline{\mathbf{U}}_1(N)\mathbf{R}_2\mathbf{U}_2(N-2) \cdots \mathbf{U}_2(N-n),$$

$$\cdots \qquad \cdots \qquad \cdots$$

$$\pi_2(N; N) = r(0; N)\mathbf{p}\overline{\mathbf{U}}_1(N)\mathbf{R}_2\mathbf{U}_2(N-2)\mathbf{U}_2(N-3) \cdots \mathbf{U}_2(0).$$

We still have to evaluate $r(0; N)$, but that is easy enough to do since the sum of all probabilities must be 1. That is,

$$\sum_{n=0}^{N} r(n; N) = r(0; N) + \pi_1(1; N)\varepsilon'_1 + \sum_{n=2}^{N} \pi_2(n; N)\varepsilon'_2 = 1.$$

We can actually write this in a compact and recursive way, just as we did previously with the matrix, $\mathbf{K}(N)$ in (5.4.6b). Recall that $\mathbf{R}_2\varepsilon'_2 = \varepsilon'_1$, and define the vector

$$\mathbf{x}_2(N) := \mathbf{p}\overline{\mathbf{U}}_1(N)\mathbf{R}_2;$$

then

$$r(1; N) = r(0; N)\mathbf{x}_2(N)\varepsilon'_2.$$

Next define the Ξ_2-space matrix,

$$\begin{aligned}\mathbf{K}_2(N) := \mathbf{I}_2 + \mathbf{U}_2(N-2) + \mathbf{U}_2(N-2)\mathbf{U}_2(N-3) + \\ \cdots + \mathbf{U}_2(N-2)\mathbf{U}_2(N-3) \cdots \mathbf{U}_2(N-n) + \\ \cdots + \mathbf{U}_2(N-2)\mathbf{U}_2(N-3) \cdots \mathbf{U}_2(0).\end{aligned} \tag{6.1.12a}$$

When we factor the terms right-multiplying $\mathbf{U}_2(N-2)$, we get

$$\mathbf{K}_2(N) = \mathbf{I}_2 + \mathbf{U}_2(N-2)\mathbf{K}_2(N-1), \tag{6.1.12b}$$

and therefore, the sum over all states reduces to

$$r(0; N)\Big[1 + \mathbf{x}_2(N)\mathbf{K}_2(N)\varepsilon'_2\Big] = 1$$

or

$$\frac{1}{r(0; N)} = \Big[1 + \mathbf{x}_2(N)\mathbf{K}_2(N)\varepsilon'_2\Big]. \tag{6.1.12c}$$

We will discuss an efficient algorithm after we have dealt with the general case, but our accomplishments so far deserve a summary theorem-algorithm.

Theorem 6.1.2: Consider an M/ME/2//*N* loop with matrices, $\mathbf{M_k}$, $\mathbf{P_k}$, $\mathbf{B_k}$, $\mathbf{V_k}$, $\mathbf{Q}_2$, and $\mathbf{R}_2$, as defined in this section. Given $N > 2$,

BEGIN PROCEDURE

$\mathbf{U}_2(0; N) = \lambda\mathbf{V}_2$.

* Then

FOR $n = 1$ TO $N - 2$:

$$\mathbf{U}_2(n) = \lambda\Big[\lambda\mathbf{I}_2 + \mathbf{B}_2 - \mathbf{U}_2(n-1)\mathbf{M}_2\mathbf{Q}_2\mathbf{R}_2\Big]^{-1};$$

END FOR.

* Next evaluate

$$\overline{\mathbf{U}}_1(N) = \lambda\Big[\lambda \mathbf{I}_2 + \mathbf{B}_2 - \mathbf{R}_2\mathbf{U}_2(N-2)\mathbf{M}_2\mathbf{Q}_2\Big]^{-1};$$

$$\mathbf{x}_1(0) = \mathbf{p};$$

$$\mathbf{x}_1(1; N) = \mathbf{x}_1(0)\overline{\mathbf{U}}_1(N);$$

$$x(1; N) = \mathbf{x}_1(1; N)\varepsilon'_1;$$

$$\mathbf{x}_2(2; N) = \mathbf{x}_1(1; N)\mathbf{R}_2\mathbf{U}_2(N-2);$$

$$x(2; N) = \mathbf{x}_2(2; N)\varepsilon'_2;$$

$$sum = 1 + x(1; N) + x(2; N).$$

FOR $n = 3$ TO N:

$$\mathbf{x}_2(n; N) = \mathbf{x}_2(n-1)\mathbf{U}_2(N-n);$$

$$x(n; N) = \mathbf{x}_2(n; N)\varepsilon'_2;$$

$$sum = sum + x(n; N);$$

END FOR.

* The steady-state probability vectors and their associated scalars are given by the following.

$$r(0; N) = \frac{1}{sum},$$

$$\pi_1(1; N) = r(0; N)\mathbf{x}_1(1; N),$$

$$r(1; N) = r(0; N)x(1; N).$$

FOR $n = 2$ TO N:

$$\pi_2(n; N) = r(0; N)\mathbf{x}_2(n; N);$$

$$r(n; N) = r(0; N)x(n; N);$$

END FOR.

* The mean throughput is given by

$$\Lambda(N) = \lambda[1 - r(N; N)].$$

END PROCEDURE

All other performance parameters can be calculated from these. ■

It does not pay to go deeply into the significance of this theorem, since we will be deriving the solutions for the general system in the next section. We point out though, that as an algorithm, the most computationally intense portion involves finding the $\mathbf{U}_2$'s in Equations (4.1.9), since that requires taking the inverse of $D(2) \times D(2)$ matrices. In studying closed systems, researchers are usually not interested in just one value of N. Rather, they must look at a whole range of values up to, say, N_{max}. In that case, the matrices required to solve for a system with N_{max} customers can be used to solve for systems with fewer customers. Put differently, if one wishes to solve for a system with $N + 1$ customers, after solving for the same system with N customers, only one more inverse need be taken [to find $\mathbf{U}_2(N+1)$].

6.2. STEADY-STATE M/G/C//N-TYPE SYSTEMS

In the preceding section we described the steady-state solution for systems with $C = 2$. The extension to systems in which there are more than two servers in S_1 is straightforward but requires some generalizations of definitions.

6.2.1. Steady-State M/ME/C//N Loop

Let us start with a definition of our state spaces. An alternative, but equivalent definition will be given in Section 6.2.2.

Definition 6.2.1

$\Xi_k := \{\, i = \{i_1, i_2, \cdots, i_k\} \mid 1 \le i_1 \le i_2 \le \cdots \le i_k \le m \,\}^{\dagger}$ for $1 \le k \le C$. Ξ_k *is the set of all internal states of* S_1 *when there are* k *active customers there.* Each k-tuple represents a state in which $k \le C$ customers are active in a subsystem, S_1, which has C identical servers in it. One customer is at phase i_1, another is at phase i_2, and so on, each in a different server. They never get in each other's way, since there are at least as many servers as there are customers. If there are more than C customers at S_1, the excess numbers must queue up outside. ◊

The number of states in Ξ_k is

$$D(k) = \begin{pmatrix} m+k-1 \\ k \end{pmatrix}. \tag{6.2.1}$$

We now make the following generalizations of previous definitions, where $N > C$. First,

$\boldsymbol{\varepsilon}'_{\mathbf{k}} :=$ *the* $D(k)$*–dimensional column vector of all* 1's.

Definition 6.2.2

$[\mathbf{M_k}]$ *is a diagonal matrix whose* ii*th component is the probability rate of leaving state* $i \in \Xi_k$, *for* $k = 1, 2, \cdots, C$. Thus if $i = \{i_1, i_2, \cdots, i_k\}$, then $[\mathbf{M_k}]_{ii} = \mu_{i_1} + \mu_{i_2} + \cdots + \mu_{i_k}$. This is the *$k$–space completion rate matrix*. ◊

Definition 6.2.3

$\boldsymbol{\pi}_{\mathbf{k}}(n; N) :=$ *probability vector of dimension* $D(k)$, *that there are* n *customers at* S_1, *and* $N-n$ *customers at* S_2. The ith component is the probability that the active customers in S_1 are collectively in state $i \in \Xi_k$. We adhere to the notation $k = n$ if $n \le C$, and $k = C$ otherwise. $r(n; N) = \boldsymbol{\pi}_{\mathbf{k}}(n; N)\boldsymbol{\varepsilon}'_{\mathbf{k}}$ is the associated scalar probability. ◊

† Note that the i_l's need not be distinct.

Definition 6.2.4

$[\mathbf{R_k}]_{ij}$:= *probability that a customer, who upon entering* S_1 *and finding it in internal state* $i \in \Xi_{k-1}$*, will go to the server and phase that puts the system in state* $j \in \Xi_k$. $\mathbf{R_k}$ is a $D(k-1) \times D(k)$-dimensional matrix with the property that $\mathbf{R_k}\boldsymbol{\varepsilon}'_\mathbf{k} = \boldsymbol{\varepsilon}'_\mathbf{k-1}$. We could let $\mathbf{p} = \mathbf{R}_1$ if we so choose. ◊

For descriptive purposes we think of the index i as representing the set $\{ i_1, i_2, \cdots, i_k \}$ (which by Definition 6.2.1, it really is), with the same for j. Then we can say that the matrix element, $[\mathbf{R_k}]_{ij}$, is 0 unless

$$i \cap j = i, \qquad \text{where } i \in \Xi_{k-1} \text{ and } j \in \Xi_k .$$

Remember, by their definition, the set j has one more member than the set i, so there must be exactly one distinct member of j (possibly appearing more than once in the set) which is not in i in order for $[\mathbf{R_k}]_{ij}$ to have a nonzero value. Then, as a direct generalization of the discussion following Definition 6.1.7, call that one element ν, and

$$[\mathbf{R_k}]_{ij} = p_\nu \quad (\text{if } i \cap j = i). \tag{6.2.2a}$$

Definition 6.2.5

$[\mathbf{Q_k}]_{ij}$:= *probability that a customer, upon leaving* S_1 *when the system was in state* $i \in \Xi_k$*, leaves the system in state* $j \in \Xi_{k-1}$ *after he exits.* This matrix is of dimension $D(k) \times D(k-1)$. If we chose, we could have let $\mathbf{q}' = \mathbf{Q}_1$. ◊

Here $[\mathbf{Q_k}]_{ij} := 0$, unless $i \cap j = j$. We now have a little generalization problem. Let ν be the left-over element. It is possible that in the set of $\{i_l\text{'s}\}$, ν appears more than once; then any one of those customers could complete service and leave. So let α_ν be the number of times ν appears in the set i; then for $i \in \Xi_k$ and $j \in \Xi_{k-1}$,

$$[\mathbf{Q_k}]_{ij} := \frac{\alpha_\nu \, \mu_\nu q_\nu}{[\mathbf{M_k}]_{ii}} \qquad (\text{for } i \cap j = j) \tag{6.2.2b}$$

Go back to Definition 6.1.8, and verify that this formula actually matches *both* conditions there, for $k = 2$.

Our last definition of this set is the transition matrix of space Ξ_k.

Definition 6.2.6

$[\mathbf{P_k}]_{ij}$:= *probability that a customer, who upon completing at some phase in* S_1*, will go to another phase in the same server in* S_1*, thereby taking the system from state* $i \in \Xi_k$ *to* $j \in \Xi_k$. $\mathbf{P_k}$ is a non-isometric transition matrix of dimension $D(k) \times D(k)$. ◊

Exactly as in the case for $k = 2$, $[\mathbf{P_k}]_{ij} := 0$, unless $i \cap j$ is a set with $k-1$ elements. Let γ be the member of i which is not in $i \cap j$, and let ν be the member of j which is not in $i \cap j$. Also, let α_γ be the number of times γ appears in the set i; then

$$[\mathbf{P_k}]_{ij} := [\mathbf{P}_1]_{\gamma\nu} \frac{\alpha_\gamma \mu_\gamma}{[\mathbf{M_k}]_{ii}}. \tag{6.2.2c}$$

As with $\mathbf{Q_k}$, this matches the discussion following Definition 6.1.6.

If the construction of these matrices seems difficult, rest assured that it can be automated for computer use. We will look at the specification problem from a different point of view in Section 6.2.2.

As a direct generalization of (6.1.2) and (6.1.3) we can write

$$\left[\mathbf{P_k} + \mathbf{Q_k}\mathbf{R_k}\right]\boldsymbol{\varepsilon'_k} = \boldsymbol{\varepsilon'_k} \tag{6.2.3a}$$

and since $\mathbf{R_k}\boldsymbol{\varepsilon'_k} = \boldsymbol{\varepsilon'_{k-1}}$

$$\mathbf{Q_k}\boldsymbol{\varepsilon'_{k-1}} = \left[\mathbf{I_k} - \mathbf{P_k}\right]\boldsymbol{\varepsilon'_k}. \tag{6.2.3b}$$

We have the natural Ξ_k-space generalizations of (6.1.5)

$$\mathbf{B_k} = \mathbf{V_k}^{-1} := \mathbf{M_k}[\mathbf{I_k} - \mathbf{P_k}], \tag{6.2.4a}$$

which together with (6.2.3b) yields

$$\mathbf{M_k}\mathbf{Q_k}\boldsymbol{\varepsilon'_{k-1}} = \mathbf{M_k}\mathbf{Q_k}\mathbf{R_k}\boldsymbol{\varepsilon'_k} = \mathbf{B_k}\boldsymbol{\varepsilon'_k}. \tag{6.2.4b}$$

There should be little difficulty in writing down the balance equations directly as generalizations of (6.1.4a), and (6.1.6). They are, for $N > C$ customers:

$$\lambda\pi_1(0; N) = \pi_1(1; N)\mathbf{B}_1\mathbf{Q}, \tag{6.2.5a}$$

$$\pi_\mathbf{1}(1; N)(\mathbf{B_1} + \lambda\mathbf{I_1}) = \pi_\mathbf{2}(2; N)\mathbf{M_2}\mathbf{Q_2} + \lambda\pi_\mathbf{1}(0; N), \tag{6.2.5b}$$

for $2 \le k \le C-1$,

$$\pi_\mathbf{k}(k; N)(\mathbf{B_k} + \lambda\mathbf{I_k}) = \pi_\mathbf{k+1}(k+1; N)\mathbf{M_{k+1}}\mathbf{Q_{k+1}} + \lambda\pi_\mathbf{k-1}(k-1; N)\mathbf{R_k}, \tag{6.2.5c}$$

$$\pi_\mathbf{c}(C; N)(\mathbf{B_c} + \lambda\mathbf{I_c}) = \pi_\mathbf{c}(C+1; N)\mathbf{M_c}\mathbf{Q_c}\mathbf{R_c} + \lambda\pi_\mathbf{c-1}(C-1; N)\mathbf{R_c}, \tag{6.2.5d}$$

for $C+1 \le n \le N-1$,

$$\pi_\mathbf{c}(n; N)(\mathbf{B_c} + \lambda\mathbf{I_c}) = \pi_\mathbf{c}(n+1; N)\mathbf{M_c}\mathbf{Q_c}\mathbf{R_c} + \lambda\pi_\mathbf{c}(n-1; N) \tag{6.2.5e}$$

and

$$\pi_\mathbf{c}(N; N)\mathbf{B_c} = \lambda\pi_\mathbf{c}(N-1; N). \tag{6.2.5f}$$

Again following the usual procedure, let

$$\mathbf{U_c}(0) := \lambda\mathbf{V_c}, \tag{6.2.6a}$$

and assume that the following is true for $n < l \le N$ [it is certainly true for $l = N$ by (6.2.5f) and (6.2.6a)]:

$$\pi_{\mathbf{c}}(l;N) = \pi_{\mathbf{c}}(l-1;N)\,\mathbf{U_c}(N-l). \tag{6.2.6b}$$

Be careful; the first index on $\pi_{\mathbf{c}}(\,\cdot\,;N)$ increases with n, while the index on $\mathbf{U_c}(\,\cdot\,)$ decreases. Then combining (6.2.6b) and (6.2.5e) gives

$$\pi_{\mathbf{c}}(n;N)[\mathbf{B_c} + \lambda\mathbf{I_c} - \mathbf{U_c}(N-n+1)\mathbf{M_c Q_c R_c}] = \lambda\pi_{\mathbf{c}}(n-1;N),$$

which implies that [compare with Equations (6.2.6b) and (6.1.8)]

$$\mathbf{U_c}(n) = \lambda[\mathbf{B_c} + \lambda\mathbf{I_c} - \mathbf{U_c}(n-1)\mathbf{M_c Q_c R_c}]^{-1} \quad \text{for } n \ge 1. \tag{6.2.6c}$$

Next, for $n = C$, we have

$$\pi_{\mathbf{c}}(C;N)[\mathbf{B_c} + \lambda\mathbf{I_c} - \mathbf{U_c}(N-C-1)\mathbf{M_c Q_c R_c}] = \lambda\pi_{\mathbf{c-1}}(C-1;N)\mathbf{R_c},$$

or

$$\pi_{\mathbf{c}}(C;N) = \pi_{\mathbf{c-1}}(C-1;N)\mathbf{R_c U_c}(N-C).$$

Note that the $\mathbf{U_c}$'s do not depend on N, in the following sense. Suppose that we were interested in a system with $C = 3$ and $N = 6$. Then to come down the ladder from $n = 6$, $[N]$, to $n = 2$, $[C-1]$, we would have to calculate the matrices, $\mathbf{U_c}(0)$, $\mathbf{U_c}(1)$, $\mathbf{U_c}(2)$, and $\mathbf{U_c}(3)$. If we then decided that we wanted to study $N = 7$, we would only have to calculate the additional matrix, $\mathbf{U_c}(4)$, since the others are the same. (However, if we wish to study $C = 4$, then *everything* is different.) Now we must consider a class of matrices that depend on N as well as n, in order to deal with the situation for $n < C-1$. This is a generalization of $\mathbf{U}_1(N)$ in Theorem 6.1.2. For instance, for $k = C-1$,

$$\pi_{\mathbf{c-1}}(C-1;N)[\mathbf{B_{c-1}} + \lambda\mathbf{I_{c-1}} - \mathbf{R_c U_c}(N-C)\mathbf{M_c Q_c}] = \lambda\pi_{\mathbf{c-2}}(C-2;N)\mathbf{R_{c-1}}.$$

The matrices we are about to define may not be the best selection for efficiency, but they provide a certain elegance that some day may prove to be useful. We define

$$\mathbf{U_c}(N \mid C) := \mathbf{U_c}(N-C), \tag{6.2.7a}$$

and

$$\mathbf{U_{c-1}}(N \mid C) := \lambda[\mathbf{B_{c-1}} + \lambda I_{C-1} - \mathbf{R_c U_c}(N \mid C)\mathbf{M_c Q_c}]^{-1},$$

which implies that

$$\pi_{\mathbf{c-1}}(C-1;N) = \pi_{\mathbf{c-2}}(C-2;N)\,\mathbf{R_{c-1} U_{c-1}}(N \mid C).$$

In general, we define the $D(k) \times D(k)$-dimensional matrices,

$$\mathbf{U_k}(N \mid C) := \lambda[\mathbf{B_k} + \lambda\mathbf{I_k} - \mathbf{R_{k+1} U_{k+1}}(N \mid C)\,\mathbf{M_{k+1} Q_{k+1}}]^{-1}. \tag{6.2.7b}$$

Then we can write

$$\pi_{\mathbf{k}}(k;N) = \pi_{\mathbf{k-1}}(k-1;N)\mathbf{R_k U_k}(N \mid C) \quad \text{for } 1 \le k < C. \tag{6.2.8}$$

Clearly, the matrices, $\{\mathbf{U_k}(N \mid C)\}$ are very different from the matrices, $\{\mathbf{U_c}(n)\}$. Take note of the subtle notational differences. This actually parallels much of Section 5.4.2.

Before collecting the foregoing formulas in a theorem, we wish to state and prove the following lemma concerning the matrices, $\mathbf{U_k}(N \mid C)$ and $\mathbf{U_c}(n)$, which is directly related to Lemma 5.4.1.

Lemma 6.2.1: Let $\mathbf{U_c}(n)$ be defined by (6.2.6a) and (6.2.6c); then

$$\mathbf{U_c}(n)\mathbf{B_c}\varepsilon'_{\mathbf{c}} = \lambda\varepsilon'_{\mathbf{c}} \quad \text{for} \quad n \geq 0. \tag{6.2.9a}$$

Furthermore, let $\mathbf{U_k}(N \mid C)$ be defined by Equations (6.2.7). Then

$$\mathbf{U_k}(N \mid C)\mathbf{B_k}\varepsilon'_{\mathbf{k}} = \lambda\varepsilon'_{\mathbf{k}} \quad \text{for} \quad 1 \leq k \leq C. \tag{6.2.9b}$$

■

Proof: First note, by postmultiplying (6.2.6a) with $\mathbf{B_c}\varepsilon'_{\mathbf{c}}$, that

$$\mathbf{U_c}(0)\mathbf{B_c}\varepsilon'_{\mathbf{c}} = \lambda\varepsilon'_{\mathbf{c}}.$$

Next, define

$$\mathbf{A_c}(n) := [\mathbf{U_c}(n)]^{-1},$$

and assume that (6.2.9a) is true for $n = 0, 1, \cdots, l-1$, then from (6.2.6c),

$$\lambda\mathbf{A_c}(l)\varepsilon'_{\mathbf{c}} = [\mathbf{B_c} + \lambda\mathbf{I_c} - \mathbf{U_c}(l-1)\mathbf{M_c}\mathbf{Q_c}\mathbf{R_c}]\varepsilon'_{\mathbf{c}}.$$

But from (6.2.4b) [for $k = C$], and by assumption,

$$\mathbf{U_c}(l-1)\mathbf{M_c}\mathbf{Q_c}\mathbf{R_c}\varepsilon'_{\mathbf{c}} = \varepsilon'_{\mathbf{c}}.$$

Thus we have

$$\lambda\mathbf{A_c}(l)\varepsilon'_{\mathbf{c}} = \mathbf{B_c}\varepsilon'_{\mathbf{c}},$$

from which (6.2.9a) follows, for $n = l$. Therefore, by induction, it is true for all n. Next observe that from (6.2.7a) and (6.2.9a), (6.2.9b) must be true for $k = C$. Now, for all relevant k, define

$$\mathbf{A_k}(N \mid C) := [\mathbf{U_k}(N \mid C)]^{-1},$$

assume that (6.2.9b) is true for $k = C, C-1, \cdots, l+1$, and note from (6.2.4b) and Definition 6.2.4 that

$$\mathbf{R_{k+1}}\mathbf{U_{k+1}}(N \mid C)\mathbf{M_{k+1}}\mathbf{Q_{k+1}}\varepsilon'_{\mathbf{k}}$$

$$= \mathbf{R_{k+1}}\mathbf{U_{k+1}}(N \mid C)\mathbf{B_{k+1}}\varepsilon'_{\mathbf{k+1}} = \lambda\mathbf{R_{k+1}}\varepsilon'_{\mathbf{k+1}} = \lambda\varepsilon'_{\mathbf{k+1}}$$

is true for all $k = C, C-1, \cdots, l$. Then (6.2.9b) must be true for $k = l$, and thus by induction, for all k. QED

Compare with Lemma 4.1.1. Also, we see that two new sets of isometric matrices have been created, since $[\lambda\mathbf{V_k}\mathbf{A_k}(N \mid C)]\varepsilon'_{\mathbf{k}} = \varepsilon'_{\mathbf{k}}$, and so on. We could have defined the $\mathbf{U}(N \mid C)$ matrices to include the $\mathbf{R_k}$'s, but the new objects would not be square matrices. By defining them the way we did, the $\mathbf{R_k}$'s and $\mathbf{Q_k}$'s remain as the only matrices that connect objects of different spaces.

At last we can write the solution vectors in terms of the single scalar, $r(0; N)$. We will state them in the form of a theorem.

Theorem 6.2.2: The steady-state probability vectors of closed M/ME/C//N loops are given below. First define the auxiliary vectors, starting with $\mathbf{x_0}(N \mid C) := \mathbf{p}$, and using (6.2.7b). For $k = 1, 2, \cdots, C$,

$$\mathbf{x_k}(N \mid C) := \mathbf{p}\,\mathbf{U_1}(N \mid C)\,\mathbf{R_2}\,\mathbf{U_2}(N \mid C) \cdots \mathbf{R_k}\,\mathbf{U_k}(N \mid C)$$

$$= \mathbf{x_{k-1}}(N \mid C)\,\mathbf{R_k}\,\mathbf{U_k}(N \mid C). \tag{6.2.10a}$$

Then starting with $\mathbf{x_c}(C; N) := \mathbf{x_c}(N \mid C)$, and using (6.2.6c), define for $n = C+1, C+2, \cdots, N$,

$$\mathbf{x_c}(n; N) := \mathbf{x_c}(N \mid C)\,\mathbf{U_c}(N-C-1)\,\mathbf{U_c}(N-C-2) \cdots \mathbf{U_c}(N-n)$$

$$= \mathbf{x_c}(n-1; N)\,\mathbf{U_c}(N-n). \tag{6.2.10b}$$

The steady-state probability vectors are given by

$$\pi_{\mathbf{k}}(k; N) = r(0; N)\,\mathbf{x_k}(N \mid C) \quad \text{for } 0 \le k \le C \tag{6.2.11a}$$

and

$$\pi_{\mathbf{c}}(n; N) = r(0; N)\,\mathbf{x_c}(n; N) \quad \text{for } C \le n \le N. \tag{6.2.11b}$$

The associated steady-state scalar probabilities are given by

$$r(n; N) = \pi_{\mathbf{k}}(n; N)\,\varepsilon'_{\mathbf{k}} \quad \text{for } 0 \le n \le N, \tag{6.2.11c}$$

where $k = n$ for $n < C$, and $k = C$ otherwise. $r(0; N)$ comes from the normalization requirement, therefore,

$$\frac{1}{r(0; N)} = \sum_{k=0}^{C-1} \mathbf{x_k}(N \mid C)\,\varepsilon'_{\mathbf{k}} + \sum_{n=C}^{N} \mathbf{x_c}(n; N)\,\varepsilon'_{\mathbf{c}}. \tag{6.2.11d}$$

Since $\mathbf{x_c}(C; N) := \mathbf{x_c}(N \mid C)$, this expression could have been placed in either sum term. The mean queue length (and from Little's theorem, the mean system time) can be calculated directly from its definition, $\bar{n} = \sum_{n=1}^{n=N} n\, r(n; N)$. ■

We realize that the contents of this theorem are very difficult to grasp, but we do want the reader to use them computationally at some time in the future. Therefore, let us pause for a moment to look at the equations for some specific values of C and N. Suppose that we let $N = 6$ and $C = 3$; then the solution vectors are given by the following set.

$$\pi_1(0; 6) = r(0; 6)\,\mathbf{p}$$

$$\pi_1(1; 6) = r(0; 6)\,\mathbf{p}\mathbf{U_1}(6 \mid 3)$$

$$\pi_2(2; 6) = r(0; 6)\,\mathbf{p}\mathbf{U_1}(6 \mid 3)\,\mathbf{R_2}\,\mathbf{U_2}(6 \mid 3)$$

$$\pi_3(3; 6) = r(0; 6)\,\mathbf{p}\mathbf{U_1}(6 \mid 3)\,\mathbf{R_2}\,\mathbf{U_2}(6 \mid 3)\,\mathbf{R_3}\,\mathbf{U_3}(3)$$

$$\pi_3(4; 6) = r(0; 6)\,\mathbf{p}\mathbf{U_1}(6 \mid 3)\,\mathbf{R_2}\,\mathbf{U_2}(6 \mid 3)\,\mathbf{R_3}\,\mathbf{U_3}(3)\,\mathbf{U_3}(2)$$

$$\pi_3(5; 6) = r(0; 6)\,\mathbf{p}\mathbf{U_1}(6 \mid 3)\,\mathbf{R_2}\,\mathbf{U_2}(6 \mid 3)\,\mathbf{R_3}\,\mathbf{U_3}(3)\,\mathbf{U_3}(2)\,\mathbf{U_3}(1)$$

$$\pi_3(6; 6) = r(0; 6)\,\mathbf{p}\mathbf{U_1}(6 \mid 3)\,\mathbf{R_2}\,\mathbf{U_2}(6 \mid 3)\,\mathbf{R_3}\,\mathbf{U_3}(3)\,\mathbf{U_3}(2)\,\mathbf{U_3}(1)\,\mathbf{U_3}(0).$$

We start with $\mathbf{U}_3(0) = \lambda\mathbf{V}_3$, and then calculate $\mathbf{U}_3(1)$, $\mathbf{U}_3(2)$, and $\mathbf{U}_3(3)$ recursively by (6.2.6c). Next, since $\mathbf{U}_3(6 \mid 3) = \mathbf{U}_3(3)$, we calculate $\mathbf{U}_2(6 \mid 3)$ and $\mathbf{U}_1(6 \mid 3)$ recursively from (6.2.7b). We can calculate $r(0; 6)$ by the procedure mentioned in the Theorem 6.2.2, or we can set up a $\mathbf{K}$ matrix, as we did in previous sections. For instance, look at

$$\begin{aligned}\mathbf{K}_3(6) := \mathbf{I}_3 + \mathbf{U}_3(3) + \mathbf{U}_3(3)\mathbf{U}_3(2) + \mathbf{U}_3(3)\mathbf{U}_3(2)\mathbf{U}_3(1) \\ + \mathbf{U}_3(3)\mathbf{U}_3(2)\mathbf{U}_3(1)\mathbf{U}_3(0) \\ = \mathbf{I}_3 + \mathbf{U}_3(3)\,[\mathbf{K}_3(5)],\end{aligned}$$

where the definition of $\mathbf{K}_3(5)$ should be clear. Next calculate the scalar

$$sum = 1 + \mathbf{p}\,\mathbf{U}_1(6 \mid 3)\boldsymbol{\varepsilon}'_1$$

and the vector (which is actually $\mathbf{x}_2(6 \mid 3)\mathbf{R}_3$),

$$\mathbf{x}_3 = \mathbf{p}\,\mathbf{U}_1(6 \mid 3)\,\mathbf{U}_2(6 \mid 3)\,\mathbf{R}_3,$$

then

$$\frac{1}{r(0;\,6)} = sum + \mathbf{x}_3\mathbf{K}_3(6)\boldsymbol{\varepsilon}'_3.$$

We hope that this has helped. If not, perhaps the reader should give this one more try.

Suppose it is desirable that the probabilities for the same system with $N = 7$ be calculated; then the new set of equations needed are

$$\boldsymbol{\pi}_1(0;\,7) = r(0;\,7)\,\mathbf{p}$$

$$\boldsymbol{\pi}_1(1;\,7) = r(0;\,7)\,\mathbf{p}\mathbf{U}_1(7 \mid 3)$$

$$\boldsymbol{\pi}_2(2;\,7) = r(0;\,7)\,\mathbf{p}\mathbf{U}_1(7 \mid 3)\,\mathbf{R}_2\mathbf{U}_2(7 \mid 3)$$

$$\boldsymbol{\pi}_3(3;\,7) = r(0;\,7)\,\mathbf{p}\mathbf{U}_1(7 \mid 3)\,\mathbf{R}_2\mathbf{U}_2(7 \mid 3)\,\mathbf{R}_3\mathbf{U}_3(4)$$

$$\boldsymbol{\pi}_3(4;\,7) = r(0;\,7)\,\mathbf{p}\mathbf{U}_1(7 \mid 3)\,\mathbf{R}_2\mathbf{U}_2(7 \mid 3)\,\mathbf{R}_3\mathbf{U}_3(4)\,\mathbf{U}_3(3)$$

$$\boldsymbol{\pi}_3(5;\,7) = r(0;\,7)\,\mathbf{p}\mathbf{U}_1(7 \mid 3)\,\mathbf{R}_2\mathbf{U}_2(7 \mid 3)\,\mathbf{R}_3\mathbf{U}_3(4)\,\mathbf{U}_3(3)\,\mathbf{U}_3(2)$$

$$\boldsymbol{\pi}_3(6;\,7) = r(0;\,7)\,\mathbf{p}\mathbf{U}_1(7 \mid 3)\,\mathbf{R}_2\mathbf{U}_2(7 \mid 3)\,\mathbf{R}_3\mathbf{U}_3(4)\,\mathbf{U}_3(3)\,\mathbf{U}_3(2)\,\mathbf{U}_3(1)$$

$$\boldsymbol{\pi}_3(7;\,7) = r(0;\,7)\,\mathbf{p}\mathbf{U}_1(7 \mid 3)\,\mathbf{R}_2\mathbf{U}_2(7 \mid 3)\,\mathbf{R}_3\mathbf{U}_3(4)\,\mathbf{U}_3(3)\,\mathbf{U}_3(2)\,\mathbf{U}_3(1)\,\mathbf{U}_3(0).$$

Comparing this to the previous set, we see that only $\mathbf{U}_3(4)$ must be calculated from (6.2.6c), but using this matrix, the matrices, $\mathbf{U}_1(7 \mid 3)$ and $\mathbf{U}_2(7 \mid 3)$ must be calculated recursively by (6.2.6c), starting from $\mathbf{U}_3(7 \mid 3) = \mathbf{U}_3(4)$. We must also calculate

$$sum = 1 + \mathbf{p}\mathbf{U}_1(7 \mid 3)\boldsymbol{\varepsilon}'_1,$$

$$\mathbf{x}_3 = \mathbf{p}\mathbf{U}_1(7 \mid 3)\mathbf{U}_2(7 \mid 3)\mathbf{R}_3,$$

and

$$\mathbf{K}_3(7) = \mathbf{I}_3 + \mathbf{U}_3(4)\,\mathbf{K}_3(6),$$

to get

$$\frac{1}{r(0;\,7)} = sum + \mathbf{x}_3 \mathbf{K}_3(7) \boldsymbol{\varepsilon}'_3 .$$

Note that $\pi_2(2;\,N)$ can be thought of as either the last of the set of vectors which do not use the $\mathbf{R}_3(n-C)$ matrices, or by multiplying it by $\mathbf{R}_3 \mathbf{I}_3$ (remember that $\mathbf{R}_3 \boldsymbol{\varepsilon}'_3 = \boldsymbol{\varepsilon}'_2$), the first among those that do. In defining $\mathbf{K}_3(N)$, we have put it in the latter class. That is where the $\mathbf{I}_3$ came from.

It might prove useful for future reference to write down the first few $\pi_\mathbf{k}$'s for arbitrary C. Let $N_c := N - C$. Then (with one equation number assigned to the whole collection)

$$\pi_\mathbf{1}(0;\,N) = r(0;\,N)\,\mathbf{p}$$

$$\pi_\mathbf{1}(1;\,N) = r(0;\,N)\,\mathbf{p}\mathbf{U}_\mathbf{1}(N \mid C)$$

$$\pi_\mathbf{2}(2;\,N) = r(0;\,N)\,\mathbf{p}\mathbf{U}_\mathbf{1}(N \mid C)\mathbf{R}_\mathbf{2}\mathbf{U}_\mathbf{2}(N \mid C)$$

$$\cdots \qquad \cdots$$

$$\pi_\mathbf{c}(C;\,N) = r(0;\,N)\,\mathbf{p}\mathbf{U}_\mathbf{1}(N \mid C)\mathbf{R}_\mathbf{2}\mathbf{U}_\mathbf{2}(N \mid C) \cdots \mathbf{R}_\mathbf{c}\mathbf{U}_\mathbf{c}(N_c) \qquad (6.2.12a)$$

$$\pi_\mathbf{c}(C+1;\,N) = r(0;\,N)\,\mathbf{p}\mathbf{U}_\mathbf{1}(N \mid C)\mathbf{R}_\mathbf{2}\mathbf{U}_\mathbf{2}(N \mid C) \cdots \mathbf{R}_\mathbf{c}\mathbf{U}_\mathbf{c}(N_c)\mathbf{U}_\mathbf{c}(N_c-1)$$

$$\pi_\mathbf{c}(C+2;\,N) = r(0;\,N)\,\mathbf{p}\mathbf{U}_\mathbf{1}(N \mid C)\mathbf{R}_\mathbf{2}\mathbf{U}_\mathbf{2}(N \mid C) \cdots \mathbf{R}_\mathbf{c}\mathbf{U}_\mathbf{c}(N_c)\mathbf{U}_\mathbf{c}(N_c-1)\mathbf{U}_\mathbf{c}(N_c-2)$$

$$\pi_\mathbf{c}(C+3;\,N) = r(0;\,N)\,\mathbf{p}\mathbf{U}_\mathbf{1}(N \mid C)\mathbf{R}_\mathbf{2}\mathbf{U}_\mathbf{2}(N \mid C) \cdots \mathbf{R}_\mathbf{c}\mathbf{U}_\mathbf{c}(N_c)\mathbf{U}_\mathbf{c}(N_c-1)\mathbf{U}_\mathbf{c}(N_c-2)\mathbf{U}_\mathbf{c}(N_c-3)$$

$$\cdots \qquad \cdots$$

The probability vector, when there are more than C customers at S_1, can be written as

$$\pi_\mathbf{c}(C+n;\,N) = \pi_\mathbf{c}(C;\,N)\,\mathbf{U}_\mathbf{c}(N_c-1)\,\mathbf{U}_\mathbf{c}(N_c-2) \cdots \mathbf{U}_\mathbf{c}(N_c-n). \qquad (6.2.12b)$$

In general, to calculate the characteristics of any system for *one more* value of N requires an inversion of one matrix of each of the dimensions $D(1) \times D(1)$, $D(2) \times D(2)$, $\cdots$, and $D(C) \times D(C)$. But since $D(C)$ is usually much larger than $D(k)$ $(k < C)$, only $\mathbf{U}_\mathbf{c}(N-C)$ is computationally significant. For most matrix inversion routines the number of instructional steps required is proportional to the cube of the dimensions of the matrix, which in our case is of order $[D(C)]^3$. Now, matrix multiplication is also of the order $[D(C)]^3$, and it would seem that we need $N-C$ of them (plus other multiplications of lower order). But (6.2.10b) tells us that we can perform our calculations by multiplying matrices on vectors, which is only of order $[D(C)]^2$. In summary, the total computational effort for evaluating an M/ME/C//N network for *all* customer populations from $N = C+1$ to some maximum, $N = N_{mx}$, is

$$O\Big((N_{mx} - C)\,[D(C)]^3\Big) + O\Big([N_{mx}]^2\,[D(C)]^2\Big).$$

The normalization matrix can be calculated recursively by

$$\mathbf{K_c}(C-1) := \mathbf{I_c}\,, \tag{6.2.13a}$$

and for $N = C, C+1, \cdots, N_{mx}$,

$$\mathbf{K_c}(N) = \mathbf{I_c} + \mathbf{U_c}(N-C)\,\mathbf{K_c}(N-1). \tag{6.2.13b}$$

With this matrix, (6.2.11d) can be rewritten in two forms, depending on whether the term corresponding to $n = C-1$ is included in the first term or the second.

$$\frac{1}{r(0;\,N)} = \sum_{k=0}^{C-1} \mathbf{x_k}(N \mid C)\boldsymbol{\varepsilon}'_{\mathbf{k}} + \mathbf{x_c}(C;\,N)\mathbf{K_c}(N-1)\boldsymbol{\varepsilon}'_{\mathbf{c}}, \tag{6.2.13c}$$

or

$$\frac{1}{r(0;\,N)} = \sum_{k=0}^{C-2} \mathbf{x_k}(N \mid C)\boldsymbol{\varepsilon}'_{\mathbf{k}} + \mathbf{x_{c-1}}(N \mid C)\mathbf{R_c}\mathbf{K_c}(N)\boldsymbol{\varepsilon}'_{\mathbf{c}}. \tag{6.2.13d}$$

It is yet to be seen which one is ultimately better for algorithmic development. However, only (6.2.13c) reduces directly to the M/ME/1//N loop for $C = 1$.

Clearly [Equation (6.2.1)],

$$D(C) = \binom{m+C-1}{C} = \frac{(m+C-1)!}{C!\,(m-1)!}$$

is the critical number that determines whether the calculation of a given system is feasible. The inversion of matrices with dimension 300 is a small effort for today's medium-sized computers, even for supermicros, particularly those with array processors or parallel multiprocessors. However, a 2000-dimensional matrix, although manageable, would be somewhat of a challenge, partly since such a matrix would require 16 megabytes of storage in main memory (remember, N_C matrices are needed, and paging them in and out could be disastrously slow).

Simple manipulation of the binomial coefficients shows that if either m or C is small, the other can be quite large without exceeding these bounds. A subsystem containing four identical servers ($C = 4$) can be solved with relative ease if each server has ≤ eight phases ($m = 8$), even with as many as 100 customers ($N_{mx} = 100$), for then $D = 330$. However, a system with $m = 10$ and $C = 5$ ($D = 2002$) would require over 200 times as much computer time. Increasing m by only one to 11 would increase D another 50% to 3003, and over another factor of 3 in computation time. The significance of this is that a small increase in C or m causes a great increase in the time (and space) required to do a calculation. Every year new computers come on the market that are bigger, faster, and cheaper than the previous year's, yet each can boast only a small increase in the soluble problem space. The skeptical reader should try some larger values for C and m, to see how easy it would be to saturate all the computers in existence. The numerical (as opposed to analytical) study of much larger systems must wait until a way is found to decompose the various matrices into smaller parts.

6.2.2. Generalized M/ME/*C*//*N* System

We mentioned in the preceding section that an alternative definition of our state spaces was available, and in fact more useful in the long run. The one we gave, however, was more concise and simpler to start with. Observe that if we have $k \le C$ customers in S_1, the C servers are identical, and the k customers are indistinguishable, we only have to know how many customers are at each phase. Consider the following set.

Definition 6.2.7

$$\Xi_k := \left\{ i = \langle\, \alpha_1, \alpha_2, \cdots, \alpha_m \,\rangle \;\middle|\; 0 \le \alpha_l \le k, \text{ and } \sum_{l=1}^{m} \alpha_l = k \right\} \quad \text{for} \quad 1 \le k \le C.$$

Ξ_k is the set of all internal states of S_1 when there are k active customers there. Each ordered m-tuple represents a state in which $k \le C$ customers are active in a subsystem, S_1, which has C identical servers in it. There are α_1 customers at phase 1, α_2 customers at phase 2, and so on, each in a different server. They never get in each other's way, since there are at least as many servers as there are customers. If there are more than C customers at S_1, the excess numbers must queue up outside. ◊

Our claim is that this definition is equivalent to Definition 6.2.1, in that there is a one-to-one mapping of the states in the two sets onto each other. We show this most easily by the following example. Suppose that $m = 5$ and that $k = 4$; then a typical state using Definition 6.2.1 would be

$$\{\, 2, 2, 4, 5 \,\} \quad (i_1 = i_2 = 2,\; i_3 = 4, \text{ and } i_4 = 5).$$

This means that one of the customers is at phase 2 of one of the servers, another customer is also at phase 2, but in another server, a third customer is at phase 4 in yet another server, and the fourth customer in at phase 5. Therefore, there are no customers at phase 1 ($\alpha_1 = 0$) in any of the servers, there are two customers at phase 2 in two of the servers ($\alpha_2 = 2$), one at phase 4 ($\alpha_4 = 1$), and one at phase 5 ($\alpha_5 = 1$). That is, the following two ordered sequences give us the same information and are therefore equivalent:

$$\{\, 2, 2, 4, 5 \,\} \equiv \langle 0, 2, 0, 1, 1 \rangle.$$

Definitions 6.2.2 to 6.2.6 are all the same, but the various matrix elements can be computed differently. For instance, Definition 6.2.2 can be changed to read: Let $i = \langle\, \alpha_1, \alpha_2, \cdots, \alpha_m \,\rangle \in \Xi_k$, then

$$[\mathbf{M_k}]_{ii} = \alpha_1\mu_1 + \alpha_2\mu_2 + \cdots + \alpha_m\mu_m = \sum_{\nu=1}^{m} \alpha_\nu\mu_\nu .$$

Note that each of the objects, $i = \langle\, \alpha_1, \alpha_2, \cdots, \alpha_m \,\rangle$, can be thought of as a vector with m components [not to be confused with our row or column vectors of dimension $D(k)$]. Thus subtraction of any two vectors, even from different spaces, is well defined, since they all have the same number of components. But to keep our notation clear, we will write the following instead of $(i-j)$. Suppose that we have $i \in \Xi_{k_1}$ with components $\{\alpha_l\}$, and $j \in \Xi_{k_2}$ with components $\{\beta_l\}$; then we write

$$[\langle i \rangle - \langle j \rangle]_l := \nu_l := \alpha_l - \beta_l ,$$

where the following sums are true:

$$\sum_{l=1}^{m} \nu_l = \sum_{l=1}^{m} \alpha_l - \sum_{l=1}^{m} \beta_l = k_1 - k_2 .$$

We do not want to get too elaborate with our notation, but we need some definiteness to calculate the other matrix elements.

Look at Definition 6.2.4. $[\mathbf{R_k}]_{ij}$ is zero unless all but one of the components of $[\langle j \rangle - \langle i \rangle]$ is zero, in which case the nonzero element would have the value 1. Let ν be the component that is not 0; then $[\mathbf{R_k}]_{ij}$ is given by (6.2.2a), the same as before.

Next look at Definition 6.2.5. $[\mathbf{Q_k}]_{ij}$ is zero unless all but one of the components of $[\langle i \rangle - \langle j \rangle]$ is zero, in which case the nonzero element would have the value 1. Let ν be the component that is not 0; then $[\mathbf{Q_k}]_{ij}$ is given by (6.2.2b), where α_ν is the νth component of $\langle i \rangle$. This is exactly the same as before.

Finally, look at Definition 6.2.6. As with the others, $[\mathbf{P_k}]_{ij} := 0$, unless $[\langle i \rangle - \langle j \rangle]$ has exactly two nonzero elements, one with the value 1 and the other with the value -1. This is nothing more than stating that only one customer can move at a time, and he can only go to one new phase. Let γ be the member of $[\langle i \rangle - \langle j \rangle]$, which is 1 (that is the phase the customer left), and let ν be the member of $[\langle i \rangle - \langle j \rangle]$ which is -1 (that is the phase to which he went). Also, let α_γ be the γth component of $\langle i \rangle$, then just as we did for the two previous matrices, $[\mathbf{P_k}]_{ij}$ is given by (6.2.2c). Once again this is identical to what we had before, including the meaning of α_ν. We can also include the possibility that $[\mathbf{P}]_{ii} \neq 0$. Let the discussion above be true for $\langle i \rangle \neq \langle j \rangle$. Then for $[\langle i \rangle = \langle j \rangle]$, we have

$$[\mathbf{P_k}]_{ii} = \frac{1}{[\mathbf{M_k}]_{ii}} \sum_{\gamma=1}^{m} [\mathbf{P_1}]_{\gamma\gamma} \alpha_\gamma \mu_\gamma .$$

Note that if all the diagonal elements of $\mathbf{P_1}$ are zero, so are all the diagonal elements of $\mathbf{P_k}$.

What, you may ask, have we gained by this notational change? Well, first of all, it is easier to program. Second, we see that in all expressions for the components of matrix elements, α_ν and μ_ν always appear together as a product. Now let us define load-dependent completion rates for each of the *phases* in a server.

Definition 6.2.8

$\mu_\nu(l)$ = *probability rate that one of the customers at phase* ν *will complete, given that there are* l *customers at that phase.* Note that there is no distinction between having k identical servers, and only one server whose phases are load dependent. ◊

How interesting. We can either think of S_1 as a subsystem with C identical servers, each with m phases, or as one server with m phases, where each phase has a completion rate that depends on the number of customers in S_1 who are at that phase.

Now comes the generalization. Why must $\mu_\nu(l)$ be equal to $l\,\mu_\nu$, where ν is one of the m phases? If we want to study an M/ME/C//N loop, it must, but if we let $\mu_\nu(l)$ be anything greater than 0, the equations we have derived remain unchanged! Given this new freedom, what have we got? This is described by the following. The word *loop* has such a limited connotation that we are changing to the word *system*, which sounds much broader in scope. Also, the distinction between *server* and *phase* has become confused, so we will now use the word *server* or *stage*, and drop *phase*, in order to conform to the terminology associated with Jackson networks. The reader is thus entitled to think of the internal components of S_1 as real things.

Definition 6.2.9

Generalized M/ME/C//N *system* := *a two subsystem loop in which* S_2 *is an exponential server (perhaps load dependent), and* S_1 *is a network of load-dependent exponential servers satisfying the following rules. No more than* C *customers can be active inside* S_1 *at a time. If there are more than* C *customers at* S_1, *the excess numbers queue up outside. If there are fewer than* C *customers present, an arriving customer enters immediately. When a customer leaves, a new one, if available enters.* A customer upon entering S_1 goes directly to server ν with probability $[\mathbf{p}]_\nu$. The probability rate of leaving a server is $\mu_\nu(l)$, where l is the number of customers at server ν. If a completion occurs at ν, then with probability $[\mathbf{P_1}]_{\nu\gamma}$ a customer goes to server γ, and with probability $[\mathbf{q}]_\nu$ leaves S_1. ◊

We summarize this with a theorem.

Theorem 6.2.3: The steady-state vectors for a generalized M/ME/C//N system, with $N > C$, are given by Theorem 6.2.2 with the matrices $\mathbf{M_k}$, $\mathbf{P_k}$, $\mathbf{R_k}$ and $\mathbf{Q_k}$ modified as follows. For $1 \le k \le C$, let

$$\langle i \rangle = \langle \alpha_1, \alpha_2, \cdots, \alpha_\nu, \cdots, \alpha_m \rangle \in \Xi_k,$$

$$\langle j \rangle = \langle \beta_1, \beta_2, \cdots, \beta_\nu, \cdots, \beta_m \rangle \in \Xi_k,$$

where the α's and β's are nonnegative integers whose sum is k.

$\mathbf{M_k}$ is a diagonal matrix with components

$$[\mathbf{M_k}]_{ii} = \sum_{\nu=1}^{m} \mu_\nu(\alpha_\nu). \tag{6.2.14a}$$

$[\mathbf{P_k}]_{ij}$, for $i, j \in \Xi_k$, is zero unless $i = j$ or $[\langle i \rangle - \langle j \rangle]$ has one 1 (at position γ) and one -1 (at position ν), the rest being 0. If this is satisfied, then

$$[\mathbf{P_k}]_{ij} = [\mathbf{P}_1]_{\gamma\nu} \frac{\mu_\gamma(\alpha_\gamma)}{[\mathbf{M_k}]_{ii}} \quad \text{for } i \neq j \tag{6.2.14b}$$

and

$$[\mathbf{P_k}]_{ii} = \frac{1}{[\mathbf{M_k}]_{ii}} \sum_{\gamma=1}^{m} [\mathbf{P_1}]_{\gamma\gamma} \mu_\gamma(\alpha_\gamma). \tag{6.2.14c}$$

$[\mathbf{R_k}]_{ij}$, for $i \in \Xi_{k-1}$, $j \in \Xi_k$, is zero unless $[\langle j \rangle - \langle i \rangle]$ has one 1 (at position ν) and the rest, 0's. If this is satisfied, then

$$[\mathbf{R_k}]_{ij} = p_\nu. \tag{6.2.14d}$$

$[\mathbf{Q_k}]_{ij}$, for $i \in \Xi_k$, $j \in \Xi_{k-1}$, is zero unless $[\langle i \rangle - \langle j \rangle]$ has one 1 (at position ν) and the rest, 0's. If this is satisfied, then

$$[\mathbf{Q_k}]_{ij} := \frac{\mu_\nu(\alpha_\nu) q_\nu}{[\mathbf{M_k}]_{ii}}. \tag{6.2.14e}$$

If $\mu_\nu(l) = l\,\mu_\nu$ for all $0 \le l \le C$ and and all $1 \le \nu \le m$, this reduces to an M/ME/C//N loop. ■

Observe that if one or more of the servers is load independent [i.e., if $\mu_\nu(l) = \mu_\nu$ for all l], queueing delays can actually occur *inside* S_1. The description just given, except for the queueing up outside S_1, is identical to that for Jackson networks. We will discuss that in the next section.

With the system described in this way, we can see how *processor sharing* queues fit in. Recall from the beginning of Section 5.4 that, using this discipline, some or all the customers at S_1 get equal access to a single server. Suppose there are k customers sharing the server, then each one must be tracked according to his progress in S_1. However, each customer can only get $(1/k)$th the resources. Now if no more than C customers are permitted to share at a time, then we have an M/G/C//N system with the following specifications. The matrices, $\mathbf{M_k}$, $\mathbf{P_k}$, $\mathbf{R_k}$, and $\mathbf{Q_k}$, $k = 1, 2, \cdots, C$, are given by Theorem 6.2.3, where $\mu_\nu(l) = l\,\mu_\nu$, (i.e., the system is *not* a generalized one). Equations (6.2.6c) and (6.2.7b) are modified by replacing $\mathbf{M_k}$ and $\mathbf{B_k}$, with $(1/k)\mathbf{M_k}$ and $(1/k)\mathbf{B_k}$, respectively. If S_1 is made up of, say, C_1 identical general servers, where $1 \le C_1 < C$, then (6.2.6c) and (6.2.7b) are only modified for $k > C_1$, in which case, use (C_1/k) instead of $(1/k)$.

For $\mathbf{U_c}(n)$, (6.2.6c) tells us that these substitutions are equivalent to replacing λ with $C\lambda/C_1$. The same cannot be said for $\mathbf{U_k}(N \mid C)$, since $\mathbf{B_k}$ and $\mathbf{M_{k+1}}$

appear together in (6.2.7b). All this leaves us with a little unsolved mystery. We know that the steady-state solution of an unconstrained network where the general servers use processor sharing is the same as a network with exponential servers. Therefore, if $C \geq N$, our steady-state formulas should collapse to the M/M/C_1//N loop. It would be nice if we could explicitly show that our matrices have this property.

We end this subject with the following summary statement. The *unrestricted* processor sharing queue, $[C \geq N]$, is simpler than the M/G/1 queue, but the *restricted* processor sharer $[C < N]$, is harder, at least for steady state conditions.

6.2.3. Relation to Jackson Networks

It cannot be emphasized too strongly that the generalized M/ME/C//N system can be applied to arbitrarily large networks, limited by their computational difficulty, containing the Jackson networks as a proper subset. In case you are not quite sure what Jackson networks are, you may consider the following theorem as their definition.

Theorem 6.2.4: The steady-state solution of a single-class Jackson network with $m+1$ load-dependent servers and C customers is the same as that for a generalized M/ME/C//C system $(N = C)$, where ME has an m-dimensional representation. The $(m+1)$st server is at S_2. Therefore, the steady-state solution vectors for both are given by

$$[\pi_{\mathbf{k}}(k;\, C)]_i = g(C) X_1(\alpha_1) X_2(\alpha_2) \cdots X_m(\alpha_m) \left(\frac{1}{\lambda} \right)^{C-k}, \qquad (6.2.15a)$$

where $\langle i \rangle = \langle \alpha_1, \alpha_2, \cdots, \alpha_\nu, \cdots, \alpha_m \rangle \in \Xi_k$, $X_\nu(0) := 1$, and

$$X_\nu(l) := \frac{1}{\mu_\nu(1)\,\mu_\nu(2) \cdots \mu_\nu(l)} = \frac{X_\nu(l-1)}{\mu_\nu(l)}. \qquad (6.2.15b)$$

$g(C)$ is a normalization constant, fixed to make the probabilities sum to 1. As written here, S_2 is load independent. That limitation is not necessary. ■

If ν is a load-independent server, then $X_\nu(l) = 1/\mu_\nu^l$, while if $\mu_\nu(l) = l\,\mu_\nu$, (6.2.15b) gives

$$X_\nu(l) = \frac{1}{l!} \left(\frac{1}{\mu_\nu} \right)^l.$$

This corresponds to a *delay* stage. We used objects similar to the X_ν's in discussing load-dependent servers in Section 2.1.4; however, the notation used there was somewhat different.

The product-form solution for Jackson networks (as given above) is already well known and simpler to set up than our matrix formulation. You can see now why we never bothered to look at algorithms for calculating the solutions in the earlier discussions. However, the product solution *is* **not** *valid* for systems for which $N > C$, that is, when there are *population constraints* on the number of customers who can be simultaneously active in S_1. In that case, our procedure *cannot* be avoided. There is a standard approximation that is used in modeling networks, but it is not known how accurate it is in general. We will give an example of this in Section 6.2.4.

As a last comment in this section, observe that it is the constraint on population activity that causes our problems to grow to *matrix* proportions. That, in turn, subtly depends on the dimensionality function, $D(k)$. Further discussion in this direction is outside the scope of this book, except to note that population constraints are special cases of *blocking*, which also lies outside this book.

6.2.4. Time-Sharing Systems with Population Constraints

The last generalization we can make to our loop without greatly increasing the mathematical complexity of our model was already alluded to in the preceding section. We can make S_2 into a load-dependent server. This slight change turns out to give a potentially powerful tool for studying the behavior of time-sharing systems, as well as other systems with population constraints. Furthermore, the computational complexity is not changed. First we will look at the changes that we must make to the formulas, and then we will look at an application.

We need only look at those formulas containing λ, which is now $\lambda(l)$, for $l = 1, 2, \cdots$. Therefore, the matrices $\mathbf{M_k}$, $\mathbf{P_k}$, $\mathbf{R_k}$, and $\mathbf{Q_k}$ are unchanged. Only the $\mathbf{U_k}(N \mid C)$'s and $\mathbf{U_c}(n)$'s must be modified. What we have to do is combine what we did in Section 5.4.1 with what we have here. The reader may go through the complete derivation alone, we only make some observations. Start with the balance equations (6.2.5) and replace each λ with $\lambda(N-n)$, where n is the first argument in $\pi_c(\boldsymbol{n}; N)$, and so on. Remember, n is the number of customers at S_1, while $\lambda(\cdot)$ depends on the number of customers at S_2. This leads to the following modified solutions [compare with (6.2.6a) and (6.2.6c)]:

$$\mathbf{U_c}(0) = \lambda(1)\,\mathbf{V_c}\,, \tag{6.2.16a}$$

$$\mathbf{U_c}(l) = \lambda(l+1)\,[\mathbf{B_c} + \lambda(l)\,\mathbf{I_c} - \mathbf{U_c}(l-1)\mathbf{M_c}\mathbf{Q_c}\mathbf{R_c}]^{-1} \quad \text{for } l \geq 1. \tag{6.2.16b}$$

Also, (6.2.7a) remains unchanged, and (6.2.7b) changes to

$$\mathbf{U_k}(N \mid C) :=$$

$$\lambda(N-k+1)\,[\mathbf{B_k} + \lambda(N-k)\,\mathbf{I_k} - \mathbf{R_{k+1}}\,\mathbf{U_{k+1}}(N \mid C)\,\mathbf{M_{k+1}}\,\mathbf{Q_{k+1}}]^{-1}. \tag{6.2.16c}$$

For $N \leq C$, Theorem 6.2.4 is changed by replacing the λ term in (6.2.15a) with $1/[\lambda(1)\lambda(2) \cdots \lambda(C-k)]$, the λ equivalent of (6.2.15b). That is it. Nothing else changes. A close look at (6.2.6b) shows that $\mathbf{U_c}(l)$ really depends on the number of customers at S_2 (remember, we started at the *top*), just as $\lambda(l)$ does, so that is all that we have to change. Aby Tehranipour [TEHR84], [TEHR90] was the first one to recognize this. Let us see what that allows us to do.

Consider a system with N customers. When a customer is at S_2, he spends some time thinking about what to do, and after a mean time of Z (exponentially distributed) joins the queue at S_1. After a mean time of $R(N)$ he leaves S_1 and returns to S_2, starting the process over again. Z is known as the *think time*, and $R(N)$ is called the *response time* for the process. The probability rate for him to leave S_2, given that he is there thinking, is $1/Z$. If there are l thinking customers at S_2, the probability rate for any one of them to leave is simply l/Z. In other words, S_2 is a load-dependent server with service rate

$$\lambda(l) = \frac{l}{Z}.$$

That is why a server with this kind of behavior is often called a *think stage* It also shows up as the description of failures in the *machine minding* problem. Here, any number of machines are running simultaneously and independently of each other, and the rate at which they break down is proportional to the number running. It is also referred to as a *delay* stage, since customers can pause somewhere (not counting their waiting in a queue), independently of each other.

The view we take here is that of computer users who sit at their terminals and think (no comments, please), or type, and every once in a while hit the *return* key, which sends their prepared *transactions* to an external computer network, which they share. It is assumed that they do nothing while they wait for the computer system's response. Drinking coffee or talking to a friend does not count as doing anything, nor does any activity, however productive, that is not related to system usage. This is then a *time-sharing* (*TS*) stage in a *time-sharing computer system.* Now, from Little's theorem, (1.1.2) we see that the mean number of customers who are *thinking* at any time [call it $\bar{l}(N)$] is related to the mean rate at which transactions are processed [call it the *throughput*, with the symbol $\Lambda(N)$] by

$$\bar{l}(N) = \Lambda(N)Z,$$

since Z is the mean time each customer spends at S_2 between transaction submittals. On the other hand, the mean number of transactions that are being processed (or waiting to be processed) at any time [call it $\bar{n}(N)$] is related to the same throughput by the following:

$$\bar{n}(N) = \Lambda(N)R(N),$$

since $R(N)$ is the mean time a transaction spends at S_2. Now since we are dealing with a closed system, we must have $\bar{n}(N) + \bar{l}(N) = N$. So if we add the two equations above together and solve for $R(N)$, we get the *fundamental formula for time-sharing systems*:

$$R(N) = \frac{N}{\Lambda(N)} - Z \qquad (6.2.17)$$

This equation is as general as Little's theorem and tells us some general things about TS systems. (Be careful, though. There are numerous counterexamples that show up just when you least expect.) For instance, when N becomes very large (i.e., when too many users try to access the same computer system simultaneously), we can expect S_1 to saturate, so the throughput reaches a limiting value, which we call

$$\Lambda := \lim_{N \to \infty} \Lambda(N).$$

Then we see, for large N,

$$R(N) \approx \frac{1}{\Lambda} N - Z.$$

In other words, $R(N)$ approaches a straight line whose slope is $1/\Lambda$ and whose y-intercept is $-Z$. At the other extreme, $R(1)$ is the amount of time it should take, on average, for a single transaction to be processed if there is no interference from other tasks. Without too much difficulty, a reasonably good performance modeler should be able to find a satisfactory value for $R(1)$, Λ, and Z. Then all one has to do to get a decent *understanding* of the performance of the particular time-sharing system is to draw a smooth curve that starts at the point $[1, R(1)]$ and asymptotically approaches the line $x/\Lambda - Z$. Figure 6.2.1 shows several possible ways to do this. Clearly, if we really know what those three parameters are, we know the ballpark we are playing in, but do we know the game we are playing? As you can see, the different curves can differ by a factor of 10 or more in the intermediate region. Since underutilized systems ($N = 1$) almost always perform well, and overloaded systems are usually quite unsatisfactory, what planners want to know is: *How many users can a system support in a satisfactory manner?* So the name of the game is finding the right middle.

As long as there are no constraints on the number of transactions that can be processed simultaneously, (i.e., when $N \le C$), Jackson networks can be used

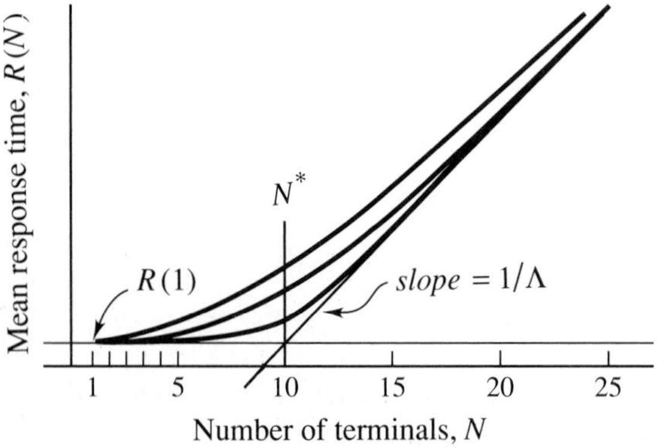

Figure 6.2.1: Response time curves for a family of time-sharing systems with the same value for minimal load $R(1)$, think-time Z, and asymptotic throughput Λ. All are bounded from below by the horizontal line, $y = R(1)$, and the asymptote, $y = x/\Lambda - Z$. The intersection of those two lines occurs at $N^* = \Lambda[Z + R(1)]$. N^* is often taken as the *number of customers that the TS system can support,* but the response times for the different systems vary enormously at that point.

quite effectively for performance modeling. However, it is well known that most systems will actually reduce their throughput if too many transactions are present, in a phenomenon known as *thrashing*. Briefly, if the amount of main memory (or cache memory) is insufficient to hold all active transactions simultaneously, then as each task is given its slice of time to use the central processor (CPU), it must first reclaim its memory space. The more jobs active, the more time is spent reclaiming main memory. To counter this, well-run computer systems will restrict the number of tasks, or transactions (our customers) who can be active simultaneously. That is, they impose a *population size constraint*, our parameter, C.

A common technique for dealing with constraints of this kind, called *decomposition*, or *aggregation*, or simply the *natural approximation*, is to *short-circuit* S_1 so that k customers return as soon as they leave. The rate at which they go around the loop is, then, $\Lambda(k)$, $k = 1, 2, \cdots, C$. Then S_1 is replaced by a load-dependent server with service rates as follows:

$$\mu(n) = \Lambda(n) \quad \text{for } n \le C$$

and

$$\mu(n) = \Lambda(C) \quad \text{for } n \ge C.$$

We have seen a simple version of this in Section 4.4.3. The technique is so compelling that many practitioners think it is exact, which it is for those systems where Jackson networks are exact. *But* it is *not* exact for systems with population constraints! (This is why the author became involved in LAQT in the first place.)

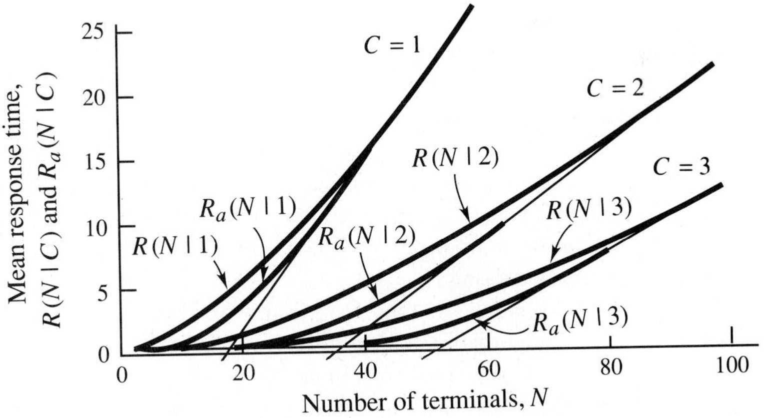

Figure 6.2.2: Response-time curves for a TS system, where the computer subsystem (S_1) is taken to be C identical servers with hyperexponential distributions. The mean service time for each is $R(1) = 0.6$ second, with coefficient of variation of $C^2 = 9.0$. The think time is $Z = 10.0$ seconds. Three different values for C (1, 2, and 3) were used. The curves marked $R(N \mid C)$ are the exact calculations, while the curves marked $R_a(N \mid C)$ come from the natural, decomposition, aggregation (whatever) approximation. For all C and for all N, the approximation lies below the correct value.

Example 6.2.1: We have calculated response times for an M/H_2/C//N loop, using the exact solution as given by Theorem 6.2.2 or Theorem 6.2.3, and its natural approximation. The calculations of $R(N)$ and $R_a(N)$ (a for approximation) versus N, for $C = 1, 2$, and 3, are given in Figure 6.2.2. As one would expect, the asymptotic slope decreases with increasing C, since $\Lambda(C)$ increases with C. Note that the natural approximation always gives the right asymptotic slope and correct value for $R(1)$ (i.e., it is in the right ballpark), but it can be off by more than a factor of 2 where results are most important, in the intermediate region (it is playing the wrong game). In this case it always yields an overly optimistic result [i.e., $R_a(N) < R(N)$]. We do not really know if this would hold true for all systems. ●

Although the decomposition method is used regularly, it is not known in general how good (or bad) an approximation it is, partly because most researchers have not been able to use the LAQT procedure appropriately. This is a pity, since LAQT can be used to explore the behavior of even more complicated queueing systems by, for instance, doing an aggregate approximation to S_2, while leaving S_1 as is. In this way one could study the interaction of two arbitrarily complicated subnetworks. Of course, we would not know how accurate that approximation is, but at the moment, we know almost nothing about such systems. An exact solution of such networks exists and falls under the more general name of *quasi birth–death* (QBD) processes. However, a full-blown calculation of that magnitude would require using matrices of the size, $D_1(C_1) \times D_2(C_2)$. One can see from our discussion on computational complexity at the end of Section 6.2.1 that it could easily become intractable.

6.3. OPEN GENERALIZED M/G/C QUEUE

The matrices $\mathbf{M_k}$, $\mathbf{P_k}$, $\mathbf{Q_k}$, $\mathbf{R_k}$, $\mathbf{B_k}$, and $\mathbf{V_k}$, for $k = 1, 2, \cdots, C$, which we tediously described and showed how to build from $\mathbf{M_1}$, $\mathbf{P_1}$, and $\mathbf{p}$ (and the load-dependence factors) in the preceding section, are the only building blocks we need for the rest of this chapter. If that already seems like too much, rest assured that we could not do it with less. We need that much information just to describe such complicated systems.

The procedure for *opening* our loop is the same as always. If the maximal service rate of S_1 is greater than λ, then as N becomes larger, the probability that S_2 will be idle goes to 0. But what *is* the maximal rate?, you ask. The answer will have to wait until the next section. For now we assume that the appropriate conditions are satisfied, in which case S_2 behaves as a Poisson source of customers to S_1. It would be expected that the limit as N goes to infinity of $\mathbf{U_c}(N)$ exists, and from (6.2.6c), satisfies the equation

$$\mathbf{U_c} := \lim_{N\to\infty} \mathbf{U_c}(N) = \lambda\left[\lambda\mathbf{I_c} + \mathbf{B_c} - \mathbf{U_c}\mathbf{M_c}\mathbf{Q_c}\mathbf{R_c}\right]^{-1}. \tag{6.3.1}$$

We ran across a formula like this in Equations (5.5.6). There is no known explicit expression for $\mathbf{U_c}$ except when $C = 1$. In that case we are dealing with the M/ME/1 queue, and we know that $\mathbf{U_1}$ is none other than our old friend $\mathbf{U}$ of

Equations (4.1.4). One way to find the numerical value for a matrix that satisfies the equation is to iterate on $\mathbf{U_c}$. That is, keep calculating $\mathbf{U_c}(C)$, $\mathbf{U_c}(C+1)$, $\mathbf{U_c}(C+2)$, and so on, until no changes are perceived. There are faster methods available if one is not interested in the sequence of finite systems [WALL69]. Anyway, suppose that $\mathbf{U_c}$ is known. Then

$$\mathbf{U_c}(\infty \mid C) := \lim_{N \to \infty} \mathbf{U_c}(N \mid C) = \mathbf{U_c},$$

and calculate $\mathbf{U_1}(\infty \mid C)$, $\mathbf{U_2}(\infty \mid C)$, $\cdots$, and $\mathbf{U_{c-1}}(\infty \mid C)$ using (6.2.7b). That is,

$$\begin{aligned}\mathbf{U_k}(\infty \mid C) &:= \lim_{N \to \infty} \mathbf{U_k}(N \mid C) \\ &= \lambda\left[\lambda \mathbf{I_k} + \mathbf{B_k} - \mathbf{R_{k+1}} \mathbf{U_{k+1}}(\infty \mid C) \mathbf{M_{k+1}} \mathbf{Q_{k+1}}\right]^{-1}. \end{aligned} \tag{6.3.2}$$

All the $\mathbf{U}$'s satisfy Lemma 6.2.1. Also, (6.2.10) of Theorem 6.2.2 remains valid with ∞ replacing N. Fortunately, (6.2.11) simplifies to

$$\mathbf{x_c}(C+n) := \lim_{N \to \infty} \mathbf{x_c}(n, N) = \mathbf{x_c}(C) \mathbf{U_c}^{n}, \tag{6.3.3a}$$

where

$$\mathbf{x_c}(C) := \mathbf{x_c}(\infty \mid C) \tag{6.3.3b}$$

Then the steady-state solution vectors are given by the following.

$$\pi_{\mathbf{k}}(k) = r(0) \mathbf{x_k}(\infty \mid C) \qquad \text{for } 0 \le k \le C \tag{6.3.4a}$$

and

$$\pi_{\mathbf{c}}(n) = r(0) \mathbf{x_c}(n) = r(0) \mathbf{x_c}(C) \mathbf{U_c}^{n-c} \qquad \text{for } C \le n. \tag{6.3.4b}$$

The associated scalar probabilities are, of course,

$$r(n) = \pi_{\mathbf{k}}(n) \varepsilon'_{\mathbf{k}}, \tag{6.3.4c}$$

where $k = n$ for $n \le C$ and $k = C$ otherwise. $r(0)$ comes from the normalization

$$\frac{1}{r(0)} = \sum_{k=0}^{C-1} \mathbf{x_k}(\infty \mid C) \varepsilon'_{\mathbf{k}} + \sum_{n=C}^{\infty} \mathbf{x_c}(n) \varepsilon'_{\mathbf{c}}.$$

We finally get a break. Since the $\mathbf{x_c}(n)$'s are matrix geometric in form, we can write a closed-form expression for the infinite sum, so

$$\frac{1}{r(0)} = \sum_{k=0}^{C-1} \mathbf{x_k}(\infty \mid C) \varepsilon'_{\mathbf{k}} + \mathbf{x_c}(C) [\mathbf{I_c} - \mathbf{U_c}]^{-1} \varepsilon'_{\mathbf{c}}. \tag{6.3.5}$$

The wonderful geometric property, which occurs so often in queueing systems is, almost assuredly, the major reason why researchers have studied open systems more than closed systems. They are easier.

6.4. TRANSIENT GENERALIZED M/ME/*C*-TYPE QUEUE

Our final topic for this chapter is, as usual, transient behavior. It is surprising, considering how complex generalized M/ME/C systems are, that what we did in previous chapters extends so easily here. It is also nice to know that everything we can do on this subject depends only on the matrices we have already created, *not* on their details. However, we will not go into it in such depth, leaving the untouched topics as exercises.

6.4.1. Queue Reduction at S_1 with No New Arrivals

First we must verify the physical meanings of our basic matrices. We were beginning to see in Chapter 5 that is was not always simple to decide what state a system is in initially. We shall have worse problems here, although sometimes we can come up with interesting answers. For instance, we know that if a system has been running a long time unobserved, we may presume that it is in its steady state. Suppose at some moment there are n customers at S_1, with k customers active. So, $k = C$ if $n \geq C$, and $k = n$ if $n \leq C$. We can assume that the system is, at that moment, in state

$$\mathbf{p_{ic}} = \frac{1}{r(n;N)} \boldsymbol{\pi}_\mathbf{c}(n;N) \quad \text{for } n \geq C, \tag{6.4.1a}$$

and

$$\mathbf{p_{ik}} = \frac{1}{r(k;N)} \boldsymbol{\pi}_\mathbf{k}(k;N) \quad \text{for } n = k \leq C. \tag{6.4.1b}$$

The subscript **ik** stands for: **k**–*space* vector for the **i***nitial* composite state. The $\boldsymbol{\pi}$'s come from Theorem 6.2.2 for the closed system and (6.3.4) for the open system. Obviously, $\mathbf{p_{ik}}\, \boldsymbol{\varepsilon}'_\mathbf{k} = 1$. See Section 4.5.2 for a discussion of what is meant by this vector.

Next define the family of $D(k)$-dimensional row vectors,

$$\mathbf{p_k} := \mathbf{p}\, \mathbf{R_2} \mathbf{R_3} \cdots \mathbf{R_k} \quad \text{for } 2 \leq k \leq C^{\dagger}. \tag{6.4.2a}$$

Because $\mathbf{R_k} \boldsymbol{\varepsilon}'_\mathbf{k} = \boldsymbol{\varepsilon}'_\mathbf{k-1}$, we see that $\mathbf{p_k}\, \boldsymbol{\varepsilon}'_\mathbf{k} = 1$ for all k. Suppose now that the system was initially idle and suddenly n customers showed up en masse. Or suppose that the customers were already there, but S_1 was inoperative and then suddenly started up. The initial vector in this case is given by

$$\mathbf{p_{ik}} = \mathbf{p_k} \quad \text{for } n = k \leq C, \tag{6.4.2b}$$

$$\mathbf{p_{ik}} = \mathbf{p_c} \quad \text{for } n \geq C, \tag{6.4.2c}$$

In the latter case there would be $n - C$ customers still waiting outside S_1. Physically, we see that the first customer enters and puts S_1 in composite state $\mathbf{p} \in \Xi_1$. Then the second customer enters and takes the subsystem from that state to $\mathbf{p}\,\mathbf{R_2}$, and so on. How simple.

Next, let us suppose that the subsystem is initially in state $i \in \Xi_k$. How long will it take before someone leaves? Assume that the entryway to S_1 is shut off so that no new customers can enter. Let us give the symbol a formal definition.

† We could have, if we liked, let $\mathbf{R_1} := \mathbf{p_1} := \mathbf{p}$.

Definition 6.4.1

$[\boldsymbol{\tau}'_{\mathbf{k}}]_i$:= *mean time until a customer leaves* S_1, *given that the subsystem was in state* $i \in \Xi_k$ $(k \le C)$ *and no new customers enter.* $\boldsymbol{\tau}'_{\mathbf{k}}$ is a $D(k)$-dimensional column vector. This describes a *collective* process, in that any one of the k customers could leave, and we do not know, or care, which. ◊

The $n = 1$ equivalent was discussed in Section 3.1.1, where we showed in (3.1.2b) that it was equal to $\mathbf{V}\boldsymbol{\varepsilon}'$. We give an extension of that derivation here. The vector equation is as follows. If there are k customers at S_1, then

$$\boldsymbol{\tau}'_{\mathbf{k}} = [\mathbf{M}_{\mathbf{k}}]^{-1}\boldsymbol{\varepsilon}'_{\mathbf{k}} + \mathbf{P}_{\mathbf{k}}\boldsymbol{\tau}'_{\mathbf{k}}.$$

In words, the mean time until someone leaves, $[\boldsymbol{\tau}'_{\mathbf{k}}]$, is equal to the sum of two terms; the time until something happens, $[1/(\mathbf{M}_{\mathbf{k}})_{ii} = (\mathbf{M}_{\mathbf{k}}^{-1}\boldsymbol{\varepsilon}'_{\mathbf{k}})_i]$, and if the event did not result in a departure, the system goes to another state, $[\mathbf{P}_{\mathbf{k}}]$, and a customer leaves from there. Notice the words we used: *the system goes to another state*. We could have said, instead, that *one of the k customers moves from one phase to another, thereby changing the state of* S_1. We solve for $\boldsymbol{\tau}'_{\mathbf{k}}$, and using (6.2.4a), with the understanding that if $n \ge C$, we are dealing with C-space objects (i.e., replace all subscripts k with C), we get

$$\boldsymbol{\tau}'_{\mathbf{k}} = \mathbf{V}_{\mathbf{k}}\,\boldsymbol{\varepsilon}'_{\mathbf{k}}. \tag{6.4.3}$$

Surprised? Of course not. In fact, we can even show that $\mathbf{B}_{\mathbf{k}}$ is the generating matrix for the distribution of the time until someone leaves.

We do have a problem, though. Whereas for a single customer, we had a natural candidate for the initial state, namely, $\mathbf{p}$, which led to the description of a pdf, now we have many candidates. For instance, if S_1 just opened up and C customers flowed in, we are dealing with (6.4.2) for $\mathbf{p}_{\mathbf{ic}}$, so the mean time until the first one leaves will be

$$\mathbf{p}\,\mathbf{R}_2 \cdots \mathbf{R}_{\mathbf{c}}\mathbf{V}_{\mathbf{c}}\boldsymbol{\varepsilon}'_{\mathbf{c}} = \mathbf{p}_{\mathbf{c}}\,\mathbf{V}_{\mathbf{c}}\,\boldsymbol{\varepsilon}'_{\mathbf{c}}.$$

Furthermore, the density function (pdf) of that time is given by

$$\mathbf{p}_{\mathbf{c}}\,[\,\mathbf{B}_{\mathbf{c}}\exp(-x\mathbf{B}_{\mathbf{c}})\,]\,\boldsymbol{\varepsilon}'_{\mathbf{c}}.$$

Other combinations are equally welcome. This does seem like the appropriate definition of interdeparture times, but when will the second customer leave? We have two possibilities, depending on whether the departed customer is replaced. We have not even begun to look at what happens if a customer enters S_1 *before* the first customer leaves. Let us postpone consideration of this last possibility, which falls into the category of first-passage times, and answer the intermediate question: *What state will the system be in immediately after the first customer leaves*? We define it as follows.

Definition 6.4.2

$[\mathbf{Y_k}]_{ij}$:= *probability that S_1 will be in internal state $j \in \Xi_{k-1}$ immediately after a departure, given that the system was initially in state $i \in \Xi_k$, and no other customers have entered.* $\mathbf{Y_k}$ is a $D(k) \times D(k-1)$-dimensional matrix, and is isometric, in that $\mathbf{Y_k}\,\varepsilon'_{\mathbf{k-1}} = \varepsilon'_{\mathbf{k}}$. By definition we are assuming that S_1 has exactly k active customers, thus $1 \le k \le C$. $\mathbf{Y}_1 = \varepsilon'$, since if $k = 1$ a departing customer leaves the empty state behind. ◊

An equation can be written down directly from the following argument. When an event occurs in S_1, either someone leaves, [$\mathbf{Q_k}$], or the internal state of the subsystem changes, [$\mathbf{P_k}$], and somebody eventually leaves, [$\mathbf{Y_k}$]. Mathematically,

$$\mathbf{Y_k} = \mathbf{Q_k} + \mathbf{P_k}\,\mathbf{Y_k},$$

and solving for $\mathbf{Y_k}$, we get [using (6.2.4a)]

$$\mathbf{Y_k} = [\mathbf{I_k} - \mathbf{P_k}]^{-1}\mathbf{Q_k} = [\mathbf{I_k} - \mathbf{P_k}]^{-1}[\mathbf{M_k}]^{-1}\mathbf{M_k}\mathbf{Q_k} = \mathbf{V_k}\,\mathbf{M_k}\,\mathbf{Q_k}. \tag{6.4.4}$$

[Compare with $\boldsymbol{\varepsilon}' = (\mathbf{I} - \mathbf{P})^{-1}\mathbf{q}'$, Equation (3.1.1b), for $k = 1$.] Both the first and third versions of $\mathbf{Y_k}$ may prove useful, and either (6.2.3b) or (6.2.4b) can be used to prove that $\mathbf{Y_k}$ is isometric.

Now we are ready to consider how long it will take for a second customer to leave S_1 after the first one left. Suppose first that no new customer enters; then that time is, for $k \le C$,

$$\mathbf{p_{ik}}\,\mathbf{Y_k}\,\mathbf{V_{k-1}}\,\varepsilon'_{\mathbf{k-1}} = \mathbf{p_{ik}}\,\mathbf{Y_k}\,\tau'_{\mathbf{k-1}}$$

where the initial vector depends on the initial conditions. The time between the second and third departures is

$$\mathbf{p_{ik}}\,\mathbf{Y_k}\,\mathbf{Y_{k-1}}V_{\mathbf{k-2}}\boldsymbol{\varepsilon}'_{\mathbf{k-2}} = \mathbf{p_{ik}}\,\mathbf{Y_k}\,\mathbf{Y_{k-1}}\tau'_{\mathbf{k-2}},$$

and so on. The successive multiplications of $\mathbf{Y_k}$'s occur often, so we provide them with their own symbol and definition.

Definition 6.4.3

$[\mathbf{Y_k}(l)]_{ij}$:= *probability that S_1 will be in internal state $j \in \Xi_l$ immediately after $k - l$ departures, given that the system was initially in state $i \in \Xi_k$, and no other customers have entered.* $\mathbf{Y_k}(l)$ is a $D(k) \times D(l)$-dimensional matrix and is isometric, in that $\mathbf{Y_k}(l)\varepsilon'_{\mathbf{l}} = \varepsilon'_{\mathbf{k}}$. We are using the term *isometric* to include any matrix (square or otherwise) or vector whose row sums are 1. By definition we are assuming that S_1 starts with exactly k active customers; thus $0 \le l < k \le C$. Also, $\mathbf{Y_k}(k-1) := \mathbf{Y_k}$. Keep alert to the fact that $\mathbf{Y_k}(\cdot)$ with an argument is *different* from $\mathbf{Y_k}$ without an argument. ◊

These matrices are easy enough to construct, since they satisfy the obvious recurrence relation, starting with $\mathbf{Y_k}(k-1) = \mathbf{Y_k}$:

$$\mathbf{Y_k}(l-1) = \mathbf{Y_k}(l)\,\mathbf{Y_l}, \tag{6.4.5a}$$

or explicitly,

$$\mathbf{Y_k}(l-1) = \mathbf{Y_k}\,\mathbf{Y_{k-1}} \cdots \mathbf{Y_{l+1}}\,\mathbf{Y_l}. \tag{6.4.5b}$$

The argument $(l-1)$ helps a little, for it tells us that the operator, $\mathbf{Y_k}(l-1)$, takes the system from k customers to $l-1$ customers.

If S_1 is an ME/C (not generalized) subsystem and $\mathbf{p_{ic}}$ is given by (6.4.2c), we are dealing with the *order statistics of C identically distributed, mutually independent random variables*. We will state this as a theorem after the following formal definition.

Definition 6.4.4

Let $X_1, X_2, \cdots, X_c$ be C identically distributed, mutually independent random variables, whose distribution functions are each generated by $<\mathbf{p}, \mathbf{B}>$. Let $Z_1 \le Z_2 \le \cdots \le Z_c$ be the size place reordering of the X's. Then the *random variable* Z_k is called the kth-order statistic. In terms of our ME/C (not generalized) subsystem, Z_1 is the time the first customer leaves S_1, leaving behind $C-1$ customers, Z_2 is the time the second one leaves, and so on. ◊

By constructing the matrices $\mathbf{M_k}$, $\mathbf{P_k}$, $\mathbf{Q_k}$, $\mathbf{R_k}$, $\mathbf{B_k}$, and $\mathbf{V_k}$, for $k = 1, 2, \cdots, C$ according to Equations (6.2.2) [i.e., $\mu_\gamma(l) = l\,\mu_\gamma(1)$ for all l and all $1 \le \gamma \le m$] we have made the X_k's mutually independent and identically distributed. Also, by selecting $\mathbf{p_{ic}}$ to be equal to $\mathbf{p_c} = \mathbf{p}\mathbf{R}_2\mathbf{R}_3 \cdots \mathbf{R_c}$ we have started service for all C customers at the same time. We now state the theorem on order statistics.

Theorem 6.4.1: Let Z_k, $k = 1, 2, \cdots, C$, be given according to Definition 6.4.4. Let S_1 be constructed as an ME/C subsystem. Then

$$E(Z_1) = \mathbf{p_c}\,\mathbf{V_c}\,\varepsilon'_{\mathbf{c}},$$

$$E(Z_2 - Z_1) = \mathbf{p_c}\,\mathbf{Y_c}\,\mathbf{V_{c-1}}\,\varepsilon'_{\mathbf{c-1}},$$

and in general,

$$E(Z_{c-k+1} - Z_{c-k}) = \mathbf{p_c}\,\mathbf{Y_c}\,\mathbf{Y_{c-1}} \cdots \mathbf{Y_{k+1}}\,\mathbf{V_k}\,\varepsilon'_{\mathbf{k}} = \mathbf{Y_c}(k)\,\mathbf{V_k}\,\varepsilon'_{\mathbf{k}}. \tag{6.4.6a}$$

We can actually say how these variables are distributed.

$$\Pr(Z_1 \ge x) = \mathbf{p_c}\exp(-x\mathbf{B_c})\,\varepsilon'_{\mathbf{c}},$$

$$\Pr(Z_2 - Z_1 \ge x) = \mathbf{p_c}\,\mathbf{Y_c}\exp(-x\mathbf{B_{c-1}})\,\varepsilon'_{\mathbf{c-1}},$$

and in general,

$$\Pr(Z_{c-k+1} - Z_{c-k} \ge x) = \mathbf{p_c}\,\mathbf{Y_c}(k)\exp(-x\mathbf{B_k})\,\varepsilon'_{\mathbf{k}}. \tag{6.4.6b}$$

Note that it is the *differences* of the Z's that are described here. The distributions of the Z's themselves are given by the convolutions of these functions. ■

This may be getting a bit obscure and abstract, so let us interject the simplest of examples. Suppose that ME is exponential (i.e., $m = 1$). Then $D(k) = 1$ for all k, $\mathbf{M_k} = \mathbf{B_k} \Rightarrow k\,\mu$, and just about everything else becomes 1. Then

$$E(Z_1) = \frac{1}{C\,\mu},$$

and in general, we get the well-known formula for the order statistics of exponential servers, namely [we can think of $E(Z_0)$ as being 0],

$$E(Z_k - Z_{k-1}) = \frac{1}{(C-k+1)\,\mu}.$$

Remember, this is the mean time *between* departures; the mean time for departures themselves is the partial sum of the interdeparture times:

$$E(Z_k) = E(Z_{k-1}) + \frac{1}{(C-k+1)\,\mu} = \frac{1}{\mu} \sum_{l=c-k+1}^{c} \frac{1}{l}.$$

In particular, the time for the last customer to leave is

$$E(Z_c) = \frac{1}{\mu} H_c := \frac{1}{\mu} \sum_{l=1}^{c} \frac{1}{l},$$

where H_c is known as the *harmonic series*. Remember, these last formulas are valid only for the M/M/C queue.

Let us return to our generalized subsystem, and suppose that $n > C$. Then when a customer leaves, another immediately takes his place, putting the system in the state

$$\mathbf{Y_c\,R_c} = \mathbf{V_c\,M_c\,Q_c\,R_c}\,.$$

This object is a singular, isometric, square matrix of dimension $D(C) \times D(C)$. We first ran across the $C = 1$ version of this in Chapter 3, namely, $\mathbf{Y}_1 \mathbf{R}_1 = \boldsymbol{\varepsilon}'\mathbf{p} = \mathbf{Q}$ (as always, not to be confused with $\mathbf{Q_1} = \mathbf{q}'$, or Q). Let us summarize this with a theorem about the time for a queue to *drain*.

Theorem 6.4.2: Consider a generalized subsystem, S_1, in which a maximum of C customers can be active simultaneously. Suppose that there are n customers at S_1, with no new arrivals possible, and at the moment the process begins, the subsystem is in state $\mathbf{p_{ik}}$. The process ends when all customers are gone. Then T_n, the *mean time for the queue to drain*, is given by the following:

$$T_n = \mathbf{p_{ik}\,V_k\,\varepsilon'_k} + \mathbf{p_{ik}\,Y_k\,V_{k-1}\,\varepsilon'_{k-1}} + \mathbf{p_{ik}\,Y_k}(k-2)\,\mathbf{V_{k-2}\,\varepsilon'_{k-2}} + \cdots + \mathbf{p_{ik}\,Y_k}(1)\,\mathbf{V}_1\,\boldsymbol{\varepsilon}'_1 \quad \text{for } n = k \le C, \tag{6.4.7a}$$

$$T_{c+1} = \mathbf{p_{ic}\,\tau'_c} + \mathbf{p_{ic}\,Y_c\,R_c\,\tau'_c} + \mathbf{p_{ic}\,Y_c\,R_c}\left[\mathbf{Y_c\,\tau'_{c-1}} + \mathbf{Y_c}(c-2)\,\boldsymbol{\tau}'_{\mathbf{c-2}} + \cdots + \mathbf{Y_c}(1)\,\boldsymbol{\tau}'_1\right]. \tag{6.4.7b}$$

In general, for $n = l + C, l > 0$,

$$T_n = \mathbf{p_{ic}} \left[\sum_{j=0}^{l} (\mathbf{Y_c}\,\mathbf{R_c})^j \right] \boldsymbol{\tau}'_\mathbf{c}$$

$$+ \mathbf{p_{ic}} (\mathbf{Y_c}\,\mathbf{R_c})^l \mathbf{Y_c} \left[\mathbf{Y_c} \boldsymbol{\tau}'_{\mathbf{c}-1} + \mathbf{Y_c}(c-2) \boldsymbol{\tau}'_{\mathbf{c}-2} + \cdots + \mathbf{Y_c}(1) \boldsymbol{\tau}'_\mathbf{1} \right]. \quad (6.4.7c)$$

The separate terms are the mean times for each successive customer to leave after the previous one has left. The τ's are given by (6.4.3) and the **Y**'s are given by (6.4.4) and (6.4.5). ■

It is left to the reader to devise a simple recursive algorithm for evaluating Equations (6.4.7). Perhaps we should think of Theorem 6.4.1 as a corollary to Theorem 6.4.2.

Some examples where *time to drain* can be important are the following.

1. A multiprogramming computer system has been in operation all day, and everyone except the operator has gone home. The operator cannot go home until all jobs are done, including those in the waiting queue. C is the maximum degree of multiprogramming, and n is the number of jobs in the system. Then $\mathbf{p_{ik}}$ is given by (6.4.1), and T_n is the mean time until the operator can go home.

2. A multiprogramming computer system has been in operation for a long time, and the operating systems people must bring it down for some reason or other. They can shut off the queue of waiting jobs but must let those in progress continue until they finish. C is the maximum degree of multiprogramming, $\mathbf{p_{ik}}$ is given by (6.4.1), and n is the number of jobs in the system when the queue is turned off. If $n = k \le C$, then T_k is the mean time until they can bring down the system. If $n > C$, then T_c is the mean time, but use (6.4.1a) with n (not C) for $\mathbf{p_{ic}}$.

3. We have $n \ge C$ identical devices, of which we would like C to be running simultaneously (hot backup), but we can survive even if all but one are broken. $\mathbf{p_{ic}}$ is given by $\mathbf{p_c}$ (6.4.2a), and T_n is the mean time until all are broken (MTTF, without repair). There are initially $n - C$ devices in *cold backup* and $C - 1$ in *hot backup*. We can generalize; failure can be defined as occurring when the number still at S_1 drops below a certain value.

4. You are driving cross-country and are in a hurry. Your car has five brand-new tires. You will have time to change but not to fix, a flat if it occurs. Equation (6.4.2a) for $C = 4$ (unless you are driving a trailer truck) is the initial state, and failure occurs when you are down to three tires (hold the steering wheel steady when this occurs). T is the sum of the first two terms in (6.4.7b).

5. Same as example 4, but now you are driving a rented a car, so the four mounted tires have already been used to an uncertain amount, but the spare is new. What is $\mathbf{p}_{ik}$ now?

Presumably the reader can think up a few more examples.

6.4.2. Semi-Markov Arrival (or Departure) Processes

In the second paragraph of Section 6.3 we stated that the maximal service rate of S_1 must be greater than λ for the steady-state M/ME/C queue to exist, without determining what the maximal rate *is*. We can, and will, do that now. We will also describe the departure process from S_1 (which is, of course, the same as the arrival process to S_2), when its queue is unboundedly large. This is the direct generalization of the renewal processes described in Section 3.3, but the name given to this process does not contain the word *renewal* in it, and instead goes by the name given in the title of this section. It is also known as a *semi-Markov point process.* We are not particularly interested in where names come from, but we give them so that the reader can have a reference point for reading the general literature. Based on our (and everyone else's) definition, this really is not a renewal process, and we hope that by the end of this section, you will see why.

In the preceding section we showed how to calculate the mean time for a customer to leave S_1, given some initial state. We also showed how to calculate the time for the second, third, and nth customers to leave. All these times are different even if the queue is long enough to guarantee that there will always be more than C customers at S_1. Fortunately, this sequence approaches a limit. That is, let t_n be the mean value of the time interval between the departures of the nth and the $(n-1)$st customers for any initial state vector, $\mathbf{p}_{ic}$. Then

$$t_n = \mathbf{p}_{ic}\,(\mathbf{Y_c}\,\mathbf{R_c})^n\,\boldsymbol{\varepsilon}'_{\mathbf{c}}, \tag{6.4.8a}$$

for all n, if the queue is unboundedly large. Let us assume that

$$\bar{t} := \lim_{n\to\infty} t_n \qquad \text{exists.} \tag{6.4.8b}$$

For the limit to exist, the matrix, $\mathbf{Y_c}\,\mathbf{R_c} = \mathbf{V_c}\,\mathbf{M_c}\,\mathbf{Q_c}\,\mathbf{R_c}$, must satisfy certain properties, some of which we already know to be true. We know, for instance, that this matrix is isometric and not invertible. Thus it has one eigenvalue equal to 1 and multiple occurrences of 0 as an eigenvalue. We know the latter because the matrix $\mathbf{Q_c}\,\mathbf{R_c}$ is of dimension $D(C)$, but since $\mathbf{Q_c}$ and $\mathbf{R_c}$ are not square matrices, $\mathbf{Q_c}\,\mathbf{R_c}$ is at most of *rank* $D(C-1)$. We will assume that all the other eigenvalues are less than 1 in magnitude. Recall from the definition of *isometric* that $\boldsymbol{\varepsilon}'_{\mathbf{c}}$ is a right eigenvector. Let's take a look. From (6.2.4b)

$$\mathbf{Y_c R_c}\,\boldsymbol{\varepsilon}'_{\mathbf{c}} = \mathbf{V_c M_c Q_c R_c}\,\boldsymbol{\varepsilon}'_{\mathbf{c}} = \mathbf{V_c B_c}\,\boldsymbol{\varepsilon}'_{\mathbf{c}} = \boldsymbol{\varepsilon}'_{\mathbf{c}}.$$

This, as we have seen several times, is an eigenvector equation, with eigenvalue 1. Next, let $\boldsymbol{\pi}_{\mathbf{c}}$ be the left eigenvector of $\mathbf{Y_c R_c}$ with eigenvalue 1, normalized so that $\boldsymbol{\pi}_{\mathbf{c}}\boldsymbol{\varepsilon}'_{\mathbf{c}} = 1$. Then we must also have

$$\boldsymbol{\pi}_{\mathbf{c}} \mathbf{Y}_{\mathbf{c}} \mathbf{R}_{\mathbf{c}} = \boldsymbol{\pi}_{\mathbf{c}}^{\dagger}. \tag{6.4.9a}$$

Now from the spectral decomposition theorem (1.3.8b), we can write

$$(\mathbf{Y}_{\mathbf{c}} \mathbf{R}_{\mathbf{c}})^n = \boldsymbol{\varepsilon}'_{\mathbf{c}} \boldsymbol{\pi}_{\mathbf{c}} + \sum_{i=2}^{D(C)} \lambda_i^n \mathbf{v}_{\mathbf{i}}' \mathbf{u}_{\mathbf{i}},$$

where the λ's, **u**'s and **v**'s are the eigenvalues and left and right eigenvectors of $\mathbf{Y}_{\mathbf{c}} \mathbf{R}_{\mathbf{c}}$, excluding 1, $\boldsymbol{\pi}_{\mathbf{c}}$, and $\boldsymbol{\varepsilon}'_{\mathbf{c}}$. Assuming that $|\lambda_i| < 1$ for $i \geq 2$, we can take the limit directly, to get

$$\lim_{n \to \infty} (\mathbf{Y}_{\mathbf{c}} \mathbf{R}_{\mathbf{c}})^n = \boldsymbol{\varepsilon}'_{\mathbf{c}} \boldsymbol{\pi}_{\mathbf{c}}. \tag{6.4.9b}$$

Then we have

$$\bar{t} = \mathbf{p}_{\mathbf{ic}} \boldsymbol{\varepsilon}'_{\mathbf{c}} \boldsymbol{\pi}_{\mathbf{c}} \mathbf{V}_{\mathbf{c}} \boldsymbol{\varepsilon}'_{\mathbf{c}} = \boldsymbol{\pi}_{\mathbf{c}} \mathbf{V}_{\mathbf{c}} \boldsymbol{\varepsilon}'_{\mathbf{c}}, \tag{6.4.10a}$$

since $\mathbf{p}_{\mathbf{ic}} \boldsymbol{\varepsilon}'_{\mathbf{c}} = 1$. Note that $\bar{t}$ is independent of the state the system was in initially, as all good Markov processes should be. This limit is the correct maximal, mean interdeparture time from S_1, so $1/\bar{t}$ is the maximal service rate of S_1. Therefore, if we define ρ_C as

$$\rho_C := \lambda \bar{t} = \lambda \left[\boldsymbol{\pi}_{\mathbf{c}} \mathbf{V}_{\mathbf{c}} \boldsymbol{\varepsilon}'_{\mathbf{c}} \right], \tag{6.4.10b}$$

then we can say that the steady-state M/ME/C queue exists as long as $\rho_C < 1$, thus finally completing our thoughts for Section 6.3. But we have not finished our thoughts for this section. If $\rho_C > 1$, then S_1 becomes a semi-Markov point process source for S_2. We are not prepared to deal fully with that possibility here, but we do have some items to mention before closing.

First we would like to give some meaning to the vector $\boldsymbol{\pi}_{\mathbf{c}}$. To do that, look at the isometric matrix [compare with (3.3.11a)]

$$\boldsymbol{P}_C := \mathbf{P}_{\mathbf{c}} + \mathbf{Q}_{\mathbf{c}} \mathbf{R}_{\mathbf{c}}. \tag{6.4.11a}$$

This matrix moderates the following process. There are C customers in S_1. The internal transitions are governed by $\mathbf{P}_{\mathbf{c}}$, but when a transition occurs that results in a customer leaving, [$\mathbf{Q}_{\mathbf{c}}$], that customer immediately returns, putting the system in a new state of Ξ_C, [$\mathbf{R}_{\mathbf{c}}$]. Therefore, $\boldsymbol{P}_C$ is the transition matrix describing a short-circuited S_1. The left eigenvector, $\mathbf{y}_{\mathbf{c}}$, defined by

$$\mathbf{y}_{\mathbf{c}} \boldsymbol{P}_C = \mathbf{y}_{\mathbf{c}} \tag{6.4.11b}$$

and $\mathbf{y}_{\mathbf{c}} \boldsymbol{\varepsilon}'_{\mathbf{c}} = 1$, is interpreted in the following way. $[\mathbf{y}_{\mathbf{c}}]_i$ is the probability that the short-circuited system will be found in state $i \in \Xi_C$ between events. Next rewrite (6.4.11a) and (6.4.11b) in the form

$$\mathbf{y}_{\mathbf{c}} \mathbf{Q}_{\mathbf{c}} \mathbf{R}_{\mathbf{c}} = \mathbf{y}_{\mathbf{c}} [\mathbf{I}_{\mathbf{c}} - \mathbf{P}_{\mathbf{c}}], \tag{6.4.11c}$$

and compare with (6.4.9a) using the first part of (6.4.4):

$$\boldsymbol{\pi}_{\mathbf{c}} = \boldsymbol{\pi}_{\mathbf{c}} [\mathbf{I}_{\mathbf{c}} - \mathbf{P}_{\mathbf{c}}]^{-1} \mathbf{Q}_{\mathbf{c}} \mathbf{R}_{\mathbf{c}}. \tag{6.4.11d}$$

† Be careful not to confuse $\boldsymbol{\pi}_{\mathbf{c}}$ with the vector $\boldsymbol{\pi}_{\mathbf{c}}(n)$ in the preceding section.

Comparing these two equations, we see that $\boldsymbol{\pi}_\mathbf{c}$ must be proportional to $\mathbf{y_c}[\mathbf{I_c} - \mathbf{P_c}]$, i.e.,

$$\boldsymbol{\pi}_\mathbf{c} \sim \mathbf{y_c}[\mathbf{I_c} - \mathbf{P_c}]$$

or normalizing so that $\mathbf{y_c}\boldsymbol{\varepsilon}'_\mathbf{c} = 1$,

$$\mathbf{y_c} = \frac{1}{\boldsymbol{\pi}_\mathbf{c}[\mathbf{I_c} - \mathbf{P_c}]^{-1}\boldsymbol{\varepsilon}'_\mathbf{c}}\boldsymbol{\pi}_\mathbf{c}[\mathbf{I_c} - \mathbf{P_c}]^{-1}.$$

Next consider the vector whose ith component is $x_i := y_i / [\mathbf{M_c}]_{ii}$. Recall that $1/[\mathbf{M_c}]_{ii}$ is the mean time the system spends in state i every time it finds itself there. Therefore, as a direct generalization of the *mean residual vector* defined in (3.3.10b),

$$\mathbf{x_c} \sim \mathbf{y_c}[\mathbf{M_c}]^{-1} \tag{6.4.12a}$$

must be proportional to the steady-state probability vector of short-circuited S_1. On the other hand, we have (substituting for $\mathbf{y_c}$)

$$\mathbf{x_c} \sim \boldsymbol{\pi}_\mathbf{c}[\mathbf{I_c} - \mathbf{P_c}]^{-1}[\mathbf{M_c}]^{-1} = \boldsymbol{\pi}_\mathbf{c}\mathbf{V_c}, \tag{6.4.12b}$$

where we set $\mathbf{x_c}\,\boldsymbol{\varepsilon}'_\mathbf{c} = \bar{t}$. Compare this with (3.3.10b) and (6.4.10a). The ith components of the three isometric vectors $\mathbf{y_c}$, $\mathbf{x_c}/\bar{t}$, and $\boldsymbol{\pi}_\mathbf{c}$ are, respectively, the steady-state probability of finding S_1 in state $i \in \Xi_C$ between events, the steady-state probability that a random observer will find S_1 in state i, and the steady-state probability that a leaving–reentering customer will put S_1 in state i. And then there is $\mathbf{p_c}$ of (6.4.2). Can anything be clearer?

We have seen that the vector-matrix pair $< \mathbf{p_{ic}}(\mathbf{Q_c}\mathbf{R_c})^n, \mathbf{B_c} >$ generates the interdeparture-time distribution for the nth customer to leave S_1 when there is an unbounded queue. We have also seen that when n is large enough, all the interdeparture distributions are the same and are generated by the pair $< \boldsymbol{\pi}_\mathbf{c}, \mathbf{B_c} >$. Can we not say, then, that this process approaches a renewal process asymptotically? The answer to this is *no*, but the explanation is very subtle. The manifest property that is missing is the following. In renewal processes, the distribution function of the interdeparture time for *two* customers is the convolution of two successive single-customer distributions. That is *not* true here.

The physical explanation is as follows. In a renewal process, when a customer leaves S_1, he leaves behind the empty state, no matter how long he was in service himself. Therefore, the next customer always starts the same way. For our semi-Markov process, the state the subsystem is in immediately after the nth departure is likely to be quite different if the nth interdeparture time is short than if the interdeparture time was long. ($\boldsymbol{\pi}_\mathbf{c}$ is the initial vector *averaged* over all possible times for *all* previous departures.) Thus the $(n+1)$st departure depends on the time of the nth departure, so a simple convolution is invalid.

The mathematical explanation has to do with the following. The generating operator for a renewal process (equivalent to $C = 1$) is our idempotent, isometric, singular friend $\mathbf{Q} = \boldsymbol{\varepsilon}'\mathbf{p}$, which has left eigenvector $\mathbf{p}$. Because this matrix is of rank 1, all other vectors are annihilated when multiplied by $\mathbf{Q}$. The convolution is generated by the vector $[\mathbf{p}, \mathbf{o}]$ and the matrix:†

$$\begin{bmatrix} \mathbf{B} & -\mathbf{BQ} \\ \mathbf{O} & \mathbf{B} \end{bmatrix}.$$

On the other hand, the convolution of two successive departures in a semi-Markov process is given by the vector $[\pi_\mathbf{c}, \mathbf{o_c}]$ and the matrix

$$\begin{bmatrix} \mathbf{B_c} & -\mathbf{M_c Q_c R_c} \\ \mathbf{O_c} & \mathbf{B_c} \end{bmatrix}.$$

It is true that $\mathbf{p_c}$ is a left eigenvector of $\mathbf{Q_c R_c}$, but this matrix does not annihilate everything else. (Some years ago, I asked Ward Whitt this question. He, of course gave me the correct answer [WHIT80], but not until now did I understand why this process does not necessarily approach a renewal process.)

This chapter should contain several interesting areas for further study by the reader. We have a few questions of our own.

1. What is the relation between the product-form solution of steady-state Jackson networks and the matrices we had to create for the M/ME/C//C loop? It would seem that the only properties necessary for the two to give the same results is that the dimension of the Ξ_k's be equal to the binomial coefficients as given in (6.2.6b).

2. There is no formal difference between a subsystem with C identical servers and the *generalized* subsystem. Since the former has some special properties, it would seem that those matrices should be capable of being broken down into smaller parts, so that the difference between the two can be seen explicitly.

3. Do there exist smaller-dimensional matrices that represent these processes equally well? We should be able to study the class of similarity transformations that leave the various equations invariant or the various results unchanged.

6.4.3. A Little Bit of Up and Down, with Arrivals

The work we did in Chapters 4 and 5 on *first-passage matrices and times*, as well as the various $\mathbf{W}$ *matrices and other properties of the busy period*, can be generalized to the networks we are treating in this chapter. We must be more careful though, since the operators, both in size and content, change from one queue length to the next. We will give a sampling of how this can be done in just two areas, first-passage times up and first-passage times down. These two topics have increasingly important applicability to *real-world problems*. *Up* is easier, so we do that first.

† See Section 3.3.1, and in particular Equations (3.3.3)].

First-Passage Processes for Queue Growth

What we are about to do is taken directly from Section 4.5.1 with the added problem that dimensions and operators change as we go up the ladder. We also have the problem of setting up new notation. As in previous chapters, we will use the symbol **H** for our isometric, first-passage matrices. We also need an auxiliary matrix for definition purposes. So, for our first definition,

Definition 6.4.5

$\mathbf{X_k}$:= *probability matrix of first passage from k to $k+1$, where $k < C$*. That is, $[\mathbf{X_k}]_{ij}$ is the probability that S_1 will be in state $j \in \Xi_{k+1}$ when its queue goes from k to $k+1$ for the first time, given that it started in state $i \in \Xi_k$. This is a $D(k) \times D(k+1)$-dimensional matrix and is isometric, since $\mathbf{X_k}\boldsymbol{\varepsilon}'_{\mathbf{k+1}} := \boldsymbol{\varepsilon}'_{\mathbf{k}}$. We are not bothering to use **u** for *up*, since $\mathbf{X_k}$ will be replaced by $\mathbf{H_{uk}}$. Note that this matrix is only defined for $k < C$. ◊

To be consistent with previous notation **X** should have been subscripted as **k+1**, since that matches the higher dimension of this rectangular matrix. However, we will soon define the related matrix **H**, which is the one we will actually use in our final formulas.

Look at the time-dependent state diagram in Figure 6.4.1. There are five different types of equations we must look at for queue growth, namely $0 < 1 < k < C < n$. N is relevant only for queue decrease. Let us start with no one at S_1. Then all that can happen is for a customer to arrive, putting S_1 into state **p**. Therefore,

$$\mathbf{X_0} = \mathbf{p}.$$

Next consider 1 customer in S_1. Now three things can happen.

1. A second customer arrives directly, with probability $\lambda[\mathbf{M_1} + \lambda\mathbf{I_1}]^{-1}$, and enters, thereby changing the state of the subsystem, $[\mathbf{R_2}]$.

2. A transition occurs in S_1, $[\mathbf{M_1}(\mathbf{M_1} + \lambda\mathbf{I_1})^{-1}]$, resulting in a customer changing phase, $[\mathbf{P_1}]$, and then eventually the queue gets to length 2, $[\mathbf{X_1}]$.

3. A transition occurs in S_1, $[\mathbf{M_1}(\mathbf{M_1} + \lambda\mathbf{I_1})^{-1}]$, resulting in a customer leaving, $[\mathbf{Q_1} = \mathbf{q}']$, and then the queue eventually grows from 0 to 1 to 2, $[\mathbf{X_0 X_1}]$.

The equation for this is

$$\mathbf{X_1} = \lambda[\mathbf{M_1} + \lambda\mathbf{I_1}]^{-1}\mathbf{R_2} + \mathbf{M_1}[\mathbf{M_1} + \lambda\mathbf{I_1}]^{-1}\mathbf{X_1} + \mathbf{M_1}[\mathbf{M_1} + \lambda\mathbf{I_1}]^{-1}\mathbf{Q_1}\mathbf{X_0}\mathbf{X_1}.$$

Premultiply both sides by $[\mathbf{M_1} + \lambda\mathbf{I_1}]$, collect all terms multiplying $\mathbf{X}_1$, and get

$$[\mathbf{M_1} + \mathbf{I_1} - \mathbf{M_1}\mathbf{P_1} - \mathbf{M_1}\mathbf{Q_1}\mathbf{R_1}]\mathbf{X_1} = \lambda\mathbf{R_2}.$$

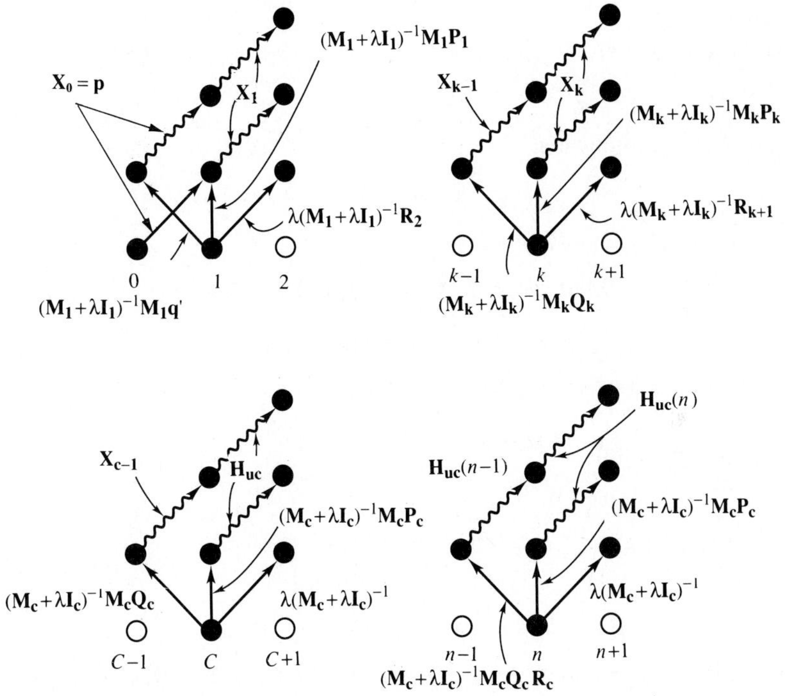

Figure 6.4.1: Time-dependent state transition diagram for *up* processes. This is applicable to both the open and closed M/ME/*C*//*N* queues, where $C < N$. There are five different sets of transition types, one each for $n = 0$, $n = 1$, $n = k < C$, $n = C, N > n > C$. $n = N$ is only relevant for *down* processes.

We have deliberately used $\mathbf{Q_1}$ and $\mathbf{R_1}$ for $\mathbf{q}'$ and $\mathbf{p}$, respectively. Now solve for $\mathbf{X_1}$, using $\mathbf{B_1} = \mathbf{M_1}(\mathbf{I_1} - \mathbf{P_1})$,

$$\mathbf{X_1} = \lambda[\lambda\mathbf{I_1} + \mathbf{B_1} - \mathbf{M_1Q_1R_1}]^{-1}\mathbf{R_2}. \tag{6.4.13a}$$

Let us make some simplifying definitions. Let $\mathbf{H_{u0}} := 1$ [a one-dimensional matrix, since $D(0) = 1$], and let

$$\mathbf{H_{u1}} := \lambda[\lambda\mathbf{I_1} + \mathbf{B_1} - \mathbf{M_1Q_1H_{u0}R_1}]^{-1}, \tag{6.4.13b}$$

then $\mathbf{X_1} = \mathbf{H_{u1}R_2}$. What is nice is that $\mathbf{H_{u1}}$ is an isometric, invertible, square matrix. We prove the isometric property by multiplying (6.4.13b) from the right with the expression in brackets [notice that we could not do this in (6.4.13a), because $\mathbf{R_2}$ is in the way] and then right-multiply by ε'_1 to get

$$\mathbf{H_{u1}}[\lambda\mathbf{I_1} + \mathbf{B_1} - \mathbf{M_1Q_1R_1}]\varepsilon'_1 = \lambda\varepsilon'_1.$$

We know from (6.2.4b) that $\mathbf{M_1Q_1R_1}\varepsilon'_1 = \mathbf{B_1}\varepsilon'_1$, so we are indeed left with

$$\mathbf{H_{u1}}\varepsilon'_1 = \varepsilon'_1.$$

Only then does it follow that

$$\mathbf{X_2}\varepsilon'_2 = \varepsilon'_1.$$

We can do the very same thing for $1 < k < C$, so

$$\mathbf{X_k} = \lambda[\mathbf{M_k} + \lambda\mathbf{I_k}]^{-1}\mathbf{R_{k+1}} + \mathbf{M_k}[\mathbf{M_k} + \lambda\mathbf{I_k}]^{-1}\mathbf{X_k}$$
$$+ \mathbf{M_k}[\mathbf{M_k} + \lambda\mathbf{I_k}]^{-1}\mathbf{Q_k}\mathbf{X_{k-1}}\mathbf{X_k}.$$

Doing the usual maneuverings, we come up with

$$\mathbf{X_k} = \lambda[\lambda\mathbf{I_k} + \mathbf{B_k} - \mathbf{M_kQ_kX_{k-1}}]^{-1}\mathbf{R_{k+1}}.$$

Next we let

$$\mathbf{X_k} = \mathbf{H_{uk}}\mathbf{R_{k+1}}; \tag{6.4.14}$$

then we have

$$\mathbf{H_{uk}} = \lambda[\lambda\mathbf{I_k} + \mathbf{B_k} - \mathbf{M_kQ_kH_{u\,k-1}R_k}]^{-1}. \tag{6.4.15a}$$

We prove that $\mathbf{H_{uk}}$ is isometric, by induction. We know that $\mathbf{H_{u1}}\varepsilon'_1 = \varepsilon'_1$, and assume that it is true for $l = 1, 2, \cdots, k-1$, then

$$\mathbf{H_{uk}}[\lambda\mathbf{I_k} + \mathbf{B_k} - \mathbf{M_kQ_kH_{u\,k-1}R_k}]\varepsilon'_k = \mathbf{H_{uk}}\varepsilon'_k = \lambda\varepsilon'_k.$$

It follows from this that $\mathbf{X_k}$ is isometric. Note that the $\mathbf{H_{uk}}$'s do not depend on C as long as k is less than C. We can interpret $[\mathbf{H_{uk}}]_{ij}$ as given in the following.

Definition 6.4.6

$[\mathbf{H_{uk}}]_{ij}$:= *probability that S_1 will be in state $j \in \Xi_k$ just before a customer arrives to raise the queue length from k to $k+1$ for the first time, given that S_1 started in state $i \in \Xi_k$.* $\mathbf{H_{uk}}$ and $\mathbf{X_k}$ are related by (6.4.14). ◊

Now we must look at the system when there are C or more customers at S_1. To do that we must first define a set of matrices for $n \geq C$.

Definition 6.4.7

$\mathbf{H_{uc}}(n)$:= *probability matrix of first passage from n to $n+1$, where $n \geq C$.* Thus $[\mathbf{H_{uc}}(n)]_{ij}$ is the probability that S_1 will be in state $j \in \Xi_c$ just after a customer arrives to raise the queue length from n to $n+1$ for the first time, given that the process started with the system in state $i \in \Xi_c$. ◊

We do not need the auxiliary **X** matrices, since the first-passage matrices of Definition 6.4.7 are already square and isometric. Put another away, from what we know of their meanings, they must be equal. After all, if the queue length at S_1 is greater than or equal to C, the system will be in the same internal state immediately before and immediately after an arrival.

Let us start with $n = C$. The first-passage matrix satisfies

$$\mathbf{H_{uc}}(C) = \lambda[\mathbf{M_c} + \lambda\mathbf{I_c}]^{-1} + \mathbf{M_c}[\mathbf{M_c} + \lambda\mathbf{I_c}]^{-1}\mathbf{P_c}\,\mathbf{H_{uc}}(C)$$

$$+ \mathbf{M_c}[\mathbf{M_c} + \lambda\mathbf{I_c}]^{-1}\mathbf{Q_c}\,\mathbf{H_{uc-1}}\,\mathbf{R_c}\,\mathbf{H_{uc}}(C).$$

When we solve explicitly for $\mathbf{H_{uc}}(C)$ we get

$$\mathbf{H_{uc}}(C) = \lambda[\lambda\mathbf{I_c} + \mathbf{B_c} - \mathbf{M_k}\mathbf{Q_c}\mathbf{H_{u\,c-1}}\,\mathbf{R_c}]^{-1}.$$

This is in exactly the same form as (6.4.15a), so we can say that

$$\mathbf{H_{uc}}(C) = \mathbf{H_{uc}},$$

but things will be a little different from now on. The defining equation when $n > C$ is the following:

$$\mathbf{H_{uc}}(n) = \lambda[\mathbf{M_c} + \lambda\mathbf{I_c}]^{-1} + \mathbf{M_c}[\mathbf{M_c} + \lambda\mathbf{I_c}]^{-1}\,\mathbf{P_c}\,\mathbf{H_{uc}}(n)$$

$$+ \mathbf{M_c}[\mathbf{M_c} + \lambda\mathbf{I_c}]^{-1}\,\mathbf{Q_c}\,\mathbf{R_c}\,\mathbf{H_{uc}}(n-1)\,\mathbf{H_{uc}}(n).$$

Notice the subtle change. In the last term, $\mathbf{R_c}$ is now to the left of the two **H**'s instead of between them. The reason is simple. When there are more than C customers at the subsystem, a departing customer can immediately be replaced, $[\mathbf{R_c}]$, and eventually a customer comes to raise the queue to $n+1$ for the first time, $[\mathbf{H_{uc}}(n)]$. If there are C or fewer customers at S_1, then eventually the subsystem has $k-1$ customers when a kth one arrives, $[\mathbf{H_{uk-1}}]$, and *then* the customer enters, $[\mathbf{R_k}]$. This, by the way, shows us the significance of having matrices that do not commute ($\mathbf{H_{uc-1}}\,\mathbf{R_c} \neq \mathbf{R_c}\,\mathbf{H_{uc}}$). This slight difference yields a somewhat different recursive equation:

$$\mathbf{H_{uc}}(n) = \lambda[\lambda\mathbf{I_c} + \mathbf{B_c} - \mathbf{M_c}\,\mathbf{Q_c}\,\mathbf{R_c}\,\mathbf{H_{uc}}(n-1)]^{-1}. \qquad (6.4.15b)$$

We leave it as an exercise to prove that these matrices are isometric.

Recall that $\mathbf{p_k}$ from Equations (6.4.2) is the state probability vector for S_1 if k customers entered the empty subsystem simultaneously. If the customers arrived randomly, the state S_1 would be in when there are *finally* k customers there for the first time is defined by (compare with Definition 4.5.4),

Definition 6.4.8

$\mathbf{p_{uk}}(n)$:= *probability vector of first passage from 0 to n.* $[\mathbf{p_{uk}}(n)]_i$ is the probability that S_1 will be in state $i \in \Xi_k$ when there are n customers there for the first time. Two conventions go with this. First, $k = n$ when $n \le C$, and $k = C$ when $n \ge C$. Second, when $k = n \le C$, we will drop the argument:

$$\mathbf{p_{uk}} := \mathbf{p_{uk}}(k) \quad \text{for } k \le C.$$

The process starts with the arrival of the first customer and ends when the queue (counting the ones in service) reaches n. The queue could have gone back to 0 any number of times before the process ends. ◊

It is plain to see that

$$\mathbf{p_{uk}} := \mathbf{p}\,\mathbf{H_{u1}}\,\mathbf{R_2}\,\mathbf{H_{u2}}\,\mathbf{R_3} \cdots \mathbf{H_{uk-1}}\,\mathbf{R_k} \qquad (\text{for } n = k \le C) \qquad (6.4.16a)$$

and

$$\mathbf{p_{uc}}(n) := \mathbf{p}\,\mathbf{H_{u1}}\,\mathbf{R_2}\,\mathbf{H_{u2}}\,\mathbf{R_3} \cdots \mathbf{H_{uc-1}}\,\mathbf{R_c}\,\mathbf{H_{uc}}\,\mathbf{H_{uc}}(C+1) \cdots \mathbf{H_{uc}}(n-1)$$

$$= \mathbf{p_{uc}}\,\mathbf{H_{uc}}\,\mathbf{H_{uc}}(C+1) \cdots \mathbf{H_{uc}}(n-1) \qquad (\text{for } n > C) \qquad (6.4.16b)$$

We can *read* these formulas physically. A first customer arrives, [$\mathbf{p}$]. There is one customer at S_1 when a second one arrives, [$\mathbf{H_{u1}}$], and enters, [$\mathbf{R_2}$]. Eventually, there are two customers at S_1 when a third one arrives, [$\mathbf{H_{u2}}$], and enters, [$\mathbf{R_3}$], and so on. Once there are C or more customers at the subsystem, [$\mathbf{H_{uc}}(n)$], the arriving customer does not enter (*no* $\mathbf{R}$). The $\mathbf{p}$'s satisfy a natural recursive equation.

$$\mathbf{p_{uk+1}} = \mathbf{p_{uk}}\,\mathbf{H_{uk}}\,\mathbf{R_{k+1}} \quad \text{for } k < C, \qquad (6.4.16c)$$

$$\mathbf{p_{uc}}(n+1) = \mathbf{p_{uc}}(n)\,\mathbf{H_{uc}}(n) \quad \text{for } n \ge C. \qquad (6.4.16d)$$

Without belaboring the point, note the difference between $\mathbf{p_c}$ and $\mathbf{p_{uc}}(n)$. In the former, even if more than C customers come simultaneously, only C of them will enter, so the vector is *the same* for all $n \ge C$. In the latter case, when that special customer who will raise the queue to $n+1$ for the first time, arrives, even though he does not enter S_1, the system is in a special state. That special state, $\mathbf{p_{uc}}(n)$, is *different* for all n.

After the following definition, we are finally ready to set up equations for the first-passage times, after which we will summarize everything in a theorem.

Definition 6.4.9

[$\boldsymbol{\tau}'_{\mathbf{uk}}(n)$] := *mean first-passage time vector for an* M/ME/C//N *loop to go from* n *to* $n+1$. The ith component is the mean time until the queue at S_1 (as usual, counting the ones in service) reaches $n+1$ customers for the first time, given that the process started in state $\{i\,;n\}$. The conventions are the same as in Definition 6.4.8, namely, $k = n$ for $n \le C$ and $k = C$ when $n \ge C$. Also,

$$\boldsymbol{\tau}'_{\mathbf{uk}} := \boldsymbol{\tau}'_{\mathbf{uk}}(k) \quad \text{for } k \le C.$$

This is a $D(k)$-dimensional column vector. ◊

Clearly, when no one is at S_1, the mean time until someone arrives is $1/\lambda$, so

$$\tau_0 = \frac{1}{\lambda}.^{\dagger} \qquad (6.4.17a)$$

† We have not used boldface since this is a one-dimensional object.

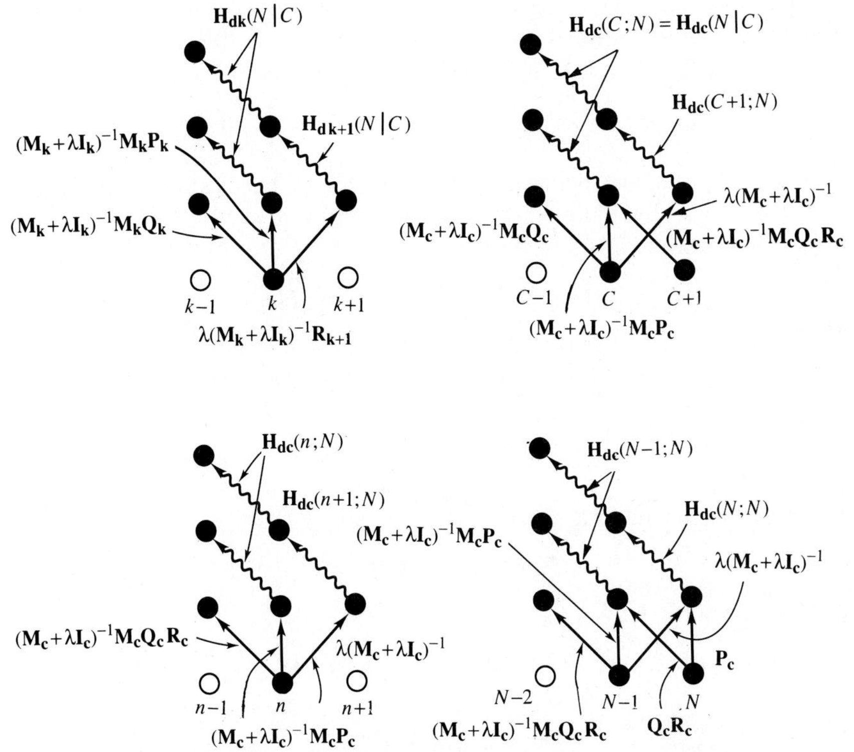

Figure 6.4.2: Time-dependent state transition diagram for *down* processes. This is applicable to the closed M/ME/C//N queue, where $C < N$. There are five different sets of transition types, one each for $0 < n = k < C$, $n = C$, $N-1 > n > C$, $n = N-1$ and $n = N$. $n = 0$ is only relevant for *up* processes.

We can see why the $(\mathbf{H_{dc}})$'s are not invertible from the presence of the term, $\mathbf{Q_c}\,\mathbf{R_c}$. Let us now define the set of auxiliary matrices suggested by this equation:

$$\mathbf{X_c}(n;N) := \left[\lambda\mathbf{I_c} + \mathbf{B_c} - \lambda\mathbf{H_{dc}}(n+1;N)\right]^{-1}. \tag{6.4.22a}$$

Then clearly,

$$\mathbf{H_{dc}}(n;N) = \mathbf{X_c}(n;N)\,\mathbf{M_c}\,\mathbf{Q_c}\,\mathbf{R_c}. \tag{6.4.22b}$$

We add to that the initial definition,

$$\mathbf{X_c}(N;N) := \mathbf{V_c}, \tag{6.4.22c}$$

and substitute (6.4.22b) back into (6.4.22a) to get for $C < n < N$,

$$\mathbf{X_c}(n;N) = \left[\lambda\mathbf{I_c} + \mathbf{B_c} - \lambda\mathbf{X_c}(n+1;N)\,\mathbf{M_c}\,\mathbf{Q_c}\,\mathbf{R_c}\right]^{-1}. \tag{6.4.22d}$$

Then (6.4.22c) and (6.4.22d) completely define the $(\mathbf{X_c})$'s, while (6.4.22b) gives all the $(\mathbf{H_{dc}})$'s from N down to $C+1$.

When there are exactly C customers at S_1, everything is as before, except that when the queue finally drops, there is no new customer waiting to enter, so there is no need for $\mathbf{R_c}$. Thus the first-passage equation comes to

$$\mathbf{H_{dc}}(N \mid C) = \left[\lambda \mathbf{I_c} + \mathbf{B_c} - \lambda \mathbf{H_{dc}}(C+1;\, N)\right]^{-1} \mathbf{M_c Q_c}\,. \tag{6.4.23a}$$

This equation is almost identical in form to (6.4.21b), but as already mentioned, it is missing $\mathbf{R_c}$. But if we let (6.4.22a) be valid for $n = C$ (up to now we only made it true down to $C+1$), then we have

$$\mathbf{H_{dc}}(N \mid C) = \mathbf{X_c}(C;\, N)\mathbf{M_c Q_c}\,. \tag{6.4.23b}$$

With fewer than C customers at S_1, a new arrival immediately enters, while a departure leaves one less customer in service. This leads to the last set of equations:

$$\mathbf{H_{dk}}(N \mid C) = \left[\lambda \mathbf{I_c} + \mathbf{B_c} - \lambda \mathbf{R_{k+1}} \mathbf{H_{d\,k+1}}(N \mid C)\right]^{-1} \mathbf{M_c Q_c}\,. \tag{6.4.23c}$$

We define our last set of $\mathbf{X}$ matrices, starting with

$$\mathbf{X_c}(N \mid C) := \mathbf{X_c}(C;\, N), \tag{6.4.24a}$$

and for all other k, $1 \le k < C$,

$$\mathbf{X_{dk}}(N \mid C) := \left[\lambda \mathbf{I_c} + \mathbf{B_c} - \lambda \mathbf{R_{k+1}} \mathbf{X_{d\,k+1}}(N \mid C)\mathbf{M_{k+1}} \mathbf{Q_{k+1}}\right]^{-1}. \tag{6.4.24b}$$

This lets us write

$$\mathbf{H_{dk}}(N \mid C) := \mathbf{X_{dk}}(N \mid C)\mathbf{M_c Q_c}\,, \qquad 1 \le k \le C, \tag{6.4.24c}$$

where we have incorporated (6.4.23b). STOP. Surely we have seen this all before. Of course, they all look alike, but exactly alike? In this case they are just like the equations governing the steady-state matrices. Compare the following sets of equations: (6.4.22c) with (6.2.6a), (6.4.22d) with (6.2.6c), (6.4.24a) with (6.2.7a), and (6.4.24b) with (6.2.7b).

Clearly, $\lambda \mathbf{X_c}(n;\, N)$ satisfies exactly the same recursive equation as $\mathbf{U_c}(N-n)$, and just as important, with the same initial condition [(6.4.22c) versus (6.2.6a)]. Therefore, we have discovered that

$$\mathbf{X_c}(n;\, N) = \frac{1}{\lambda}\mathbf{U_c}(N-n) \quad \text{for } C \le n \le N \tag{6.4.25a}$$

and

$$\mathbf{X_k}(N \mid C) = \frac{1}{\lambda}\mathbf{U_k}(N \mid C) \quad \text{for } 1 \le k \le C. \tag{6.4.25b}$$

We now have everything we want to know about the $\mathbf{H}$'s.

Lemma 6.4.4: The matrices described in Definitions 6.4.10 and 6.4.11 are isometric and are related to the steady-state matrices by the following equations:

$$\mathbf{H_{dc}}(n;\, N) = \frac{1}{\lambda}\mathbf{U_c}(N-n)\mathbf{M_c}\,\mathbf{Q_c}\,\mathbf{R_c} \quad \text{for } C < n \le N \tag{6.4.26a}$$

and

$$\mathbf{H_{dk}}(N \mid C) = \frac{1}{\lambda}\mathbf{U_k}(N \mid C)\mathbf{M_k}\,\mathbf{Q_k} \qquad \text{for } 1 \le k \le C. \tag{6.4.26b}$$

The **U**'s are given by Equations (6.2.6) and (6.2.7). ■

Proof: The isometric property follows directly from (6.2.4b). QED

We saw something like this in Section 4.5.3, but $\mathbf{H_d}(n; N)$ [the $C = 1$ equivalent to $\mathbf{H_{dc}}(n; N)$], turned out to be the matrix **Q**, so the relation equivalent to (6.4.26a) is trivial.

Note from (6.2.9) that it is $[(1/\lambda)\mathbf{U_c}(N-n)\mathbf{B_c}]$ that is isometric, and not $\mathbf{U_c}(N-n)$ itself. Indeed, it is this product that has the physical interpretation, for we can rewrite (6.4.26a) as

$$\mathbf{H_{dc}}(n; N) = \frac{1}{\lambda}\mathbf{U_c}(N-n)\mathbf{B_c}\,\mathbf{V_c}\,\mathbf{M_c}\,\mathbf{Q_c}\,\mathbf{R_c} = \left[\frac{1}{\lambda}\mathbf{U_c}(N-n)\mathbf{B_c}\right]\mathbf{Y_c}\,\mathbf{R_c},$$

where $\mathbf{Y_c}$ comes from Definition 6.4.2 and (6.4.4). From their definition we see that the $[\mathbf{Y_c}\,\mathbf{R_c}]$ portion carries a process to a departure and subsequent entry without any intervening arrivals. Therefore, the ijth component of the portion in brackets must be the *probability that* S_1 *is in state* $j \in \Xi_c$, *with* n *customers, immediately after the last arrival but before the departure that finally lowers the queue to* $n-1$ *for the first time* (given that the system was originally in state $i \in \Xi_c$ with n customers). This is a rather complicated interpretation, but it need not be understood for the development of our formulas.

Perhaps we should have given a special symbol to the isometric product and used it in our exposition, and maybe we will in the future. For now we will stop using the **X**'s and express the first-passage times in terms of the already familiar **H**'s. First we give the *down* equivalent of $\tau'_{\mathbf{uk}}(n)$ in Definition 6.4.9.

Definition 6.4.12

$[\tau'_{\mathbf{dk}}(n; N \mid C)] :=$ *mean first-passage time vector for a generalized* M/ME/C//N *loop to go from* n *to* $n-1$. *The* i*th component is the mean time until the queue at* S_1 *(as usual, counting the ones in service) reaches* $n-1$ *customers for the first time, given that the process started in state* $\{i; n\}$. The conventions are the same as in Definition 6.4.9, namely, $k = n$ for $n \le C$ and $k = C$ when $n \ge C$. Also,

$$\tau'_{\mathbf{dk}}(k; N \mid C) := \tau'_{\mathbf{dk}}(k; N) \qquad \text{for } k \le C.$$

This is a $D(k)$-dimensional column vector, and it depends on N. ◊

When all customers are at S_1, there are no arrivals, so the mean time to drop is simply

$$\tau'_{\mathbf{dc}}(N; N) := \mathbf{V_c}\,\varepsilon'_{\mathbf{c}}. \tag{6.4.27a}$$

Otherwise, we must go down, across and up, giving

$$\boldsymbol{\tau}'_{\mathbf{dc}}(n;N) = (\lambda\mathbf{I_c} + \mathbf{M_c})^{-1}\boldsymbol{\varepsilon}'_{\mathbf{c}} + (\lambda\mathbf{I_c} + \mathbf{M_c})^{-1}\mathbf{M_c}\,\mathbf{P_c}\,\boldsymbol{\tau}'_{\mathbf{dc}}(n;N)$$
$$+ \lambda(\lambda\mathbf{I_c} + \mathbf{M_c})^{-1}[\boldsymbol{\tau}'_{\mathbf{dc}}(n+1;N) + \mathbf{H_{dc}}(n+1;N)\,\boldsymbol{\tau}'_{\mathbf{dc}}(n;N)],$$

which rearranges to yield

$$[\lambda\mathbf{I_c} + \mathbf{M_c} - \mathbf{M_c}\mathbf{P_c} - \lambda\mathbf{H_{dc}}(n+1;N)]\boldsymbol{\tau}'_{\mathbf{dc}}(n;N) = \boldsymbol{\varepsilon}'_{\mathbf{c}} + \lambda\boldsymbol{\tau}'_{\mathbf{dc}}(n+1;N).$$

Use (6.4.26a) for $\mathbf{H_{dc}}(n+1;N)$ and compare with (6.2.6c) to get for $C \le n < N$,

$$\lambda\boldsymbol{\tau}'_{\mathbf{dc}}(n;N) = \mathbf{U_c}(N-n)\,\boldsymbol{\varepsilon}'_{\mathbf{c}} + \lambda\,\mathbf{U_c}(N-n)\,\boldsymbol{\tau}'_{\mathbf{dc}}(n+1;N). \qquad (6.4.27b)$$

This equation (as noted) is valid for $n = C$. There is another expression that is more useful, which we include in the theorem below.

For $k < C$ we have a somewhat different set of equations.

$$\boldsymbol{\tau}'_{\mathbf{dk}}(N \mid C) = (\lambda\mathbf{I_k} + \mathbf{M_k})^{-1}\boldsymbol{\varepsilon}'_{\mathbf{k}} + (\lambda\mathbf{I_k} + \mathbf{M_k})^{-1}\mathbf{M_k}\,\mathbf{P_k}\,\boldsymbol{\tau}'_{\mathbf{dk}}(N \mid C)$$
$$+ \lambda(\lambda\mathbf{I_k} + \mathbf{M_k})^{-1}\mathbf{R_{k+1}}[\boldsymbol{\tau}'_{\mathbf{dk+1}}(N \mid C) + \mathbf{H_{dk+1}}(N \mid C)\,\boldsymbol{\tau}'_{\mathbf{dk}}(N \mid C)].$$

Just as we did above, regroup terms, use (6.4.26b), compare with (6.2.7b), and get for $1 \le k < C$,

$$\lambda\boldsymbol{\tau}'_{\mathbf{dk}}(N \mid C) = \mathbf{U_k}(N \mid C)\,\boldsymbol{\varepsilon}'_{\mathbf{k}} + \lambda\,\mathbf{U_k}(N \mid C)\,\mathbf{R_{k+1}}\,\boldsymbol{\tau}'_{\mathbf{dk+1}}(N \mid C). \qquad (6.4.27c)$$

The summary theorem now follows.

Theorem 6.4.5: Given a generalized M/ME/C//N loop, the first-passage matrices and vectors for decreasing length, as given in Definitions 6.4.10 to 6.4.12, satisfy the following formulas.

(a) The first-passage matrices are given by [Equation (6.4.26)]

$$\mathbf{H_{dc}}(n;N) = \frac{1}{\lambda}\mathbf{U_c}(N-n)\,\mathbf{M_c}\,\mathbf{Q_c}\,\mathbf{R_c} \qquad \text{for } C < n \le N$$

and

$$\mathbf{H_{dk}}(N \mid C) = \frac{1}{\lambda}\mathbf{U_k}(N \mid C)\,\mathbf{M_k}\,\mathbf{Q_k} \qquad \text{for } 1 \le k \le C.$$

(b) The first-passage vectors are given by [Equation (6.4.27)]

$$\boldsymbol{\tau}'_{\mathbf{dc}}(N;N) := \mathbf{V_c}\boldsymbol{\varepsilon}'_{\mathbf{c}}.$$

For $C \le n < N$,

$$\boldsymbol{\tau}'_{\mathbf{dc}}(n;N) = \frac{1}{\lambda}\mathbf{U_c}(N-n)\,\boldsymbol{\varepsilon}'_{\mathbf{c}} + \mathbf{U_c}(N-n)\,\boldsymbol{\tau}'_{\mathbf{dc}}(n+1;N),$$

and for $1 \le k < C$,

$$\boldsymbol{\tau}'_{\mathbf{dk}}(N \mid C) = \frac{1}{\lambda}\mathbf{U_k}(N \mid C)\,\boldsymbol{\varepsilon}'_{\mathbf{k}} + \mathbf{U_k}(N \mid C)\,\mathbf{R_{k+1}}\,\boldsymbol{\tau}'_{\mathbf{dk+1}}(N \mid C).$$

(c) The formula, (6.4.27b), has a more useful form. For $0 < l \le N - C$,

$$\lambda\boldsymbol{\tau}'_{\mathbf{dc}}(N-l;N) = \mathbf{U_c}(l)\,\mathbf{K_c}(C+l-1)\,\boldsymbol{\varepsilon}'_{\mathbf{c}} = [\mathbf{K_c}(C+l) - \mathbf{I_c}]\,\boldsymbol{\varepsilon}'_{\mathbf{c}} \qquad (6.4.28a)$$

and for $C < n < N$ (where $l = N - n$)

$$\lambda \tau'_{\mathbf{dc}}(n; N) = \mathbf{U_c}(N-n)\,\mathbf{K_c}(N-n+C-1)\,\boldsymbol{\varepsilon}'_{\mathbf{c}} \qquad (6.4.28b)$$

$$= [\mathbf{K_c}(N-n+C) - \mathbf{I_c}]\,\boldsymbol{\varepsilon}'_{\mathbf{c}}.$$

$\mathbf{K_c}(n)$ is the steady-state normalization matrix defined by (6.2.13). The middle form of both versions is also valid for $l = 0$ (or $n = C$). The right-hand side of (6.4.28a) does not contain N explicitly. Therefore, $\tau'_{\mathbf{dc}}(N-l; N)$ depends only on l, the number of customers at S_2. ■

Proof: Substitute (6.4.28b) directly into (6.4.27b), and use (6.2.13). QED

Equations (6.4.28) look just like (4.5.14a), the result for the generalized M/ME/1//N loop, except that things change once the queue length drops below C. This theorem is quite interesting, since it lets us find out about transient behavior using no more information than is needed for the steady-state solution.

What can we do with these? Well, first of all, by definition

$\mathbf{p}\tau'_{\mathbf{1}}(N \mid C) =$ *mean busy−period time of a generalized* M/ME/C//N *system.*

For anything else, we require more information. Let us suppose for definiteness that there are $N > C$ customers at S_1, and that the system is in some internal state represented by $\mathbf{p_{ic}}$, the same initial vectors we discussed in Equations (6.4.1), (6.4.2), and (6.4.16). Then the mean time for the queue to drop by 1 is given by

$$t_d(n;\, 1;\, N) := \mathbf{p_{ic}}\,\tau'_{\mathbf{dc}}(n; N).$$

The time it takes to drop by one more is

$$t_d(n;\, 2;\, N) := \mathbf{p_{ic}}\,\mathbf{H_{dc}}(n; N)\,\tau'_{\mathbf{dc}}(n-1; N).$$

In general, the state the system will be in after the queue has dropped by $l \geq 1$ customers is

$$\mathbf{p_{dc}}(n;\, l;\, N) := \mathbf{p_{ic}}\,\mathbf{H_{dc}}(n; N)\,\mathbf{H_{dc}}(n-1; N) \cdots \mathbf{H_{dc}}(n-l+1; N), \qquad (6.4.29a)$$

and if we let, by definition,

$$\mathbf{p_{dc}}(n;\, 0;\, N) := \mathbf{p_{ic}}, \qquad (6.4.29b)$$

then

$$\mathbf{p_{dc}}(n;\, l+1;\, N) := \mathbf{p_{dc}}(n;\, l;\, N)\,\mathbf{H_{dc}}(n-l; N) \quad \text{for } l > 0. \qquad (6.4.29c)$$

This is (more or less) the *down* equivalent of Definition 6.4.8, except that we can start with any length and in any initial state. We do not have an official name for it, so we will not give it an official definition designation. In any case, the mean time to drop from l to $l-1$, given the constraints above, is

$$t_d(n;\, l+1;\, N) = \mathbf{p_{dc}}(n;\, l;\, N)\,\tau'_{\mathbf{dc}}(n-l; N). \qquad (6.4.30)$$

If the queue should drop below C, use (6.4.26b) instead of (6.4.26a). Remember, these objects are the times to drop by *one more* customer. The total time it takes to drop from n to $n-l$ is the sum,

$$t_d(n \to n-l; N) := \sum_{j=0}^{l-1} t_d(n; j; N). \tag{6.4.31}$$

Now, this object has a name – several names, in fact. If, for instance, we have $l = n$, we have the *n-busy period*, starting with initial condition, $\mathbf{p_{ic}}$. Potentially, the most important interpretation of this is the MTTF, with backup and repair. It is important enough to give it a definition.

Definition 6.4.13

MTTF for a C-parallel, $(N-C)$-backup system, with exponential repair times is an M/ME/C//N *loop with the initial state given by* (6.4.2) *for $k = N$*. The system starts with N brand new identical devices. Their individual failure times are generated by $<\mathbf{p};\mathbf{B}>$. C devices are started simultaneously and the rest are kept in *cold backup*. When one device breaks, it is immediately replaced by one of the backups and is sent to a single *repairman* who takes exponential time (with mean, $1/\lambda$) to pick up, repair, and return a device that is as good as new. Failure occurs when the number of devices that are functional (counting the number that are running) drops to a prespecified number, say, $\phi \geq 0$. The mean time for this process is given by $t_d(N \to \phi; N)$. ◊

Consider some of the variations we can perform on this.

1. The system has been running for a long time, and presently there are n functional devices (and $N-n$ devices in repair). The initial vector is given by (6.4.1) for $k = n$, and MTTF is given by $t_d(n \to \phi; N)$.

2. We are starting with new devices, but only n are available at the moment, with $N-n$ still awaiting single delivery at rate λ. The initial vector is given by (6.4.2) for $k = n$, and MTTF is $t_d(n \to \phi; N)$.

3. The system was originally as given in Definition 6.4.12, but two devices have already failed and one has been repaired. The initial vector is given by (assuming that $N \geq C$)

$$\mathbf{p_{ic}} = \mathbf{p_c}\,\mathbf{H_{dc}}(N; N)\,\mathbf{H_{dc}}(N-1; N)\,\mathbf{H_{uc}}(N-2),$$

and MTTF is given by $t_d(N-1 \to \phi; N)$. Notice the *up* first-passage matrix.

The last example can be very useful for *dynamic updating* the MTTF. Every time a device fails, postmultiply the initial vector by the appropriate $\mathbf{H_{dk}}$ to create an updated initial vector for the MTTF starting now. Similarly, every time a device is returned from repair, postmultiply by the appropriate $\mathbf{H_{uk}}$. The

MTTF's change accordingly. By virtue of the fact that the **H**'s do not commute, we see that two failures followed by one repair gives different results from one failure followed by one repair and then another failure. Dynamic updating can help us even if we do not know the initial state. Pick any initial vector, then update it regularly. Eventually, your poor guess will be forgotten, and the updated vector will converge to the correct updated initial vector. – etc., etc., etc.

We could go on indefinitely enumerating systems that can be described this way, but we will do only one more analysis before giving up on this chapter. Consider what happens when $N \to \infty$. Ah yes, of course, the open system. We have actually done this already, since the $\mathbf{H_{dk}}$'s are known in terms of the $\mathbf{U_k}$'s. Remember, though, that the limit exists only if ρ_C given by (6.4.10b) is less than 1. From (6.3.1),

$$\mathbf{H_{dc}} := \lim_{N\to\infty} \mathbf{H_{dc}}(n; N) := \frac{1}{\lambda}\mathbf{U_c M_c Q_c R_c} \quad \text{for } n \geq C \tag{6.4.32a}$$

and

$$\mathbf{H_{dk}}(\infty \mid C) := \lim_{N\to\infty} \mathbf{H_{dk}}(N \mid C) = \frac{1}{\lambda}\mathbf{U_k}(\infty \mid C)\mathbf{M_c Q_c}. \tag{6.4.32b}$$

So we do not have to cascade the **H**'s down from infinity to find out what they are. In fact, they are all the same for $n \geq C$. All we have to do is solve for $\mathbf{U_c}$ in (6.3.1). Similarly, we can find the first-passage times,

$$\boldsymbol{\tau}'_{\mathbf{dc}} := \lim_{N\to\infty} \boldsymbol{\tau}'_{\mathbf{dc}}(n; N) = \frac{1}{\lambda}[\mathbf{K_c}(\infty) - \mathbf{I_c}]\boldsymbol{\varepsilon}'_{\mathbf{c}}.$$

But from its recursive definition, (6.2.13b), $\mathbf{K_c}(\infty)$ can be shown to be

$$\mathbf{K_c} := \mathbf{K_c}(\infty) = [\mathbf{I_c} - \mathbf{U_c}]^{-1} \tag{6.4.33}$$

[compare with (4.2.3a)]. Therefore,

$$\boldsymbol{\tau}'_{\mathbf{dc}} := \frac{1}{\lambda}\mathbf{U_c}[\mathbf{I_c} - \mathbf{U_c}]^{-1}\boldsymbol{\varepsilon}'_{\mathbf{c}}. \tag{6.4.34}$$

How interesting (we think lots of things are interesting), these vectors are independent of n, just like the M/ME/1 queue in (4.5.16a). Some thought would lead us to believe that this is reasonable. Does this mean that the time for the queue to drop by n is simply n multiplied by the time it takes to drop by 1, just as it is for the M/G/1 queue, based on (4.5.16b)? The answer is *NO*, because the departing customer leaves the system in a different state from that which it was in at the previous departure. Equation (6.4.29a) still holds true, and becomes

$$\mathbf{p_{dc}}(n-l) := \lim_{N\to\infty} \mathbf{p_{dc}}(n; l; N) := \mathbf{p_{ic}}\,\mathbf{H_{dc}}^{n-l} \tag{6.4.35a}$$

and

$$t_d(n-l+1) := \lim_{N\to\infty} t_d(n; l+1; N) = \mathbf{p_{dc}}(n-l)\boldsymbol{\tau}'_{\mathbf{dc}} = \mathbf{p_{ic}}\,\mathbf{H_{dc}}^{n-l}\boldsymbol{\tau}'_{\mathbf{dc}}. \tag{6.4.35b}$$

Each step takes a different amount of time than the previous one.

There are many useful applications of this set of equations. Here are some that come to mind. Let $\bar{n}$ be the mean number of customers at S_1 in a steady-state system.

1. A computer system has been in operation for a long time, when suddenly n jobs arrive in a bunch, while the Poisson arrivals continue at the same rate. How long will it take before the system settles back down to its steady state? Use (6.4.1) as the initial vector, with $k = \bar{n}$, but use n for $n - l + 1$ in (6.4.35b). We call this the *rush-hour* approximation.

2. A computer system has been running for a long time, with an arrival rate of λ_1. After 5 P.M., the arrival rate drops to λ_2. How long will it take to reach its new steady state? When can a part of the subsystem be taken off-line (reduce C)?

3. The system has been down for a while, and when it starts up there are $n > \bar{n}$ jobs in the queue. How long will it take for the system to settle down?

6.5. CONCLUSIONS

We have seen that there are innumerable problems that can be explored using M/ME/*C* queues. They are more general than single-class Jackson networks. In fact, the formulas as derived here apply to more general systems than the ones we called *generalized* M/ME/*C*//*N* *systems*. The equations depend on the *defined* properties of the input matrices (i.e., $\mathbf{M_k}$, $\mathbf{Q_k}$, $\mathbf{R_k}$, and $\mathbf{P_k}$) and not how they were constructed. Although we did describe how to construct them, we did not make use of those properties. In other words, almost any quasi-birth-death process may be analyzed in this way. Since our formalism covers a larger class of problems than we had intended, this leads to the question as to whether the matrices can be given more detailed properties that can be incorporated to yield more specific results.

Most of the material laid out here remains unexplored, even though they are now computationally manageable. It is hoped that this chapter, in particular, will help stimulate such activity. The two groups who this author feels would be most interested in this material are researchers in computer performance and systems reliability. Yet their interests tend to be at opposite ends of the ρ scale. That is, performance modelers usually assume that the system can handle the load ($\rho_c < 1$). Otherwise, throw away the system. Therefore, they are interested in steady-state solutions (probably overly so), and even open systems, particularly since systems with thousands of terminals now exist. On the other hand, reliability researchers usually assume that it takes less time to fix an object than it took to break it ($\rho_c > 1$). Therefore, except for questions of *inventory*, open systems are uninteresting. Furthermore, the steady state tells us nothing about MTTF. Yet the underlying formalism is identical for both groups. So it is important that the queueing theory practitioners in each camp understand clearly the difference of their goals when they communicate with each other.

CHAPTER 7

G / G / 1 / / N LOOP

"Those who cannot remember the past are condemned to repeat it" – George Santayana.

We are finally facing up to giving structure to S_2. In many ways, this is the hardest queueing system for which analytic results are known. The mathematics required at present to describe such systems is too complicated for one to get reasonable insight from the formulas themselves. Furthermore, we must now specify *two* nonexponential functions, finding that the system behavior depends not merely on ρ, the ratio of their mean service times, and their second moments, or variances, but to a great extent on the parameter $C(x)$, which is the probability that the customer in service at S_1 will finish before the customer at S_2. This parameter, in turn, depends on x, the difference between the times when the two customers started service.

As long as S_2 was exponentially distributed, $C(x)$ (for $x \geq 0$) reduced to the Laplace transform of $b_1(t)$, and everything came out to be reasonably manageable, as described in previous chapters. For the G/G/1//N queue, things get messy (messier?). In matrix representations, this shows up in the difficulty one has in describing two different servers that are simultaneously active. This involves taking the *direct product* of two independent vector spaces. Presently, we discuss one such way to do this, the *Kronecker product*, and then go on to find the steady-state solution of the ME/ME/1//*N* loop. We do not continue on to the open queue, since we have not found how to get an explicit solution for that case. We do, however, discuss how this might be done eventually. In the final section we discuss some transient behavior, by looking at the mean time to failure for a system with small N.

This material is taken in large part from the Ph.D. thesis by Appie van de Liefvoort [LIEF82], most of which was also published in [LIEF86]. But first we will look at $C(x \geq 0)$, without relying on any direct product representation.

7.1. BASIS-FREE EXPRESSION FOR $\Pr[X_1 < X_2]$

Let us consider two subsystems, S_i, $i = 1, 2$, each represented by $\langle \mathbf{p_i}, \mathbf{B_i} \rangle$ with dimension m_i. Let X_i be the random variables for the service times of the two servers. Now suppose that S_2 started service x units of time before S_1, but has not finished when S_1 begins. Then $C(x)$ is defined to be the probability that S_1 will finish before S_2. That is,

$$C(x) = \Pr[X_1 < X_2 + x] \tag{7.1.1a}$$

From elementary probability theory we can write, for $x \geq 0$,

$$C(x) = \frac{\int_0^\infty R_2(x+t)\, b_1(t)\, dt}{R_2(x)}. \tag{7.1.1b}$$

We will now naively try to find an operator expression for $C(x)$, without first explicitly defining what the product of two operators from different spaces looks like, component by component. We do know, however, that since $\mathbf{B_1}$ operates only on vectors describing S_1, and $\mathbf{B_2}$ operates only on vectors describing S_2, they cannot have any affect on each other. After all, S_1 and S_2 are completely independent of each other (i.e., what happens in one subsystem cannot directly affect what happens in the other). We thus assert the *independence principle*, which states: *All operations on S_1 vectors commute with all operations on S_2 vectors.* In particular, this means that

$$\mathbf{B_1 B_2} = \mathbf{B_2 B_1}, \quad \mathbf{p_1 p_2} = \mathbf{p_2 p_1}, \quad \varepsilon'_1 \varepsilon'_2 = \varepsilon'_2 \varepsilon'_1,$$

and so on. We will refer to the vector space made up of all linear combinations of the vectors that describe the internal state of S_i as *space i* or *i-space*.

Before going on, we mention that the behavior of customers in S_1 and that of customers in S_2 eventually become correlated to each other as the two subsystems exchange customers, despite the independence principle. We will see later that the exchange of customers requires that operators from the different spaces be added together, and through these, the commutativity property is lost. For instance, let $\mathbf{X_i}$ and $\mathbf{Y_i}$ be operators on vectors in space i. Then, of course, $\mathbf{X_1 X_2} = \mathbf{X_2 X_1}$, $\mathbf{X_1 Y_2} = \mathbf{Y_2 X_1}$, and so on. But suppose that $\mathbf{Y_i}$ does not commute with $\mathbf{X_i}$. Then the operator

$$\mathbf{Z} := \mathbf{Y_1} + \mathbf{X_1} + \mathbf{Y_2} + \mathbf{X_2}$$

does not commute with any of them.

Now let us replace R_i and b_i in (7.1.1b) with their matrix equivalents, and then see what happens.

$$\mathbf{p_2}\left[\exp(-x\mathbf{B_2})\right]\varepsilon'_2\, C(x)$$
$$= \int_0^\infty \mathbf{p_2}\left[\exp\{-(x+t)\mathbf{B_2}\}\right]\varepsilon'_2\, \mathbf{p_1}\left[\mathbf{B_1}\exp(-t\mathbf{B_1})\right]\varepsilon'_1\, dt.$$

Next apply the commutativity rule as many times as necessary, and recognize that $\exp[-(x+t)\mathbf{B_2}] = \exp(-x\mathbf{B_2})\exp(-t\mathbf{B_2})$, to get

$$\mathbf{p_2}\left[\exp(-x\mathbf{B_2})\right]\varepsilon'_2 C(x)$$
$$= \mathbf{p_1 p_2 B_1}\exp(-x\mathbf{B_2})\int_0^\infty \left[\exp(-t\mathbf{B_2})\exp(-t\mathbf{B_1})\, dt\right]\varepsilon'_1\varepsilon'_2.$$

Since $\mathbf{B_1}$ and $\mathbf{B_2}$ commute, we have $\exp[-t(\mathbf{B_1}+\mathbf{B_2})] = \exp(-t\mathbf{B_1})\exp(-t\mathbf{B_2})$, so

$$\mathbf{p_2}\left[\exp(-x\mathbf{B_2})\right]\varepsilon'_2 C(x)$$
$$= \mathbf{p_1 p_2 B_1}\exp(-x\mathbf{B_2})\left[\int_0^\infty \exp[-t(\mathbf{B_1}+\mathbf{B_2})]\, dt\right]\varepsilon'_1\varepsilon'_2$$
$$= \mathbf{p_1 p_2}\left[\exp(-x\mathbf{B_2})\mathbf{B_1}(\mathbf{B_1}+\mathbf{B_2})^{-1}\right]\varepsilon'_1\varepsilon'_2.$$

We have made use of the fact that for any invertible matrix,

$$\int_0^\infty \exp(-t\mathbf{X})\,dt = \mathbf{X}^{-1}.$$

Last, define the following vectors (without asking until the next section what their components look like, or even their dimensions):

$$\mathbf{p} := \mathbf{p}_1\mathbf{p}_2 \quad \text{and} \quad \boldsymbol{\varepsilon}' := \varepsilon'_1\varepsilon'_2,$$

then the original definition of Ψ is the same as it was before, namely

$$\Psi[\mathbf{X}] = \mathbf{p}[\mathbf{X}]\boldsymbol{\varepsilon}' = \mathbf{p}_1\mathbf{p}_2[\mathbf{X}]\varepsilon'_1\varepsilon'_2, \tag{7.1.2a}$$

where $\mathbf{X}$ can be anything like $\mathbf{B}_1$, $\mathbf{B}_2$, $\mathbf{B}_1\mathbf{B}_2$, $\mathbf{B}_1 + \mathbf{B}_2$, or any combination of such things. For instance,

$$\Psi[\mathbf{B}_1] = \mathbf{p}_1\mathbf{p}_2[\mathbf{B}_1]\varepsilon'_1\varepsilon'_2 = \mathbf{p}_1[\mathbf{B}_1]\varepsilon'_1\mathbf{p}_2\varepsilon'_2 = \mathbf{p}_1[\mathbf{B}_1]\varepsilon'_1 \tag{7.1.2b}$$

(just as before),

$$\Psi[\mathbf{B}_1\mathbf{B}_2] = \mathbf{p}_1[\mathbf{B}_1]\varepsilon'_1\mathbf{p}_2[\mathbf{B}_2]\varepsilon'_2 = \Psi[\mathbf{B}_1]\Psi[\mathbf{B}_2], \tag{7.1.2c}$$

and

$$\Psi[\mathbf{B}_1 + \mathbf{B}_2] = \Psi[\mathbf{B}_1] + \Psi[\mathbf{B}_2]. \tag{7.1.2d}$$

Then we can write (remembering that $\mathbf{V}_i$ is still $\mathbf{B}_i^{-1}$)

$$C(x) = \frac{\Psi[\exp(-x\mathbf{B}_2)(\mathbf{I} + \mathbf{V}_1\mathbf{B}_2)^{-1}]}{\Psi[\exp(-x\mathbf{B}_2)]}. \tag{7.1.3a}$$

In particular, the probability that the customer in S_1 will finish before the one in S_2, given that they started at the same time, is

$$C(0) = \int_0^\infty R_2(t)b_1(t)\,dt = \Psi[\mathbf{B}_1(\mathbf{B}_1 + \mathbf{B}_2)^{-1}] = \Psi[(\mathbf{I} + \mathbf{V}_1\mathbf{B}_2)^{-1}]. \tag{7.1.3b}$$

Note that when S_2 is exponential (i.e., one-dimensional), $\mathbf{B}_2 = \lambda$ and $C(0) = \Psi[(\mathbf{I} + \lambda\mathbf{V}_1)^{-1}]$, which from (3.1.10) and Theorem 3.1.1 is indeed the Laplace transform, $[B_1^*(\lambda)]$ of $b_1(x)$, as stated in the second paragraph of this chapter. Some authors have used the symbolic notation for $C(0)$ in general,

$$C(0) = B_1^*(\mathbf{B}_2),$$

although it is not clear what it means, except in terms of (7.1.3b)†.

Now, you are dying to ask, "What can $(\mathbf{B}_1 + \mathbf{B}_2)$ [or $(\mathbf{I} + \mathbf{V}_1\mathbf{B}_2)$] mean? How can one add two matrices from two different spaces together? After all, they may not even have the same dimensions. Why, that is like adding *apples and oranges*!" (and indeed it is). We can still delay giving the full

† Asmussen uses a very similar notation in his book [ASMU87], but be warned that his meaning is completely different.

answer if we avoid having to use $(\mathbf{B}_1 + \mathbf{B}_2)$ directly. For instance, we can formally expand the expression for $C(0)$ in a Maclaurin series, as follows:

$$C(0) = \Psi\left[\mathbf{I} + \sum_{k=1}^{\infty} (-1)^k \mathbf{V}_1{}^k \mathbf{B}_2{}^k\right] = 1 + \sum_{k=1}^{\infty} (-1)^k \Psi[\mathbf{V}_1{}^k]\Psi[\mathbf{B}_2{}^k].$$

[(Remember that if two operators, $\mathbf{A}$ and $\mathbf{B}$, commute, then $(\mathbf{AB})^k = \mathbf{A}^k \mathbf{B}^k$.] We know what $\Psi[\mathbf{V}_1{}^k]$ and $\Psi[\mathbf{B}_2{}^k]$ are from (3.1.8b) and (3.1.9). Therefore, if the series converges, we can write

$$C(0) = \sum_{k=0}^{\infty} \frac{\mathrm{E}(X_1^k)}{k!} R_2^{(k)}(0). \tag{7.1.3c}$$

We could have gotten this expression directly from the integral form for $C(0)$ in (7.1.3b), but still, it does show that the matrix forms in that equation have real meaning. The power of our formalism is utilized only when we can use $(\mathbf{B}_1 + \mathbf{B}_2)^{-1}$ directly. So, without further delay, we finally show how this is done.

7.2. DIRECT PRODUCTS OF VECTOR SPACES

Equations involving matrices that operate on vectors in different spaces are not uncommon, although they are usually restricted to combinations of square matrices of order m with matrices or order 1, the scalars. In this case, no problems arise, since there is a natural embedding of the scalars into the matrices of order m: The scalars are isomorphic to the diagonal matrices whose nonzero elements are all equal. Because of this embedding, one does not hesitate to write $a = a \cdot \mathbf{I}$, even though this equality does not make any sense technically. In this chapter we are dealing with two sets of matrices of order *greater than* 1. Before equations containing these objects can be evaluated, the matrices must be replaced by their images under an embedding into a *direct-product space*, much like the scalar a is replaced by $a \cdot \mathbf{I}$ before the expression $\mathbf{A} + a \cdot \mathbf{I}$ can be evaluated.

7.2.1. Kronecker Products

The Kronecker product is one way to represent the embedding (or combining) of two disjoint operator spaces. In particular, if $\mathbf{K}_1$ is an $m_1 \times n_1$ matrix operating on objects in space 1, and $\mathbf{K}_2$ is an $m_2 \times n_2$ matrix of space 2, the Kronecker product of $\mathbf{K}_1$ and $\mathbf{K}_2$, denoted by $\mathbf{K}_1 \otimes \mathbf{K}_2$, is the matrix of size $(m_1 m_2) \times (n_1 n_2)$ that is obtained by multiplying each element of $\mathbf{K}_1$ [designated as $(\mathbf{K}_1)_{ij}$] by the full matrix, $\mathbf{K}_2$. $\mathbf{K}_1 \otimes \mathbf{K}_2$ can be regarded as an $m_1 \times n_1$ matrix whose elements are themselves matrices of size $m_2 \times n_2$. For instance, let $\mathbf{K}_1$ be 2×3; then

$$\mathbf{K} := \mathbf{K}_1 \otimes \mathbf{K}_2 = \begin{bmatrix} (\mathbf{K}_1)_{11}\,\mathbf{K}_2 & (\mathbf{K}_1)_{12}\,\mathbf{K}_2 & (\mathbf{K}_1)_{13}\,\mathbf{K}_2 \\ (\mathbf{K}_1)_{21}\,\mathbf{K}_2 & (\mathbf{K}_1)_{22}\,\mathbf{K}_2 & (\mathbf{K}_1)_{23}\,\mathbf{K}_2 \end{bmatrix}. \tag{7.2.1}$$

Note that the Kronecker product is *not* commutative or symmetric. That is, $\mathbf{K_1} \otimes \mathbf{K_2} \neq \mathbf{K_2} \otimes \mathbf{K_1}$, although the two representations are equivalent. What we are doing, in essence, is creating a supermatrix, $\mathbf{K}$, with elements, $K_{\underline{k}\underline{l}}$, where $\underline{k}$ and $\underline{l}$ are themselves ordered pairs. That is,

$$\underline{k} = (k_1, k_2) \in \{(k_1, k_2) \mid k_1 \in \Xi_1, k_2 \in \Xi_2\},$$

where Ξ_i is the set of internal states of S_i. In order to write down $\mathbf{K}$ in a rectangular array, it is necessary to give a linear ordering to the pairs (k_1, k_2). Equation (7.2.1) implies one such ordering; $\underline{\mathbf{K}} := \mathbf{K_2} \otimes \mathbf{K_1}$ would give a different ordering. $\mathbf{K}$ and $\underline{\mathbf{K}}$ are the same size and have the same elements, but they are arranged differently. With this definition, the following multiplication rule is valid. Let $\mathbf{K_i}$ and $\mathbf{L_i}$ be any two arrays in space i for which $\mathbf{K_i L_i}$ is defined; then

$$\mathbf{KL} = [\mathbf{K_1} \otimes \mathbf{K_2}] \cdot [\mathbf{L_1} \otimes \mathbf{L_2}] = \mathbf{K_1 L_1} \otimes \mathbf{K_2 L_2}. \tag{7.2.2}$$

Note that $\mathbf{KL} = \mathbf{LK}$ if and only if $\mathbf{K_1 L_1} = \mathbf{L_1 K_1}$ and $\mathbf{K_2 L_2} = \mathbf{L_2 K_2}$.

To keep our ordering of elements consistent, we adhere to the following conventions. What we must do is embed the row vectors, column vectors, and square matrices of each space (e.g., $\mathbf{p_i}$, $\mathbf{B_i}$, and $\boldsymbol{\varepsilon}'_\mathbf{i}$) into the product space. As implied by the nonequality of $\mathbf{K}$ and $\underline{\mathbf{K}}$, this cannot be done in a symmetric manner. We will use the single symbol, $\hat{}$ (called a *caret* or *hat*), to designate this mapping. Thus

$$\hat{\mathbf{A}}_\mathbf{1} := \mathbf{A_1} \otimes \mathbf{I_2}$$

and

$$\hat{\mathbf{A}}_\mathbf{2} := \mathbf{I_1} \otimes \mathbf{A_2},$$

where $\mathbf{I_i}$ is the identity matrix of dimensions $m_i \times m_i$, and $\mathbf{A_i}$ is any matrix of that dimension. Both $\hat{\mathbf{A}}_\mathbf{1}$ and $\hat{\mathbf{A}}_\mathbf{2}$ are of dimension $(m_1 m_2) \times (m_1 m_2)$. The subscripts, 1 or 2, on all matrices and vectors denote the space they come from, even after the embedding. Matrices without any subscript are assumed to be in the product space already. The special matrix

$$\mathbf{I} := \mathbf{I_1} \otimes \mathbf{I_2}$$

is the identity matrix of the product space.

From (7.2.2) we have the nice property that

$$\hat{\mathbf{A}}_\mathbf{1} \cdot \hat{\mathbf{A}}_\mathbf{2} = \mathbf{A_1} \otimes \mathbf{A_2} = \hat{\mathbf{A}}_\mathbf{2} \cdot \hat{\mathbf{A}}_\mathbf{1}.$$

This is just the property we needed to satisfy our *independence principle*. The embedded matrices commute, even though the Kronecker product of the two matrices does not. We also can see that the *hat* and *inverse* operators commute. Let $\mathbf{R_1} := \mathbf{A_1}^{-1}$; then

$$\hat{\mathbf{R}}_\mathbf{1} = \mathbf{R_1} \otimes \mathbf{I_2} = \mathbf{A_1}^{-1} \otimes \mathbf{I_2} = (\mathbf{A_1} \otimes \mathbf{I_2})^{-1} = (\hat{\mathbf{A}}_\mathbf{1})^{-1}.$$

Our next project is to embed the various vectors into the product space. Let $\mathbf{a_i}$ be any row vector in space i, and $\mathbf{b}'_\mathbf{i}$ be any column vector. Then

$$\hat{\mathbf{a}}_1 := \mathbf{a}_1 \otimes \mathbf{I}_2, \quad \hat{\mathbf{b}}'_1 := \mathbf{b}'_1 \otimes \mathbf{I}_2, \tag{7.2.3a}$$

and

$$\hat{\mathbf{a}}_2 := \mathbf{I}_1 \otimes \mathbf{a}_2, \quad \hat{\mathbf{b}}'_2 := \mathbf{I}_1 \otimes \mathbf{b}'_2. \tag{7.2.3b}$$

This seems simple enough, but these objects are *not* vectors in the product space. They are, in fact, rectangular matrices of the following dimensions (read "*the dimensions of* · " for "Dim[·]"):

$$\mathrm{Dim}[\hat{\mathbf{a}}_1] = (m_2) \times (m_1 m_2),$$

$$\mathrm{Dim}[\hat{\mathbf{b}}'_1] = (m_1 m_2) \times (m_2),$$

$$\mathrm{Dim}[\hat{\mathbf{a}}_2] = (m_1) \times (m_1 m_2),$$

and

$$\mathrm{Dim}[\hat{\mathbf{b}}'_2] = (m_1 m_2) \times (m_1).$$

We know that the simple *dot product*, $\mathbf{a}_2 \mathbf{b}'_2$ is a scalar, call it c. But

$$\hat{\mathbf{a}}_2 \hat{\mathbf{b}}'_2 = [\mathbf{I}_1 \otimes \mathbf{a}_2] \cdot [\mathbf{I}_1 \otimes \mathbf{b}'_2] = \mathbf{I}_1 \mathbf{I}_1 \otimes \mathbf{a}_2 \mathbf{b}'_2 = c\, \mathbf{I}_1 \otimes 1 = c\, \mathbf{I}_1. \tag{7.2.3c}$$

What, then, are the appropriate vectors for the product space? Just as we found that the Kronecker product of two square matrices is a square matrix, we can see that the Kronecker product of two row vectors is a row vector with $(m_1 m_2)$ components (i.e., $\mathbf{a}_1 \otimes \mathbf{a}_2$ is a row vector); similarly for column vectors. In particular, we define the two special vectors

$$\mathbf{p} := \mathbf{p}_1 \otimes \mathbf{p}_2 \tag{7.2.4a}$$

and

$$\boldsymbol{\varepsilon}' := \varepsilon'_1 \otimes \varepsilon'_2. \tag{7.2.4b}$$

Yes, $\boldsymbol{\varepsilon}'$ is an $(m_1 m_2)$-dimensional column vector of all 1's. Also,

$$\mathbf{p} \cdot \boldsymbol{\varepsilon}' = [\mathbf{p}_1 \otimes \mathbf{p}_2] \cdot [\varepsilon'_1 \otimes \varepsilon'_2] = \mathbf{p}_1 \varepsilon'_1 \otimes \mathbf{p}_2 \varepsilon'_2 = 1 \otimes 1 = 1. \tag{7.2.4c}$$

Well, strictly speaking, $1 \otimes 1$ is not exactly the same as "1," but they have the same effect on everything. After all, what is a 1 by 1 matrix whose only element is 1?

The embedded vectors $\hat{\mathbf{a}}_i$ and $\hat{\mathbf{b}}'_i$ are needed, but we must be careful how they are used. For instance,

$$\hat{\mathbf{p}}_1 \cdot \hat{\mathbf{p}}_2 \neq \mathbf{p}.$$

In fact, it is not defined, since the object $\mathbf{I}_2 \mathbf{p}_2$ has no meaning. What we must use, instead, is

$$\mathbf{p} = \mathbf{p}_1 \hat{\mathbf{p}}_2 = \mathbf{p}_2 \hat{\mathbf{p}}_1.$$

In general, we have the following lemma.

Lemma 7.2.1: Let $\mathbf{a}_i$, $\mathbf{b}'_i$, and $\mathbf{A}_i$ be objects from space i $(i = 1, 2)$. Then the following are all vectors in the product space.

Row vectors:†

$$\mathbf{a}_1\hat{\mathbf{a}}_2 = \mathbf{a}_2\hat{\mathbf{a}}_1 = \mathbf{a}_1 \otimes \mathbf{a}_2 \neq \mathbf{a}_2 \otimes \mathbf{a}_1;$$

$$\mathbf{a}_2\hat{\mathbf{a}}_1\hat{\mathbf{A}}_2 = \mathbf{a}_1\hat{\mathbf{a}}_2\hat{\mathbf{A}}_2 = \mathbf{a}_2\mathbf{A}_2\hat{\mathbf{a}}_1;$$

$$\mathbf{a}_1\mathbf{A}_1\hat{\mathbf{a}}_2 = \mathbf{a}_1\hat{\mathbf{a}}_2\hat{\mathbf{A}}_1 = \mathbf{a}_2\hat{\mathbf{a}}_1\hat{\mathbf{A}}_1.$$

Column vectors:†

$$\hat{\mathbf{b}}'_1\mathbf{b}'_2 = \hat{\mathbf{b}}'_2\mathbf{b}'_1 = \mathbf{b}'_1 \otimes \mathbf{b}'_2 \neq \mathbf{b}'_2 \otimes \mathbf{b}'_1;$$

$$\hat{\mathbf{b}}'_1\mathbf{A}_2\mathbf{b}'_2 = \hat{\mathbf{A}}_2\hat{\mathbf{b}}'_1\mathbf{b}'_2 = \hat{\mathbf{A}}_2\hat{\mathbf{b}}'_2\mathbf{b}'_1;$$

$$\hat{\mathbf{A}}_1\hat{\mathbf{b}}'_2\mathbf{b}'_1 = \hat{\mathbf{b}}'_2\mathbf{A}_1\mathbf{b}'_1 = \hat{\mathbf{A}}_1\hat{\mathbf{b}}'_1\mathbf{b}'_2.$$

Mixed vectors:

$$\hat{\mathbf{a}}_1 \cdot \hat{\mathbf{b}}'_1 = [\mathbf{a}_1 \cdot \mathbf{b}'_1]\mathbf{I}_2$$

$$\hat{\mathbf{a}}_2 \cdot \hat{\mathbf{b}}'_2 = [\mathbf{a}_2 \cdot \mathbf{b}'_2]\mathbf{I}_1,$$

where $[\mathbf{a}_i \cdot \mathbf{b}_i]$ is a scalar. The first and fourth sets of these equations come close to satisfying our commutativity property, so with some care we can assume that our independence principle applies. ■

7.2.2. Ψ Projections onto Subspaces

We next deal with *projections*, or *deflations.* Here we *project* or *deflate* a square matrix in the product space to one in space i. We already deflated matrices to scalars by use of the $\Psi[\cdot]$ operators. We generalize that here. First define (or rather define again)

$$\Psi[\mathbf{X}] := \mathbf{p}\mathbf{X}\boldsymbol{\varepsilon}',$$

where $\mathbf{X}$ is any square matrix in the product space and $\mathbf{p}$ and $\boldsymbol{\varepsilon}'$ are given by (7.2.4). This is clearly a scalar, with the same properties that we wanted it to have in (7.1.2a). The two new projections are those that deflate $\mathbf{X}$ to space i. Define the following:

$$\Psi_2[\mathbf{X}] := \hat{\mathbf{p}}_1[\mathbf{X}]\hat{\boldsymbol{\varepsilon}}'_1, \tag{7.2.5a}$$

$$\Psi_1[\mathbf{X}] := \hat{\mathbf{p}}_2[\mathbf{X}]\hat{\boldsymbol{\varepsilon}}'_2. \tag{7.2.5b}$$

Note the apparent mismatch between the subscripts on the Ψ's and on the vectors. This is correct, for $\Psi_2[\mathbf{X}]$ *is* a matrix in space 2. In a sense, the vectors $\mathbf{p}_1$ and $\boldsymbol{\varepsilon}'_1$ have deflated the dependence of $\mathbf{X}$ on space 1 to a scalar. This is quite clear if $\mathbf{X}$ is itself an embedding of an operator in space 2. Suppose that $\mathbf{X} = \hat{\mathbf{X}}_2 = \mathbf{I}_1 \otimes \mathbf{X}_2$. Then

† For embedded row vectors, the *hatted* object must be on the right, while for embedded column vectors, the *hatted* object must be on the left.

$$\Psi_2[\mathbf{X}] = \hat{\mathbf{p}}_1[\mathbf{I}_1 \otimes \mathbf{X}_2]\hat{\varepsilon}'_1 = (\mathbf{p}_1 \otimes \mathbf{I}_2)\cdot[\mathbf{I}_1 \otimes \mathbf{X}_2]\cdot(\varepsilon'_1 \otimes \mathbf{I}_2)$$
$$= [\mathbf{p}_1 \mathbf{I}_1 \varepsilon'_1] \otimes (\mathbf{I}_2 \mathbf{X}_2 \mathbf{I}_2) = 1 \otimes \mathbf{X}_2 = \mathbf{X}_2 .$$

Thus we see that the projections $\Psi_i[\cdot]$ are inverses of the embeddings $\hat{\cdot}_i$, in an operator sense, and in fact satisfy the properties,

$$\hat{\Psi}_i[\hat{\Psi}_i[\mathbf{X}]] = \hat{\Psi}_i[\mathbf{X}], \tag{7.2.5c}$$

which can be proven after some effort by direct substitution. The most important single property of projection operators is that they are idempotent. That is, successive operations yield the same result, which indeed is shown by (7.2.5c).

We look further at $\Psi_1[\mathbf{X}]$, with the intention of reducing it to a scalar. Then

$$\mathbf{p}_1 \Psi_1[\mathbf{X}]\varepsilon'_1 = \mathbf{p}_1 \hat{\mathbf{p}}_2[\mathbf{X}]\hat{\varepsilon}'_2 \varepsilon'_1 = \mathbf{p}[\mathbf{X}]\boldsymbol{\varepsilon}' = \Psi[\mathbf{X}]. \tag{7.2.6a}$$

Similarly,

$$\mathbf{p}_2 \Psi_2[\mathbf{X}]\varepsilon'_2 = \Psi[\mathbf{X}]. \tag{7.2.6b}$$

Thus the order in which one deflates **X** is immaterial. Now, we could have written $\Psi_2[\Psi_1[\mathbf{X}]] = \Psi_1[\Psi_2[\mathbf{X}]]$, but the outer Ψ_2 (or Ψ_1) implies that this is an object in 2-space (or 1-space), when in fact it is a scalar. Therefore, we will use the notation

$$\Psi[\Psi_1[\mathbf{X}]] = \Psi[\Psi_2[\mathbf{X}]] = \Psi[\mathbf{X}], \tag{7.2.6c}$$

since this notation unambiguously says that whatever is inside the brackets is reduced to a scalar.

Before going on to the ME/ME/1//*N* queue, we conclude this section with three important lemmas.

Lemma 7.2.2: [*Eigenvalues and eigenvectors of* $(\mathbf{A}_1 \otimes \mathbf{B}_2)$ *and* $(\hat{\mathbf{A}}_1 + \hat{\mathbf{B}}_2)$]
Let

$$\alpha_1, \alpha_2, \cdots, \alpha_{m_1}$$

be the eigenvalues of $\mathbf{A}_1$, with corresponding left eigenvectors

$$\mathbf{a}_1, \mathbf{a}_2, \cdots, \mathbf{a}_{\mathbf{m}_1}.$$

Furthermore, let

$$\beta_1, \beta_2, \cdots, \beta_{m_2}$$

be the eigenvalues of $\mathbf{B}_2$, with corresponding left eigenvectors

$$\mathbf{b}_1, \mathbf{b}_2, \cdots, \mathbf{b}_{\mathbf{m}_2}.$$

Then the eigenvalues of $\mathbf{A}_1 \otimes \mathbf{B}_2$ $(\hat{\mathbf{A}}_1 \hat{\mathbf{B}}_2)$ are the $(m_1 \cdot m_2)$ products,

$$\alpha_k \, \beta_l$$

(i.e., any eigenvalue of $\mathbf{A}_1$ times any eigenvalue of $\mathbf{B}_2$ is an eigenvalue of their Kronecker product). The corresponding left eigenvector is

$$\mathbf{a_k} \otimes \mathbf{b_l}.$$

Similarly, the eigenvalues of $\hat{\mathbf{A}}_1 + \hat{\mathbf{B}}_2$ are the $(m_1 \cdot m_2)$ sums,

$$\alpha_k + \beta_l,$$

with the same eigenvector, $\mathbf{a_k} \otimes \mathbf{b_l}$. The right eigenvectors are similarly constructed from the right eigenvectors of $\hat{\mathbf{A}}_1$ and $\hat{\mathbf{B}}_2$. ■

We do not make any use of this lemma in this book, but it may be significant in future research, perhaps in conjunction with the next lemma.

Lemma 7.2.3: Remember that $\mathbf{Q_i} = \boldsymbol{\varepsilon}'_\mathbf{i}\mathbf{p_i}$. Thus

$$\hat{\mathbf{Q}}_1 = [\boldsymbol{\varepsilon}'_1\mathbf{p}_1] \otimes \mathbf{I}_2 = [\boldsymbol{\varepsilon}'_1 \otimes \mathbf{I}_2]\cdot[\mathbf{p}_1 \otimes \mathbf{I}_2] = \hat{\boldsymbol{\varepsilon}}'_1\,\hat{\mathbf{p}}_1$$

and

$$\hat{\mathbf{Q}}_2 = \mathbf{I}_1 \otimes [\boldsymbol{\varepsilon}'_2\mathbf{p}_2] = \hat{\boldsymbol{\varepsilon}}'_2\,\hat{\mathbf{p}}_2.$$

The product space $\mathbf{Q}$ is idempotent [i.e., $\mathbf{Q}^2 = \mathbf{Q}$] and of rank† 1, and satisfies the following:

$$\mathbf{Q} := \boldsymbol{\varepsilon}'\mathbf{p} = [\boldsymbol{\varepsilon}'_1 \otimes \boldsymbol{\varepsilon}'_2]\cdot[\mathbf{p}_1 \otimes \mathbf{p}_2] = \boldsymbol{\varepsilon}'_1\mathbf{p}_1 \otimes \boldsymbol{\varepsilon}'_2\mathbf{p}_2 = \mathbf{Q}_1 \otimes \mathbf{Q}_2 = \hat{\mathbf{Q}}_1\hat{\mathbf{Q}}_2.$$

Both $\hat{\mathbf{Q}}_\mathbf{i}$'s are also idempotent but they are not of rank 1. Instead, $\hat{\mathbf{Q}}_1$ is of rank m_2, and $\hat{\mathbf{Q}}_2$ is of rank m_1. Furthermore, there exist m_2 (left and right) eigenvectors of $\hat{\mathbf{Q}}_1$ with eigenvalue 1, and $m_2(m_1 - 1)$ (left and right) eigenvectors with eigenvalue, 0. Furthermore, the matrix $(\mathbf{I} - \hat{\mathbf{Q}}_1)$ is also idempotent, with rank $m_2(m_1 - 1)$, satisfying the *null* or *orthogonality* equation,

$$(\mathbf{I} - \hat{\mathbf{Q}}_1)\hat{\mathbf{Q}}_1 = \mathbf{O},$$

with an identical result for $\hat{\mathbf{Q}}_2$. In fact, *every* idempotent matrix satisfies the null equation. It follows directly that the following are true:

$$\hat{\mathbf{Q}}_1\hat{\boldsymbol{\varepsilon}}'_1 = \hat{\boldsymbol{\varepsilon}}'_1, \quad \hat{\mathbf{Q}}_1\boldsymbol{\varepsilon}' = \boldsymbol{\varepsilon}', \quad \hat{\mathbf{p}}_1\hat{\mathbf{Q}}_1 = \hat{\mathbf{p}}_1, \quad \mathbf{p}\hat{\mathbf{Q}}_1 = \mathbf{p}.$$

Recall that $\hat{\boldsymbol{\varepsilon}}'_1$ is not a vector, but the first of the equations above tells us that each of its m_2 columns is a right eigenvector of $\hat{\mathbf{Q}}_1$ with eigenvalue 1. Since $\hat{\mathbf{Q}}_1$ is of rank m_2, there are no other unit eigenvectors. The equivalent can be said of the m_2 rows of $\hat{\mathbf{p}}_1$. The duals of all these statements are valid for $\hat{\mathbf{Q}}_2$. ■

Also, from (7.2.3c) we know that $\hat{\mathbf{p}}_1\hat{\boldsymbol{\varepsilon}}'_1 = \mathbf{I}_2$ and $\hat{\mathbf{p}}_2\hat{\boldsymbol{\varepsilon}}'_2 = \mathbf{I}_1$.

Lemma 7.2.4: Let $\mathbf{F}$ and $\mathbf{G}$ be any matrices in the product space, and let $\Psi_2[\mathbf{I} - \mathbf{GF}]$ be nonsingular. Then the matrix $\mathbf{I} - \mathbf{F}\hat{\mathbf{Q}}_1\mathbf{G}$ is nonsingular, and its inverse is

$$(\mathbf{I} - \mathbf{F}\hat{\mathbf{Q}}_1\mathbf{G})^{-1} = \mathbf{I} + \mathbf{F}\left(\hat{\Psi}_2[\mathbf{I} - \mathbf{GF}]\right)^{-1}\hat{\mathbf{Q}}_1\mathbf{G}. \tag{7.2.7}$$

† Recall that the *rank* of a finite dimensional matrix is equal to the number of its non-zero eigenvalues. $\mathbf{Q}$ has one eigenvalue equal to 1, and all the rest are equal to 0; thus it is of rank 1.

Interchanging the indices 1 and 2 gives the dual result. We have used the notation

$$\hat{\Psi}_2[\mathbf{X}] := \mathbf{I}_1 \otimes \Psi_2[\mathbf{X}].$$

The proof, though tedious, is by direct multiplication. ■

Note that this lemma is a direct generalization of Lemma 4.2.1, and Equations (4.2.2).

Before going on, the reader should be sure that the material of this section is fairly familiar. However, a specific example of embedded matrices is deferred until we have the explicit solution for the ME/ME/1//*N* loop. Perhaps the best strategy would be to read everything, up through the example, as best one can, and then go back to the beginning of this section.

7.3. STEADY-STATE ME/ME/1//*N* LOOP

We have set up a rather elaborate mathematical apparatus and will present a considerable number of formulas before this chapter is completed. If the reader feels that the concrete results we give appear small in comparison, be encouraged. We are presenting more formulas than necessary in the hope that they will help some reader to discover further significant results. We will touch on this at the end of the chapter.

7.3.1. Balance Equations

Let us consider the usual two-server loop as given in Figure 7.3.1. Each subsystem S_i, $i = 1, 2$, can only have one active customer at a time, and the queueing

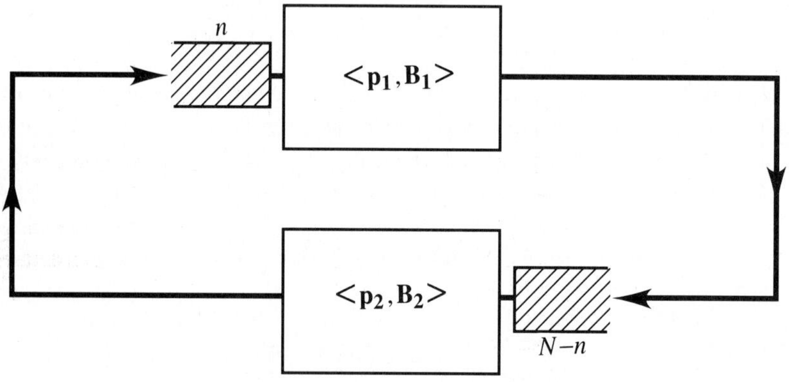

Figure 7.3.1: Closed loop of two matrix exponential servers. There are n customers at S_1, and the one being served is at phase $k_1 \in \Xi_1$. $N - n$ customers are at S_2, with the active one being at phase $k_2 \in \Xi_2$. Thus the system is in state $\{\underline{k}; n; N\}$, where $\underline{k} = (k_1, k_2) \in \Xi$.

discipline is FCFS. Both S_1 and S_2 are nonexponential and represented by $<\mathbf{p_i}, \mathbf{B_i}>$, with dimension $m_i > 1$, and the associated objects, $\mathbf{M_i}$, $\mathbf{V_i}$, $\varepsilon'_\mathbf{i}$, $\mathbf{Q_i}$, $\mathbf{P_i}$, and $\mathbf{q'_i} = (\mathbf{I_i} - \mathbf{P_i})\varepsilon'_\mathbf{i}$. As before, the diagonal elements of $\mathbf{M_1}$ are denoted by μ_k ($k = 1$ to m_1), and as a generalization of previous chapters, the diagonal elements of $\mathbf{M_2}$ are denoted by λ_k ($k = 1$ to m_2). N is the number of customers in the system, and n is the number at S_1 (with $N-n$ customers at S_2). If neither subsystem is empty, we must know where both of the active customers are to specify the system completely. Let Ξ_i be the set of phases associated with S_i, where $|\Xi_i| = m_i$. We extend the notation further.

Definition 7.3.1

$\{\underline{k}; n; N\}$ *corresponds to one possible state of an* ME/ME/1//N *loop, for* $0 < n < N$. *N is the total number of customers in the system, n is the number of customers at S_1, including the one in service, and $\underline{k}$ stands for the ordered pair (k_1, k_2), where $k_i \in \Xi_i$.* We say that the system is in the state $\{\underline{k}; n; N\}$. $\Xi := \Xi_1 \otimes \Xi_2 := \{(k_1, k_2) \mid k_i \in \Xi_i, i = 1, 2\}$. Since only one customer can be active at a time in S_i, Ξ is the set of all *internal states* of the system as a whole. As long as neither queue is empty, we can say that the system is in internal state $\underline{k} \in \Xi$, or that the active customers are at phases k_1 and k_2 in their respective subsystems. ◊

Clearly, $|\Xi| = (m_1 \cdot m_2)$, but this full space is relevant only if $n \neq 0$ and $n \neq N$. In those two cases, the state space collapses to Ξ_2 or Ξ_1, respectively. With this understanding, we define the steady-state probability vectors.

Definition 7.3.2

$[\mathbf{\Pi}(n; N)]_{\underline{k}}$:= *steady-state probability that there are n* $(0 < n < N)$ *customers at S_1 and $N - n$ customers at S_2, where $\underline{k} = (k_1, k_2) \in \Xi$ (i.e., the active customer at S_i is at phase k_i).* $[\mathbf{\Pi}(n; N)]$ is an $(m_1 \cdot m_2)$-dimensional row vector whose components are ordered according to the Kronecker product convention implied in (7.2.1), which also corresponds to the lexicographical ordering

$$\mathbf{\Pi}(n; N) = [\Pi_{(1,1)}, \Pi_{(1,2)}, \cdots, \Pi_{(1,m_2)}, \Pi_{(2,1)}, \cdots, \Pi_{(m_1,m_2)}].$$

[We have suppressed the components' dependence on $(n;N)$.] The associated scalar probability is denoted by

$$r(n; N) = \mathbf{\Pi}(n; N)\boldsymbol{\varepsilon}' \tag{7.3.1}$$

The steady-state probability vector for $n = 0$ is a vector in 2-space, since no one is at S_1, and is denoted by $\boldsymbol{\pi}_2(0; N)$. Similarly, the probability vector for $n = N$ is denoted by $\boldsymbol{\pi}_1(N; N)$. For convenience [in analogy with what we did in (4.1.1)], we define

$$\mathbf{\Pi}(0; N) := \mathbf{p_1} \otimes \boldsymbol{\pi}_2(0; N) \tag{7.3.2a}$$

and

$$\Pi(N\,;N) := \pi_1(N\,;N) \otimes \mathbf{p}_2. \tag{7.3.2b}$$

With these definitions, (7.3.1) is valid for all n. ◊

We are now ready to set up the balance equations. The process is a straightforward extension of Section 4.1.1. The complications arise in rewriting the equations as matrix equations of objects in the product space. Recall that the balance equations are derived from the fact that the sum of probability rates of arrows entering a given state are equal to the sum of those leaving. The arrows are shown in Figure 7.3.2 for an arbitrary state $\{(k,s);n;N\}$, where $(k,s) \in \Xi$

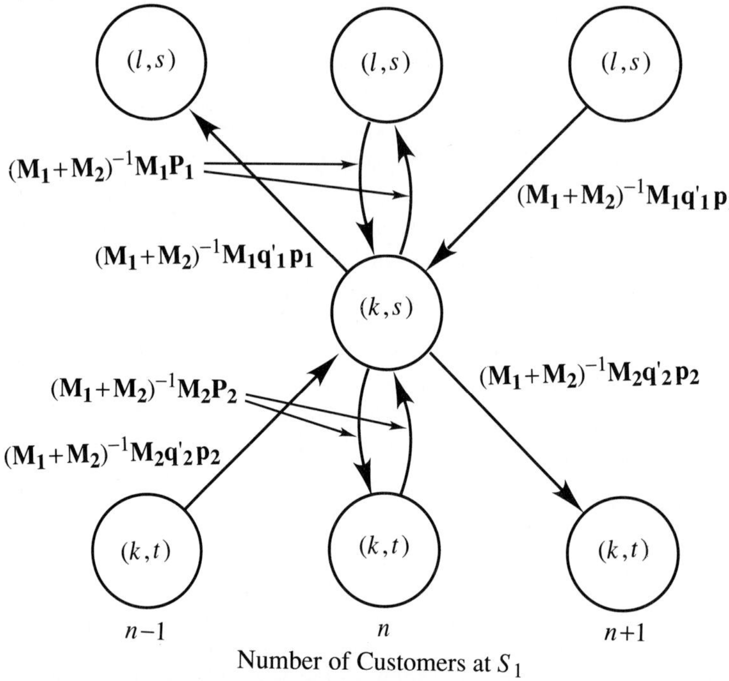

Figure 7.3.2: Steady-state transition diagram for the $\{(k,s);n;N\}$th state of an ME/ME/1//N closed loop. Arrows coming from or going to (k,s) from above correspond to events that occurred in S_1, while arrows below (k,s) correspond to events that occurred in S_2. Vertical lines are internal transitions. There are no horizontal arrows, since exactly one internal state must change [transitions such as $(\mathbf{P}_1)_{kk}$ and $(\mathbf{q}'_1)_k\,(\mathbf{p}_1)_k$ are to be visualized as changes]. The expression next to each arrow is the probability that the corresponding event will occur, given that the system is in the state designated by the node at the arrow's tail. Thus the sum of all arrows leaving node $\{(k,s);n;N\}$ equals 1. (This includes the sum over all $l \in \Xi_1$ for *up* arrows, and the sum over all $t \in \Xi_2$ for *down* arrows.)

(i.e., $k \in \Xi_1$ and $s \in \Xi_2$) and $0 < n < N$. The probability rate of an arrow is, in turn, equal to the steady-state probability that the system is in the state designated by its tail, times the probability rate of leaving that state, times the probability that an arrow will occur – given that the system is in the state of the tail.

For instance, the probability rate of the arrow going from $\{(k,s);n;N\}$ to $\{(k,t);n{+}1;N\}$ is

$$\left[[\mathbf{\Pi}(n\,;N\,)]_{(k,s)}\right]\times\left[\lambda_s+\mu_k\right]\times\left[\frac{\lambda_s}{\lambda_s+\mu_k}(\mathbf{q}'_2)_s\,(\mathbf{p}_2)_t\right]$$

$$=[\mathbf{\Pi}(n\,;N\,)]_{(k,s)}\lambda_s\,(\mathbf{q}'_2)_s\,(\mathbf{p}_2)_t.$$

This particular arrow corresponds to the following process: The customer at phase s in S_2 finishes there and leaves, $[(\mathbf{q}'_2)_s]$, going to S_1 and raising its queue length to $n+1$. Simultaneously, the next customer in the queue enters S_2, and goes to phase t, $[(\mathbf{p}_2)_t]$. There is one arrow for each phase in S_2, so we must sum over t.

When doing the same for the other seven types of arrows, we get the balance equations for $0<n<N$. Since the sum of the probabilities of the four arrows leaving the state sum to 1, the left-hand side of this equation is simple.

$$[\mathbf{\Pi}(n\,;N\,)]_{(k,s)}\,(\lambda_s+\mu_k)$$
$$=\sum_{l\,\in\,\Xi_1}[\mathbf{\Pi}(n{+}1\,;N\,)]_{(l,s)}\,\mu_l\,(\mathbf{q}'_1)_l\,(\mathbf{p}_1)_k+\sum_{l\,\in\,\Xi_1}[\mathbf{\Pi}(n\,;N\,)]_{(l,s)}\,\mu_l\,(\mathbf{P}_1)_{lk}$$
$$+\sum_{t\,\in\,\Xi_2}[\mathbf{\Pi}(n\,;N\,)]_{(k,t)}\,\lambda_t\,(\mathbf{P}_2)_{ts}+\sum_{t\,\in\,\Xi_2}[\mathbf{\Pi}(n-1\,;N\,)]_{(k,t)}\,\lambda_t\,(\mathbf{q}'_2)_s\,(\mathbf{p}_2)_t\,.$$

Let us clean this up. By our definitions, $(\mathbf{M}_1)_{kk}=\mu_k$ and $(\mathbf{M}_2)_{ss}=\lambda_s$. Therefore, we have $\mu_l\,(\mathbf{P}_1)_{lk}=[\mathbf{M}_1\mathbf{P}_1]_{lk}$, and $\mu_l\,(\mathbf{q}'_1)_l\,(\mathbf{p}_1)_k=[\mathbf{M}_1\mathbf{q}'_1\mathbf{p}_1]_{lk}$, with comparable expressions for objects with subscript 2. Next, from (4.1.3a), we know that

$$\mathbf{M}_i\mathbf{q}'_i=\mathbf{B}_i\boldsymbol{\varepsilon}'_i\quad\text{and}\quad\mathbf{M}_i\mathbf{q}'_i\mathbf{p}_i=\mathbf{B}_i\mathbf{Q}_i\,.$$

Last, the reader should verify that

$$\lambda_s+\mu_k=\left[\hat{\mathbf{M}}_1+\hat{\mathbf{M}}_2\right]_{(k,s)(k,s)}.$$

The last problem is to convert the summations in one space to a sum over indices in the product space. We can do this here because each term is made up of a $\mathbf{\Pi}$ times an object that is in only one of the subspaces. For instance, for any matrix $\mathbf{X}_1$, we can do the following:

$$\sum_{l\,\in\,\Xi_1}\Pi_{(l,s)}(\mathbf{X}_1)_{lk}=\sum_{l\,\in\,\Xi_1}\sum_{t\,\in\,\Xi_2}\Pi_{(l,t)}(\mathbf{X}_1)_{lk}\,(\mathbf{I}_2)_{ts}=\sum_{\underline{l}\,\in\,\Xi}\Pi_{\underline{l}}\,(\hat{\mathbf{X}}_1)_{\underline{lk}}=[\mathbf{\Pi}\hat{\mathbf{X}}_1]_{\underline{k}},$$

where

$$\underline{l}:=(l\,,t)\in\Xi\quad\text{and}\quad\underline{k}:=(k\,,s)\in\Xi.$$

After some practice, the reader can become comfortable with this.

Since the balance equations are valid for all $\underline{k}\in\Xi$, we can put all the pieces together and write the following vector balance equation:

$$\mathbf{\Pi}(n\,;N\,)\left[\hat{\mathbf{M}}_1+\hat{\mathbf{M}}_2\right]=\mathbf{\Pi}(n{+}1\,;N\,)\hat{\mathbf{B}}_1\hat{\mathbf{Q}}_1$$
$$+\mathbf{\Pi}(n\,;N\,)\hat{\mathbf{M}}_1\hat{\mathbf{P}}_1+\mathbf{\Pi}(n\,;N\,)\hat{\mathbf{M}}_2\hat{\mathbf{P}}_2+\mathbf{\Pi}(n-1\,;N\,)\hat{\mathbf{B}}_2\hat{\mathbf{Q}}_2.$$

Now we take all the terms with $\mathbf{\Pi}(n\,;N)$ to the left side, and recognize that $\hat{\mathbf{M}}_i - \hat{\mathbf{M}}_i\hat{\mathbf{P}}_i = \hat{\mathbf{B}}_i$, and get, finally,

$$\mathbf{\Pi}(n\,;N)\left[\hat{\mathbf{B}}_1 + \hat{\mathbf{B}}_2\right] = \mathbf{\Pi}(n{+}1\,;N)\hat{\mathbf{B}}_1\hat{\mathbf{Q}}_1 + \mathbf{\Pi}(n{-}1\,;N)\hat{\mathbf{B}}_2\hat{\mathbf{Q}}_2. \tag{7.3.3a}$$

With similar but simpler procedures, we can find the balance equations for $n = 0$ and $n = N$. We summarize these equations with the next theorem.

Theorem 7.3.1: The steady-state balance equations for an ME/ME/1//*N* loop are given by (7.3.3a) for $n = 1, 2, \cdots, N-1$, and

$$\mathbf{\Pi}(0;N)\hat{\mathbf{B}}_2 = \mathbf{\Pi}(1;N)\hat{\mathbf{B}}_1\hat{\mathbf{Q}}_1 \tag{7.3.3b}$$

and

$$\mathbf{\Pi}(N;N)\hat{\mathbf{B}}_1 = \mathbf{\Pi}(N{-}1;N)\hat{\mathbf{B}}_2\hat{\mathbf{Q}}_2. \tag{7.3.3c}$$

Because S_1 and S_2 now play symmetric roles, these equations are invariant to the interchanges $n \Leftrightarrow N-n$, together with $(\cdot)_1 \Leftrightarrow (\cdot)_2$. ■

Observe the similarity between these equations and the balance equations of Equations (4.1.3) for the M/ME/1//*N* queue.

7.3.2. Steady-State Solution

Setting up the balance equations, even in such an elegant form as given in Theorem 7.3.1, in no way guarantees that an explicit solution can be found. As we saw in Chapter 6, often one must be satisfied with a recursive algebraic solution. In attempting to solve finite difference equations, of which this is an example, one is almost always denied even a recursive solution, and as with the solution of differential equations, one must settle for a brute force numerical solution. We know this much in general: that if by luck, ingenuity, or stroke of genius one could find a formula that satisfies all the balance equations, it would be the unique nontrivial solution. Appie van de Liefvoort did just that. We now present the solution and outline the proof that it is correct.

The formula has not been used very much, and no one knows just which of the several forms that it can take will ultimately prove to be the most useful. Thus we will present several expressions for the same variables. We start with the usual set of definitions. We begin with an operator we used in Section 7.1. Let

$$\mathbf{D} := \left[\hat{\mathbf{B}}_1 + \hat{\mathbf{B}}_2\right]^{-1}. \tag{7.3.4}$$

Recall that this operator is the generator of first-finishing probabilities, and in simpler days (when one of the **B**'s was exponential) was the generator of the Laplace transform. We next introduce two matrices that look suspiciously like the **A** matrix of (4.1.4a) but are not.

$$\mathbf{S}^+ := \hat{\mathbf{B}}_1 + \hat{\mathbf{B}}_2 - \hat{\mathbf{B}}_2\hat{\mathbf{Q}}_1 = \mathbf{D}^{-1}\left[\mathbf{I} - \mathbf{D}\hat{\mathbf{B}}_2\hat{\mathbf{Q}}_1\right], \tag{7.3.5a}$$

$$\mathbf{S}^- := \hat{\mathbf{B}}_1 + \hat{\mathbf{B}}_2 - \hat{\mathbf{B}}_1\hat{\mathbf{Q}}_2 = \mathbf{D}^{-1}\left[\mathbf{I} - \mathbf{D}\hat{\mathbf{B}}_1\hat{\mathbf{Q}}_2\right]. \tag{7.3.5b}$$

Note that these matrices cannot be decomposed into a product of matrices from the two separate spaces. Also, they do not commute with each other if both S_1 and S_2 are nontrivial (i.e., have dimension greater than 1). If S_2 is one-dimensional, $\hat{\mathbf{B}}_2 \Rightarrow \lambda\mathbf{I}$, $\hat{\mathbf{Q}}_2 \Rightarrow \mathbf{I}$, and $\mathbf{S}^-$ reduces to $\lambda\mathbf{I}$, while $\mathbf{S}^+ \Rightarrow \lambda\mathbf{A}$.

We next collect several formulas that $\mathbf{S}^\pm$ satisfy in relation to the $\boldsymbol{\varepsilon}'_i$'s.

Lemma 7.3.2: The following relations hold:

$$\mathbf{S}^+\hat{\boldsymbol{\varepsilon}}'_1 = \hat{\mathbf{B}}_1\hat{\boldsymbol{\varepsilon}}'_1 \quad \text{and} \quad \mathbf{S}^-\hat{\boldsymbol{\varepsilon}}'_2 = \hat{\mathbf{B}}_2\hat{\boldsymbol{\varepsilon}}'_2, \tag{7.3.6a}$$

or equivalently [see (7.2.4b) and Lemma 7.2.3],

$$\mathbf{S}^+\hat{\mathbf{Q}}_1 = \hat{\mathbf{B}}_1\hat{\mathbf{Q}}_1 \quad \text{and} \quad \mathbf{S}^-\hat{\mathbf{Q}}_2 = \hat{\mathbf{B}}_2\hat{\mathbf{Q}}_2. \tag{7.3.6b}$$

We can also see that

$$\mathbf{S}^+\boldsymbol{\varepsilon}' = \hat{\mathbf{B}}_1\boldsymbol{\varepsilon}' \quad \text{and} \quad \mathbf{S}^-\boldsymbol{\varepsilon}' = \hat{\mathbf{B}}_2\boldsymbol{\varepsilon}', \tag{7.3.6c}$$

or equivalently,

$$\mathbf{S}^+\mathbf{Q} = \hat{\mathbf{B}}_1\mathbf{Q} \quad \text{and} \quad \mathbf{S}^-\mathbf{Q} = \hat{\mathbf{B}}_2\mathbf{Q}. \tag{7.3.6d}$$

The proofs follow directly just by carrying out the algebra. ■

Keep in mind that whereas Equations (7.3.6c) are vector equations, Equations (7.3.6a) are not, since $\boldsymbol{\varepsilon}'$ is a vector in the product space, but $\hat{\boldsymbol{\varepsilon}}'_1$ and $\hat{\boldsymbol{\varepsilon}}'_2$ are rectangular matrices of dimensions $(m_1 \cdot m_2) \times (m_2)$ and $(m_1 \cdot m_2) \times (m_1)$, respectively.

Since $\hat{\mathbf{Q}}_1$ commutes with $\hat{\mathbf{B}}_2$, and so on, there is an equivalent set of equations for $\hat{\mathbf{p}}_i$, and so on, namely

Lemma 7.3.3: The following relations hold:

$$\hat{\mathbf{p}}_1\mathbf{S}^+ = \hat{\mathbf{p}}_1\hat{\mathbf{B}}_1 \quad \text{and} \quad \hat{\mathbf{p}}_2\mathbf{S}^- = \hat{\mathbf{p}}_2\hat{\mathbf{B}}_2, \tag{7.3.7a}$$

or equivalently [see (7.2.4a) and Lemma 7.2.3],

$$\hat{\mathbf{Q}}_1\mathbf{S}^+ = \hat{\mathbf{Q}}_1\hat{\mathbf{B}}_1 \quad \text{and} \quad \hat{\mathbf{Q}}_2\mathbf{S}^- = \hat{\mathbf{Q}}_2\hat{\mathbf{B}}_2. \tag{7.3.7b}$$

We can also see that

$$\mathbf{p}\mathbf{S}^+ = \mathbf{p}\hat{\mathbf{B}}_1 \quad \text{and} \quad \mathbf{p}\mathbf{S}^- = \mathbf{p}\hat{\mathbf{B}}_2, \tag{7.3.7c}$$

or equivalently,

$$\mathbf{Q}\mathbf{S}^+ = \mathbf{Q}\hat{\mathbf{B}}_1 \quad \text{and} \quad \mathbf{Q}\mathbf{S}^- = \mathbf{Q}\hat{\mathbf{B}}_2. \tag{7.3.7d}$$

The proofs are identical to those in Lemma 7.3.2. As in the previous lemma, Equations (7.3.7c) are vector equations, but Equations (7.3.7a) are not. ■

We will need the inverses of $\mathbf{S}^{\pm}$, which if they exist, we call $\mathbf{T}^{\pm}$. Therefore, by Lemma 7.2.4 and (7.3.5a),

$$\mathbf{T}^{+} := (\mathbf{S}^{+})^{-1} := \left(\mathbf{I} - \mathbf{D}\hat{\mathbf{B}}_2\hat{\mathbf{Q}}_1\right)^{-1}\mathbf{D} = \left(\mathbf{I} + \mathbf{D}\hat{\mathbf{B}}_2\hat{\mathbf{Q}}_1\hat{\Psi}_2^{-1}[\mathbf{D}\hat{\mathbf{B}}_1]\right)\mathbf{D}, \quad (7.3.8a)$$

where we have used $\hat{\Psi}_1^{-1}[\cdot]$ † for $\left(\hat{\Psi}_1[\cdot]\right)^{-1}$. Similarly, we have

$$\mathbf{T}^{-} := (\mathbf{S}^{-})^{-1} := \left(\mathbf{I} - \mathbf{D}\hat{\mathbf{B}}_1\hat{\mathbf{Q}}_2\right)^{-1}\mathbf{D} = \left(\mathbf{I} + \mathbf{D}\hat{\mathbf{B}}_1\hat{\mathbf{Q}}_2\hat{\Psi}_1^{-1}[\mathbf{D}\hat{\mathbf{B}}_2]\right)\mathbf{D}. \quad (7.3.8b)$$

These equations are rather complicated, but it seems clear that $\mathbf{T}^{+}$ and $\mathbf{T}^{-}$ exist as long as $\Psi_2[\mathbf{D}\hat{\mathbf{B}}_1]$ and $\Psi_1[\mathbf{D}\hat{\mathbf{B}}_2]$ are nonsingular. We shall assume this to be true, since to assume otherwise would end our exploration instantly.

We now introduce the pivotal matrix of this chapter, the one that is the true generalization of the $\mathbf{U}$ in Equations (4.1.4), and the one that provides the geometric solution to the ME/ME/1//*N* loop. We use the same symbol for it.

$$\mathbf{U} := \mathbf{T}^{+}\mathbf{S}^{-} = \left[\mathbf{I} - \mathbf{D}\hat{\mathbf{B}}_2\hat{\mathbf{Q}}_1\right]^{-1}\left[\mathbf{I} - \mathbf{D}\hat{\mathbf{B}}_1\hat{\mathbf{Q}}_2\right]. \quad (7.3.9a)$$

Using (7.3.8a) and (7.3.5b), $\mathbf{U}$ can also be expressed as

$$\mathbf{U} = \left(\mathbf{I} + \mathbf{D}\hat{\mathbf{B}}_2\hat{\mathbf{Q}}_1\hat{\Psi}_2^{-1}[\mathbf{D}\hat{\mathbf{B}}_1]\right)\left[\mathbf{I} - \mathbf{D}\hat{\mathbf{B}}_1\hat{\mathbf{Q}}_2\right]. \quad (7.3.9b)$$

From Equations (7.3.5) and the discussion following them, we see that $\mathbf{U}$ reduces precisely to (4.1.4b) when S_2 is one-dimensional, while

$$\mathbf{U}^{-1} = \mathbf{T}^{-}\mathbf{S}^{+} = \left[\mathbf{I} - \mathbf{D}\hat{\mathbf{B}}_1\hat{\mathbf{Q}}_2\right]^{-1}\left[\mathbf{I} - \mathbf{D}\hat{\mathbf{B}}_2\hat{\mathbf{Q}}_1\right] \quad (7.3.9c)$$

reduces to $\mathbf{A}$ in (4.1.4a). Actually, another matrix, defined by

$$\mathbf{R} := \mathbf{S}^{-}\mathbf{T}^{+} = \left[\mathbf{I} - \mathbf{D}\hat{\mathbf{B}}_1\hat{\mathbf{Q}}_2\right]\left[\mathbf{I} - \mathbf{D}\hat{\mathbf{B}}_2\hat{\mathbf{Q}}_1\right]^{-1}, \quad (7.3.9d)$$

could equally be the pivotal matrix.

We need one final set of equations before presenting the major goal of this chapter, which as usual we precede with a definition:

$$\mathbf{E} := (\mathbf{I} - \hat{\mathbf{Q}}_1)(\mathbf{I} - \hat{\mathbf{Q}}_2) = (\mathbf{I} - \hat{\mathbf{Q}}_2)(\mathbf{I} - \hat{\mathbf{Q}}_1). \quad (7.3.10a)$$

$\mathbf{E}$ is idempotent, with rank $(m_1 - 1)\cdot(m_2 - 1)$. As with all idempotent matrices, $(\mathbf{I} - \mathbf{E})$ is also idempotent, with rank $(m_1 \cdot m_2) - (m_1 - 1)\cdot(m_2 - 1)$, which equals $m_1 + m_2 - 1$.

Lemma 7.3.4: For any matrix $\mathbf{X}$ in the product space,

$$(\mathbf{I} - \mathbf{X}\hat{\mathbf{Q}}_i)(\mathbf{I} - \hat{\mathbf{Q}}_i) = (\mathbf{I} - \hat{\mathbf{Q}}_i). \quad (7.3.10b)$$

† This is as good a place as any to caution the reader that, in general, $\Psi_i[\mathbf{F}^{-1}]$ does not equal $\Psi_i^{-1}[\mathbf{F}]$.

Therefore, since $\hat{\mathbf{Q}}_1$ and $\hat{\mathbf{Q}}_2$ commute,

$$\mathbf{S}^+\mathbf{E} = (\hat{\mathbf{B}}_1 + \hat{\mathbf{B}}_2)\mathbf{E} = \mathbf{S}^-\mathbf{E} \quad \text{and} \quad \mathbf{E}\mathbf{S}^+ = \mathbf{E}(\hat{\mathbf{B}}_1 + \hat{\mathbf{B}}_2) = \mathbf{E}\mathbf{S}^-, \tag{7.3.10c}$$

so the following identities hold:

$$\mathbf{U}\mathbf{E} = \mathbf{E} \quad \text{or} \quad (\mathbf{I} - \mathbf{U})\mathbf{E} = \mathbf{O}, \tag{7.3.10d}$$

$$\mathbf{E}\mathbf{R} = \mathbf{E} \quad \text{or} \quad \mathbf{E}(\mathbf{I} - \mathbf{R}) = \mathbf{O}, \tag{7.3.10e}$$

$$\mathbf{T}^{\pm}(\hat{\mathbf{B}}_1 + \hat{\mathbf{B}}_2)\mathbf{E} = \mathbf{E} = \mathbf{E}(\hat{\mathbf{B}}_1 + \hat{\mathbf{B}}_2)\mathbf{T}^{\pm}. \tag{7.3.10f}$$

Also,

$$\mathbf{S}^-\hat{\mathbf{Q}}_1 + \mathbf{S}^+\hat{\mathbf{Q}}_2 = [\hat{\mathbf{B}}_1 + \hat{\mathbf{B}}_2]\,[\mathbf{I} - \mathbf{E}] \tag{7.3.10g}$$

and

$$\hat{\mathbf{Q}}_1\mathbf{S}^- + \hat{\mathbf{Q}}_2\mathbf{S}^+ = [\mathbf{I} - \mathbf{E}]\,[\hat{\mathbf{B}}_1 + \hat{\mathbf{B}}_2]. \tag{7.3.10h}$$

The proofs are by direct substitution for **U**, **R**, $\mathbf{S}^{\pm}$, and $\mathbf{T}^{\pm}$. ■

Equations (7.3.10) will be used to prove our theorem, and they contain information that may be of critical importance for future research. Equations (7.3.10d) and (7.3.10e) look like eigenvalue equations, and in fact, imply that there are as many eigenvectors of **U** (and **R**) with eigenvalue 1 as the rank of **E**. Therefore, $(\mathbf{I} - \mathbf{U})$ has at most $m_1 + m_2 - 1$ nonzero eigenvalues. In the previous chapters, $m_2 = 1$, so $(\mathbf{I} - \mathbf{U})$ had m_1 nonzero eigenvalues, which was all of them, and so was invertible. But now that $m_2 > 1$, this is no longer true. The same statement can be made for **R**. This will inhibit our ability to find an explicit solution to the open ME/ME/1 queue.

The main theorem is now stated and proved.

Theorem 7.3.5: Given a closed ME/ME/1//N loop, with $N > 1$, the steady-state vector and scalar probabilities are described by

$$\mathbf{\Pi}(n\,;N) = \mathbf{\Pi}(0;N)\hat{\mathbf{B}}_2\mathbf{U}^n\,\mathbf{T}^-, \quad n = 1, 2, \cdots, N-1, \tag{7.3.11a}$$

and

$$\mathbf{\Pi}(N\,;N) = \mathbf{\Pi}(0;N)\hat{\mathbf{B}}_2\mathbf{U}^{N-1}\hat{\mathbf{V}}_1. \tag{7.3.11b}$$

Equation (7.3.11a) can also be written in the form

$$\mathbf{\Pi}(n\,;N) = \mathbf{\Pi}(0;N)\hat{\mathbf{B}}_2\mathbf{T}^-\mathbf{R}^{n-1}. \tag{7.3.11c}$$

This form would seem to be preferred, since the geometric factor appears at the right of the expression, allowing one to use

$$\mathbf{\Pi}(n\,;N) = \mathbf{\Pi}(n-1;N)\mathbf{R}, \quad n = 1, 2, \cdots, N-1$$

(which we actually do in the algorithm below). It remains to be seen which will be more significant in the long run. ■

Proof: First we show that Equations (7.3.11) are symmetric in S_1 and S_2. We do this by expressing everything in terms of $\mathbf{\Pi}(N;N)$ and interchanging 1 and 2. The interchange, in turn, causes + and − to interchange; thus $\mathbf{U}$ goes to $\mathbf{U}^{-1}$, which we shall call $\mathbf{A}$. From (7.3.11b) we get

$$\mathbf{\Pi}(0;N) = \mathbf{\Pi}(N;N)\hat{\mathbf{B}}_1\mathbf{A}^{N-1}\hat{\mathbf{V}}_2.$$

Similarly, from (7.3.11a) and the above, we get

$$\begin{aligned}\mathbf{\Pi}(n;N) &= \mathbf{\Pi}(N;N)\hat{\mathbf{B}}_1\mathbf{A}^{N-1}\hat{\mathbf{V}}_2\hat{\mathbf{B}}_2\mathbf{U}^n\mathbf{T}^- \\ &= \mathbf{\Pi}(N;N)\hat{\mathbf{B}}_1\mathbf{A}^{N-1}\mathbf{U}^n\mathbf{T}^- = \mathbf{\Pi}(N;N)\hat{\mathbf{B}}_1\mathbf{A}^{N-n}\mathbf{U}\mathbf{T}^-,\end{aligned}$$

or by expressing the first argument in terms of the number of customers at S_2, namely, $k = N - n$, and observing that $\mathbf{U}\mathbf{T}^- = \mathbf{T}^+$, we obtain

$$\mathbf{\Pi}(N-k;N) = \mathbf{\Pi}(N-0;N)\hat{\mathbf{B}}_1\mathbf{A}^k\mathbf{T}^+.$$

Clearly, these equations look just like the original ones after one makes the required interchanges. This means, then, that if we satisfy (7.3.3b), by duality we automatically satisfy (7.3.3c).

To satisfy (7.3.3b), take (7.3.11a) for $n=1$, recall again that $\mathbf{U}\mathbf{T}^- = \mathbf{T}^+$, and manipulate to get

$$\mathbf{\Pi}(1;N)\mathbf{S}^+ = \mathbf{\Pi}(0;N)\hat{\mathbf{B}}_2.$$

Now multiply both sides by $\hat{\mathbf{Q}}_1$, use (7.3.2), (7.3.6b), and Lemma 7.2.1 to get

$$\mathbf{\Pi}(1;N)\hat{\mathbf{B}}_1\hat{\mathbf{Q}}_1 = [\mathbf{p}_1 \otimes (\pi_2(0;N)\mathbf{B}_2)]\cdot[\mathbf{Q}_1 \otimes \mathbf{I}_2] = \mathbf{\Pi}(0;N)\hat{\mathbf{B}}_2.$$

In the last step we used (7.2.2) and the fact that $\mathbf{p}_1\mathbf{Q}_1 = \mathbf{p}_1$. This is indeed the same as (7.3.3b).

Last, we substitute (7.3.6b) and (7.3.11c) into the right-hand side of (7.3.3a) and get

$$\begin{aligned}\mathbf{\Pi}(n;N)[\mathbf{S}^-\mathbf{T}^+\mathbf{S}^+\hat{\mathbf{Q}}_1 + \mathbf{S}^+\mathbf{T}^-\mathbf{S}^-\hat{\mathbf{Q}}_2] &= \mathbf{\Pi}(n;N)[\mathbf{S}^-\hat{\mathbf{Q}}_1 + \mathbf{S}^+\hat{\mathbf{Q}}_2] \\ &= \mathbf{\Pi}(n;N)[\hat{\mathbf{B}}_1 + \hat{\mathbf{B}}_2] - \mathbf{\Pi}(n;N)[\hat{\mathbf{B}}_1 + \hat{\mathbf{B}}_2]\mathbf{E},\end{aligned}$$

where we used (7.3.10d). But from (7.3.10b) and (7.3.10c), we see that the last term vanishes.

$$\mathbf{\Pi}(n;N)[\hat{\mathbf{B}}_1 + \hat{\mathbf{B}}_2]\mathbf{E} = \mathbf{\Pi}(0;N)\hat{\mathbf{B}}_2\mathbf{U}^n\mathbf{T}^-[\hat{\mathbf{B}}_1 + \hat{\mathbf{B}}_2]\mathbf{E}$$

$$= \mathbf{\Pi}(0;N)\hat{\mathbf{B}}_2\mathbf{U}^n\mathbf{E} = \mathbf{p}_1 \otimes [\pi_2(0;N)\mathbf{B}_2]\mathbf{E} = [\pi_2(0;N)\mathbf{B}_2][\hat{\mathbf{p}}_1\mathbf{E}] = \mathbf{o},$$

by virtue of the fact that the last term in brackets is 0. The term that remains is equal to the left-hand side of (7.3.3a). QED

This was surely a tedious proof, but we are not yet finished. Note that the solution is given in terms of $\pi_2(0;N)$ [or $\pi_1(N;N)$]. This is an m_2 [or m_1] component object, and we can only fix one constant with the normalization requirement that the sum of all probabilities be 1. Naturally, when m_2 was 1 there was no problem, but now there is. Equation (7.3.11b) is a matrix equation with $(m_1 \cdot m_2)$ components, but only $(m_1 + m_2)$ unknowns. Therefore, these unknowns must be related in some way. We rewrite the equation using (7.3.2).

$$\pi_1(N;N)\mathbf{B}_1 \otimes \mathbf{p}_2 = \mathbf{p}_1 \otimes \pi_2(0;N)\mathbf{B}_2\mathbf{U}^{N-1}.$$

Next rewrite this equation using Lemma 7.2.1 (and momentarily drop the dependence of the π's on n and N),

$$\mathbf{p}_2\hat{\pi}_1\hat{\mathbf{B}}_1 = \pi_2\mathbf{B}_2\hat{\mathbf{p}}_1\mathbf{U}^{N-1}.$$

Next postmultiply both sides with $\hat{\varepsilon}'_1$ to get

$$\mathbf{p}_2 \cdot \left(\hat{\pi}_1\hat{\mathbf{B}}_1\hat{\varepsilon}'_1\right) = \pi_2\mathbf{B}_2\hat{\Psi}_2[\mathbf{U}^{N-1}]. \tag{7.3.12a}$$

The expression in parentheses is the inner product of two 1-space hatted vectors and is therefore equal to a scalar times $\mathbf{I}_2$ (see Lemma 7.2.1). Define that scalar to be

$$\beta(N) := \pi_1\mathbf{B}_1\varepsilon'_1.$$

Observe that the right-hand expression of (7.3.12a), just like the other side, depends only on 2-space objects; therefore, we can remove the *hats*, and the equation becomes

$$\beta(N)\mathbf{p}_2 = \pi_2\mathbf{B}_2\Psi_2[\mathbf{U}^{N-1}].$$

Solving for π_2, we get

$$\pi_2(0;N) = \beta(N)\mathbf{p}_2\Psi_2^{-1}[\mathbf{U}^{N-1}]\mathbf{V}_2. \tag{7.3.12b}$$

By similar manipulations, it can be shown that π_2 also satisfies the following eigenvector equation:

$$\pi_2 = \pi_2\mathbf{B}_2\Psi_2\left[\mathbf{U}^{N-1}\hat{\mathbf{Q}}_2\mathbf{A}^{N-1}\right]\mathbf{V}_2. \tag{7.3.12c}$$

Either of these last two equations can be used to find π_2.

It is not understood what the significance is to having *two* defining equations for π_2. Clearly, both must yield the same vector, to within a constant [which we can take to be $\beta(N)$]. This constant can be determined by normalization as follows:

$$1 = \sum_{n=0}^{N} r(n;N) = \sum_{n=0}^{N} \mathbf{\Pi}(n;N)\boldsymbol{\varepsilon}' = \mathbf{\Pi}(0;N)\boldsymbol{\varepsilon}'$$
$$+\mathbf{\Pi}(0;N)\hat{\mathbf{B}}_2[\mathbf{U} + \mathbf{U}^2 + \cdots + \mathbf{U}^{N-1}]\mathbf{T}^{-}\boldsymbol{\varepsilon}' + \mathbf{\Pi}(0;N)\hat{\mathbf{B}}_2\mathbf{U}^{N-1}\hat{\mathbf{V}}_1\boldsymbol{\varepsilon}'.$$

These expressions are so cumbersome to work with that one can get discouraged from going on. But only with continued use will we be able to find simpler formulas. For now, we follow what we did for the M/G/1//N queue in (4.1.6d), and define

$$\mathbf{K}(N) := \mathbf{I} + [\mathbf{U} + \mathbf{U}^2 + \cdots + \mathbf{U}^{N-1}]\mathbf{T}^-\hat{\mathbf{B}}_2 + \mathbf{U}^{N-1}\hat{\mathbf{V}}_1\hat{\mathbf{B}}_2.$$

Then

$$1 = \mathbf{\Pi}(0;N)\hat{\mathbf{B}}_2\mathbf{K}(N)\hat{\mathbf{V}}_2\boldsymbol{\varepsilon}' = \pi_2(0;N)\mathbf{B}_2\hat{\mathbf{p}}_1\mathbf{K}(N)\hat{\varepsilon}'_1\mathbf{V}_2\varepsilon'_2,$$

and using (7.3.12b), we obtain

$$\begin{aligned} 1 &= \beta(N)\mathbf{p}_2\Psi_2^{-1}[\mathbf{U}^{N-1}]\mathbf{V}_2\mathbf{B}_2\Psi_2[\mathbf{K}(N)]\mathbf{V}_2\varepsilon'_2 \\ &= \beta(N)\Psi\Big[\Psi_2^{-1}[\mathbf{U}^{N-1}]\Psi_2[\mathbf{K}(N)]\mathbf{V}_2\Big]. \end{aligned} \tag{7.3.13}$$

We summarize this maze of formulas with the following theorem.

Theorem 7.3.6: Explicit expressions for the steady-state vector probabilities for a closed ME/ME/1//N loop, with $N > 1$ and $0 < n < N$, are

$$\mathbf{K}(N) = \mathbf{I} + [\mathbf{U} + \mathbf{U}^2 + \cdots + \mathbf{U}^{N-1}]\mathbf{T}^-\hat{\mathbf{B}}_2 + \mathbf{U}^{N-1}\hat{\mathbf{V}}_1\hat{\mathbf{B}}_2. \tag{7.3.14a}$$

$$\beta(N)^{-1} = \Psi\Big[\Psi_2^{-1}[\mathbf{U}^{N-1}]\Psi_2[\mathbf{K}(N)]\mathbf{V}_2\Big], \tag{7.3.14b}$$

$$\pi_2(0;N) = \beta(N)\mathbf{p}_2\Psi_2^{-1}[\mathbf{U}^{N-1}]\mathbf{V}_2, \tag{7.3.14c}$$

$$\mathbf{\Pi}(n;N) = \beta(N)\mathbf{p}_2\Psi_2^{-1}[\mathbf{U}^{N-1}]\hat{\mathbf{p}}_1\mathbf{U}^{n-1}\mathbf{T}^+, \tag{7.3.14d}$$

$$\mathbf{\Pi}(N;N) = \beta(N)\mathbf{p}_2\Psi_2^{-1}[\mathbf{U}^{N-1}]\hat{\mathbf{p}}_1\mathbf{U}^{N-1}\hat{\mathbf{V}}_1. \tag{7.3.14e}$$

The associated scalar probabilities are given by

$$r(0;N) = \pi_2(0;N)\cdot\varepsilon'_2 = \beta(N)\Psi\Big[\Psi_2^{-1}[\mathbf{U}^{N-1}]\mathbf{V}_2\Big], \tag{7.3.14f}$$

$$r(n;N) = \mathbf{\Pi}(n;N)\cdot\boldsymbol{\varepsilon}' = \beta(N)\Psi\Big[\Psi_2^{-1}[\mathbf{U}^{N-1}]\cdot\Psi_2[\mathbf{U}^{n-1}\mathbf{T}^+]\Big], \tag{7.3.14g}$$

$$r(N;N) = \mathbf{\Pi}(N;N)\cdot\boldsymbol{\varepsilon}' = \beta(N)\Psi\Big[\Psi_2^{-1}[\mathbf{U}^{N-1}]\cdot\Psi_2[\mathbf{U}^{N-1}\hat{\mathbf{V}}_1]\Big]. \tag{7.3.14h}$$

Surely we will find something better someday. ■

7.3.3. Outline of an Efficient Algorithm

The formulas we have derived for the ME/ME/1//N queue appear rather intimidating, but they actually can be calculated systematically and efficiently. We aid the reader to recognize the relative ease with which this can be done by giving an algorithm for the calculation of the steady-state scalar probabilities, the mean queue length, and the throughput, parametrically on the number of customers, N, in the system. We make no claims that this is the most efficient possible, but it could be worse. We state without proof that ($\mathbf{\Pi}(n;N)\hat{\mathbf{B}}_1\varepsilon'$) is the equilibrium flow from external state n to $n+1$. Therefore [see (1.1.1), (2.1.5a), and (4.1.8)],

$$\Lambda(N) = \sum_{n=1}^{N}\mathbf{\Pi}(n;N)\hat{\mathbf{B}}_1\boldsymbol{\varepsilon}'. \tag{7.3.16}$$

In general, computational costs can be greatly reduced by working with matrix-vector products rather that matrix-matrix products. Therefore, we introduce the vector $\mathbf{k}'(N)$ of dimension $(m_1 \cdot m_2)$:

$$\mathbf{k}'(N) := \mathbf{K}(N)\,\hat{\mathbf{V}}_2\boldsymbol{\varepsilon}',$$

which has the following properties:

$$\mathbf{k}'(1) = (\hat{\mathbf{V}}_1 + \hat{\mathbf{V}}_2)\cdot\boldsymbol{\varepsilon}' = [\mathbf{V}_1\varepsilon'_1] \otimes \varepsilon'_2 + \varepsilon'_1 \otimes [\mathbf{V}_2\varepsilon'_2],$$

$$\mathbf{k}'(N) = t_2\boldsymbol{\varepsilon}' + \mathbf{U}\cdot\mathbf{k}'(N-1) \quad \text{for} \quad N > 1,$$

and

$$\hat{\mathbf{p}}_1\cdot\mathbf{k}'(N) = \Psi_2\big[\mathbf{K}(N)\big]\cdot\mathbf{V}_2\cdot\boldsymbol{\varepsilon}',$$

where $t_2 = \Psi\big[\mathbf{V}_2\big]$ is the mean service time of S_2 and $\hat{\mathbf{p}}_1\cdot\mathbf{k}'(N)$ is a column vector in 2-space. Next, the *block vectors* of dimension $(m_1 \cdot m_2) \times m_2$, defined by

$$\mathbf{u}'(N) := \mathbf{U}^N\,\hat{\varepsilon}'_1$$

satisfy the recursive relations, starting with $\mathbf{u}'(0) = \hat{\varepsilon}'_1$,

$$\mathbf{u}'(N) := \mathbf{U}\cdot\mathbf{u}'(N-1) \quad \text{for} \quad N > 0.$$

Finally, to avoid needless repetitions (and perhaps to clarify), the symbols $\hat{\mathbf{y}}_1$, $\mathbf{X}$, $\mathbf{x}'$, and $\mathbf{z}_2$ are used to keep intermediate results for $\hat{\mathbf{p}}_1\mathbf{T}^+$, $\mathbf{S}^-\hat{\mathbf{V}}_1$, $\hat{\mathbf{B}}_1\boldsymbol{\varepsilon}'$, and $\mathbf{p}_2\Psi_2^{-1}[\mathbf{U}^N]$, respectively. It is important in understanding the equations, as well as in coding the algorithms, to keep track of the dimensions of each of these objects. Therefore, we have given them symbols that match their dimensions. But be warned that the subscripts on these temporary variables in no way imply that they commute with objects from the other space. Their dimensions are (as before, read "*the dimensions of* $[\cdot]$" for "Dim$[\cdot]$"):

$$\mathrm{Dim}[\hat{\mathbf{y}}_1] = \mathrm{Dim}[\hat{\mathbf{p}}_1\mathbf{T}^+] = \mathrm{Dim}[\hat{\mathbf{p}}_1] = m_2 \times (m_1 \cdot m_2),$$

$$\mathrm{Dim}[\mathbf{X}] = \mathrm{Dim}[\mathbf{S}^-\hat{\mathbf{V}}_1] = \mathrm{Dim}[\mathbf{S}^-] = (m_1 \cdot m_2) \times (m_1 \cdot m_2),$$

$$\mathrm{Dim}[\hat{\mathbf{B}}_1\boldsymbol{\varepsilon}'] = \mathrm{Dim}[\mathbf{x}'] = \mathrm{Dim}[\boldsymbol{\varepsilon}'] = (m_1 \cdot m_2) \times 1,$$

$$\mathrm{Dim}\Big[\mathbf{p}_2\Psi_2^{-1}[\mathbf{U}^n]\Big] = \mathrm{Dim}[\mathbf{z}_2] = \mathrm{Dim}[\mathbf{p}_2] = 1 \times m_2.$$

Algorithm for Calculating Properties of ME/ME/1//N Loops

* *Initialization*
* *Assume that* $\mathbf{p}_1, \mathbf{P}_1, \mathbf{M}_1, \mathbf{p}_2, \mathbf{P}_2,$ *and* $\mathbf{M}_2$ *are given, then:*

BEGIN PROCEDURE

FOR $i = 1$ TO 2, DO

$\mathbf{B}_i \leftarrow \mathbf{M}_i(\mathbf{I}_i - \mathbf{P}_i)$

$\mathbf{V}_i \leftarrow$ *inverse* $[\mathbf{B}_i]$

$t_i \leftarrow \mathbf{p}_i\mathbf{V}_i\varepsilon'_i$

END FOR

$\mathbf{S}^- \leftarrow \hat{\mathbf{B}}_1 + \hat{\mathbf{B}}_2 - \hat{\mathbf{B}}_1\hat{\mathbf{Q}}_2$

$\mathbf{S}^+ \leftarrow \hat{\mathbf{B}}_1 + \hat{\mathbf{B}}_2 - \hat{\mathbf{B}}_2\hat{\mathbf{Q}}_1$

$\mathbf{T}^+ \leftarrow \textit{inverse}\,[\mathbf{S}^+]$

$\mathbf{U} \leftarrow \mathbf{T}^+\mathbf{S}^-$

$\mathbf{R} \leftarrow \mathbf{S}^-\mathbf{T}^+$

$\mathbf{k}'(1) \leftarrow [\mathbf{V}_1\varepsilon'_1] \otimes \varepsilon'_2 + \varepsilon'_1 \otimes [\mathbf{V}_2\varepsilon'_2]$

$\mathbf{u}'(0) \leftarrow \hat{\varepsilon}'_1$

$\hat{\mathbf{y}}_1 \leftarrow \hat{\mathbf{p}}_1\mathbf{T}^+$

$\mathbf{X} \leftarrow \hat{\mathbf{V}}_1\hat{\mathbf{B}}_2 + \mathbf{I} - \hat{\mathbf{Q}}_2 \quad \left(= \mathbf{S}^-\hat{\mathbf{V}}_1 = \hat{\mathbf{V}}_1\mathbf{S}^-\right)$

$\mathbf{x}' \leftarrow \hat{\mathbf{B}}_1\boldsymbol{\varepsilon}'$

* *For a parametric study, where the number of customers, N, varies from 2*
* *to a certain limit, N_m, the normalization constant and the initial vector*
* *need to be calculated. Also, the vector $\Pi(1;N)$ can be calculated, and*
* *initial values for mean queue length and throughput must be set.*

FOR $N = 2$ TO N_m, DO

$\mathbf{k}'(N) \leftarrow t_2\boldsymbol{\varepsilon}' + \mathbf{U}\mathbf{k}'(N-1)$

$\mathbf{u}'(N-1) \leftarrow \mathbf{U}\mathbf{u}'(N-2)$

$\mathbf{z}_2 \leftarrow \mathbf{p}_2 \cdot \textit{inverse}\,[\hat{\mathbf{p}}_1\mathbf{u}'(N-1)]$

$[\beta(N)]^{-1} \leftarrow \mathbf{z}_2\hat{\mathbf{p}}_1\mathbf{k}'(N)$

$\pi_2(0;N) \leftarrow \beta(N)\mathbf{z}_2\mathbf{V}_2$

$r(0;N) \leftarrow \pi_2(0;N)\varepsilon'_2$

$\Pi(1;N) \leftarrow \beta(N)\mathbf{z}_2 \cdot \hat{\mathbf{y}}_1$

$r(1;N) \leftarrow \Pi(1;N)\boldsymbol{\varepsilon}'$

$\bar{q} \leftarrow r(1;N)$

$\Lambda \leftarrow \Pi(1;N) \cdot \mathbf{x}'$

* *Each steady-state probability vector can now be calculated iteratively.*

FOR $i = 2$ TO $N-1$, DO

$\Pi(i;N) \leftarrow \Pi(i-1;N) \cdot \mathbf{R}$

$r(i;N) \leftarrow \Pi(i;N) \cdot \boldsymbol{\varepsilon}'$

$\bar{q} \leftarrow \bar{q} + i \cdot r(i;N)$

$\Lambda \leftarrow \Lambda + \Pi(i;N) \cdot \mathbf{x}'$

END FOR

* *The last terms need to be calculated separately.*

$\Pi(N;N) \leftarrow \Pi(N-1;N) \cdot \mathbf{X}$

$r(N;N) \leftarrow \mathbf{\Pi}(N;N)\cdot\boldsymbol{\varepsilon}'$

$\bar{q} \leftarrow \bar{q} + N\cdot r(N;N)$

$\Lambda \leftarrow \Lambda + \mathbf{\Pi}(N;N)\cdot\mathbf{x}'$

END FOR

END PROCEDURE

It is worth discussing the computational complexity of this algorithm. Let $T(N_m)$ be the number of multiplications and divisions for this procedure. Then

$$T(N_m) = 3m_1^\alpha\cdot m_2^\alpha + m_1^2\cdot m_2^2 + m_1^\alpha + m_2^\alpha + 2m_1^2 + m_2^2 + m_1\cdot m_2$$
$$+ (N_m - 1)\cdot(m_1^2\cdot m_2^\alpha 2m_1^2\cdot m_2^2 + 2m_1\cdot m_2^2 + m_2^\alpha + m_2^2 + 4m_1\cdot m_2 + 2m_2 + 1)$$
$$+ \tfrac{1}{2}(N_m-1)\cdot(N_m-2)\cdot(m_1^2 m_2^2 + m_1\cdot m_2 + 1)$$
$$= \mathrm{O}\left(3m_1^\alpha\cdot m_2^\alpha + N_m m_1^2\cdot m_2^\alpha + \tfrac{1}{2}N_m^2 m_1^2\cdot m_2^2\right),$$

where m^α is the complexity for multiplying two $m \times m$ matrices. For the special case where $\alpha = 3$, the order of complexity reduces to

$$T(N_m) = \mathrm{O}\left(3m_1^3\cdot m_2^3 + N_m m_1^2\cdot m_2^3 + \tfrac{1}{2}N_m^2 m_1^2\cdot m_2^2\right).$$

In most cases, the complexity is likely to be dominated by the N_m^2 term. If one is interested in $\bar{q}$ and Λ but not in $r(n;N)$, $\bar{q}$ and Λ can be calculated in a way analogous to the way β is calculated [using $\mathbf{k}'(N)$], thereby eliminating the inner loop on i, and consequently, reducing the complexity by an order of N_m. In such a case, it would take no more to compute the performance for all N from 1 to N_m than it would to calculate the performance for N_m alone.

7.3.4. An Example

We now present an example of the simplest nontrivial loop, just to see what the specific matrices look like. Let both S_1 and S_2 be Erlangian-2 servers, with parameters μ and λ, respectively. Then $\rho = \lambda/\mu$ and

$$\mathbf{p_i} = [1\ 0], \quad \mathbf{M_1} = \mu\mathbf{I_1}, \quad \mathbf{M_2} = \lambda\mathbf{I_2},$$

$$\mathbf{q'_i} = \begin{bmatrix} 0 \\ 1 \end{bmatrix}, \quad \mathbf{P_i} = \begin{bmatrix} 0 & 1 \\ 0 & 0 \end{bmatrix}, \quad \mathbf{Q_i} = \begin{bmatrix} 1 & 0 \\ 1 & 0 \end{bmatrix},$$

$$\mathbf{B_1} = \mu\begin{bmatrix} 1 & -1 \\ 0 & 1 \end{bmatrix}, \quad \mathbf{B_2} = \lambda\begin{bmatrix} 1 & -1 \\ 0 & 1 \end{bmatrix}, \quad \mathbf{V_1} = \frac{1}{\mu}\begin{bmatrix} 1 & 1 \\ 0 & 1 \end{bmatrix}, \quad \mathbf{V_2} = \frac{1}{\lambda}\begin{bmatrix} 1 & 1 \\ 0 & 1 \end{bmatrix}.$$

The embedded matrices [from (7.2.1)] are given by

$$\hat{\mathbf{p}}_1 = \begin{bmatrix} 1 & 0 & 0 & 0 \\ 0 & 1 & 0 & 0 \end{bmatrix}, \quad \hat{\mathbf{p}}_2 = \begin{bmatrix} 1 & 0 & 0 & 0 \\ 0 & 0 & 1 & 0 \end{bmatrix},$$

$$\mathbf{p} = \mathbf{p}_1 \otimes \mathbf{p}_2 = \mathbf{p}_1 \hat{\mathbf{p}}_2 = \mathbf{p}_2 \hat{\mathbf{p}}_1 = [1\ 0\ 0\ 0],$$

$$\hat{\mathbf{Q}}_1 = \begin{bmatrix} 1 & 0 & 0 & 0 \\ 0 & 1 & 0 & 0 \\ 1 & 0 & 0 & 0 \\ 0 & 1 & 0 & 0 \end{bmatrix}, \quad \hat{\mathbf{Q}}_2 = \begin{bmatrix} 1 & 0 & 0 & 0 \\ 1 & 0 & 0 & 0 \\ 0 & 0 & 1 & 0 \\ 0 & 0 & 1 & 0 \end{bmatrix}, \quad \mathbf{Q} = \boldsymbol{\varepsilon}'\mathbf{p} = \begin{bmatrix} 1 & 0 & 0 & 0 \\ 1 & 0 & 0 & 0 \\ 1 & 0 & 0 & 0 \\ 1 & 0 & 0 & 0 \end{bmatrix},$$

where $\mathbf{Q} = \hat{\mathbf{Q}}_1 \hat{\mathbf{Q}}_2 = \mathbf{Q}_1 \otimes \mathbf{Q}_2$, and

$$\mathbf{E} = (\mathbf{I} - \hat{\mathbf{Q}}_1)(\mathbf{I} - \hat{\mathbf{Q}}_2) = \begin{bmatrix} 0 & 0 & 0 & 0 \\ 0 & 0 & 0 & 0 \\ 0 & 0 & 0 & 0 \\ 1 & -1 & -1 & 1 \end{bmatrix}.$$

Note that both $\hat{\mathbf{Q}}_1$ and $\hat{\mathbf{Q}}_2$ are of rank 2, $[m_i]$, since they each have two linearly independent rows when considered as vectors. On the other hand, $\mathbf{Q}$ is of rank 1, as it should be, since all four of its rows are the same. $\mathbf{E}$ is also of rank 1 $[(m_1 - 1) \cdot (m_2 - 1) = 1]$, since all of its columns are proportional to each other.

$$\hat{\mathbf{B}}_1 = \mu \begin{bmatrix} 1 & 0 & -1 & 0 \\ 0 & 1 & 0 & -1 \\ 0 & 0 & 1 & 0 \\ 0 & 0 & 0 & 1 \end{bmatrix}, \quad \hat{\mathbf{B}}_2 = \lambda \begin{bmatrix} 1 & -1 & 0 & 0 \\ 0 & 1 & 0 & 0 \\ 0 & 0 & 1 & -1 \\ 0 & 0 & 0 & 1 \end{bmatrix},$$

$$\hat{\mathbf{V}}_1 = \frac{1}{\mu} \begin{bmatrix} 1 & 0 & 1 & 0 \\ 0 & 1 & 0 & 1 \\ 0 & 0 & 1 & 0 \\ 0 & 0 & 0 & 1 \end{bmatrix}, \quad \hat{\mathbf{V}}_2 = \frac{1}{\lambda} \begin{bmatrix} 1 & 1 & 0 & 0 \\ 0 & 1 & 0 & 0 \\ 0 & 0 & 1 & 1 \\ 0 & 0 & 0 & 1 \end{bmatrix}.$$

Some composite matrices in the product space follow.

$$\mathbf{D}^{-1} = \hat{\mathbf{B}}_1 + \hat{\mathbf{B}}_2 = \mu \begin{bmatrix} 1+\rho & -\rho & -1 & 0 \\ 0 & 1+\rho & 0 & -1 \\ 0 & 0 & 1+\rho & -\rho \\ 0 & 0 & 0 & 1+\rho \end{bmatrix},$$

$$\mu\mathbf{D} = \frac{1}{(1+\rho)^3} \begin{bmatrix} (1+\rho)^2 & \rho(1+\rho) & (1+\rho) & 2\rho \\ 0 & (1+\rho)^2 & 0 & (1+\rho) \\ 0 & 0 & (1+\rho)^2 & \rho(1+\rho) \\ 0 & 0 & 0 & (1+\rho)^2 \end{bmatrix}.$$

Recall that in Section 7.1 we used this operator to find the expression for $C(0)$, the probability that S_1 would finish before S_2, given that they started at the same time. In our example, this turns out to be [using (7.1.3b)]

$$C(0) = \Psi[\hat{\mathbf{B}}_1\mathbf{D}] = \frac{1+3\rho}{(1+\rho)^3}. \tag{7.3.15}$$

Exercise 7.3.1: You are to compare $C(0)$ for three different cases as a function of ρ. The cases are (a) exponential – exponential, from Equation (2.1.1b) (call it C_{11}), (b) exponential – Erlangian-2 (C_{12}), and (c) Equation (7.3.16) (call it C_{22}). First verify that

$$\Psi_2[\hat{\mathbf{B}}_1\mathbf{D}] = \frac{1}{(1+\rho)^3}\begin{bmatrix} 1+\rho & 2\rho \\ 0 & 1+\rho \end{bmatrix}.$$

Then verify that C_{22} is correct [use (7.2.6)], and find similar expressions for the other two. Prove that the following inequalities hold:

$$C_{12} < C_{11} < C_{22} \quad \text{for} \quad \rho < 1,$$

$$C_{12} < C_{22} < C_{11} \quad \text{for} \quad 1 < \rho < \frac{1+\sqrt{17}}{2},$$

$$C_{22} < C_{12} < C_{11} \quad \text{for} \quad \frac{1+\sqrt{17}}{2} < \rho.$$

Remember that $\rho = \bar{x}_1/\bar{x}_2$, which for C_{12} is *not* λ/μ.

Continuing with matrices in the product space,

$$\mathbf{S}^+ = \mu\begin{bmatrix} 1 & 0 & -1 & 0 \\ 0 & 1 & 0 & -1 \\ -\rho & \rho & 1+\rho & -\rho \\ 0 & -\rho & 0 & 1+\rho \end{bmatrix}, \quad \mathbf{S}^- = \mu\begin{bmatrix} \rho & -\rho & 0 & 0 \\ -1 & 1+\rho & 1 & -1 \\ 0 & 0 & \rho & -\rho \\ 0 & 0 & -1 & 1+\rho \end{bmatrix}.$$

The determinants of $S^{\pm}$ are μ^4 and λ^4, respectively, so these matrices are non-singular. Their inverses are

$$\mathbf{T}^+ = \frac{1}{\mu}\begin{bmatrix} 1+\rho & -\rho & 1 & 0 \\ 0 & 1+\rho & 0 & 1 \\ \rho & -\rho & 1 & 0 \\ 0 & \rho & 0 & 1 \end{bmatrix}, \quad \mathbf{T}^- = \frac{1}{\mu\rho^2}\begin{bmatrix} 1+\rho & \rho & -1 & 0 \\ 1 & \rho & -1 & 0 \\ 0 & 0 & 1+\rho & \rho \\ 0 & 0 & 1 & \rho \end{bmatrix}.$$

The most important matrix, **U**, is not so simple looking:

$$\mathbf{U} = \mathbf{T}^{+}\mathbf{S}^{-} = \begin{bmatrix} \rho(2+\rho) & -2\rho(1+\rho) & 0 & 0 \\ -(1+\rho) & (1+\rho)^2 & \rho & 0 \\ \rho(1+\rho) & -\rho(1+2\rho) & 0 & 0 \\ -\rho & \rho(1+\rho) & \rho-1 & 1 \end{bmatrix},$$

from which we can get

$$\Psi_2[\mathbf{U}] = \begin{bmatrix} \rho(2+\rho) & -2\rho(1+\rho) \\ -1 & (1+\rho)^2 \end{bmatrix}.$$

From either of these equations one can calculate $\Psi[\mathbf{U}] = \Psi[\Psi_2[\mathbf{U}]] = -\rho^2$. The characteristic equation for **U** is

$$\phi(\alpha) = |\,\mathbf{U} - \alpha\mathbf{I}\,| = \left(\alpha - 1\right)\left(\alpha - \rho^2\right)\left(\alpha^2 - [1+4\rho+\rho^2]\alpha + \rho^2\right).$$

Therefore, the eigenvalues of **U** are

$$\alpha_1 = 1, \qquad \alpha_2 = \rho^2,$$

$$\alpha_3 = \frac{(\rho^2+4\rho+1) + Z(1+\rho)}{2},$$

and

$$\alpha_4 = \frac{(\rho^2+4\rho+1) - Z(1+\rho)}{2},$$

where $Z^2 = \rho^2 + 6\rho + 1$. Sadly, we see that one of the eigenvalues is 1. Therefore, $\mathbf{I} - \mathbf{U}$ has no inverse. We knew this would happen from Lemma 7.3.2 and the discussion following it. We said that there are at most $m_1 + m_2 - 1$ roots, which do *not* equal 1. In our case, $m_1 = m_2 = 2$, so there are at most three. In fact, there are exactly three if $\rho \neq 1$. The other difficulty we have is that α_3 is greater than 1 for all ρ. Therefore, some matrix elements of $\mathbf{U}^n$ must become unboundedly large as n increases. For what it is worth, the four eigenvalues satisfy the following inequality.

$$\alpha_4 < \alpha_2 < \alpha_1 = 1 < \alpha_3 \quad \text{for} \quad \rho < 1$$

and

$$\alpha_4 < \alpha_1 = 1 < \alpha_2 < \alpha_3 \quad \text{for} \quad \rho > 1,$$

The matrix **R** can do no better, since it too has a unit eigenvalue. It presumably also has an eigenvalue greater than 1. We leave it as an exercise for the reader to analyze **R** in the way that we just analyzed **U**.

Where do we go from here? Despite all these formulas, we cannot go to the open system. In Section 4.2.1 we successfully took the limit of $\mathbf{K}(N)$ as N went to infinity, because $(\mathbf{I} - \mathbf{U})^{-1}$ existed. It does not here. In Section 5.1.1 we were able to take the limit because we were able to isolate a unique eigenvalue and its associated left and right eigenvectors. So far we have not been successful in finding an appropriate generalization of this. We know this much: Victor Wallace proved that *all* open quasi birth-death processes of a certain type (of which the ME/ME/1 queue is a special case) must have a matrix geometric solution [WALL69]. It is just that neither **U** nor **R** appears to be that matrix. But an isometric transformation of **U** or **R** in the product space may well yield the correct matrix. We shall not go into detail here, but it should be possible to find a transformation that yields a matrix for which the eigenvectors belonging to the eigenvalue 1 drop out of the solution. The solution, whatever it turns out to be, almost surely will reflect the characteristics of both the M/ME/1 and ME/M/1 queues, since the ME/ME/1 queue is the generalization of both. (See, however, [RAMA90] for an iterative solution.) We have presented far more formulas than are necessary, in the hope that they will help some reader to discover how this can be done.

We close out the chapter with a short look at mean first-passage times for the queue at S_1 to drop. Extensions to other transient properties are left to the reader's ingenuity.

7.4. A MODICUM OF TRANSIENT BEHAVIOR

We shall not go into too much detail of transient behavior for two nonexponential servers, not so much because it is so hard, but because it looks so much like what we already did in Section 4.5. All objects are in the product space, so should be wearing hats. To make things simple for us, we will revert to the naive approach of Section 7.1. We will only cover the first-passage processes to drop by 1, and thereby reproduce (4.5.29a) for this more general case. The reader should review Section 4.5 before continuing.

First recall Definition 4.5.7. The matrix $[\mathbf{H_d}(n\,;N)]_{\underline{kl}}$ is identical in meaning, except that now

$$\underline{l},\ \underline{k} \in \Xi = \Xi_1 \otimes \Xi_2.$$

As in Chapter 4, after a single event occurs, the queue at S_1 can grow by one, decrease by one, or stay the same. The difference now is that there are two ways that it can stay the same, either by a transition in S_1, $[\mathbf{P_1}]$, or a transition in S_2, $[\mathbf{P_2}]$. Thus $\mathbf{H_d}$ satisfies the following:

$$\mathbf{H_d}(n\,;N) = [\mathbf{M_1} + \mathbf{M_2}]^{-1}[\mathbf{M_1P_1} + \mathbf{M_2P_2}]\mathbf{H_d}(n\,;N)$$
$$+ [\mathbf{M_1} + \mathbf{M_2}]^{-1}\mathbf{M_1 q'_1 p_1} + [\mathbf{M_1} + \mathbf{M_2}]^{-1}\mathbf{M_2 q'_2 p_2}\mathbf{H_d}(n+1\,;N)\,\mathbf{H_d}(n\,;N).$$

The quantity

$$[(\mathbf{M_1} + \mathbf{M_2})^{-1}\mathbf{M_1}]_{\underline{kk}} = \frac{\mu_{k_1}}{\mu_{k_1} + \lambda_{k_2}},$$

is the probability that the next event, when it occurs, will be in S_1, given that the system is in state $\underline{k} = (k_1, k_2) \in \Xi$. Therefore, the first term, $[\mathbf{P_1}]$, corresponds to an internal transition in S_1, with the eventual drop, $[\mathbf{H_d}(n\,;N)]$, to $n-1$.

Similarly, the second term, $[\mathbf{P_2 H_d}(n\,;N)]$, is an internal transition in S_2. The third term corresponds to a transition in S_1 that results in a departure, $[\mathbf{q'_1}]$, followed immediately by the entry of the next customer, $[\mathbf{p_1}]$, at S_1. In this process, nothing more need happen, since S_1 now has $n-1$ customers. The last term corresponds to a departure from S_2, $[\mathbf{q'_2}]$, immediately followed by the entry of the next customer, $[\mathbf{p_2}]$, and then the eventual drop from $n+1$ to n, $[\mathbf{H_d}(n{+}1\,;N)]$, followed eventually by a drop to $n-1$, $[\mathbf{H_d}(n\,;N)]$.

Recall the following, which we have seen so many times: $\mathbf{B_i} = \mathbf{M_i}(\mathbf{I}-\mathbf{P_i})^{\dagger}$ and $\mathbf{M_i q'_i} = \mathbf{B_i Q_i}$, for $i = 1, 2$. Then multiply both sides of the equation by $[\mathbf{M_1}+\mathbf{M_2}]$, regroup terms and for $0 < n < N$, come up with

$$\mathbf{H_d}(n\,;N) = [\mathbf{B_1}+\mathbf{B_2}-\mathbf{B_2 Q_2 H_d}(n{+}1\,;N)]^{-1}\mathbf{B_1 Q_1}. \tag{7.4.1a}$$

We must still get an equation for $\mathbf{H_d}(N\,;N)$, since there is no way to go up. There is one other difficulty with this state. Since there is no customer at S_2, $[\mathbf{H_d}(N\,;N)]_{i\underline{l}}$ generates a transition from $i \in \Xi_1$ to $\underline{l} \in \Xi$, (i.e., it is not a square matrix). Thus

$$\mathbf{H_d}(N\,;N) = \mathbf{P_1 H_d}(N\,;N) + \mathbf{q'_1 p_1 p_2}$$

(the customer who leaves S_1 immediately enters S_2, $[\mathbf{p_2}]$, since there was no one there before his arrival), or

$$\mathbf{H_d}(N\,;N) = [\mathbf{I}-\mathbf{P_1}]^{-1}\mathbf{q'_1 p_1 p_2} = \mathbf{Q_1 p_2}, \tag{7.4.1b}$$

since $[\mathbf{I}-\mathbf{P_1}]^{-1}\mathbf{q'_1} = \boldsymbol{\varepsilon}'_1$. Alternatively (as we have been doing all along), we can make believe that when S_2 is empty, it is in state $\mathbf{p_2}$, but cannot do anything. Then we must wipe out that state, $[\boldsymbol{\varepsilon}'_2]$, before reentering it, $[\mathbf{p_2}]$. This leads to the simpler formula,

$$\mathbf{H_d}(N\,;N) = \mathbf{Q_1 Q_2} = \mathbf{Q}. \tag{7.4.1c}$$

In either case, (7.4.1a) leads to

$$\mathbf{H_d}(N{-}1\,;N) = [\mathbf{B_1}+\mathbf{B_2}-\mathbf{B_2 Q}]^{-1}\mathbf{B_1 Q_1}.$$

Next we define $[\boldsymbol{\tau}'_{\mathbf{d}}(n\,;N)]_{\underline{k}}$ in the way we did in Definition 4.5.8 where now $\underline{k} \in \Xi$. This means that the time for the queue at S_1 to drop by 1 depends on the internal state of S_2 as well as S_1. First, let all the customers be at S_1,

$$\boldsymbol{\tau}'_{\mathbf{d}}(N\,;N) = \mathbf{M_1^{-1}}\boldsymbol{\varepsilon}'_1 + \mathbf{P_1}\boldsymbol{\tau}'_{\mathbf{d}}(N\,;N),$$

leading to

$$\boldsymbol{\tau}'_{\mathbf{d}}(N\,;N) = \mathbf{V_1}\boldsymbol{\varepsilon}', \tag{7.4.2a}$$

which is identical to (4.5.11a), as it should be, since S_2 plays no role in the process. For $n < N$,

$$\boldsymbol{\tau}'_{\mathbf{d}}(n\,;N) = (\mathbf{M_1}+\mathbf{M_2})^{-1}\boldsymbol{\varepsilon}' + (\mathbf{M_1}+\mathbf{M_2})^{-1}(\mathbf{M_1 P_1}+\mathbf{M_2 P_2})\boldsymbol{\tau}'_{\mathbf{d}}(n\,;N)$$
$$+ (\mathbf{M_1}+\mathbf{M_2})^{-1}\mathbf{M_2 q'_2 p_2}[\boldsymbol{\tau}'_{\mathbf{d}}(n{+}1\,;N) + \mathbf{H_d}(n{+}1\,;N)\boldsymbol{\tau}'_{\mathbf{d}}(n\,;N)],$$

† Keep in mind that **I** is the identity matrix of the product space and that $\mathbf{M_1}, \mathbf{B_2}$, etc., are all embedded (*hatted*) into that space.

which regroups, and rearranges to

$$\boldsymbol{\tau}'_{\mathbf{d}}(n\,;N)$$
$$= [\mathbf{B}_1 + \mathbf{B}_2 - \mathbf{B}_2\mathbf{Q}_2\mathbf{H}_{\mathbf{d}}(n{+}1\,,N\,)]^{-1}[\boldsymbol{\varepsilon}' + \mathbf{B}_2\mathbf{Q}_2\boldsymbol{\tau}'_{\mathbf{d}}(n{+}1\,;N\,)] \qquad (7.4.2b)$$

[compare with (4.5.11b)]. Equations (7.4.1) and (7.4.2) are sufficient for finding all times recursively, by following the discussion in Section 4.5.3. Let us look in particular at the special case $N = 2$. We have discussed its significance in Section 4.5.4, but now we generalize to two nonexponential servers. First,

$$t_d(2;\,2) = \mathbf{p}\boldsymbol{\tau}'_{\mathbf{d}}(2;\,2) = T_1 = \Psi[\mathbf{V}_1]$$

and

$$\boldsymbol{\tau}'_{\mathbf{d}}(1;\,2) = [\mathbf{B}_1 + \mathbf{B}_2 - \mathbf{B}_2\mathbf{Q}]^{-1}[\boldsymbol{\varepsilon}' + \mathbf{B}_2\mathbf{Q}_2\mathbf{V}_1\boldsymbol{\varepsilon}'].$$

Note that $\mathbf{V}_1$ commutes with $\mathbf{Q}_2$, and $\mathbf{Q}_2\boldsymbol{\varepsilon}' = \boldsymbol{\varepsilon}'$, so the expression in the second set of brackets becomes

$$[\mathbf{I} + \mathbf{B}_2\mathbf{V}_1]\boldsymbol{\varepsilon}' = [\mathbf{B}_1 + \mathbf{B}_2]\mathbf{V}_1\boldsymbol{\varepsilon}' = \mathbf{D}^{-1}\mathbf{V}_1\boldsymbol{\varepsilon}',$$

where we have used (7.3.4) as the definition for $\mathbf{D}$. Lemma 4.2.1 can be applied to the expression in the first set of brackets (not Lemma 7.2.4) since $\mathbf{Q}$ itself appears rather than $\mathbf{Q}_\mathbf{i}$. We get

$$[\mathbf{B}_1 + \mathbf{B}_2 - \mathbf{B}_2\mathbf{Q}]^{-1} = [\mathbf{D}^{-1}(\mathbf{I} - \mathbf{D}\mathbf{B}_2\mathbf{Q})]^{-1}$$
$$= [\mathbf{I} - \mathbf{D}\mathbf{B}_2\mathbf{Q}]^{-1}\mathbf{D} = \left(\mathbf{I} + \frac{1}{1 - \Psi[\mathbf{D}\mathbf{B}_2]}\mathbf{D}\mathbf{B}_2\mathbf{Q}\right)\mathbf{D}.$$

Therefore,

$$\boldsymbol{\tau}'_{\mathbf{d}}(1;\,2) = \left(\mathbf{I} + \frac{1}{1 - \Psi[\mathbf{D}\mathbf{B}_2]}\mathbf{D}\mathbf{B}_2\mathbf{Q}\right)\mathbf{D}\mathbf{D}^{-1}\mathbf{V}_1\boldsymbol{\varepsilon}'$$
$$= \mathbf{V}_1\boldsymbol{\varepsilon}' + \frac{T_1}{1 - \Psi[\mathbf{D}\mathbf{B}_2]}\mathbf{D}\mathbf{B}_2\boldsymbol{\varepsilon}'.$$

Finally,

$$t_d(1;\,2) = \mathbf{p}\boldsymbol{\tau}'_{\mathbf{d}}(1;\,2) = T_1 + \frac{T_1\Psi[\mathbf{D}\mathbf{B}_1]}{1 - \Psi[\mathbf{D}\mathbf{B}_1]} = \frac{T_1}{\Psi[\mathbf{D}\mathbf{B}_1]},$$

since $1 - \Psi[\mathbf{D}\mathbf{B}_2] = \Psi[\mathbf{I} - (\mathbf{B}_1 + \mathbf{B}_2)^{-1}\mathbf{B}_2] = \Psi[\mathbf{D}\mathbf{B}_1]$. Then, following Section 4.5.4,

$$MTTF\,(2) = t_d(2;\,2) + t_d(1;\,2) = T_1\left(1 + \frac{1}{\Psi[\mathbf{D}\mathbf{B}_1]}\right). \qquad (7.4.3)$$

This equation is identical to (4.5.29a) when one takes into account the slight difference of notation. In Chapter 4 we used $\mathbf{D} = (\mathbf{I} + \lambda\mathbf{V})^{-1}$, and since $\lambda \Rightarrow \mathbf{B}_2$ in this chapter, we have $\mathbf{D}$(Chapter 4) $\Rightarrow \mathbf{B}_1 \mathbf{D}$(Chapter 7). We have already shown that $\Psi[\mathbf{D}\mathbf{B}_1] = C(0)$ [Equation (7.1.3b)] is the probability that the customer in S_1 will finish before the customer in S_2, given that they started at the same time. In the context of MTTF, this is the probability that the second device will break before the first is fixed, and is true for any pair of distributions. See Section 4.5.4 for a more thorough discussion.

We hope that the reader will be able to solve for other transient properties using the material expounded upon in previous chapters. We have seen by this example that certain transient events can be computed more easily than can the steady-state solution for the same system, particularly if N is small. This should become a useful and practical tool.

Exercise 7.4.1: Take the results you got from Exercise 7.3.1 (C_{11}, C_{12}, and C_{22}), and use (7.4.3) to calculate MTTF as a function of ρ for all three systems. Assume that $T_1 = 1$ and let ρ take on the values, 0_+, 0.1, 0.5, 0.8, 1.0, 1.25, 2.0, 10.0, 100, and 1000.

CHAPTER 8
LINEAR ALGEBRAIC APPROACH

"A theory should be as simple as possible, but no simpler" - Albert Einstein

In the previous chapters we saw that matrix relations continually occur, independently of probabilistic interpretations. Surely this is not an accident. We will now show that there exists a linear algebraic formulation which is not merely an algorithmic or computational aid but a complete formal procedure for dealing with nonexponential queues. We will do this without resorting to any particular basis set, a common technique in linear algebra. We will first show that most, if not all, the equations in this book are invariant to the isometric transformations we introduced in Section 3.2.3. This invariance property implies that a basis-free formulation is possible.

In some sense this chapter will serve as a review of the first five chapters of the book, but now all the properties of the servers and queues are in terms of linear operators that modify the state vector of the system when things happen. This change in viewpoint may appear self-evident to some readers, and if so, fine, but it is important to mention, nonetheless. We do not claim that we are doing this in the best or most efficient way to set up the algebraic structure. Surely some readers can do better. We merely wish to show that it can be done. Therefore, questions that may arise should not be considered to be weaknesses of the theory but issues to be cleared up or clarified, which may actually lead to new insights.

8.1. ISOMETRIC TRANSFORMATIONS

In Section 3.2.3 (Theorem 3.2.1), we showed that if $<\mathbf{p}, \mathbf{B}>$ is a faithful representation of a given distribution function, so is $<\mathbf{pS}^{-1}, \mathbf{SBS}^{-1}>$, where $\mathbf{S}$ is any isometric, invertible matrix. That is, $B(t)$, $R(t)$, $b(t)$, and $B^*(s)$ remain unchanged by such transformations. These isometric transformations go beyond the description of a single server. We now extend this idea to any row vector $\mathbf{u}$, column vector $\mathbf{v}'$, and square matrix $\mathbf{X}$, and define the following mapping.

Definition 8.1.1

Let $\mathbf{S}$ be any isometric, nonsingular matrix; then the following mapping (or similarity transformation) is called an *isometric transformation*:

$$\tilde{\mathbf{u}} := \mathbf{uS}^{-1}, \quad \tilde{\mathbf{v}}' = \mathbf{Sv}', \quad \text{and} \quad \tilde{\mathbf{X}} := \mathbf{SXS}^{-1},$$

for every row vector, column vector, and matrix of interest. Because $\mathbf{S}$ is isometric, $\boldsymbol{\varepsilon}'$ does not change under any transformation. Note that

$$\tilde{\mathbf{u}}\,\tilde{\mathbf{v}}' = \mathbf{u}\mathbf{S}^{-1}\mathbf{S}\mathbf{v}' = \mathbf{u}\mathbf{v}' \quad \text{and} \quad \tilde{\Psi}[\tilde{\mathbf{X}}] := \tilde{\mathbf{p}}\tilde{\mathbf{X}}\boldsymbol{\varepsilon}' = \Psi[\mathbf{X}].$$

These equations show us that inner products and $\Psi[\cdot]$ operations remain unchanged (are invariant) under isometric transformations. In general, we say that an *equation is invariant* if it is identical in form for the transformed objects as it is for the original objects. ◊

We showed in Section 4.1.2 that the steady-state solutions of M/G/1 queues (both open and closed) depend on the matrix $\mathbf{A}$ and its inverse $\mathbf{U}$, defined by

$$\mathbf{A} := \mathbf{U}^{-1} := \mathbf{I} + \frac{1}{\lambda}\mathbf{B} - \mathbf{Q}, \tag{8.1.1}$$

where λ is the service rate of S_2 (or the Poisson arrival rate to S_1), and $\mathbf{Q} := \varepsilon'\mathbf{p}$. Since $\mathbf{S}\boldsymbol{\varepsilon}' = \boldsymbol{\varepsilon}'$, it follows from their definitions that

$$\tilde{\mathbf{Q}} = \mathbf{S}\mathbf{Q}\mathbf{S}^{-1} = \mathbf{S}\boldsymbol{\varepsilon}'\mathbf{p}\mathbf{S}^{-1} = \boldsymbol{\varepsilon}'\tilde{\mathbf{p}}$$

and

$$\tilde{\mathbf{A}} = \mathbf{S}\mathbf{A}\mathbf{S}^{-1} = \mathbf{I} + \frac{1}{\lambda}\tilde{\mathbf{B}} - \tilde{\mathbf{Q}}.$$

Thus the equations defining $\mathbf{Q}$ and $\mathbf{A}$ are *invariant* to isometric transformations. From Theorem 4.1.2, the steady-state vector and scalar probabilities for the open queue are

$$\pi(n) = (1-\rho)\mathbf{p}\mathbf{U}^n \tag{8.1.2a}$$

and

$$r(n) = \pi(n)\boldsymbol{\varepsilon}' = (1-\rho)\Psi[\mathbf{U}^n]. \tag{8.1.2b}$$

Any isometric transformation, $\mathbf{S}$, will produce the following:

$$\tilde{\pi}(n) := \pi(n)\mathbf{S}^{-1} = (1-\rho)\mathbf{p}\mathbf{S}^{-1}\mathbf{S}\mathbf{U}^n\mathbf{S}^{-1} = (1-\rho)\tilde{\mathbf{p}}\,\tilde{\mathbf{U}}^n$$

and (note that $\mathbf{S}\mathbf{U}^2\mathbf{S}^{-1} = \mathbf{S}\mathbf{U}\mathbf{S}^{-1}\mathbf{S}\mathbf{U}\mathbf{S}^{-1} = \tilde{\mathbf{U}}^{-2}$, and so on for all n)

$$\tilde{r}(n) := \tilde{\pi}(n)\boldsymbol{\varepsilon}' = (1-\rho)\tilde{\Psi}[\tilde{\mathbf{U}}^n] = (1-\rho)\Psi[\mathbf{U}^n] = r(n).$$

Clearly, the $\pi(n)$'s and the $r(n)$'s are also invariant to isometric transformations. In particular, we see that for scalars [e.g., $r(n)$], *invariance* means *no change*.

We mentioned in Section 4.4 that the standard algorithm for evaluating the steady-state probabilities for the M/G/1 queue requires the *exponential moments* defined by

$$\alpha_n(s) := \int_0^\infty \frac{(sx)^n}{n!} e^{-sx} b(x)\,dx.$$

We then showed that [Equations (4.4.1)]

$$\alpha_n(s) = \Psi[(s\mathbf{V}\mathbf{D})^n\mathbf{D}],$$

where $\mathbf{D} := (\mathbf{I} + s\mathbf{V})^{-1}$. Clearly, these expressions are invariant to isometric transformations, so the results one gets with one representation will be identical to the results one gets with any similar representation. For instance [using $(\mathbf{SXS}^{-1})^{-1} = (\mathbf{S}^{-1})^{-1}\mathbf{X}^{-1}\mathbf{S}^{-1} = \mathbf{SX}^{-1}\mathbf{S}^{-1}$],

$$\tilde{\mathbf{D}} := \mathbf{SDS}^{-1} = \mathbf{S}[\mathbf{I} + s\mathbf{V}]^{-1}\mathbf{S}^{-1} = [\mathbf{S}(\mathbf{I} + s\mathbf{V})\mathbf{S}^{-1}]^{-1}$$
$$= [\mathbf{I} + s\mathbf{SVS}^{-1}]^{-1} = [\mathbf{I} + s\tilde{\mathbf{V}}]^{-1}.$$

We mention that the various formulas describing the M/ME/1 queue are very similar to those for the M/M/1 queue and would generalize trivially were it not for the fact that $\mathbf{Q}$ and $\mathbf{B}$ do not commute! On the other hand, things would have been a lot worse if $\mathbf{Q}$ did not have rank one, which is the case for M/G/*C*//*N* systems.

8.2. LINEAR ALGEBRAIC FORMULATION

We are now ready to formulate queueing processes without resorting to individual components or phases. First we shall look at a single isolated (general) server, S. To do this we change the meaning of our notation somewhat. In general, one can start with a set of independent *basis vectors* (the equivalent of our phases) and then generate the entire vector space by taking all possible linear combinations of the basis vectors. Alternatively, we can start with an abstract vector space, and then, if we need one, select a basis set. We did the former in previous chapters. We will do the latter here.

8.2.1. Description of a Single Server

Let $\mathbf{r}$ be a *vector* in some discrete (in our case, finite-dimensional) vector space, Ξ, which contains all we know about S. Previously, we considered Ξ to be the set of phases of S and then constructed the vectors from them. Now, we let Ξ be the set of *all* vectors. In doing this we are actually tightening up our mathematics. Keep in mind, though, that not every vector in Ξ has physical meaning. Let the *length* of any vector in Ξ be its *dot product* with a special, unique vector from the adjoint space $\Xi'^{\dagger}$, denoted by $\boldsymbol{\varepsilon}'$. Then, for one thing,

$$R := \mathbf{r}\boldsymbol{\varepsilon}' = \text{probability } \textit{that S is busy}.$$

R is a measurable quantity, but the components of $\mathbf{r}$ need not be. From an outside observer's point of view, S can only be in one of two *external states*: either it is busy or it is not. What goes on inside is hidden from view until S stops, in which case $\mathbf{r}$ becomes the null vector, $\mathbf{o}$. (Of course, if an observer really can look inside, S *must* truly be a *phase* distribution.)

† Technically, objects in Ξ' are *linear functionals* which map vectors in Ξ into the complex numbers. It is well known that this is also a vector space (see, e.g., [HALM55]) and is isomorphic to (i.e., has the same dimension as) Ξ. When one is working with an explicit basis, one thinks of row and column vectors, while the scalar mapping is the dot product.

From the basic Markov property, only one thing can happen at a time, and it can only depend on the state the system is in when it happens. Also, a transition that does not change the *length* of **r** is not directly observable. Let **P** be a linear operator on Ξ which moderates internal transitions, while **q**' moderates completion of service. That is, given that something has occurred, **rq**' is the probability that service ended, **rP** is the new state S is in if service has not ended, and **rP**ε' is the probability that service did not end. Since this is all that can occur, we must have

$$\mathbf{rq}' + \mathbf{rP}\boldsymbol{\varepsilon}' = \mathbf{r}\boldsymbol{\varepsilon}'.$$

This equation must be true for all $\mathbf{r} \in \Xi$ which have physical meaning, so it follows that $\mathbf{q}' + \mathbf{P}\boldsymbol{\varepsilon}' = \boldsymbol{\varepsilon}'$. Put differently, this equation can be rewritten as

$$\mathbf{r}[\mathbf{q}' + \mathbf{P}\boldsymbol{\varepsilon}' - \boldsymbol{\varepsilon}'] = 0.$$

Then if it is true for m linearly independent **r**'s, the term in brackets must be identically equal to **o**'. (m is the dimension of vector space Ξ.) Thus **q**' and **P** are related by the relation

$$\mathbf{q}' = (\mathbf{I} - \mathbf{P})\boldsymbol{\varepsilon}'. \tag{8.2.1}$$

The time scale for the behavior of S comes in through the operator **T**, where $\mathbf{rT}\boldsymbol{\varepsilon}'$ is the mean time to the next event. Also, let $\tau' \in \Xi'$ be a linear functional such that $\mathbf{r}\tau'$ is the mean time until service terminates, given that S is initially busy, and described by state vector, **r**, where $\mathbf{r}\varepsilon' = 1$. Then we can write

$$\mathbf{r}\tau' = \mathbf{rT}\boldsymbol{\varepsilon}' + \mathbf{rP}\tau'.$$

In words, the time for service to complete is made up of two parts. First there is the time until the next event, [$\mathbf{rT}\boldsymbol{\varepsilon}'$], and if that event was not a termination, [$\mathbf{rP}\boldsymbol{\varepsilon}'$], then S changes its internal state, [**rP**], and completes service from there, [τ']. Since this equation is valid for all $\mathbf{r} \in \Xi$, we can once again discard **r** and solve for τ' to get

$$\tau' = (\mathbf{I} - \mathbf{P})^{-1}\mathbf{T}\boldsymbol{\varepsilon}' = \mathbf{V}\boldsymbol{\varepsilon}', \tag{8.2.2}$$

where $\mathbf{V} := (\mathbf{I} - \mathbf{P})^{-1}\mathbf{T}$.

Next, let **M** be a linear operator on our vector space that moderates the occurrence of events. Then $\mathbf{rM}\varepsilon'$ is the instantaneous rate for something to happen. A physical interpretation of what this means is as follows. Suppose that there exists a basis set for Ξ in which **M** is diagonal. (It might be quite interesting to explore systems in which this were not possible, although it is not clear what that would mean.) Each basis vector $\mathbf{u_i}$ is referred to as a *phase* or *pure state*, and μ_i, the eigenvalue of **M** that goes with $\mathbf{u_i}$ is the formal "probability rate" at which the system leaves state i. The time to leave state i is "exponentially distributed" with exponent μ_i. Note that the μ's need not be real, so this interpretation may not have physical meaning. As we shall see presently, this will not lead to any contradictions, as long as physical (i.e., observable) quantities do not depend on the individual components of the vectors in Ξ.

Based on our assumptions about S, only two types of things can happen. We have just seen that either the system changes its internal state according to the linear operator $\mathbf{P}$, or it stops (i.e., the customer leaves), according to the adjoint vector $\mathbf{q}'$. That is, $\mathbf{rMq}'$ is the probability rate that S will have an event that results in a departure. We now examine how the internal status of S evolves in time. Let $\mathbf{r}(t)$ contain that information and have initial value $\mathbf{p} := \mathbf{r}(0)$ such that $\mathbf{p}\boldsymbol{\varepsilon}' = 1$. In other words, we assume that whenever S first starts service (at $t = 0$) it will always be represented by the *initial* or *entrance vector* $\mathbf{p}$. Then we have

$$R(t) = \mathbf{r}(t)\boldsymbol{\varepsilon}' = \textit{probability that S is still busy at time } t,$$

and $R(0) = 1$. Now, in some small time interval δ,

$$\mathbf{r}(t)\,\delta\,\mathbf{M} + O(\delta^2) = \textit{probability that something will happen.}$$

Then either nothing happens in the interval $[\mathbf{r}(t)(\mathbf{I} - \delta\mathbf{M})]$, or the event results in an internal transition $[\mathbf{r}(t)\delta\mathbf{MP}]$, or there is a departure [no term needed]. Thus we have

$$\mathbf{r}(t+\delta) = \mathbf{r}(t)(\mathbf{I} - \delta\mathbf{M}) + \mathbf{r}(t)\delta\mathbf{MP} + \mathbf{O}(\delta^2). \tag{8.2.3a}$$

In the usual way, bring $\mathbf{r}(t)$ to the left-hand side of the equation, divide both sides by δ, and take the limit as δ goes to 0, to get, with the aid of the definition, $\mathbf{B} := \mathbf{M}(\mathbf{I} - \mathbf{P})$,

$$\frac{d\mathbf{r}(t)}{dt} = -\mathbf{r}(t)\mathbf{M}(\mathbf{I} - \mathbf{P}) = -\mathbf{r}(t)\mathbf{B}. \tag{8.2.3b}$$

Even on an abstract vector space, the solution of this differential equation is simple as long as we understand that $\exp(-t\mathbf{B})$ stands for its Maclaurin's series expansion. Now since $\mathbf{p} = \mathbf{r}(0)$, we get

$$\mathbf{r}(t) = \mathbf{p}\exp(-t\mathbf{B}), \tag{8.2.3c}$$

and by postmultiplying with $\boldsymbol{\varepsilon}'$, we have

$$R(t) = \mathbf{r}(t)\boldsymbol{\varepsilon}' = \Psi\left[\exp(-t\mathbf{B})\right]. \tag{8.2.3d}$$

Since $b(t) = -R'(t)$, Equation (3.1.7d) directly follows. In fact all of the equations in Theorem 3.1.1 follow from this if we recognize that $\mathbf{B} = \mathbf{V}^{-1}$, which in turn is true if and only if $\mathbf{T} = \mathbf{M}^{-1}$. Their derivation is almost identical to that which we gave in Chapter 3, except that here we never impose any physical meanings or constraints on the individual components of the matrices.

Equation (8.2.3d) is the primary one that places constraints on $\mathbf{p}$ and $\mathbf{B}$. Since $R(t)$ is an observable function, it must satisfy the following:

$$t_2 > t_1 \geq 0 \quad \text{implies that} \quad R(t_1) \geq R(t_2) \geq 0. \tag{8.2.4}$$

If Ξ is finite-dimensional, this constraint is no more or less than requiring that $1 - R(t)$ be a *matrix exponential* probability distribution function.

In a base-free description, one might ask what the dimensionality of Ξ might be. This has a straightforward answer when one notes that $\boldsymbol{\varepsilon}'$ is a unique, invariant vector. Therefore, we define the dimension of Ξ' to be the smallest integer for which the family of vectors

$$\boldsymbol{\varepsilon}',\ \mathbf{B}\boldsymbol{\varepsilon}',\ \mathbf{B}^2\boldsymbol{\varepsilon}',\ \mathbf{B}^3\boldsymbol{\varepsilon}',\ \cdots,\mathbf{B}^n\boldsymbol{\varepsilon}',\ \cdots \tag{8.2.5a}$$

is linearly independent. This is a base-free property, even though one usually uses some basis set representation to find that integer. Ξ and Ξ' must have the same dimension (call it m), and because of (8.2.5a), there must exist m linearly independent vectors, $\{\mathbf{r_j}\}$, in Ξ for which

$$\mathbf{r_j}\boldsymbol{\varepsilon}' = 1, \quad j = 1, 2, \cdots, m. \tag{8.2.5b}$$

These (or any independent linear combination of them) can be used as the basis set for Ξ. If we so desired, we could pick an appropriate linear combination that makes $\mathbf{M}$ a diagonal matrix, as discussed in the paragraph following (8.2.2). Note that *all* bases which satisfy (8.2.5b) *must* be related to each other by some isometric transformation. Let $\{\mathbf{r_j}\}$ and $\{\tilde{\mathbf{r}}_\mathbf{j}\}$ be two bases for Ξ. Then there exists a matrix (or linear transformation) $\mathbf{S}$ such that

$$\mathbf{r_j}\mathbf{S} = \tilde{\mathbf{r}}_\mathbf{j} \quad \text{for} \quad j = 1, 2, \cdots, m, \tag{8.2.5c}$$

and $\mathbf{S}\boldsymbol{\varepsilon}' = \boldsymbol{\varepsilon}'$.

Having said this, we see that all of the above depend on four independent objects, $\mathbf{p}$, $\mathbf{M}$, $\mathbf{P}$, and $\boldsymbol{\varepsilon}'$. There might be a smaller set by combining $\mathbf{M}$ and $\mathbf{P}$ in $\mathbf{B}$. We use $\mathbf{M}$ in describing the interaction of two servers, but even in that more complicated system $\mathbf{M}$ can be absorbed into $\mathbf{B}$, so it remains to be seen if $\mathbf{M}$ is a fundamental object. In any case, we use the expression, $S = \langle \mathbf{p}, \mathbf{B} \rangle$ to mean that S can be represented by $\langle \mathbf{p}, \mathbf{B} \rangle$ (i.e., that the pdf for S satisfies Theorem 3.1.1).

8.2.2. Residual Vector and Related Properties

From now on we will use the terminology of queueing theory to describe the behavior of servers. Thus S is busy if there is a customer there, and becomes idle when the customer finishes service and leaves. Suppose that instead of leaving forever, the customer immediately returns and starts up again. This parallels what we did in Section 3.3.3. The equation governing this process is directly related to (8.2.3a), except that we must add the term previously ignored, namely that the customer upon leaving [$\mathbf{q}'$] immediately reenters [$\mathbf{p}$]. Thus

$$\mathbf{r}(t+\delta) = \mathbf{r}(t)(\mathbf{I} - \delta\mathbf{M}) + \mathbf{r}(t)\delta\mathbf{MP} + \mathbf{r}(t)\delta\mathbf{Mq}'p + O(\delta^2).$$

Note that $\mathbf{Mq}'p = \mathbf{M}(\mathbf{I} - \mathbf{P})\boldsymbol{\varepsilon}'\mathbf{p} = \mathbf{BQ}$. In the usual way, we get the following differential equation, which is a special case of the Chapman–Kolmogorov equations, (1.3.2b):

$$\frac{d\mathbf{r}(t)}{dt} = -\mathbf{r}(t)[\mathbf{B}(\mathbf{I} - \mathbf{Q})]. \tag{8.2.6}$$

Since $(\mathbf{I}-\mathbf{Q})\boldsymbol{\varepsilon}' = \mathbf{o}'$, we know that a steady-state solution vector, $\boldsymbol{\pi}_\mathbf{r} := \lim_{t\to\infty}\mathbf{r}(t)$ exists and satisfies the eigenvector equation:

$$\boldsymbol{\pi}_\mathbf{r}\mathbf{B}(\mathbf{I}-\mathbf{Q}) = \mathbf{o}.$$

In Section 3.3 this vector was shown to be [Equation (3.3.10b)],

$$\boldsymbol{\pi}_\mathbf{r} = \frac{\mathbf{pV}}{\mathbf{pV}\boldsymbol{\varepsilon}'} = \frac{1}{\bar{x}}\mathbf{pV} \tag{8.2.7a}$$

(with $\boldsymbol{\pi}_\mathbf{r}\boldsymbol{\varepsilon}' = \mathbf{1}$). We can say that $\langle \boldsymbol{\pi}_\mathbf{r}, \mathbf{B} \rangle$ generates the residual process, including for instance, the mean residual time,

$$\overline{x_r} = \boldsymbol{\pi}_\mathbf{r}\mathbf{V}\boldsymbol{\varepsilon}' = \frac{\mathbf{pV}^2\boldsymbol{\varepsilon}'}{\bar{x}} = \frac{\overline{x^2}}{2\,\bar{x}}. \tag{8.2.7b}$$

This is the mean time for service to complete if it is not known when service began.

8.2.3. Networks of Servers

Let us consider a *true* closed queueing network, as shown in Figure 8.2.1. The probability of going from S_i to S_j is a real number that is greater than or equal to

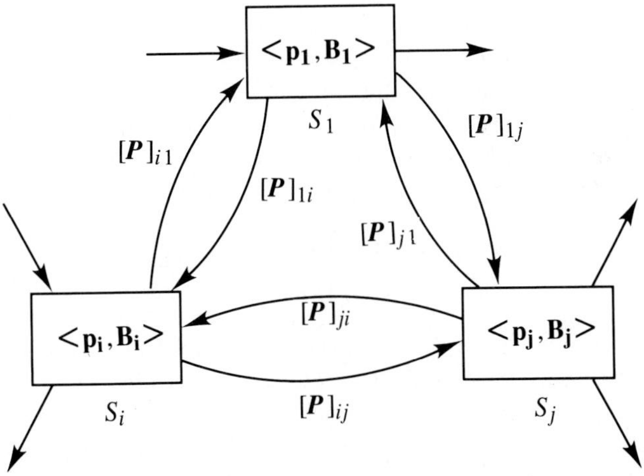

Figure 8.2.1: A true network of m nonexponential servers. Each server can be represented by the vector-matrix pair $\langle \mathbf{p_j}, \mathbf{B_j} \rangle$. S_j has its own queue, with n_j customers in it. $\boldsymbol{P}$ is the transition matrix whose ijth component is the probability that a customer, upon leaving S_i, will go to S_j.

0, and the sum over j is equal to 1. Let that number be $[\boldsymbol{P}]_{ij}$. Then $[\boldsymbol{P}]_{ij} \geq 0$, and

$$\sum_{j=1}^{m} [\boldsymbol{P}]_{ij} = 1 \quad \text{for } all\ i.$$

In other words, $\boldsymbol{P}$ is a stochastic or transition matrix in the usual sense. Each server is represented by a vector-matrix pair, $<\mathbf{p_j}, \mathbf{B_j}>$ (or perhaps the matrices $\mathbf{M_j}$ and $\mathbf{P_j}$). Server S_j has its own queue, containing n_j customers. The *external state* of the system is denoted by the n-tuple

$$\underline{n} = < n_1, n_2, \cdots, n_m >, \qquad \sum_{j=1}^{m} n_j = N,$$

where N is the total number of customers in the system. The S_j's are completely independent of each other, and interact only through the exchange of customers, but only *one* thing can happen at a time. For a given $\underline{n}$, the composite *internal state* of the system must be described by a vector that is in the direct product space of the Ξ's. Let $\mathbf{r_j} \in \Xi_j$ be a state vector for S_j. Then a state vector for the system as a whole, $\mathbf{r}(\underline{n})$ is an element of the *product space* Ξ, or

$$\mathbf{r}(\underline{n}) \in \Xi = \Xi_1 \otimes \Xi_2 \otimes \cdots \otimes \Xi_m := \{\, \mathbf{r} = \mathbf{r_1} \otimes \mathbf{r_2} \otimes \cdots \otimes \mathbf{r_m} \mid \mathbf{r_j} \in \Xi_j \,\}.$$

Understand that this is a base-free statement. There are many ways of giving implementations of this space, one of which is the family of *Kronecker products* that we discussed in Chapter 7 [GRAH81]. However, there are numerous advantages of avoiding a specific implementation, at least until comprehensive numerical calculations must be performed. First of all, one can naively manipulate various operators from the different spaces without being in error. For instance, the statement that the servers act independently of each other implies that the various operators on each subspace commute with each other. One can formally write $\mathbf{B_1}\,\mathbf{B_2} = \mathbf{B_2}\,\mathbf{B_1}$. But the Kronecker product of the two matrix representations, $\mathbf{B_1} \otimes \mathbf{B_2}$, is an entirely different animal from $\mathbf{B_2} \otimes \mathbf{B_1}$, so commutativity does not have a meaning unless one embeds both matrices into the product space by defining a *hat* [$\hat{\cdot}$] operator, as in the following:

$$\hat{\mathbf{B}}_1 := \mathbf{B_1} \otimes \mathbf{I_2}$$

and

$$\hat{\mathbf{B}}_2 := \mathbf{I_1} \otimes \mathbf{B_2}.$$

Note that the asymmetry of these two formulas obscures the symmetric relation that the different servers naturally have with each other.

If all the servers have one-dimensional representations, Ξ is one-dimensional and we have a Jackson network, [JACK63], [GORD67], and we have nothing new to contribute. If at least one of the servers needs a higher-dimensional representation, we are into LAQT. If, in particular, exactly one subspace, say Ξ_1, is multidimensional, Ξ is isomorphic to Ξ_1 and implementations are tractable. If two or more spaces are multidimensional, one can no longer avoid the problems inherent in product space arithmetic. In any case, the instantaneous rate for something to happen is generated by

$$\mathbf{M} := \mathbf{M_1} + \mathbf{M_2} + \cdots + \mathbf{M_m}.$$

The probability that the next event will occur in S_j is generated by

$$\mathbf{M_j}\,\mathbf{M}^{-1} = \mathbf{M}^{-1}\,\mathbf{M_j},$$

where we have noted that **M** commutes with each of its components. Strictly speaking, one must first embed each matrix into Ξ, and *then* add, but formally this just means putting a *hat* on everything.

8.3. SYSTEMS WITH TWO SERVERS

We have no intention at this time of trying to continue our discussion of many-server systems. Thus let us let $m = 2$ hereafter. Our purpose is to show that many of the known results of queueing theory that have matrix formulations (beyond those we discussed in Section 8.1) are invariant to isometric transformations and can be written in a base-free way. We will enumerate some results concerning the G/M/1 and M/G/1 queues and then look momentarily at the G/G/1 queue. Finally, we will look at some transient behavior in M/G/1 systems, noting that the procedure is completely generalizable.

In a closed loop, S_1 and S_2 play exactly equivalent roles. But as we have mentioned numerous times before, if the number of customers in the system is so large that one or the other has no likelihood of ever being idle, *that* subsystem is equivalent to a source of customers to the other. Clearly, for subsystems where only one customer can be served at a time, the one with the smaller maximal throughput will be that subsystem, or server. By convention, we have assumed that S_2 has the longer mean service time for M/G/1 and $G_2/G_1/1$ queues. But for G/M/1 queues S_1 has the longer service time. Let G_i describe the pdf type of server S_i; then we are looking at $G_2/G_1/1//N$ loops, and their open extensions [i.e., $G_2/G_1/1//(N\rightarrow\infty)$ is equivalent to $G_2/G_1/1$)].

8.3.1. G/M/1 Queue

We have already shown that the steady-state M/G/1 queue is invariant. The same matrix which governs that system [the matrix **A** of (8.1.1)] also has relevance to the open G/M/1 queue, except that now, as in Chapter 5, $\rho = 1/\zeta = \lambda\bar{x} > 1$. For instance, let s and $\hat{\mathbf{u}}$ satisfy the eigenvector equation:

$$\hat{\mathbf{u}}\mathbf{A} = s\,\hat{\mathbf{u}}, \tag{8.3.1a}$$

where s is the smallest positive eigenvalue of **A**, and $\hat{\mathbf{u}}\boldsymbol{\varepsilon}' = 1$. We know that $s < 1$ if and only if $\zeta < 1$ (ζ is the utilization factor now). In Theorem 5.1.1, we showed the following:

$$r(n) = (1-s)\zeta s^{n-1}, \quad n > 0 \tag{8.3.1b}$$

and

$$r(0) = 1 - \zeta. \tag{8.3.1c}$$

Note that the eigenvalues are an invariant property of any matrix. That is, if **X** and $\tilde{\mathbf{X}}$ are related by an isometric transformation, they have the same set of eigenvalues. Also, recall from (5.1.6b) that

$$\hat{\mathbf{u}} = \lambda\mathbf{p}\mathbf{V}[\mathbf{I} + \lambda(1-s)\mathbf{V}]^{-1}. \tag{8.3.1d}$$

It follows from Corollary 5.1.1 that $\Psi[(\mathbf{I}+\lambda(1-s)\mathbf{V})^{-1}] = B^*[\lambda(1-s)] = s$. So we even get the famous relation between the Laplace transform and s without ever knowing what a Riemann–Stieltjes integral is, and from a base-free matrix algebraic formulation.

Next recall two other distributions related to the G/M/1 queue. The first is the interdeparture time distribution we gave in Section 5.2.2, which is generated by $<\mathbf{p}_{2d}, \mathbf{B}_{2d}>$ where

$$\mathbf{p}_{2d} := [\, s\hat{\mathbf{u}}\, ,\ 1-s\ ]; \quad \mathbf{B}_{2d} := \begin{bmatrix} \mathbf{B} & \mathbf{B}\boldsymbol{\varepsilon}' \\ \mathbf{o} & \lambda \end{bmatrix} \Big\} \ (m+1). \tag{8.3.2}$$

The second distribution describes the arrival time conditioned by departures, which is generated by $<\hat{\mathbf{u}}, \mathbf{B}>$. This is rather interesting, for it tells us that the generator of the arrival process is in composite state $\hat{\mathbf{u}}$ at the moment a customer leaves the G/M/1 queue, thus giving us a meaning of the eigenvector of $\mathbf{B}$ belonging to the smallest eigenvalue, s.

The last process we mention here is the system time for the M/G/1 queue. It is generated by the vector-matrix pair (Section 4.2.3) $<\mathbf{p}_s, \mathbf{B}_s>$, where

$$\mathbf{B}_s := \mathbf{B} - \lambda\mathbf{Q} \tag{8.3.3a}$$

and

$$\mathbf{p}_s := (1-\rho)\,\mathbf{p}(\mathbf{I}-\mathbf{U})^{-1}. \tag{8.3.3b}$$

It is clear that all three distributions $<\mathbf{p}_d, \mathbf{B}_d>$, $<\hat{\mathbf{u}}, \mathbf{B}>$, and $<\mathbf{p}_s, \mathbf{B}_s>$, are invariant to isometric transformations.

8.3.2. Two Nonexponential Servers

As we have already seen, if the representation of a nonexponential server is m-dimensional, the space required to describe its interaction with exponential servers is also m-dimensional (i.e., there is no increase in dimensionality)†. However, if two servers are nonexponential, one needs a space of $m_1 \cdot m_2$ dimensions. There is no way out of this increase in complexity; it simply reflects the amount of information needed to describe the dynamics of such complex systems. (There is an interesting exception, which we will mention in the concluding remarks.)

We first recall the steady-state solution for the closed G/G/1//N loop from Chapter 7. The operators $\mathbf{B}_i$ and $\mathbf{Q}_i$ are defined in the same way for server i as

† This, by the way, indicates that the concept of an *absorbing state* interferes with a self-consistent matrix formulation of queueing theory, since then one requires an $(m+1)$–dimensional description.

was done in the previous sections. Remember that operators with different subscripts (belonging to different subspaces) automatically commute. When we need a matrix representation of sums of their products, we embed them in the product space, which formally means putting a $\hat{\ }$ on them. Repeating Equations (7.3.5), we have

$$\mathbf{S}^+ := \mathbf{B}_1 + \mathbf{B}_2 - \mathbf{B}_2\mathbf{Q}_1, \tag{8.3.4a}$$

$$\mathbf{S}^- := \mathbf{B}_2 + \mathbf{B}_1 - \mathbf{B}_1\mathbf{Q}_2, \tag{8.3.4b}$$

$$\mathbf{T}^\pm := (\mathbf{S}^\pm)^{-1}, \tag{8.3.4c}$$

and

$$\mathbf{U} := \mathbf{T}^+\mathbf{S}^-. \tag{8.3.4d}$$

It would seem that $\mathbf{S}^+$ and $\mathbf{S}^-$ are the generalizations of $\mathbf{A}$ in (8.1.1) for the M/G/1 queue, but it is not quite that simple. Instead, $\mathbf{U}$ is the direct generalization of the '$\mathbf{U}$' for the M/G/1 queue, with no real analog for $\mathbf{A}$, since now both servers play symmetric roles in the theory, and $\mathbf{U}^{-1} = \mathbf{T}^-\mathbf{S}^+$. The steady-state solution is given by Theorem 7.3.2:

$$\pi(n, N) = \pi(0, N)\mathbf{B}_2\mathbf{U}^n\mathbf{T}^-, \quad 1 \le n = \le N-1, \tag{8.3.5a}$$

$$\pi(N, N) = \pi(0, N)\mathbf{B}_2\mathbf{U}^{N-1}\mathbf{V}_1, \tag{8.3.5b}$$

and

$$r(n, N) = \pi(n, N)\boldsymbol{\varepsilon}'. \tag{8.3.5c}$$

All of these equations are invariant to isometric transformations in the two subspaces, since a transformation in one subspace automatically commutes with matrices in the other space. An interesting research problem would be to study isometric transformations over the product space.

Recall that $\mathbf{Q_i}$ and $\mathbf{B_i}$ do not commute with each other if Ξ_i has dimension greater than 1 (nonexponential). This is what made the M/G/1 queue harder than the M/M/1 queue. But now we have the added problem that $\mathbf{S}^+$ and $\mathbf{S}^-$ do not commute with each other if both Ξ_1 and Ξ_2 are multidimensional, which is what makes the G/G/1 queue harder than the M/G/1 queue.

8.3.3. Review of Transient Behavior

In the previous chapters, we assumed that there existed a basis set of *pure* vectors, and that the system could be in one of those *pure states* initially. By doing so, we appeared to be saying that such states (which we called *phases*) have physical meaning individually. The formulation we are presenting in this chapter treats all vectors on an equal footing. Note that in deriving (8.2.1) and (8.2.2), we talked about operations (linear transformations) on an arbitrary state vector $\mathbf{r}$, and then since our intermediate equations were true for all state vectors, we threw $\mathbf{r}$ away. From a rigorous mathematical point of view we said that if an equation of the form $\mathbf{rX} = \mathbf{o}$, is true for m linearly independent vectors, where m

is the dimension of the vector space, then it must be true for all $\mathbf{r} \in \Xi$, and further, $\mathbf{X} = \mathbf{O}$. This means that we can pick *any* m linearly independent vectors from Ξ, and treat them as though they are *pure* states, even though they may have no independent physical meaning. For instance, we could pick the set discussed in (8.2.5).

The argument goes something like this. Let $\mathbf{r_1}, \mathbf{r_2}, \cdots, \mathbf{r_m}$ be a basis for Ξ. Then any physical vector $\mathbf{r}$ can be written as a linear combination of these basis vectors,

$$\mathbf{r} = \sum_{j=1}^{m} r_j \mathbf{r_j}.$$

Every linear operator $\mathbf{X}$ transforms every vector in Ξ to some other vector in Ξ. In particular, $\mathbf{r_j X} \in \Xi$, and thus can be written as a linear combination of the $\{\mathbf{r_j}\}$'s. That is,

$$\mathbf{r_j X} = \sum_{k=1}^{m} X_{jk} \mathbf{r_k},$$

and thus

$$\mathbf{rX} = \sum_{j=1}^{m} r_j \mathbf{r_j X} = \sum_{j,k=1}^{m} r_j X_{jk} \mathbf{r_k}.$$

In words, if the set of scalars, $\{r_1, r_2, \cdots, r_m\}$, describes $\mathbf{r}$ in terms of the basis set $\{\mathbf{r_i}\}$, then $\{\sum r_j X_{j1}, \sum r_j X_{j2}, \cdots, \sum r_j X_{jm}\}$ describes $\mathbf{rX}$ in the same basis, and X_{jk} is a matrix representation of transformation operator $\mathbf{X}$ in the basis $\{\mathbf{r_j}\}$. Note that similarity transformations (which include our isometric transformations) are those that change the basis set.

We can see that dealing with components of vectors and matrices is equivalent to dealing with abstract vectors and transformations. We can do it either way, without implying that the components themselves have any meaning. It is somewhat easier to speak in terms of components. Thus we have been using the notation "$\{i, n\}$" to mean that "S_1 *is in state* $\mathbf{r_i}$ *with queue length* n."

In Section 4.5 we considered the process of a queue rising in length. For $n \le N$, we defined the matrix

$$\mathbf{H_u}(n) := \textit{probability matrix of first passage from } n \textit{ to } n+1. \qquad (8.3.6a)$$

That is, we said that $[\mathbf{H_u}(n)]_{ij}$ is the probability that S_1 will be in state (phase) j (or $\mathbf{r_j}$) when its queue goes from n to $n+1$ for the first time, given that it started in state i (or $\mathbf{r_i}$) with n customers. Now we would say that if S_1 was initially described by state vector $\mathbf{r}$, with n customers, then when its queue goes from n to $n+1$ for the first time, it will be described by state vector $\mathbf{rH_u}(n)$. After a while the two viewpoints seem to be synonymous; one no longer notices the difference (are you there yet, dear reader?). By its definition from either viewpoint, the following must be true since the queue must eventually reach every length.

$$\mathbf{H_u}(n)\boldsymbol{\varepsilon}' = \boldsymbol{\varepsilon}' \quad \text{for} \quad 1 \le n < N. \qquad (8.3.6b)$$

The $\mathbf{H_u}(n)$'s are isometric.

In Section 4.5.1 we derived the recursive equations that the $\mathbf{H_u}(n)$'s must satisfy, namely

$$\mathbf{H_u}(n) = \lambda[\lambda\mathbf{I} + \mathbf{B} - \mathbf{BQH_u}(n-1)]^{-1}. \tag{8.3.7a}$$

From the definition of $\mathbf{A}$, this can also be written as

$$\mathbf{H_u}(n) = [\mathbf{A} + \mathbf{Q} - \mathbf{AQH_u}(n-1)]^{-1}. \tag{8.3.7b}$$

As with all recursive relations, we must start somewhere, which we did by noting that

$$\mathbf{H_u}(0) = \mathbf{p}, \tag{8.3.8a}$$

and thus

$$\mathbf{H_u}(1) = \lambda[\lambda\mathbf{I} + \mathbf{B} - \mathbf{BQ}]^{-1} = [\mathbf{A} + \mathbf{Q} - \mathbf{AQ}]^{-1}. \tag{8.3.8b}$$

It is easy to show that $\mathbf{H_u}(1)\boldsymbol{\varepsilon}' = \boldsymbol{\varepsilon}'$, and by induction, using (8.3.7b), prove that (8.3.6b) is true for all n. Note that in general, the $\mathbf{H_u}$'s are all different, although they do approach a limit for large n.

From these matrices we found the probability matrices of first passage from n to $n+j$, for any n and j. For instance, the probability matrix (it is actually a vector) of first passage from $0 \to n$ is

$$\mathbf{p_u}(n) := \mathbf{pH_u}(1)\mathbf{H_u}(2) \cdots \mathbf{H_u}(n-1). \tag{8.3.9}$$

These objects may not appear to be very interesting in their own right, but they are needed for calculating first-passage times, as is shown in the next paragraph.

By arguments similar to the preceding, we derived the mean time for the queue to grow from n to $n+1$ for the first time. First we defined the vector, $\boldsymbol{\tau}'_\mathbf{u}(n)$, whose ith component is $[\boldsymbol{\tau}'_\mathbf{u}(n)]_i$:= *mean first-passage time from n to n+1, having started in state* $\{i, n\}$. It then followed that

$$\boldsymbol{\tau}'_\mathbf{u}(n) = \frac{1}{\lambda}\boldsymbol{\varepsilon}' + \mathbf{H_u}(n)\mathbf{BQ}\boldsymbol{\tau}'_\mathbf{u}(n-1), \quad \text{with} \quad \boldsymbol{\tau}'_\mathbf{u}(0) := \frac{1}{\lambda}\boldsymbol{\varepsilon}'. \tag{8.3.10}$$

The sets of equations (8.3.7) and (8.3.10) are all that is needed to compute all the vector times. One can then calculate (whether the system is open or closed, irrespective of whether ρ is less than, equal to, or greater than 1) such things as:

1. The mean first-passage time of going from n to $n+1$, given that the customer in service has just begun $[\mathbf{p}\boldsymbol{\tau}'_\mathbf{u}(n)]$.

2. The mean first-passage time from n to $n+1$, given that the queue was originally empty {see (8.3.9)} $[t_u(n) := \mathbf{p_u}(n)\boldsymbol{\tau}'_\mathbf{u}(n)]$.

3. The mean first-passage time, given that a customer has just arrived and found n customers already there (see Theorem 4.5.2 and its corollaries) $[\boldsymbol{\pi}(n)\boldsymbol{\tau}'_\mathbf{u}(n)/r(n)]$.

One can even calculate in an efficient way the mean time for a queue to grow to n for the first time given that a customer has just arrived at an empty queue, namely,

$$t(1 \rightarrow n) := \sum_{k=1}^{n-1} t_u(k). \tag{8.3.11}$$

Note that this is not the same as the first excursion to n during a busy period (although that too is calculable), since this process allows the queue to empty any number of times before finally reaching its goal.

In like manner one can derive analogous expressions for M/G/C, G/G/1, and even more general systems. The most significant point in this discussion is that all the formulas are expressible in a base-free formulation invariant to isometric transformations. Thus explicit appeal to a "component" interpretation is unnecessary.

8.4. CONCLUDING REMARKS

We hope we have shown that an approach which is linear algebraic from beginning to end has great potential for covering material which hitherto has been ignored because the difficulties involved. The ubiquitousness of such an approach appears to depend on the invariance of formulas to isometric transformations. If this is so, one must be prepared to deal with representations that are distinctly *not* phase distributions. Only then can one study the purely algebraic properties of various systems using a paradigm that is different from what we have been locked into for 50 years or more. Two such research problems are described below.

1. Consider a G/G/1//N queue. In preparing such a system at say $t = 0$, one must initialize both S_1 and S_2. This would require specifying $m_1 + m_2$ quantities. That is, we have a *sum−space* description. But as the system evolves in time, the components from each subspace become correlated with those in the other, thus forcing a complete *product−space* description ($m_1 \cdot m_2$ components). However, as Van de Liefvoort has shown [LIEF86] [LIEF89], the key matrix for the steady-state solution, **U** from (8.3.4d), has $m_1 \cdot m_2 - m_1 - m_2 + 1$ eigenvectors with eigenvalue 1, all of which can be thrown away when calculating the $r(n, N)$'s, *IF* one can find an appropriate isometric transformation in the product space (such a transformation exists, finding a general form for it is the problem). This means that there exists (at least) one sum-space representation of steady-state G/G/1 queues, one that mixes the components of the two subspaces.

2. In describing M/G/C//N-type systems ($N > C$), one must work in spaces which have $D := \binom{m+C-1}{C}$ components. The steady-state solutions can then be written in terms of matrices that have this dimension. However, when $N \leq C$, the solution is known to be the product-form solution of Jackson networks! What is the relationship between the two? And as in question 1, does there exist a representation of dimension less than D which can be used?

BIBLIOGRAPHY

The following list is not meant to be exhaustive. Most of the references contain background information, or are of historical significance, or are a source for further reading. Some are included because that is where I learned something that I thought was interesting or important. For the frontiers of research see [NEUT89] who has 70 pages (!) of references to research papers related to LAQT. My favorite queueing theory text (excluding this one, of course) is [KLEI75], while I use [TRIV82] as the prerequisite book. I also like [ALLE90] for that purpose. I browsed through [COOP81] a short while ago and was startled at how much material was in there that I had not thought about. The field is so broad that almost any well-written book will have unique ideas for pondering. For research background support, there is still no equal to [FELL68] and [FELL71]. Sometimes one gets the feeling that any new idea one might have is already in there, but perhaps in a different form. So think of this as nothing more than my personal list.

[ABRA64] Milton Abramowitz, and Irene A. Stegun, *Handbook of Mathematical Functions*, U.S. Gov't. Printing Office, Washington, D.C., 1964.

[ALLE90] Arnold O. Allen, *Probability, Statistics, and Queueing Theory, with Computer Science Applications*, 2nd Ed., Academic Press, New York, 1990.

[ASMU87] S. Asmussen, *Applied Probability and Queues*, John Wiley, New York, 1987.

[BASK75] Forest Baskett, K. Mani Chandy, Richard R. Muntz and Fernando G. Palacios, "Open, Closed, and Mixed Networks of Queues with Different Classes of Customers," *Journal of the ACM*, **22**, No. 2, 248–260, 1975.

[BURK56] Paul J. Burke, "The Output of a Queueing System," *Operations Research*, **4**, 699–704, 1956.

[BUZE73] Jeffrey P. Buzen, "Computational Algorithms for Closed Queueing Networks with Exponential Servers," *Communications of the ACM*, September 1973.

[CARR79] John L. Carroll, *A Study of Closed Queueing Networks with Population Size Constraints*, Ph.D. Thesis, University of Nebraska, Lincoln, NE, 1979.

[CARR82] John L. Carroll, Lester Lipsky, and Appie van de Liefvoort, "Solutions of M/G/1//*N*–Type Loops with Extension to M/G/1 and GI/M/1 Queues," *Operations Research*, **30**, 490–514, 1982.

[COHE69] Jacob W. Cohen, *The Single Server Queue*, North Holland, Amsterdam, 1969.

[COOP81] Robert B. Cooper, *Introduction to Queueing Theory*, 2nd Ed., Elsevier North Holland, New York, 1981.

[COX55] D. R. Cox, "Use of Complex Probabilities in the Theory of Stochastic Processes," *Proceedings of the Cambridge Philosophical Society,* **51**, 313–319, 1955.

[COX62] D. R. Cox, *Renewal Theory,* Methuen, London, 1962.

[DING91] Yiping Ding, *On Performance Control of Real-Time Systems*, Ph.D. Thesis, University of Connecticut, Storrs, CT, July 1991.

[DENN78] Peter J. Denning and Jeffrey P. Buzen, "The Operational Analysis of Queueing Network Models," *Computing Surveys,* **10**, No. 3, 225–261, 1978.

[ERLA17] A. K. Erlang, "Solution of Some Problems in the Theory of Probabilities of Significance in Automatic Telephone Exchanges," *The Post Office Electrical Engineer's Journal,* **10**, 189–97, 1917–1918.

[FELL68] William Feller, *An Introduction to Probability Theory and Its Applications*, Vol. I, John Wiley, New York, 1957.

[FELL71] William Feller, *An Introduction to Probability Theory and Its Applications*, Vol. II, John Wiley, New York, 1971.

[GORD67] W. J. Gordon and Gordon F. Newell, "Closed Queueing Systems with Exponential Servers," *Operations Research,* **15**, 254–265, 1967.

[GRAH81] A. Graham, *Kronecker Products and Matrix Calculus*, Ellis Horwood, Chichester, England, 1981.

[HALM55] Paul R. Halmos, *Finite Dimensional Vector Spaces*, Princeton University Press, Princeton, NJ, 1955.

[JACK63] James R. Jackson, "Jobshop-like Queueing Systems," *Management Science,* **10**, 131–142, 1963.

[KEND64] David G. Kendall, "Some Recent Work and Further Problems in the Theory of Queues," *Theory of Probability and Its Applications,* **1**, 1–15, 1964.

[KHIN32] A. Y. Khinchin, "Mathematical Theory of Stationary Queues," *Mat. Sbornik,* **39**, 73–84, 1932.

[KING72] John F. C. Kingman, *Regenerative Phenomena*, John Wiley, New York, 1972.

[KLEI75] Leonard Kleinrock, *Queueing Systems,* Volume I: *Theory*, John Wiley, New York, 1975.

[LAZO84] Edward D. Lazowska, John Zahorjan, G. Scott Graham, and Kenneth C. Sevcik, *Quantitative System Performance - Computer System Analysis Using Queueing Network Models*, Prentice Hall, Englewood Cliffs, NJ, 1984.

[LIEF82] Appie van de Liefvoort, *An Algebraic Approach to the Steady-state Solution of* G/G/1//*N-Type Loops*, Ph.D. Thesis, University of Nebraska, Lincoln, 1982.

[LIEF86] Appie van de Liefvoort and Lester Lipsky, "A Matrix-Algebraic Solution to Two K_m Servers in a Loop," *Journal of the ACM,* **33**, 1, 207–223, 1986.

[LIEF87] Appie van de Liefvoort, "A Sum-Space Characterization of G/G/1//N-Type Queues," Technical report TR-87-5, Computer Science Department, University of Kansas, Lawrence, April 1987.

[LIEF90] Appie van de Liefvoort, "The Waiting-Time Distribution and its Moments of the PH/PH/1 Queue," *Operations Research Letters,* **9**, 261–269, 1990.

[LIPS77] Lester Lipsky and James D. Church, "Applications of a Queueing Network Model for a Computer System", *Computing Surveys*, **9**, 205–221, September 1977.

[LIPS83] Lester Lipsky and Zhixi Fang, "A Note on the Persistance of the Time-Dependent Solution of an M/M/1//N Queue," Technical Report, Department of Computer Science, University of Nebraska, Lincoln, August 1982.

[LIPS85a] Lester Lipsky, "Explicit Solutions of M/G/C//N-type Queueing Loops with Generalizations," *Operations Research,* **33**, 911–927, 1985.

[LIPS85b] Lester Lipsky and V. Ramaswami, "A Unique Minimal Representation of Coxian Service Centers," (1985), Technical Report, Department of Computer Science, University of Nebraska, Lincoln, November 1985.

[LIPS86] Lester Lipsky and Zhixi Fang, "Classification of Functions With Rational Laplace Transforms," *Summer Computer Simulation Conference,* Las Vegas, NV, July 1986.

[LIPS90] Lester lipsky, Dilip Tagare and Edward Bigos, "Evaluation of Queueing System Parameters Using Linear Algebraic Queueing Theory – An Implementation," *ACM SIGSMALL/PC Symposium on Small Systems*, Washington, DC, March 1990.

[LITT61] J. D. C. Little, "A Proof of the Queueing Formula $L = \lambda W$," *Operations Research,* **9**, 383–387, 1961.

[MARK07] Andrei A. Markov, "Extension of the Limit Theorems of Probability Theory to a Sum of Variables Connected in a Chain," *The Notes of the Imperial Academy of Science of St. Petersburg*, Series VIII, Physio-Mathematical College, Vol. XXII, 9, December 1907.

[MELA90] B. Melamed and Ward Whitt, "On Arrivals That See Time Averages," *Operations Research,* **38**, 156–172, 1990.

[MOLL89] Michael K. Molloy, *Fundamentals of Performance Modeling*, Macmillan, New York, 1989.

[MORS58] Philip M. Morse, *Queues, Inventories and Maintenance*, John Wiley, New York, 1958.

[NEUT75] Marcel F. Neuts, "Probability Distributions of Phase Type," in *Liber Amicorum Prof. Emeritus H. Florin*, Department of Mathematics, University of Louvain, Belgium, 173–206, 1975.

[NEUT77] Marcel F. Neuts, "The Mythology of the Steady State," *Joint National ORSA-TIMS Meeting*, Atlanta, GA, November 1977.

[NEUT81] Marcel F. Neuts, *Matrix-Geometric Solutions in Stochastic Models* – An Algorithmic Approach, Johns Hopkins University Press, Baltimore, 1981.

[NEUT82] Marcel F. Neuts, "Explicit Steady-State Solutions to Some Elementary Queueing Models," *Operations Research,* **30**, 480–489, 1982.

[NEUT89] Marcel F. Neuts, *Structured Stochastic Matrices of M/G/1 Type and Their Applications*, Marcel Dekker, New York, 1989.

[POLL30] F. Pollaczek, "Uber eine Aufgabe der Wahrscheinlichkeitstheorie, I und II," *Mathematische Zeitschrift,* **32**, 64–100, 729–750, 1930.

[RAMA90] V. Ramaswami, "Nonlinear Matrix Equations in Applied Probability – Solution Techniques and Open Questions," *SIAM Review*, Forthcoming.

[TAKA62] Lajos Takacs, *Introduction to the Theory of Queues,* Oxford University Press, New York, 1962.

[TEHR83] Aby Tehranipour, *Explicit Solutions of Generalized* M/G/*C*//*N Systems Including an Analysis of Their Transient Behavior*, Ph.D. Thesis, University of Nebraska, Lincoln, December 1983.

[TEHR89] Aby Tehranipour, Appie van de Liefvoort, and Lester Lipsky, "Residual Lifetimes as a Function of Queue Length for M/G/1//*N* Loops," *Joint ACM-IEEE Workshop on Applied Computing '89*, Stillwater, OK, March 1989.

[TEHR90] Aby Tehranipour and Lester Lipsky, "The Generalized M/G/1//*N*-Queue as a Model for Time-Sharing Systems," *ACM-IEEE Joint Symposium on Applied Computing '90*, Fayetteville, AR, April 1990.

[TRIV82] Kishor S. Trivedi, *Probability and Statistics with Reliability, Queueing, and Computer Science Applications*, Prentice Hall, Englewood Cliffs, NJ, 1982.

[WALL69] Victor L. Wallace, *The Solution of Quasi-Birth and Death Processes Arising from Multiple Access Computer Systems,* Ph.D. Thesis, University of Michigan, Ann Arbor, March 1969.

[WALL72] Victor L. Wallace, "Toward an Algebraic Theory of Markovian Networks," *Proceedings of the Symposium on Computer Communications Networks and Teletraffic*, 397–408, 1972.

[WHIT80] Ward Whitt, "Approximating a Point Process by a Renewal Process, I: Two Basic Methods," *Operations Research,* **30**, 125–147, 1982.

[WOLF82] R. W. Wolff, "Poisson Arrivals See Time Averages," *Operations Research,* **30**, 223–231, 1982.

LIST OF SYMBOLS

$\mathbf{A} = \mathbf{I} + \lambda^{-1}\mathbf{B} - \mathbf{Q}$.	123
$\mathbf{a}(n; N)$ – S.s. Arrival Prob. vector at S_1.	131
$a(n; N) = \mathbf{a}(n; N)\boldsymbol{\varepsilon}'$ – Scalar prob. associated with $\mathbf{a}(n; N)$.	131
$\mathbf{a_2}(k; N)$ – S.s. Arrival Prob. vector at S_2.	212
$\mathbf{a_2}(k; N \mid C)$ – S.s. Arrival Prob. vector at S_2 (ME/M/C//N).	257
$a_2(k; N) = \mathbf{a_2}(k; N)\boldsymbol{\varepsilon}'$ – Scalar prob. associated with $\mathbf{a_2}(k; N)$.	212
$\mathbf{B} := \mathbf{M}(\mathbf{I} - \mathbf{P})$ – Service rate matrix.	68
$B(t)$ – Probability Distribution Function – PDF.	13
$B^*(s)$ – Laplace transform of $b(t)$.	72
$B_d(t;N)$ – PDF for Inter-**D**eparture times (M/ME/1//N).	145
$B_s(t)$ – PDF for **S**ystem time (M/ME/1).	141
$b(t)$ – Probability density function – pdf.	13
C – Number of servers at S_1	4
C^2 – Coefficient of variation.	14
$\mathbf{d}(n; N)$ – S.s. Prob. vector as seen by departing customer.	131
$d(n; N) = \mathbf{d}(n; N)\boldsymbol{\varepsilon}'$ – Scalar prob. associated with $\mathbf{d}(n; N)$.	131
$\mathbf{d_2}(k; N)$ – S.s. Prob. vector as seen by customer departing S_2.	212
$d_2(k; N) = \mathbf{d_2}(k; N)\boldsymbol{\varepsilon}'$ – Scalar prob. associated with $\mathbf{d_2}(k; N)$.	212
$\mathbf{d_2}(k; N \mid C)$ – S.s. Prob. vector for departure from S_2 (ME/M/C//N).	257
$\mathbf{H_d}(n;N)$ – Prob. mx. of f.p. from n to $n-1$.	188
$\mathbf{H_{2d}}(k;N)$ – Prob. mx. of f.p. at S_2 from k to $k-1$.	265
$\mathbf{H_{dc}}(n;N)$ – Prob. mx. of f.p. from n to $n-1$, where $N \geq n > C$.	320
$\mathbf{H_{dk}}(N \mid C)$ – Prob. mx. of f.p. from k to $k-1$, where $C \geq k > 0$.	320
$\mathbf{H_u}(n)$ – Prob. mx. of f.p. from n to $n+1$.	174
$\mathbf{H_{2u}}(k)$ – Prob. mx. of f.p. at S_2 from k to $k+1$.	259
$\mathbf{H_u}(n \rightarrow n+l)$ – Prob. mx. of f.p. from n to $n+l$.	178
$\mathbf{H_{uc}}(n)$ – Prob. mx. of f.p. from n to $n+1$, where $n \geq C$.	314
$\mathbf{H_{uk}}$ – Related to $\mathbf{X_k}$ by $\mathbf{X_k} = \mathbf{H_{uk}}\,\mathbf{R_{k+1}}$, $k < C$.	314
$\{\, i; n; N \,\}$ – A state of an M/ME/1//N loop (also ME/M/1//N).	121, 212
$\mathbf{M}$ – Completion rate matrix.	18
$\mathbf{M_k}$ – Completion rate matrix for k *active* servers (M/ME/C).	272, 282
$N_i(t)$ – Number of departures from S_i in interval, t.	170
$\mathbf{P}$ – Transition matrix.	17, 66
$\boldsymbol{P}$ – Transition matrix between subsystems	365
$\mathbf{P_k}$ – Transition mx. for k *active* cust. (M/ME/C).	272, 283
$P_i(N)$ – S.s. Prob that S_i is busy in a system with N cust.	37

ABBREVIATIONS

b-p – Busy period.
f-p – First-passage.
f.p. – First passage.
LAQT – Linear Algebraic Queueing Theory.
ME – Matrix exponential function.
MTTF – Mean time to failure.
mx. – Matrix.
ODE – Ordinary differential equation.
PDF – Probability Distribution Function.
pdf – Probability density function.
PH – Phase distribution.
RLT – Rational Laplace transform.
RT – Relaxation Time.
s.s. – steady-state.
TS – Time-sharing.
w.o. – Without.

INDEX